D0207388

Developmental Neurobiology

To E.A.B.

Developmental Neurobiology

Lynne M. Bianchi

GS Garland Science
Taylor & Francis Group
NEW YORK AND LONDON

Vice President: Denise Schanck
Senior Development Editor: Monica Toledo
Senior Digital Project Editor: Natasha Wolfe
Senior Production Editor: Georgina Lucas
Text Editor: Kathleen Vickers
Illustrator: Nigel Orme
Text and Cover Design: Matthew McClements, Blink Studio, Ltd.
Copyeditor: John Murdzek
Proofreader: Susan Wood
Indexer: Simon Yapp at Indexing Specialists
Permissions Coordinator: Sheri Gilbert

Lynne M. Bianchi is Professor of Neuroscience and Pre-Medical Program Director at Oberlin College. She received her Ph.D. in Anatomy and Cell Biology from the University at Buffalo School of Medicine and Biomedical Sciences. She joined Oberlin College, a liberal arts college with one of the first and longest-running undergraduate neuroscience programs in the United States, in 1998. Her research interests focus on neuron–target interactions and the role of nerve growth factors in the developing auditory system.

Cover image shows a light micrograph of a mouse embryo, approximately 10.5 days post-fertilisation. The specimen was stained with a fluorescent marker that highlights the presence of precursor cells to nerve tissue then chemically treated to make it optically transparent. Image courtesy of RPS/Jim Swoger/BNPS.

© 2018 by Garland Science, Taylor & Francis Group, LLC

This book contains information obtained from authentic and highly regarded sources. Every effort has been made to trace copyright holders and to obtain their permission for the use of copyright material. Reprinted material is quoted with permission, and sources are indicated. A wide variety of references are listed. Reasonable efforts have been made to publish reliable data and information, but the author and the publisher cannot assume responsibility for the validity of all materials or for the consequences of their use.

All rights reserved. No part of this publication may be reproduced, stored in a retrieval system or transmitted in any form or by any means—graphic, electronic, or mechanical, including photocopying, recording, taping, or information storage and retrieval systems—without permission of the copyright holder.

ISBN 9780815344827

Library of Congress Cataloging-in-Publication Data
Names: Bianchi, Lynne, author.
Title: Developmental neurobiology / Lynne M. Bianchi.
Description: New York, NY: Garland Science, Taylor & Francis Group, LLC, 2018.
Identifiers: LCCN 2017034851 | ISBN 9780815344827
Subjects: LCSH: Developmental neurobiology.
Classification: LCC QP363.5 .B563 2018 | DDC 612.6/4018--dc23
LC record available at https://lccn.loc.gov/2017034851

Published by Garland Science, Taylor & Francis Group, LLC, an informa business, 711 Third Avenue, New York, NY 10017, USA, and 3 Park Square, Milton Park, Abingdon, OX14 4RN, UK.

Printed in the United States of America
15 14 13 12 11 10 9 8 7 6 5 4 3 2 1

GS Garland Science
Taylor & Francis Group

Visit our web site at http://www.garlandscience.com

Preface

No one goes into science because they love to memorize facts; they go into science because they love the process of discovery and problem solving. The field of developmental neurobiology is filled with numerous examples of creativity and insight that highlight the exciting process of scientific discovery. As an instructor, it is a pleasure to be able to discuss the motivation and experimental methods behind such studies. Whether studies were done 125 years ago or 5 weeks ago, there is always something intriguing to discuss—from the very first stages of neural induction in early embryogenesis to the refinement of synaptic connections during postnatal development.

One goal of this book is to provide historical background on topics to help students gain a perspective on how ideas have evolved over time. As instructors, it is sometimes tempting to focus only on the latest material. However, somewhere along the way, I noticed that students did not always fully grasp why a new discovery was so remarkable and realized that many had not yet heard about the earlier work that suggested another outcome, and were therefore unable to appreciate the excitement generated by the newer findings. Thus, I have found providing such background to be beneficial to students. As one reviews earlier studies, one comes to appreciate how the experiments were done, what information influenced how certain hypotheses were formed, and how unexpected findings have shifted the focus of research efforts over time. While students will have to memorize some detailed facts for a course, I hope that reading how the facts were generated will lead to an appreciation of why those details are so important for understanding how the nervous system develops.

A challenge often encountered by instructors teaching developmental neurobiology is that, at many institutions, the course is an elective course for undergraduate or first-year graduate students. Therefore, it is not unusual for students to enter the class with different academic backgrounds. Instructors need to balance providing enough information so students without much advanced biology and neuroscience can keep up, without also losing the students who have had the more advanced coursework. In writing this book, I kept those differing levels of student experience in mind. My goal was to provide sufficient background information in each chapter so that all students will be able to follow the more detailed and specific concepts as they are covered. This organization also gives advanced students a review of material and allows instructors to skim over background information when appropriate for a given class.

The opportunity to teach developmental neurobiology is always a welcome experience because there are so many topics to discuss that an instructor never runs out of material. However, when organizing the course or planning a single lecture, an instructor is required to select specific content to cover in the available time, knowing that other material must be set aside. It is never an easy task. I've chosen to particularly highlight experiments that had a major impact on the field or changed how investigators approached a particular question. These examples are not the only experiments that have shaped the field of developmental neurobiology, but they are provided to illustrate the types of work that have been done.

To start, Chapter 1 provides an overview of concepts that will be important for material covered in subsequent chapters. The chapter begins with a review of basic cell biology and anatomy of major structures in the nervous system, and then describes the embryonic development and staging criteria used for common vertebrate and invertebrate animal models, as well as for humans. Images from atlases of the different species are provided so that students have a reference for understanding studies discussed in later chapters. Chapter 1 concludes with a discussion of experimental methods commonly used by investigators and frequently discussed in subsequent chapters.

Chapters 2–10 focus on selected stages of neural development. As with any subtopic in developmental neurobiology, it is difficult to provide a comprehensive overview of every neural population and so examples that highlight major developmental mechanisms were selected, though there are certainly other equally important examples that could have been used. Chapter 2 describes the process of neural induction beginning with the discovery of the organizer through current discoveries identifying subtle differences in induction mechanism across different vertebrate and invertebrate animal models. Chapters 3 and 4 cover segmentation and patterning along the anterior–posterior and dorsal–ventral axes, respectively. The topics have been separated into two chapters because the volume of information on each has advanced to the point where covering all the material in a single chapter can become overwhelming to both the instructor and the student. Chapter 5 discusses how cells migrate to their proper location in the developing central and peripheral nervous systems, while Chapter 6 covers the cellular determination of selected neural and sensory cells. Chapter 7 explains mechanisms that guide axons to their proper target cells, and Chapter 8 discusses how target cells influence neuronal survival and the various signaling pathways that intersect to mediate neuronal survival and death. Chapters 9 and 10 cover synapse formation and reorganization at the neuromuscular junction and central nervous system, respectively. Both chapters discuss how synapses are formed at each region and how synapses are later eliminated or reorganized in early postnatal development. Rather than separate chapters based on synapse formation and synapse elimination/reorganization, the chapters are separated by the type of synapse to provide a sense of what happens at particular synapses over time in a given region of the nervous system.

For many experimental examples discussed in the book, the names of lead investigators are indicated so that students can refer to the literature and read the original papers. In several instances an investigator's name is listed with the very broad label "and colleagues." In some cases, the colleagues were a few other individuals working on the project in a single lab. In many cases, however, "and colleagues" represents the contributions of several, if not dozens, of researchers over the course of many years or, in some cases, decades. While not specifically named in the text, the contributions of the colleagues cannot be underestimated. The research of current investigators is also highlighted in boxes to provide examples of how careers in developmental neurobiology begin and evolve. Many of these boxes were written by recent graduates of Oberlin College who are now pursuing careers in scientific research or medicine, and illustrate some of the many career paths available.

Writing a book takes a remarkably long time, particularly because it has to be done in the moments that can be found outside of time dedicated to other academic responsibilities. I greatly appreciate the support and encouragement of my colleagues and friends throughout this process. I also thank the many colleagues who provided background information on various studies described in the text. The staff at the Oberlin College Archives and Science Library were extremely helpful in providing the many materials needed for preparing this book, and they were very patient when

I kept books out for extended periods of time. It is the students at Oberlin College who motivated me to begin and continue this project, and I am thankful for the many great conversations I have had with so many of them over the years.

I thank Janet Foltin for initially contacting me and assuring me that writing such a book was possible. The staff at Garland Science has made writing a textbook a very smooth process. I greatly appreciate the careful and thoughtful editing of Kathleen Vickers during the early stages of the project. I am especially grateful to Monica Toledo for her commitment to this book. She kept me on track, reviewed the text and illustrations to make sure everything fit together, and recruited reviewers and compiled their reviews for me. She also taught me a lot about the publishing process along the way. I also thank Nigel Orme for his hard work and ability to turn my sketches into clear illustrations that convey the ideas I was trying to get across and Matthew McClements for his cover and text designs. I greatly appreciate the time and effort of the many reviewers who read early drafts of the chapters. The thoughtful and detailed reviews they provided were extremely helpful and have certainly enhanced the content of the book. And, finally, I want to acknowledge and thank my husband and children for all of their support, good humor, and incredible patience during the processes of completing this book. I hope they enjoy reading it as much as I have enjoyed writing it.

ACKNOWLEDGMENTS

The author and publisher of *Developmental Neurobiology* gratefully acknowledge the contributions of the following scientists and instructors for their advice and critique in the development of this book: Coleen Atkins (University of Miami); Karen Atkinson-Leadbeater (Mount Royal University); Eric Birgbauer (Winthrop University); Jennifer Bonner (Skidmore College); Martha Bosma (University of Washington); Sara Marie Clark (Tulane University); Elizabeth Debski (University of Kentucky); Mirella Dottori (University of Melbourne); Mark Emerson (The City College of New York); Erika Fanselow (University of Pittsburgh); Deni S. Galileo (University of Delaware); Suzanna Lesko Gribble (University of Pittsburgh); Jenny Gunnersen (University of Melbourne); Elizabeth Hogan (Canisius College); Alexander Jaworski (Brown University); John Chua Jia En (National University of Singapore); Raj Ladher (National Centre for Biological Sciences); Stephen D. Meriney (University of Pittsburgh); Mary Wines-Samuelson (University of Rochester); and Richard E. Zigmond (Case Western Reserve University).

RESOURCES FOR INSTRUCTORS

The figures from *Developmental Neurobiology* are available in two convenient formats: PowerPoint® and JPEG, which have been optimized for display. Please email science@garland.com to access the resources.

Contents

Chapter 1
An Introduction to the Field of Developmental Neurobiology 1

CELLULAR STRUCTURES AND ANATOMICAL REGIONS OF THE NERVOUS SYSTEM 4

The central and peripheral nervous systems are comprised of neurons and glia 4

The nervous system is organized around three axes 7

ORIGINS OF CNS AND PNS REGIONS 8

The vertebrate neural tube is the origin of many neural structures 9

Future vertebrate CNS regions are identified at early stages of neural development 10

Timing of developmental events in various vertebrates 11

Anatomical regions and the timing of developmental events are mapped in invertebrate nervous systems 15

The *Drosophila* CNS and PNS arise from distinct areas of ectoderm 15

Cell lineages can be mapped in *C. elegans* 18

GENE REGULATION IN THE DEVELOPING NERVOUS SYSTEM 20

Experimental techniques are used to label genes and proteins in the developingnervous system 22

Altering development as a way to understand normal processes 22

Summary 26

Further Reading 26

Chapter 2
Neural Induction 29

THE ESTABLISHMENT OF NEURAL TISSUE DURING EMBRYOGENESIS 29

Gastrulation creates new cell and tissue interactions that influence neural induction 30

EARLY DISCOVERIES IN THE STUDY OF NEURAL INDUCTION 33

Amphibian models were used in early neuroembryology research and remain popular today 34

A region of the dorsal blastopore lip organizes the amphibian body axis and induces the formation of neural tissue 34

The search for the organizer's neural inducer took decades of research 35

New tissue culture methods and cell-specific markers advanced the search for neural inducers 36

NEURAL INDUCTION: THE NEXT PHASE OF DISCOVERIES 37

The search for mesoderm inducers revealed that neural induction might involve removal of animal cap-derived signals 37

Mutation of the activin receptor prevents the formation of ectoderm and mesoderm but results in the formation of neural tissue 38

Modern molecular methods led to the identification of three novel neural inducers 39

NOGGIN, FOLLISTATIN, AND CHORDIN PREVENT EPIDERMAL INDUCTION 42

Studies of epidermal induction revealed the mechanism for neural induction 42

The discovery of neural inducers in the fruit fly *Drosophila* contributed to a new model for epidermal and neural induction 42

BMP signaling pathways are regulated by SMADs 46

Additional signaling pathways may influence neural induction in some contexts 46

Species differences may determine which additional pathways are needed for neural induction 47

Summary 48

Further Reading 49

Chapter 3
Segmentation of the Anterior–Posterior Axis 51

NEURAL TUBE FORMATION 52

Early segmentation in the neural tube helps establish subsequent neural anatomical organization 54

Temporal–spatial changes in the signals required to induce head and tail structures 56

Activating, transforming, and inhibitory signals interact to pattern the A/P axis 56

SPECIFICATION OF FOREBRAIN REGIONS 57

Signals from extraembryonic tissues pattern forebrain areas 57

Forebrain segments are characterized by different patterns of gene expression 58

Signals prevent Wnt activity in forebrain regions 58

REGIONALIZATION OF THE MESENCEPHALON AND METENCEPHALON REGIONS 60
Intrinsic signals pattern the midbrain–anterior hindbrain 61
Multiple signals interact to pattern structures anterior and posterior to the isthmus 62
FGF is required for development of the cerebellum 63
FGF isoforms and intracellular signaling pathways influence cerebellar and midbrain development 63
FGF and Wnt interact to pattern the A/P axis 64

RHOMBOMERES: SEGMENTS OF THE HINDBRAIN 65
Cells usually do not migrate between adjacent rhombomeres 65
Some of the signals responsible for establishing and maintaining hindbrain segments have been identified 67

GENES THAT REGULATE HINDBRAIN SEGMENTATION 68
The body plan of *Drosophila* is a good model for studying the roles specific genes play in segmentation 68
The homeotic genes that are active in establishing segment identity are conserved across species 69
A unique set of expressed *Hox* genes defines the patterning and cell development in each rhombomere 71
Retinoic acid regulates *Hox* gene expression 73
The RA-degrading enzyme Cyp26 helps regulate *Hox* gene activity in the hindbrain 75
RA and FGF signaling interactions differentially pattern posterior rhombomeres and spinal cord 76
Cdx transcription factors are needed to regulate *Hox* gene expression in the spinal cord 76

Summary 78

Further Reading 79

Chapter 4
Patterning along the Dorsal–Ventral Axis 81

ANATOMICAL LANDMARKS AND SIGNALING CENTERS IN THE POSTERIOR VERTEBRATE NEURAL TUBE 82
The sulcus limitans is an anatomical landmark that separates sensory and motor regions 83
The roof plate and floor plate influence gene expression patterns to delineate cell groupings in the dorsal and ventral neural tube 83

VENTRAL SIGNALS AND MOTOR NEURON PATTERNING IN THE POSTERIOR NEURAL TUBE 85
The notochord is required to specify ventral structures 85
Sonic hedgehog (Shh) is necessary for floor plate and motor neuron induction 86
Shh concentration differences regulate induction of ventral neuron subtypes 89
Genes are activated or repressed by the Shh gradient 90
Shh binds to and regulates patched receptor expression 91
RA and FGF signals are also used in ventral patterning 95

DORSAL PATTERNING IN THE POSTERIOR NEURAL TUBE 95
TGFβ-related molecules help pattern the dorsal neural tube 96
Roof plate signals pattern a subset of dorsal interneurons 97
BMP-related signals pattern class A interneurons 97
BMP-like signaling pathways are regulated by SMADS 99
Wnt signaling through the β-catenin pathway influences development in the dorsal neural tube 100
Gradients of BMP and Shh antagonize each other to form D/V regions of the neural tube 102

D/V PATTERNING IN THE ANTERIOR NEURAL TUBE 104
Roof plate signals pattern the anterior D/V axis by interacting with the Shh signaling pathway 105
Zic mediates D/V axis specification by integrating dorsal and ventral signaling pathways 106
The location of cells along the A/P axis influences their response to ventral Shh signals 107
Analysis of birth defects reveals roles that D/V patterning molecules play in normal development 108

Summary 109

Further Reading 109

Chapter 5
Proliferation and Migration of Neurons 111

NEUROGENESIS AND GLIOGENESIS 111
Scientists debated whether neurons and glia arise from two separate cell populations 112
Precursor cell nuclei travel between the apical and basal surfaces 113
Interkinetic movements are linked to stages of the cell cycle 114
The plane of cell division and patterns of protein distribution determine whether a cell proliferates or migrates 115
Distinct proteins are concentrated at the apical and basal poles of progenitor cells 116
The rate of proliferation and the length of the cell cycle change over time 118

CELLULAR MIGRATION IN THE CENTRAL NERVOUS SYSTEM 122
In the neocortex, newly generated neurons form transient layers 123
Most neurons travel along radial glial cells to reach the cortical plate 124
Cells in the cortical plate are layered in an inside-out pattern 126
Changes in cortical migration patterns lead to clinical syndromes in humans 127
The *Reeler* mutation displays an inverted cell migration pattern 128
Cajal–Retzius cells release the protein Reelin, a stop signal for migrating neurons 128
Cortical interneurons reach target areas by tangential migration 130
Cell migration patterns in the cerebellum reflect its distinctive organization 131
Cerebellar neurons arise from two zones of proliferation 132
Granule cell migration from external to internal layers of the cerebellar cortex is facilitated by astrotactin and neuregulin 134
Mutant mice provide clues to the process of neuronal migration in the cerebellum 136

MIGRATION IN THE PERIPHERAL NERVOUS SYSTEM: EXAMPLES FROM NEURAL CREST CELLS 136
Neural crest cells emerge from the neural plate border 137
Neural crest cells from different axial levels contribute to specific cell populations 138
Cranial neural crest forms structures in the head 139
Multiple mechanisms are used to direct neural crest migration 140
Trunk neural crest cells are directed by permissive and inhibitory cues 141
Melanocytes take a different migratory route than other neural crest cells 143

Summary 144

Further Reading 144

Chapter 6
Cell Determination and Early Differentiation 147

LATERAL INHIBITION AND NOTCH RECEPTOR SIGNALING 148
Lateral inhibition designates future neurons in *Drosophila* neurogenic regions 148
Lateral inhibition designates stripes of neural precursors in the vertebrate spinal cord 150

CELLULAR DETERMINATION IN THE INVERTEBRATE NERVOUS SYSTEM 151
Cells of the *Drosophila* PNS arise along epidermal regions and develop in response to differing levels of Notch signaling activity 151
Ganglion mother cells give rise to *Drosophila* CNS neurons 153
Apical and basal polarity proteins are differentially segregated in GMCs 153
Cell location and the temporal expression of transcription factors influence cellular determination 154

MECHANISMS UNDERLYING FATE DETERMINATION IN VERTEBRATE CNS NEURONS 155
Changes in transcription factor expression mediate the progressive development of cerebellar granule cells 156
Temporal cues help mediate the fate of cerebral cortical neurons 157
Epigenetic factors influence determination and differentiation in vertebrate neurons 159

DETERMINATION AND DIFFERENTIATION OF NEURAL-CREST-DERIVED NEURONS 161
Environmental cues influence the fate of parasympathetic and sympathetic neurons 161
Sympathetic neurons can change neurotransmitter production later in development 163

DETERMINATION OF MYELINATING GLIA IN THE PERIPHERAL AND CENTRAL NERVOUS SYSTEM 164
Neuregulin influences determination of myelinating Schwann cells in the PNS 166
Precursor cells in the optic nerve are used to study oligodendrocyte development 167
Internal clocks establish when oligodendrocytes will start to form 168

DEVELOPMENT OF SPECIALIZED SENSORY CELLS 170
Cell–cell contact regulates cell fate in the compound eye of *Drosophila* 170
Cell–cell contacts and gene expression patterns establish R1–R7 photoreceptor cell types 173
Cells of the vertebrate inner ear arise from the otic vesicle 175
Notch signaling specifies hair cells in the organ of Corti 176
Cells of the vertebrate retina are derived from the optic cup 178
The vertebrate retina cells are generated in a specific order and are organized in a precise pattern 180
Temporal identity factors play a role in vertebrate retinal development 181

Summary 181

Further Reading 182

Chapter 7 Neurite Outgrowth, Axonal Path-finding, and Initial Target Selection 185

GROWTH CONE MOTILITY AND PATHFINDING 185

Early neurobiologists identify the growth cone as the motile end of a nerve fiber 186

In vitro and *in vivo* experiments confirm neurite outgrowth from neuronal cell bodies 186

Substrate binding influences cytoskeletal structures to promote growth cone motility 187

Actin-binding proteins regulate actin polymerization and depolymerization 189

Rho family GTPases influence cytoskeletal dynamics 190

GROWTH CONE SUBSTRATE PREFERENCES *IN VITRO* AND *IN VIVO* 191

In vitro studies confirm that growth cones actively select a favorable substrate for extension 191

Extracellular matrix molecules and growth cone receptors interact to direct neurite extension 192

Roles of pioneer axons and axonal fasciculation in target selection 193

Research in invertebrate models leads to the labeled pathway hypothesis 196

Fasciclins are expressed on axonal surfaces 197

Vertebrate motor neurons rely on local guidance cues 197

Several molecules help direct motor axons to muscles 200

INTERMEDIATE, MIDLINE TARGETS FOR SPINAL COMMISSURAL AXONS 202

In vertebrate embryos, the axons of commissural interneurons are attracted to the floor plate 203

Laminin-like midline guidance cues are found in invertebrate and vertebrate animal models 204

Homologous receptors mediate midline attractive and repulsive guidance cues 205

Slit proteins provide additional guidance cues to axons at the midline 206

Slit proteins repel commissural axons away from the midline by activating Robo receptors 207

Robo signaling is regulated by additional proteins expressed on commissural axons 208

Shh phosphorylates zip code binding proteins to increase local translation of actin and direct growth of vertebrate commissural axons at the midline 209

THE RETINOTECTAL SYSTEM AND THE CHEMOAFFINITY HYPOTHESIS 211

Early scientists focus on studies of physical cues and neural activity in regulating axon-target recognition 211

Amphibian retinal ganglion cell axons regenerate to reestablish neural connections 212

Retinotectal maps are found in normal and experimental conditions 214

Some experimental evidence contradicts the chemoaffinity hypothesis 215

A "stripe assay" reveals growth preferences for temporal retinal axons 215

Retinotectal chemoaffinity cues are finally identified in the 1990s 218

Eph/ephrin signaling proves to be more complex than originally thought 220

Axonal self-avoidance as a mechanism for chemoaffinity 222

Summary 223

Further Reading 224

Chapter 8 Neuronal Survival and Programmed Cell Death 227

GROWTH FACTORS REGULATE NEURONAL SURVIVAL 227

The death of nerve cells was not initially recognized as a normal developmental event 228

Studies reveal that target tissue size affects the number of neurons that survive 228

Some tumor tissues mimic the effect of extra limb buds on nerve fiber growth 229

In vitro studies led to a bioassay method to study nerve growth factors 231

The factor released by sarcoma 180 is found to be a protein 231

Nerve growth factor is identified in salivary glands 233

Recognition that not all neuronal populations respond to NGF leads to the discovery of brain-derived neurotrophic factor 234

Discoveries of other NGF-related growth factors rapidly followed 236

NGF SIGNALING MECHANISMS AND NEUROTROPHIN RECEPTORS 237

NGF undergoes retrograde transport from the nerve terminal to the cell body 237

NGF receptors are first identified in the PC12 cell line 238

Activation of Trk receptors stimulates multiple intracellular signaling pathways 240

Interaction of full-length Trk receptors with truncated Trk receptors or p75NTR further influences cell survival 243

Other growth factors also regulate neuronal survival and outgrowth 244

Ciliary neurotrophic factor is isolated based on an assay for developing ciliary ganglion neurons 245

The CNTF receptor requires multiple components to function 246
Growth factors unrelated to CNTF promote survival of developing CG and motor neurons 247

PROGRAMMED CELL DEATH DURING NEURAL DEVELOPMENT 248
Studies reveal cell death is an active process dependent on protein synthesis 250
Cell death genes are identified in *C. elegans* 251
Homologs of the *C. elegans ced* and *egl* genes contribute to the mammalian apoptotic pathway 252
p75NTR and precursor forms of neurotrophins help mediate neuronal death during development 254

Summary 255

Further Reading 256

Chapter 9 Synaptic Formation and Reorganization Part I: The Neuromuscular Junction 259

CHEMICAL SYNAPSE DEVELOPMENT IN THE PERIPHERAL AND CENTRAL NERVOUS SYSTEMS 260
Reciprocal signaling by presynaptic and postsynaptic cells results in the development of unique synaptic elements 261

THE VERTEBRATE NEUROMUSCULAR JUNCTION AS A MODEL FOR SYNAPSE FORMATION 262
At the NMJ, the presynaptic motor axon releases acetylcholine to depolarize the postsynaptic muscle cell 263
The distribution of AChRs has been mapped in developing muscle fibers 264
The density of innervation to muscle fibers changes during vertebrate development 266
The synaptic basal lamina is a site of NMJ organizing signals 267
AChRs cluster opposite presynaptic nerve terminals in response to agrin released by motor neurons 269
The agrin hypothesis is revised based on additional observations 271
The receptor components MuSK and Lrp4 mediate agrin signaling 272
Rapsyn links AChRs to the cytoskeleton 276
AChR subunits are synthesized in nuclei adjacent to the nerve terminal 276
Perisynaptic Schwann cells play roles in NMJ synapse formation and maintenance 279
The synaptic basal lamina concentrates laminins needed for presynaptic development and alignment with postjunctional folds 280

MODELS OF SYNAPTIC ELIMINATION IN THE NMJ 282
The relative levels of neuromuscular activity determine which terminal branches remain at the endplate 283
BDNF and pro-BDNF are candidates for the protective and punishment signals 283
Perisynaptic Schwann cells influence the stability of synaptic connections 285

Summary 285

Further Reading 287

Chapter 10 Synaptic Formation and Reorganization Part II: Synapses in the Central Nervous System 289

EXCITATORY AND INHIBITORY NEURONS IN THE CENTRAL NERVOUS SYSTEM 290
Many presynaptic and postsynaptic elements are similar in excitatory and inhibitory synapses 292
The postsynaptic density is an organelle found in excitatory, but not inhibitory, neurons 293
Cell adhesion molecules mediate the initial stabilization of synaptic contacts 294
Neurexins and neuroligins also induce formation of synaptic elements and stabilize synaptic contacts 295
Reciprocal signals regulate pre- and postsynaptic development 296
Dendritic spines are highly motile and actively seek presynaptic partners 297
BDNF influences dendritic spine motility and synaptogenesis 298
Eph/ephrin bidirectional signaling mediates presynaptic development 299
Eph/ephrin signaling initiates multiple intracellular pathways to regulate the formation of postsynaptic spine and shaft synapses 301
Wnt proteins influence pre- and postsynaptic specializations in the CNS 304
Different Wnts regulate postsynaptic development at excitatory and inhibitory synapses 305
Glial cells contribute to CNS synaptogenesis 306

SYNAPSE ELIMINATION AND REORGANIZATION IN THE CNS 307
The vertebrate visual system is a popular model to study synapse elimination and reorganization 307
Spontaneous waves of retinal activity stabilize selected synapses in LGN layers 308

Competition between neurons determines which synaptic connections are stabilized 309
Neural activity resulting from early visual experience establishes ocular dominance columns in the primary visual cortex 310
Homeostatic plasticity contributes to synaptic activity 312
Intrinsic and environmental cues continue to influence synapse organization at all ages 313

Summary 316
Further Reading 317

Glossary 319
Index 330

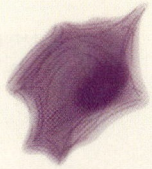

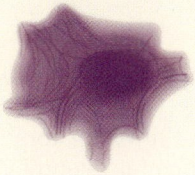

1 An Introduction to the Field of Developmental Neurobiology

Developmental neurobiology is an area of study that seeks to understand the formation of one of the most complex biological systems—the nervous system. Fundamental questions about how the nerve cells of the human nervous system initially form and extend cellular processes to make the billions of necessary connections with such precision have intrigued scientists and non-scientists for centuries. Experimentally, it is one of the most exciting fields to work in, as the questions are addressed using a variety of methods that range from the classical approaches of tissue manipulations to the most sophisticated molecular, genetic, and imaging techniques available today. It is no wonder that developmental neurobiology is a field populated by researchers with backgrounds in fields as diverse as anatomy, biochemistry, cellular and molecular biology, computational sciences, embryology, genetics, medicine, physics, physiology, and psychology (**Box 1.1**).

Identifying the paths that nerve fibers take as they extend from the brain and spinal cord to target areas throughout the body has been of interest to scientists for centuries (**Figure 1.1**), but significant advances in understanding how such pathways form during development did not occur until the mid- to late nineteenth century. During this period, detailed descriptions of the microscopic anatomy of neural tissue were described for the first time. These new findings were made possible by several technical advances arising during that period. One such technological development was the introduction of the microtome, an instrument that provides a means to cut tissues into very thin slices. Another was the increasing availability of microscopes with improved optics that allowed for better visualization of these thinner tissue slices. Additionally, scientists continued to test and refine techniques for fixing (preserving) and staining tissues, so that by the end of the nineteenth century, several improved methods for visualizing the cellular composition of tissues were available. These innovations led to discoveries that were part of the "great age of cellular biology," laying the foundation for many fundamental concepts that we now take for granted.

Several of the first explanations of how the nervous system formed and extended nerve fibers were based on these early microscopic observations.

Box 1.1 Pathways to developmental neurobiology

Individual investigators have come to the field of developmental neurobiology by following many different paths. Some, such as Hans Spemann and Rita Levi-Montalcini, began their careers studying medicine, but ultimately decided to focus on research instead. Both began research careers in the general area of zoology and gradually, as they undertook one project and then another, began to focus on questions pertaining to neurodevelopment. Some investigators, such as the biochemist Stanley Cohen, were recruited to help address a particular question in the developing nervous system, and later focused research efforts on topics beyond the nervous system. Spemann's work provided pivotal insights into how neural tissue is first formed in the early stages of embryogenesis (Chapter 2) and Levi-Montalcini and Cohen identified the first protein to promote the survival of developing neurons (Chapter 8).

Researchers also come from a variety of backgrounds. Some who study developmental neurobiology were the first in their families to attend college, whereas others descend from families comprised of several scientists and physicians. Some completed their undergraduate studies at large universities, whereas others began their studies at small colleges. Some came to college expecting to study science, while others began with different majors and uncertain career goals. Roger Sperry, for example, whose early work addressed how neurons are able to extend nerve fibers to the correct target cell, graduated from Oberlin College in 1935 with a major in English. He later earned a master's degree in psychology and studied zoology at Oberlin College prior to beginning his doctoral studies. In addition to his influential contributions to developmental neurobiology, Sperry won the Nobel Prize for Physiology or Medicine in 1981 for his work on split-brain patients that revealed how the two hemispheres of the brain communicate with one another. Yet, this very successful career could have easily taken a different path. On his college application, when asked about his future career plans, one of his suggestions was college athletic coach due to his interests and talents in various sports (**Figure 1**). It is certain that none of the scientists whose work is featured throughout this book had any idea as undergraduate students where their careers would take them, what questions they might address in the future, how long their careers in science would last, or how many other scientists they would influence.

Figure 1 Roger Sperry as the captain of the basketball team. Like many college students, Roger Sperry was unsure of what career he would pursue. Prior to entering Oberlin College, he expressed interest in science and athletic coaching. As an undergraduate, Sperry majored in English and was captain of the basketball team. After graduating in 1935, he remained at Oberlin to complete a master's degree in Psychology (1937), then took additional courses in zoology to prepare for his doctoral studies at the University of Chicago. (Courtesy of Oberlin College Archives.)

Among the most influential scientists of that period was Santiago Ramón y Cajal, whose work is described in subsequent chapters. What is remarkable about the work of Cajal and his contemporaries is that their descriptions of how neurons grew and behaved in an embryonic environment were all formulated based on images of fixed tissues. By careful observation at different stages of development, the researchers were able to formulate reasonable hypotheses about how cell growth and movement would take place. While not every hypothesis put forth in the late nineteenth century was found to be accurate, a surprisingly large number of the ideas were later found to be correct or very nearly so.

By the early twentieth century scientists had developed a variety of surgical, histological, electrophysiological, and tissue culture techniques that advanced studies in the area of developmental neurobiology. Major scientific milestones in the field often paralleled advances in other areas.

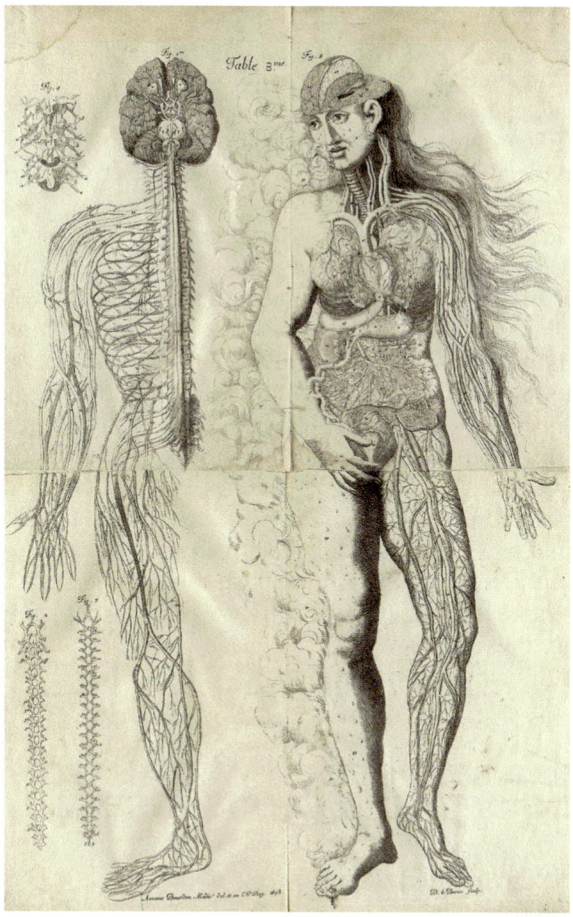

Figure 1.1 Early illustration of the nervous system. Scientists have long been interested in understanding the paths nerve fibers take from the brain and spinal cord to target regions throughout the body. This illustration was completed by the physician Amé Bourdon in 1678. (Image courtesy of U.S. National Library of Medicine, Historical Anatomies.)

For example, in the mid-twentieth century, the electron microscope made it possible to view cellular organelles and led to the conclusive identification of the synapse as the site of connection between two nerve cells. Advances in electronics were similarly influential. As recording equipment became more precise, researchers were able to detect the tiny electrical impulses produced by nerve cells. Additionally, instrumentation was developed that provided a means for making precision microelectrodes to record activity outside of cells, inside of cells, and even on a restricted patch of cell membrane. Because of these technical advances, researchers can measure isolated ionic currents and neural activity and monitor changes that occur as the nervous system develops. Also in the mid-twentieth century, the discovery of **DNA (deoxyribonucleic acid)**, the various forms of **RNA (ribonucleic acid)**, amino acid structures, and the genetic code ushered in the entirely new field of molecular biology, which provided insight into the importance of regulated gene and protein expression during neural development. Technological advances continue to be made in imaging, electronics, molecular biology, and genetics. As in the past, researchers commonly combine the available techniques to get a fuller picture of what happens as neurons progress through various developmental stages.

Discoveries regarding development of the nervous system are made using a variety of animal models, including flies, worms, frogs, fish, chicks, and mice, to track normal developmental events as well as manipulate developing systems and evaluate the impact of such changes on neural development. It is now recognized that many developmental mechanisms are highly conserved among species, and scientists working with the fruit fly (*Drosophila manangaster*), the nematode worm (*Caenorhabditis elegans*), and other invertebrate species often are the first to discover genes and signaling molecules that regulate a particular aspect of nervous system formation in multiple species.

This book describes many of the primary mechanisms by which the nervous system develops, from the initial specification of neural tissue to the refinement of neural connections during early postnatal periods. Each chapter highlights some of the experiments that were key to advancing understanding of a particular stage of neural development. These many experiments highlight the remarkable creativity and insight of the early neurobiologists who made so many major contributions, even without benefit of the more sophisticated techniques available today. One also quickly appreciates how some questions simply could not be answered until suitable technical approaches became available. It is likely that many of the techniques that are considered advanced today will appear crude to scientists in the future. Yet as in the past, the discoveries made today will add to the foundation of knowledge that will be used by future scientists, and together these discoveries will elucidate the mechanisms that govern the formation of the nervous system.

CELLULAR STRUCTURES AND ANATOMICAL REGIONS OF THE NERVOUS SYSTEM

In order to help orient readers to topics discussed in subsequent chapters, the following sections provide a brief overview of some major cellular, anatomical, and developmental features found in vertebrate and invertebrate animal models discussed in this text. The cellular composition and anatomical organization of key neural structures are described first, followed by information on specific developmental stages documented in the chick, mouse, human, fish, frog, fly, and worm nervous systems. These descriptions focus on the timing of shared developmental milestones, including gastrulation, neural plate and neural tube formation, and early brain segmentation. For more detailed explanations of these topics, refer to the references at the end of the chapter.

The central and peripheral nervous systems are comprised of neurons and glia

The vertebrate nervous system is divided into two main regions, the **central nervous system** (**CNS**) and the **peripheral nervous system** (**PNS**). The vertebrate CNS is comprised of the brain and spinal cord, while the PNS consists of collections of neurons called **ganglia** that lie outside of the CNS. The vertebrate PNS includes the neurons of the spinal sensory (dorsal root) ganglia, cranial nerve ganglia, the enteric ganglia, and ganglia of the autonomic nervous system (ANS). The invertebrate nervous system is also divided into CNS and PNS regions; however, different terminology is used for the various CNS and PNS structures, as described in this chapter.

The cells of the CNS and PNS are the **neurons** and **glia**. A vertebrate neuron consists of a cell body and cellular processes called **axons** and **dendrites** (**Figure 1.2**A). Each neuron has only one axon but may have several dendrites. The axon is typically longer than other neural processes, has a uniform diameter, and ends in specialized regions called

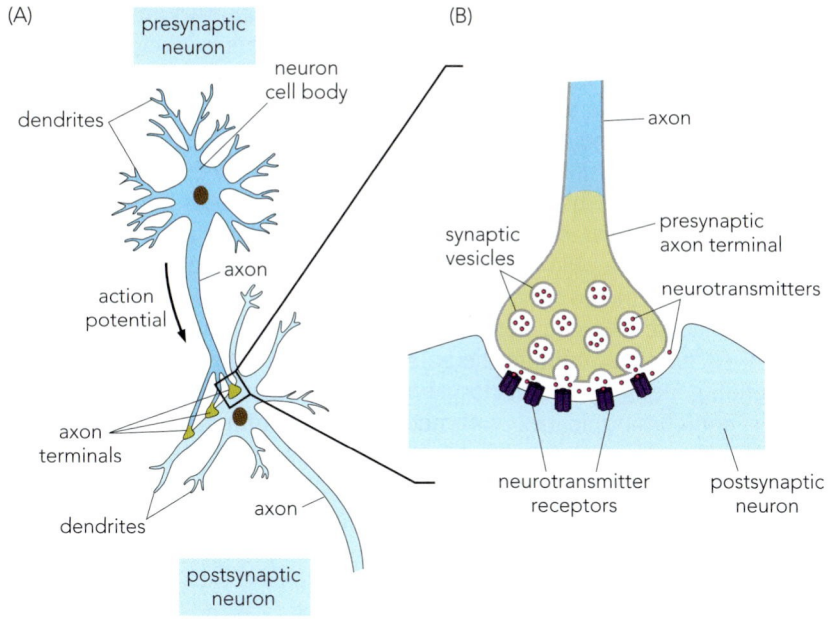

Figure 1.2 Neurons release neurotransmitters to communicate with other cells. (A) Vertebrate neurons consist of a cell body and cellular processes called axons and dendrites. The majority of neurons in the vertebrate nervous system signal to other cells by conducting the electrical activity of an action potential down the axon to stimulate the release of a neurotransmitter from the axon terminals and, in this example, to the dendrites and cell body of the postsynaptic neuron. (B) The neurotransmitter is released from synaptic vesicles then diffuses across the synaptic cleft to bind to specific receptors on the postsynaptic cell. Depending on the neurotransmitter and receptor pair, the binding will either increase or decrease the likelihood that an action potential will occur in the postsynaptic cell.

axon terminals. In contrast, the dendrites tend to be shorter, branch extensively, and have tapered ends. In some circumstances, the term **neurite** is used to refer to either axons or dendrites. For example, when viewing neuronal processes at early stages of neural development or in tissue culture preparations, it is often difficult to conclusively identify a process as an axon or a dendrite and therefore the term *neurite* is used.

Neurons primarily communicate with one another through electrical signals (**action potentials**) that are conducted along the length of the axon to initiate the release of chemical signals (**neurotransmitters**) from **synaptic vesicles** that accumulate in the axon terminals. The release of neurotransmitter occurs at the **synapse**, a small gap or cleft between the axon terminal of one neuron (the **presynaptic cell**) and the cell body or processes of another (the **postsynaptic cell**). The neurotransmitter diffuses across the **synaptic cleft** to bind to receptors on the postsynaptic cell, which may be another neuron or a muscle cell (Figure 1.2B). Neurotransmitter–receptor pairs that *increase* the likelihood that an action potential will occur are found at **excitatory synapses**. In contrast, neurotransmitter–receptor pairs that *reduce* the likelihood of an action potential firing are found at **inhibitory synapses**. A small percentage of vertebrate neurons and some invertebrate neurons communicate through **gap junctions**—channels that are formed between two cells that are in direct contact with each other. In vertebrates, the chemical synapses and gap junction synapses can work together to enhance neural transmission.

The nervous system is also comprised of a number of distinct cell types called **glia**. Originally called neuroglia in the mid 1800s, these cells were thought to be connective tissue—the "glue"—needed to support the structures of the nervous system. For over a century and half, glia were thought to be limited to this role. However, it is now clear that glia serve a number of important functions in the nervous system and in some cases participate in cell signaling. The glia in the vertebrate CNS are the oligodendrocytes, astrocytes, microglia, and ependymal cells. **Oligodendrocytes** extend cellular processes that form the myelin around axons in the CNS. Each oligodendrocyte extends processes to wrap around several nearby axons. **Myelin** provides a type of insulation that speeds the propagation of action potentials. Thus, action potentials are conducted faster along myelinated axons than along unmyelinated axons. **Astrocytes** are star-shaped cells that perform many functions in the central nervous system, such as maintaining the balance of ions in the extracellular fluid surrounding neurons, interacting with cells that form the blood–brain barrier, and communicating with neurons. **Microglia** are the smallest of the glia cell types and generally function as the immune cells of the brain to remove debris and pathogens in the CNS. Microglia may also interact with signals from the immune system to modify the stability of synaptic connections during development and in neurodegenerative conditions (**Figure 1.3**A). **Ependymal** cells line the ventricles of the CNS, where they produce cerebral spinal fluid (CSF).

In the vertebrate PNS, glial cells consist of the Schwann cells and satellite cells. Most **Schwann cells** function similarly to oligodendrocytes. However, each Schwann cell wraps around only one axon and does not extend processes to nearby axons. The **satellite cells** surround neuronal

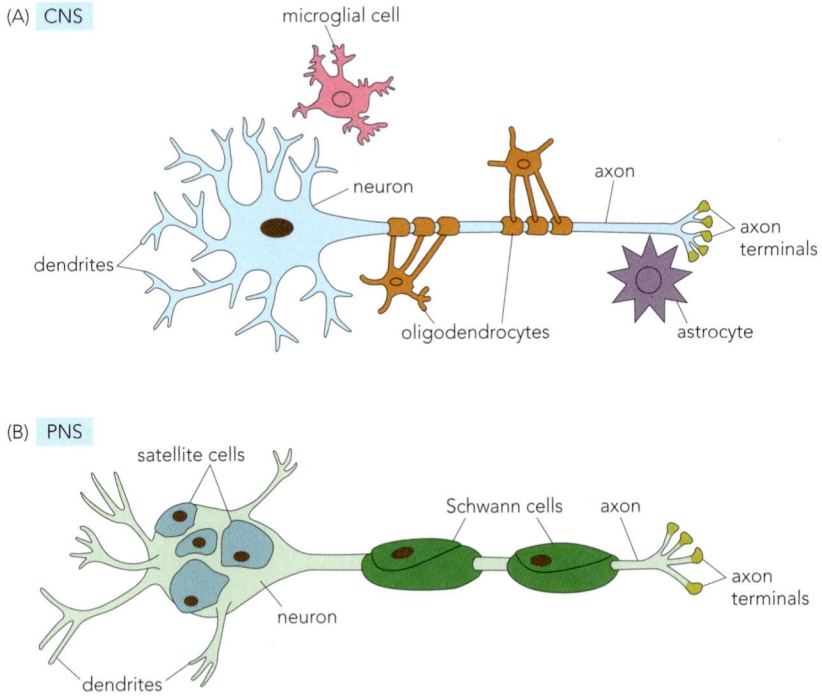

Figure 1.3 The vertebrate nervous system is comprised of neurons and glia. Neurons in the vertebrate nervous system are characterized by a single axon and many dendrites. Axons are generally longer and of uniform diameter, while the dendrites tend to be shorter, with tapered ends. Glial cells surround the neurons and perform diverse functions. (A) Neurons in the central nervous system (CNS) are surrounded by numerous glia, including astrocytes, microglia, and the myelinating oligodendrocytes that wrap around the axons. (B) In the peripheral nervous system (PNS), the cell bodies of neurons are surrounded by glial satellite cells, whereas the axons are wrapped by myelinating Schwann cells.

Timing of developmental events in various vertebrates

The formation of the blastula, gastrula, neural plate, neural tube, and brain vesicles occurs at specific times in embryonic development in each of the animal models studied. While the sequence of developmental events is consistent across all vertebrate species, the actual time that these structures arise varies. Developmental age is reported as the number of hours, days, or weeks post fertilization or by staging criteria established for each species. The stages are based on various morphological criteria, including embryo length and the presence of key developmental features, such as the number of **somites** (the blocks of mesoderm that line either side of the neuraxis). Such staging corrects for any variations that might arise from genetic or environmental influences.

Human development is often referred to in terms of weeks of gestation or Carnegie stages—stages first defined in the early twentieth century by Franklin Mall and George Streeter, both of whom worked at the Carnegie Institute in Washington, DC. Mice are described in terms of embryonic (E) days or days post coital (d.p.c.) and are often staged using criteria established by Karl Theiler. Development of chick embryos is reported based on the hours or days post fertilization or by the duration of incubation. Chick embryos are staged using criteria published by Viktor Hamburger and Howard Hamilton.

The development of the chick embryo begins *in utero* and continues after the egg is laid (about 20 hours after fertilization). *In utero* development involves the formation of the blastoderm at 10–11 hours after fertilization. The onset of gastrulation begins about the time the egg is laid and subsequent developmental events are easily monitored by cutting a small hole, or window, in the egg. Thus, the chick embryo is an extremely useful model for viewing neural development. Another useful feature of the chick egg is that development of a fertilized egg can be halted for several days if the eggs are maintained at room temperature. Development resumes when the eggs are placed in an incubator at 37.5°C. **Figure 1.8** shows several of the stages first published by Hamburger and Hamilton in 1951. At these stages of development, hours refer to the number of hours the eggs were incubated, rather than the hours post fertilization. The first somites and the head neural folds are visible at Hamburger and Hamilton (HH) stage 7 (23–26 hours of incubation). In the chick, the three primary brain vesicles are detected at HH10 (33–38 hours) and the five secondary vesicles are observed after 40–45 hours of incubation (HH11). The embryo turns to the side beginning at HH13 (48–52 hours) and the enlargement and refinement of the brain vesicles is easily viewed in the translucent embryo from HH 14–21 (about 2–3 days after incubation).

Figure 1.9 compares the development in humans and mice using Carnegie and Theiler criteria, respectively. In humans, the blastula is detected in the uterus at four days after fertilization (Carnegie stage 3, CS3) and gastrulation begins at day 16 (CS7). The neural plate and neural folds become evident at day 18 (CS8). The three primary vesicles form during the third and fourth week of gestation (days 20–24; CS8–11) and the five secondary vesicles become visible during the fifth week of gestation (CS14).

In mice, the blastula stage embryo is formed 3 to 4 days post coitus (d.p.c.), corresponding to Theiler stages 4–5 (TS04–05). Gastrulation begins at 6.5–7.5 d.p.c. (TS09) and the neural plate forms at 7.5–8 d.p.c. (TS11). The three primary vesicles are visible at 8–8.5 days of embryogenesis (TS12–13) and the five secondary vesicles are detected at days 9–10 (TS15–16). The development of a newborn mouse (TS27; 19–20 d.p.c.) is similar to that of a 9-week human (CS23). Human development continues for about 38 weeks (9 months).

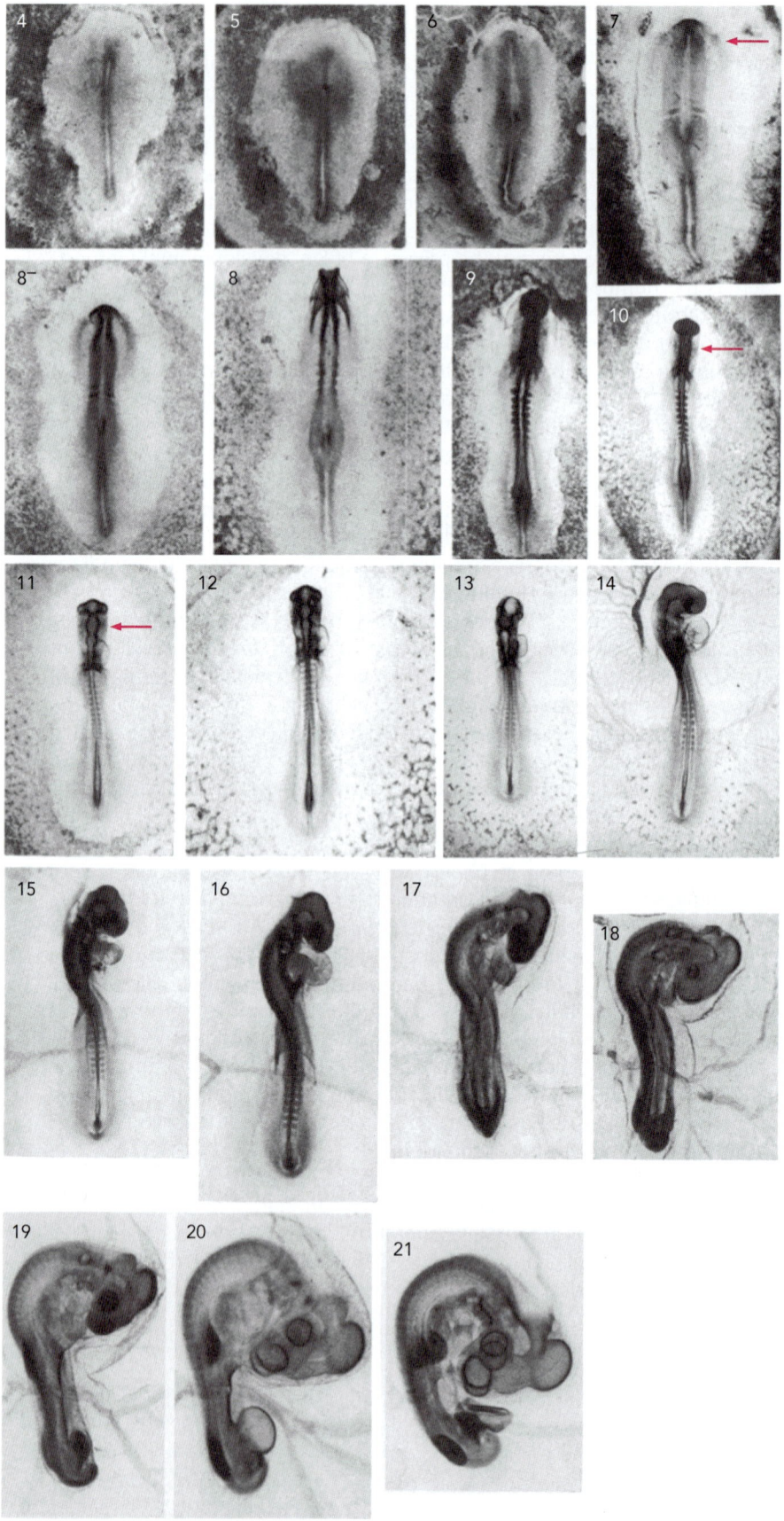

Figure 1.8 Chick development is easily viewed throughout embryogenesis. Images of developing chick embryos reveal some of the key events in neural development. Numbers in the corners of the images indicate the stage of development as determined by Hamburger and Hamilton (HH). The neural folds are first identified at 23–26 hours (HH stage 7, arrow). The primary brain vesicles are visible at HH 10 (arrow) and the secondary vesicles at HH 11 (arrow). The embryo begins to turn to the side at HH 13 and further development of the brain regions is observed through the embryo's translucent body until HH 21 (embryonic day 4). (From Hamburger V & Hamilton HL [1992] *Dev Dyn* 195:231–272.)

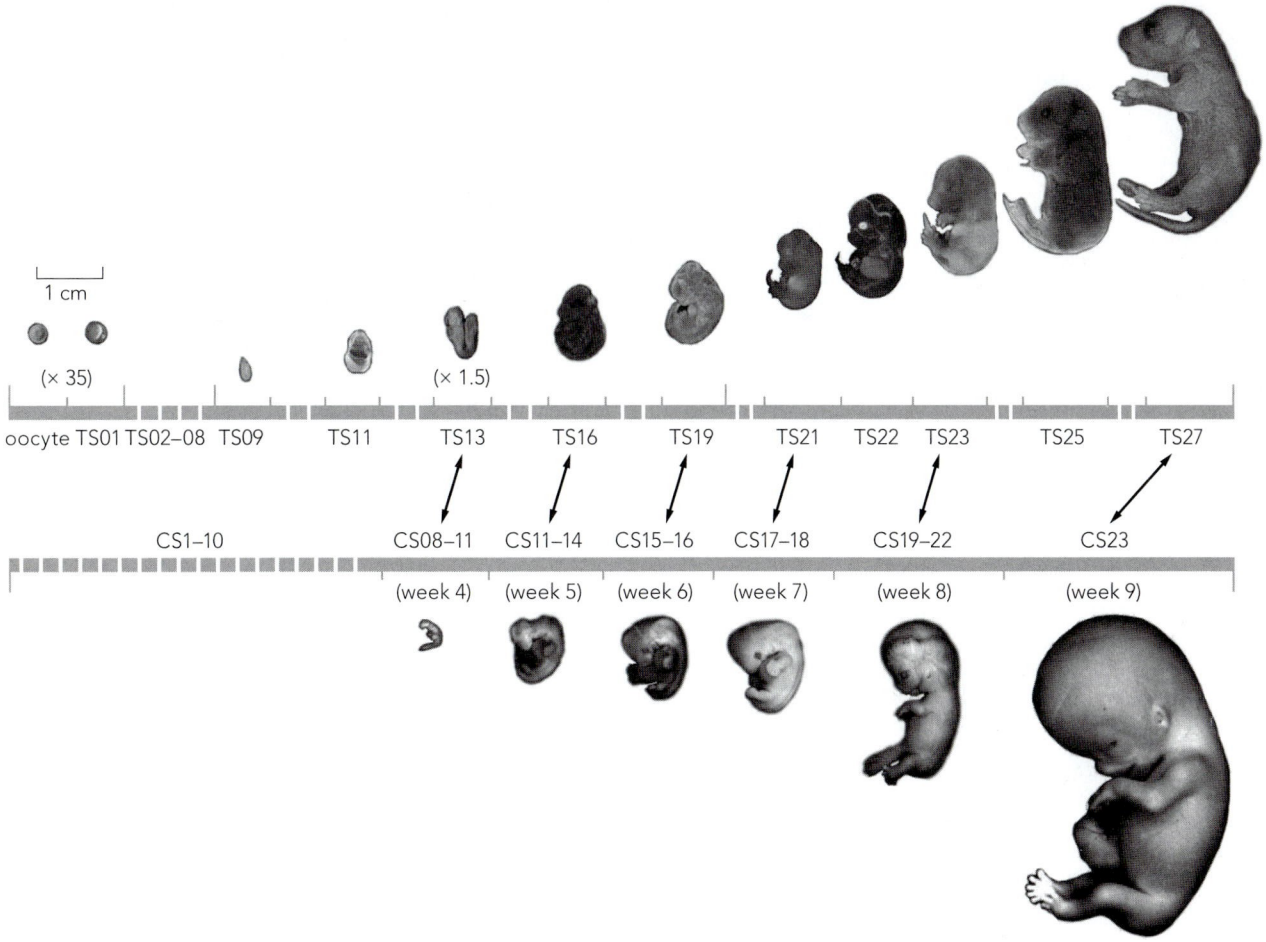

Figure 1.9 Comparison of stages of mouse and human embryonic development. Morphological criteria are used to identify the stages of embryonic development. Mouse development is staged by the criteria of Theiler, whereas human development is marked by Carnegie stages. As shown in the diagram, Theiler stage 13 (TS13) is equivalent to Carnegie stage 11 (CS11). At these stages, the three primary vesicles are observed. TS16 is equivalent to CS11–14, the stages when the five secondary brain vesicles are formed. TS27 is a newborn mouse, which is at a similar stage of development as a nine-week human (CS23). Arrows indicate times at which human and mouse development are at a similar stage. (From Xue L, Cai J-Y, Ma J et al. [2013] *BMC Genomics* 14:568.)

Fish and amphibians have also been very popular vertebrate animal models in studies of neural development. The stages of development in the zebrafish have been documented by Monte Westerfield, Charles Kimmel, and colleagues. The cells of the zygote begin to divide about 40 minutes after fertilization and are easily viewed above the yolk. The zebrafish blastula forms at a little over two hours post fertilization, when 128 cells, the blastomeres, are present. The blastula-stage embryo continues to develop through multiple stages during the first five hours after fertilization and the blastoderm is identifiable a little over 4.5 hours post fertilization (h.p.f.; **Figure 1.10**). After the blastula-stage is complete, epiboly and gastrulation, the movement and thinning of cell layers, begins just over 5 h.p.f. At these stages, the embryo begins to curl around the central yolk. Development is measured in terms of the percentage of **epiboly**, indicating the percent of the yolk that is surrounded by the blastoderm. At 50% epiboly, gastrulation begins. At just over 6 hours, the embryonic shield, a key structure in the process of neural induction (Chapter 2), is present. At 90% epiboly, 9 h.p.f., the neural plate is visible. At the completion of epiboly and gastrulation, somites are detected (10 h.p.f.), and by 16 hours, there are 14 somites and the three primary brain vesicles are observed. At 24 hours, the five secondary brain vesicles are present. At the same time,

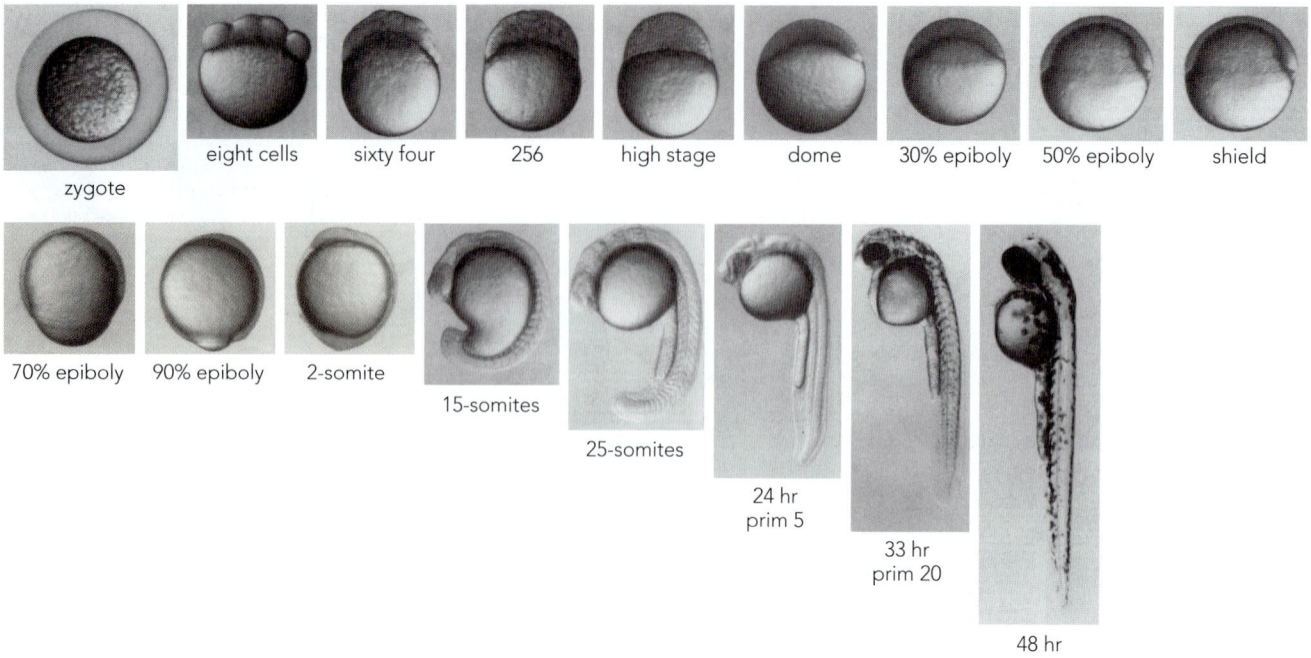

Figure 1.10 Examples of zebrafish development from zygote to hatching. Cells of the zebrafish zygote begin to divide about 40 minutes after fertilization. The resulting blastomeres continue to divide as the blastodisc forms above the yolk, as seen in the eight-cell stage shown in this figure. The blastula stage of development begins at the 128-256 cell stage (2.25-2.5 hours post fertilization, h.p.f.) then progresses through multiple stages. The high stage, for example, indicates the period that the blastodisc is located "high" on the yolk. The blastula stage continues until a little over 4 h.p.f. (dome stage). By about 4.5 hours, epiboly can be measured, indicating the percentage of the yolk surface that is surrounded by the embryo. At 50% epiboly gastrulation begins and at 90% epiboly, the neural plate is present. Somites are first visible by 10 h.p.f., and the divisions of the brain vesicles are first observed at the 14-16 somite stage. The embryo begins to straighten away from the yolk at 24 h.p.f. (prim 5), when development is measured by the myotome number that the tip of the primoridum (prim) of the lateral line organ reaches. By 48 h.p.f., the embryo hatches. (From Westerfield M [1993] The Zebrafish Handbook, 2nd ed, University of Oregon Press.)

the embryo begins to straighten away from the yolk sac. The embryos are now measured by indicating which myotome (the segment of the somite that later gives rise to muscle) that the tip of primordium (prim) of the lateral line organ (a sensory organ found in aquatic vertebrates) reaches. Thus, prim 5 indicates that the tip of the primordium of the lateral line reaches the fifth myotome. The embryo hatches at about 48 hours and enters the larval stages by 72 hours post fertilization. The larval stages last up to 29 days. Juvenile zebrafish form at day 30 and become adults by day 90.

Frogs are another frequently used vertebrate animal model in studies of neural development. Among the most commonly used frogs are *Xenopus laevis*. These frogs provide many advantages for researchers, including the ability to induce frogs to produce eggs year round and the one-year cycle needed to complete development from a fertilized egg to an adult frog. The description that follows refers to the timing of developmental events in *Xenopus* based on the time post fertilization and the staging criteria of Pieter Nieuwkoop and Jacob Faber. The timing of these events may be slightly different in other frogs. In *Xenopus*, blastula-stage embryos are noted by four hours after fertilization (stage 7). Gastrulation begins approximately 7–10 hours post fertilization (stages 10–12) and leads to the formation of the neurula-stage embryo, the stage when the neural tissue begins to form (12 and 13.5 hours post fertilization; **Figure 1.11**). The neurula stage (stages 13–21) continues until the early tailbud-stage embryo forms 24–32 hours post fertilization (stages 22–28). Primary brain vesicles appear around 24 hours after fertilization (stage 22) and the five secondary vesicles about 32 h.p.f (stage 28). The embryo develops into a tadpole by 96 hours (stages 45–50), before undergoing metamorphosis (stages 51–65) and reaching the adult stage approximately 12 months later (stage 66).

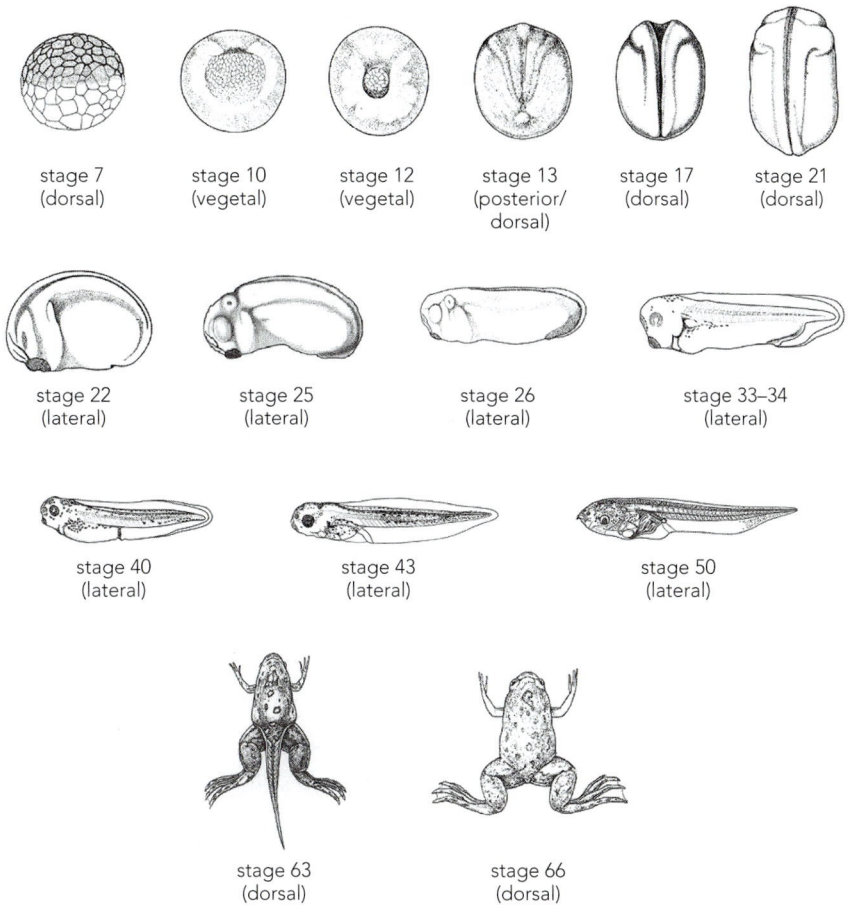

Figure 1.11 Development of the frog from egg through adult stages. Examples from the criteria established by Nieuwkoop and Faber in 1967 identify some of the key stages of development in the frog *Xenopus laevis*. Images show embryos from posterior-dorsal, lateral, and dorsal views. By stage 7 (4 hours after fertilization), the blastula-stage embryo is present. Gastrulation begins 7–10 hours after fertilization (stages 10–12). The neurula-stage embryo, when the neural structures first form, continues from 12 to 22.5 hours after fertilization (stages 13–21). The embryo then enters the tailbud stage at 24–36 hours post fertilization (stages 22–44). The brain vesicles are visible beginning 24 hours after fertilization (stage 22). The embryo develops into a tadpole by 96 hours (stages 45–50), before undergoing metamorphosis (stages 51-65). An adult frog is formed approximately 12 months later (stage 66).

Anatomical regions and the timing of developmental events are mapped in invertebrate nervous systems

Several invertebrate animal models, particularly the fruit fly *Drosophila melanogaster* and the round worm *Caenorhabditis elegans* (*C. elegans*), are also utilized in a number of pivotal studies described in subsequent chapters. *Drosophila* became a popular animal model for research in areas of genetics and developmental biology beginning with the pioneering work of Thomas Hunt Morgan in the early twentieth century. The work of Seymour Benzer and colleagues in the 1960s helped make *Drosophila* an animal model of ongoing interest to developmental neurobiologists. *C. elegans* also became a popular model beginning in the 1960s, largely through the work of Sydney Brenner's lab. These animal models are easily bred, have a short life cycle from the time of fertilization to the adult form, and exhibit naturally occurring and experimentally induced mutations that provide a means to test how specific genes regulate development of the various cells within the nervous system. Like the vertebrate nervous system, the nervous system of invertebrates arises from the ectoderm. However, there are significant differences in how and where neural structures arise in these animal models.

The *Drosophila* CNS and PNS arise from distinct areas of ectoderm

When *Drosophila* are maintained at 25°C, embryogenesis occurs over a period of approximately 22 hours and adult flies are formed within 9–12 days (**Figure 1.12**). This allows for the generation of large numbers of animals

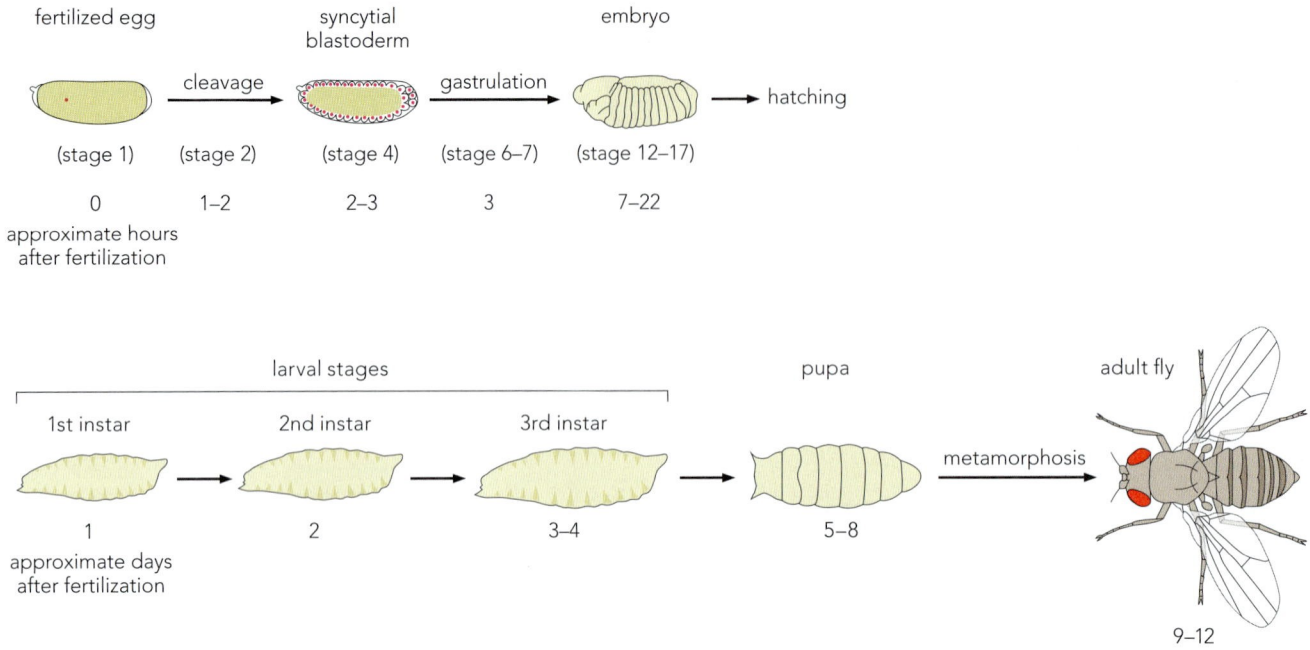

Figure 1.12 Development of the fruit fly *Drosophila melanogaster.* The fertilized *Drosophila* egg (stage 1) undergoes cleavage (stage 2) and forms a syncytial blastoderm 2–3 hours after fertilization (stage 4) followed by a cellular blastoderm (stage 5, not shown). The embryo then begins gastrulation (stages 6-7) and forms the late stage embryo 7–22 hours after fertilization (stages 12-17). The embryo enters the first larval stage by 24 hours after fertilization, and continues through three larval instar stages before forming a pupa at 5–8 days post fertilization. Metamorphosis takes place and the adult fly emerges 9–12 days after fertilization.

to evaluate in a short period of time. Staging of *Drosophila* embryos is often noted using the criteria of Volker Hartenstein and Jose Campos-Ortega.

In *Drosophila*, the cytoplasm of the fertilized egg contains many nuclei that divide rapidly (stages 1–2) prior to migrating to the outer cortex of the cell to form a syncytial blastoderm (stage 4). Each nucleus is then surrounded by a cell membrane to form the cellular blastoderm (stage 5). Gastrulation (stages 6–7) occurs within 3 hours of fertilization, and by the end of the first day (stages 16–17), the embryo hatches to enter the first larval stage called the first instar. *Drosophila* larvae progress through three instar stages, with each stage lasting one to two days (Figure 1.12). An epidermal-derived hardened shell called a cuticle surrounds each instar stage larva. At the end of each instar stage, the cuticle sheds to accommodate the growth of the larva. A new cuticle is then produced for the larger larva. Following the third instar stage, the cuticle contributes to extracellular case that surrounds the prepupa. During the pupal stage, metamorphosis occurs. Adult tissues that are derived from ectoderm, such as the nervous system, arise from pockets of epithelium formed during the larval stages. These pockets of tissue, called **imaginal discs**, attach to the inside of the larval epidermis and later evert during metamorphosis to form adult structures of the head, thorax, legs, and wings. Unlike most other larval-stage organs, the components of the gut and nervous system persist in the adult fly. Cells of the nervous system proliferate during the larval stages and begin to differentiate in the pupal stage.

In *Drosophila*, the nervous system arises from ventral ectoderm (the ventral neurogenic ectoderm), rather than the dorsal ectoderm as in vertebrates. This ectoderm gives rise to the neuroblasts (beginning at stage 9, about 4 hours after fertilization) that form the adult brain and ventral nerve cord, a structure with functions similar to the vertebrate spinal cord (**Figure 1.13**). During development, **neuroblasts** segregate from the surrounding ectoderm, then move inside the embryo along

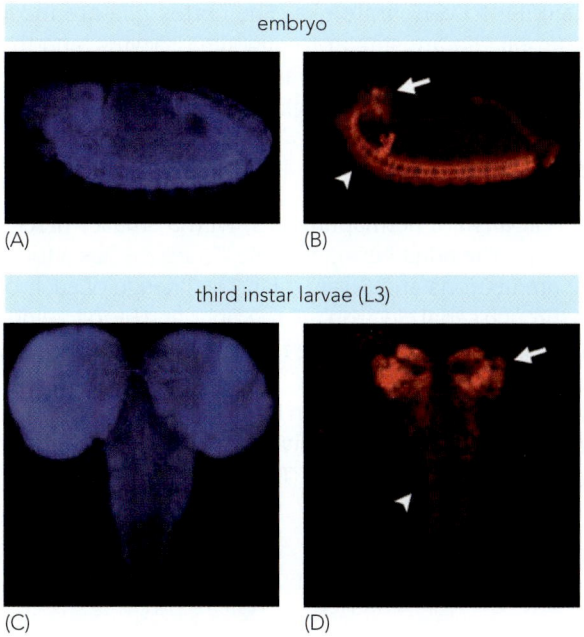

Figure 1.13 The *Drosophila* central nervous system (CNS) forms in the embryo and continues to develop during the larval stages. The outline of the *Drosophila* embryo (A) and third instar larval stage CNS (C) are shown in blue. (B) CNS structures of the embryo are identified in red. The arrow indicates the brain region and the arrowhead indicates the ventral nerve cord. (D) The brain (arrow) and ventral nerve cord (arrowhead) continue to develop and enlarge during the third instar larval stage. In this panel, the areas of red reveal sites of synaptic connections. (From Diaper DC & Hirth F [2014] In Brain Development Methods and Protocols [SG Sprecher ed], pp 3–17. Humana Press.)

the anteroposterior axis. As development progresses through the larval stages, the number of neurons increases and definitive CNS regions form. In the adult, the anterior-most region of the CNS is divided further, reminiscent of subdivisions found in the mammalian brain (**Figure 1.14**). These anterior brain regions are formed from three pairs of ganglia, with each pair controlling specific functions: the protocerebrum (forebrain, largely associated with visual regions), the deutocerebrum (midbrain, largely associated with sensory information from the antennae), and the tritocerebrum (hindbrain, primarily integrates information from the protocerebrum and deutocerebrum; linked to the ventral nerve cord). The more posterior ganglia of the CNS are part of the ventral nerve cord (Figure 1.14). These include the subesophageal ganglia (associated with head and neck regions), the thoracic ganglia (associated with legs and wings structures), and the abdominal ganglia (associated with abdominal structures).

Many of the neurons of the PNS arise from **sensory organ progenitors (SOPs)** located in the surface ectoderm. The SOPs ultimately develop into the mechanosensory, chemosensory, and chordotonal organs of the fly. PNS neurons generally develop later than the cells of the CNS. Mechanisms regulating the formation of various CNS and PNS structures in *Drosophila* are described in Chapters 2, 3, 4, and 6.

There are four types of glial cells in *Drosophila* that are designated cortex, surface, neuropil, and peripheral glia. Many of the functions of these

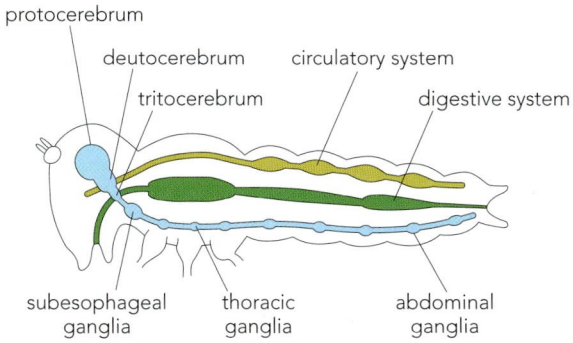

Figure 1.14 The *Drosophila* central nervous system is comprised of pairs of ganglia. The three pairs of ganglia that make up the *Drosophila* brain are divided into protocerebrum (forebrain), deutocerebrum (midbrain), and tritocerebrum (hindbrain). These ganglia connect to the ventral nerve cord comprised of subesophageal, thoracic, and abdominal ganglia. The nervous system (blue) is shown relative to the digestive (green) and circulatory (yellow) systems. (Adapted from Agricultural and Life Sciences, General Entomology, North Carolina State University.)

glia appear similar to vertebrate glia. The cortex glia are most like astrocytes, the neuropil glia are similar to oligodendrocytes, and the peripheral glia function similar to Schwann cells. The cortex, surface, and neuropil glia can also function like the microglia found in the vertebrate CNS.

Cell lineages can be mapped in *C. elegans*

Most adult *C. elegans* are hermaphrodites, with a smaller percentage of the adults being male. The adult hermaphrodite *C. elegans* has a total of 959 cells of which 302 are neurons and 56 are glial. The lineage of each cell has been documented through serial electron micrographs and by following the progeny and fate of individual cells through the translucent body of the tiny worm.

The timing of developmental events established for *C. elegans* maintained at 22°Celsius are shown in **Figure 1.15**. The egg cell is fertilized inside the worm and cells begin to divide in a specific sequence beginning about 40 minutes after fertilization. The eggs are laid at the gastrulation

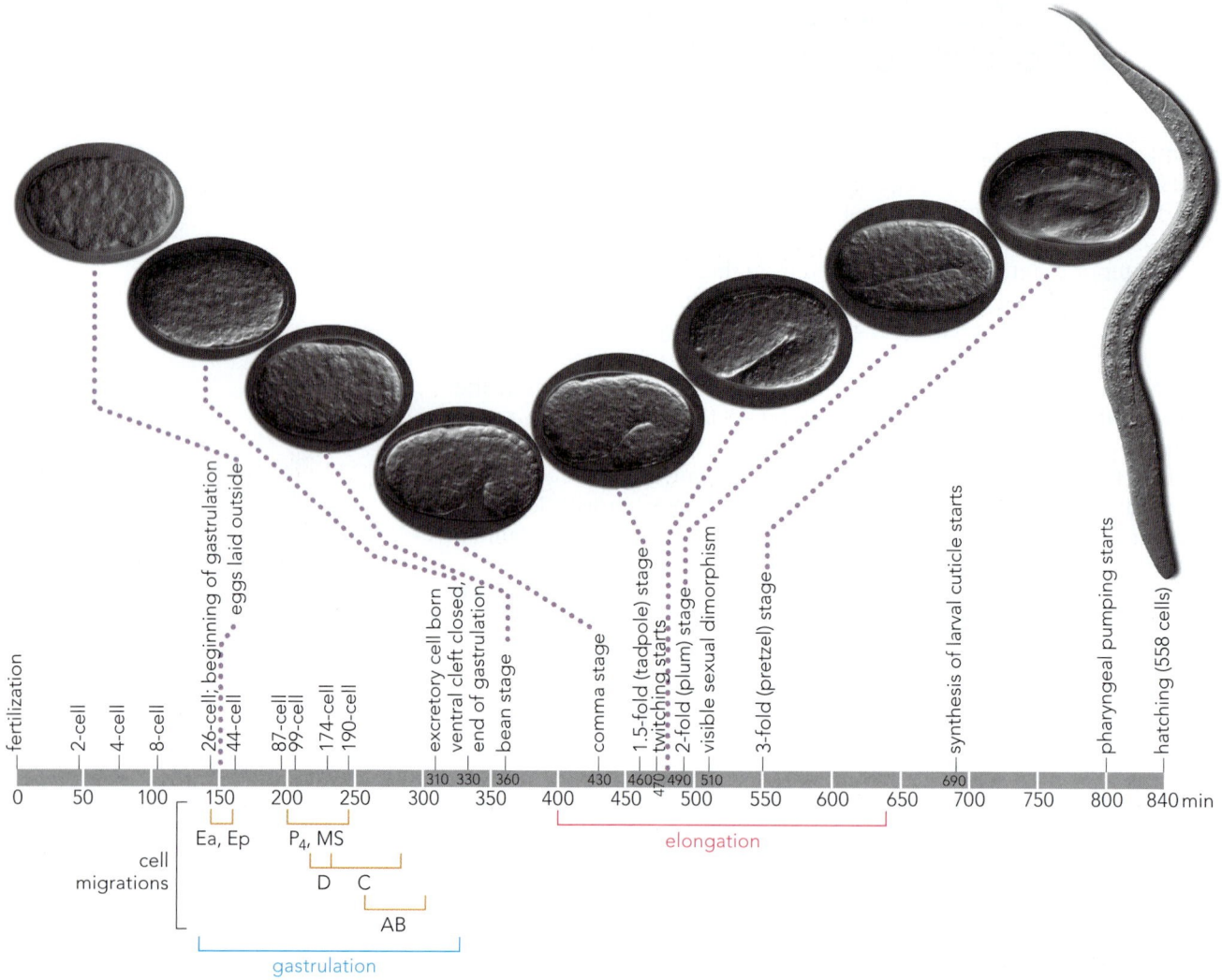

Figure 1.15 An adult *C. elegans* can form within three days of fertilization. A timeline of developmental events through hatching for *C.elegans* maintained at 22 degrees Celsius. The egg is fertilized and begins to divide in the worm. Gastrulation begins when the egg is laid, about 150 minutes after fertilization, when there are 26 cells present. Gastrulation continues until 330 minutes (5.5 hours) after fertilization, when 421 cells are present. The timing of the migration of founder cells during gastrulation is indicated below the timeline. As the worm continues to elongate and become thinner, the worm folds over itself 1.5 times (tadpole stage), then two times (plum stage) and finally three times (pretzel stage). Worms hatch 14 hours after fertilization, when there are 558 cells. The resulting larvae then progresses through four larval stages (L1–L4) before forming an adult worm. Under favorable environmental conditions, the adult worm emerges about 56 hours after fertilization. (Adapted from The Worm Atlas.)

stage, approximately 150 minutes after fertilization, when there are about 26 cells present. Gastrulation is initiated as cells move inward to form the gut and muscle tissues, while the **hypodermis**, the equivalent of ectoderm in vertebrates, remains as the outermost layer. Cells of the hypodermis that subsequently move to the inside of the embryo give rise to the majority of the neurons. The remaining cells of the hypodermis migrate over the surface of the embryo to form the epidermis. Gastrulation continues until the number of cells increase to 421 (about 5.5 hours after the first cleavage). During the final stages of embryogenesis, the shape of the embryo changes from a spherical structure to the elongated shape of the adult. As elongation continues, the worm begins to fold over first 1.5 times (tadpole stage), then 2 times (plum stage), and eventually 3 times (pretzel stage). Hatching occurs about 14 hours after fertilization, when there are 558 cells. The resulting larva then progresses through the four larval stages (L1–L4) before forming an adult worm. During the larval stages, additional neurons are produced, with the majority born during the late L1 stage. The length of time in each larval stage depends, in part, on environmental conditions such as temperature, food supply, and population density. Under favorable conditions, it takes less than 2.5 days for a *C. elegans* to complete the life cycle from fertilized egg to adult worm.

The cell fate options available to a particular cell in *C. elegans* are established early in development with the asymmetric division of the zygote into the AB and P founder cells. A series of cell divisions then results in a total of six founder cells that are designated AB, P, E, MS, C, and D (Figure 1.15). Each founder cell gives rise to progeny in a specific pattern. The initial division of the zygote yields an AB cell located at the anterior pole and the P_1 cell at the posterior pole of the embryo (**Figure 1.16**A). Next, the AB cell divides to produce two daughter cells, the anterior AB.a and the

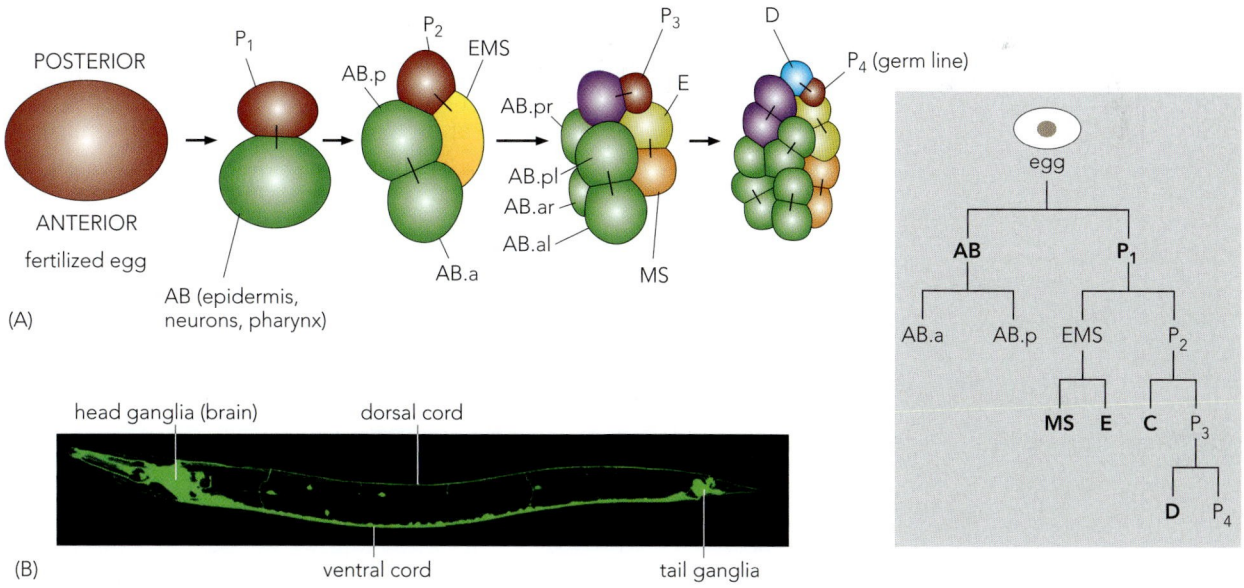

Figure 1.16 Cells in *C. elegans* divide in a precise order. (A) The asymmetric division of the fertilized egg leads to formation of a larger AB founder cell at the anterior pole of the developing embryo and a smaller P_1 founder cell at the posterior pole. The cells continue to divide until a total of six founder cells are produced (AB, P, E, MS, C, and D; bold letters, inset). As shown in the diagram, each dividing cell produces cells in a specific location. The AB cell (green) gives rise to the AB.a cell at a more anterior site and the AB.p at a more posterior site. These cells divide to produce additional daughter cells designated as AB.al and AB.ar, the left- and right-handed daughter cells of AB.a. Similarly AB.p divides to produce AB.pl and AB.pr. The P_1 cell (red) establishes the germ line (red cells) and various somatic cells (yellow, orange, purple, and blue). The AB founder cell gives rise to most neurons in *C. elegans*. Descendants of MS and C give rise to a small number of the neurons in *C. elegans*. (B) Regions of the worm nervous system are stained with green fluorescent protein to outline regions of the brain (head ganglia), the tail ganglia, and the dorsal and ventral nerve cords. (A, adapted from Alberts B, Johnson A, Lewis J et al. [2008] Molecular Biology of the Cell, 5th ed. Garland Science. B, courtesy of Harold Hutter.)

posterior AB.p. Shortly after the AB cell divides, the P_1 cell divides, producing the more posterior P_2 cell and the more anterior EMS cell (Figure 1.16A).

As more AB cells are generated, the location of a cell relative to its sister cell is specified along the anterior–posterior axis. Thus, the AB.a1 cell is the "left-handed" daughter cell of an anterior AB cell, whereas AB.pr is the "right-handed" daughter of the posterior AB daughter cell (Figure 1.16A). The EMS cell divides to produce E and MS cells, whereas P_2 produces C and P_3. P_3 then divides, producing D and P_4, the cell that establishes the germ line. Thus, the P_1 founder cell functions like a stem cell in that it gives rise to both somatic and germ cells.

Each founder cell produces progeny that go on to contribute to specific tissues. However, not all cells of a given body system arise from a single founder cell type. For example, most of the 302 neurons arise from descendants of the AB founder cell, but some arise from the MS and C cells that descend from the P_1 founder cell. In all cases, however, the individual neuronal types always arise from the same precursor and are always found in the same location in the body.

C. elegans has a somatic and pharyngeal nervous system. The somatic nervous system contains 282 neurons found in the head and tail ganglia as well as in the ventral and dorsal nerve cords (Figure 1.16B). These regions contain sensory, motor, and interneurons. The head region also contains numerous sense organs called sensilla that are comprised of free nerve endings and glial sheath and socket cells. The pharyngeal nervous system contains 20 neurons. The pharynx in *C. elegans* is segregated from the rest of the tissues of the body by a unique basement membrane and functions largely independent of the other parts of the worm.

There are 56 glial cells in *C. elegans* that are designated into three categories: sheath, socket, and glial-like nerve ring (GLR). The 24 sheath and 26 socket glia are derived from ectoderm, whereas the six GLR cells are derived from mesoderm. While the functions of glia in *C. elegans* are not as well characterized as in other animal models, they appear to assist in synaptic signaling and play roles in the development, maintenance, and activity of their associated synapses. The GLR cells seem to be specifically associated with signaling that regulates motor movement. Unlike vertebrates and *Drosophila*, axons in *C. elegans* are not myelinated, so glia do not wrap around axons to speed neural conduction.

Despite the variations in anatomy and cellular organization among the different animal models, many of the genes and signaling pathways are conserved across species, allowing discoveries in one animal model to impact discoveries in another. This is particularly helpful when a technique is more readily applied to a simpler invertebrate animal model than a more complex vertebrate model.

GENE REGULATION IN THE DEVELOPING NERVOUS SYSTEM

In all animals, each neuron found in the nervous system must selectively express specific cellular components, such as neurotransmitters, ion channels, cell surface receptors, cytoskeletal elements, and other proteins. The regulated production of these specialized proteins gives individual neurons their unique characteristics and allows them to perform specific functions in the nervous system. Neurons, like other cells, produce only the proteins required at a particular stage of development. In order to selectively produce these proteins, individual genes must be turned on (expressed) or turned off (repressed) at the correct stage of development. The process of turning genes on or off is called **gene regulation**.

A **gene** is a segment of DNA, the double-stranded, helical molecule synthesized in the nucleus of all cells. While DNA is the same in every cell of the body, the genes that are expressed in an individual cell at a particular stage of development will determine which messenger RNA (mRNA) nucleotides are transcribed from the DNA template and therefore which amino acids are translated into proteins (**Figure 1.17**). Among the numerous proteins a cell produces are **transcription factors**—proteins that bind to specific DNA sequences to enhance or suppress expression of a gene.

In many instances, gene expression and protein production are influenced by extracellular signals. The extracellular signal is typically a **ligand** that binds to a cell surface **receptor** protein to initiate intracellular signal transduction pathways. Cell signaling or **signal transduction** is the process by which signals originating outside a cell are conveyed to cytoplasmic components or the nucleus to influence cell behavior. Because the ligand is often thought of as the first messenger in a signal transduction pathway, the subsequent intracellular events are often called second messenger pathways. The activation of various signal transduction pathways regulates cellular events such as survival, death, growth, differentiation, movement, and intracellular communication.

Figure 1.18 outlines an example of a signal transduction pathway, or cascade, where an extracellular signal (ligand) binds to a cell surface receptor. Once the ligand binds to the receptor, subsequent signaling molecules are activated inside the cell. There are often several sequential signaling molecules influenced before the final cellular response is achieved. A signal is said to activate a target **downstream** when it influences the next molecule in the signal transduction pathway. The signal transduction pathway eventually regulates effector proteins that serve a variety of different cellular functions. Common effector proteins are ion channels, metabolic enzymes, cytoskeletal proteins, and gene regulatory proteins (Figure 1.18). Several examples of specific signal transduction pathways utilized during neural development are detailed in subsequent chapters.

The structure of each ligand and receptor is unique so that a given ligand only binds to corresponding receptors. This allows for binding specificity.

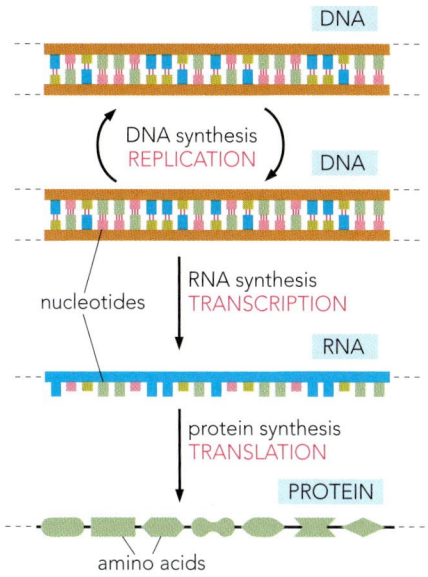

Figure 1.17 Gene regulation determines which proteins are produced in a cell. DNA synthesis (replication) takes place in the nucleus of the cell, where the template strand of DNA synthesizes a complementary strand of DNA. A DNA template strand can also initiate synthesis (transcription) of a complementary strand of RNA. During protein synthesis (translation) the sequence of nucleotides in a strand of messenger RNA determines the order of amino acids and therefore the resulting protein structure. (Adapted from Alberts B, Johnson A, Lewis J et al. [2015] Molecular Biology of the Cell, 6th ed. Garland Science.)

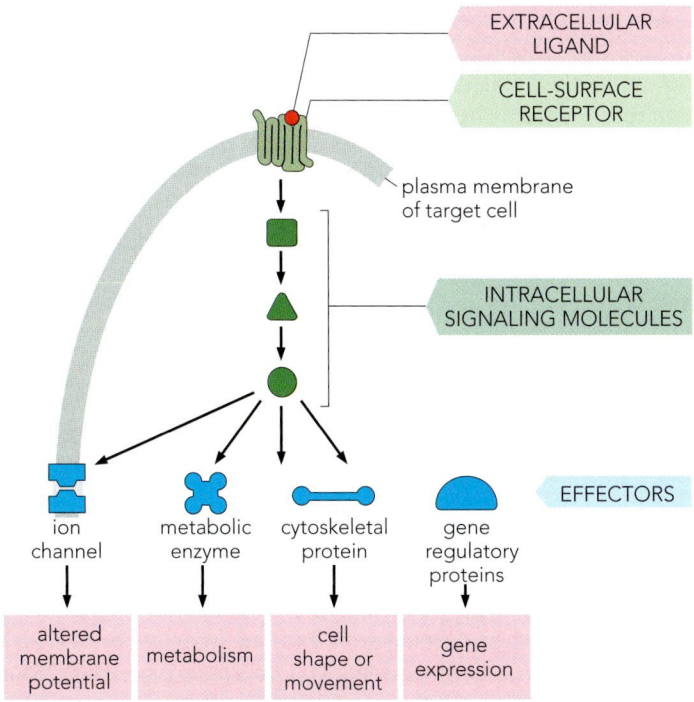

Figure 1.18 Signal transduction pathways transfer extracellular information to the cell. When an extracellular signal (ligand) binds to its corresponding receptor located on the surface of a cell, one of many intracellular signal transduction cascades can be initiated. Each step in the cascade stimulates the next molecule in the pathway until an effector protein is influenced. Examples of effector proteins include ion channels, such as those that alter a neuron's membrane potential, metabolic enzymes that impact cellular metabolism, cytoskeletal proteins that influence cell shape and movement, and gene regulatory proteins, such as transcription factors, that influence whether a gene is expressed or repressed. (Adapted from Luo L [2016] Principles of Neurobiology. Garland Science.)

Binding specificity combined with the ability of cells to regulate the expression of the multitude of ligands and receptors ensure that specific signaling pathways are only available to a cell when needed. Thus, cells become specialized so they only respond to required signals at each developmental stage and ignore other signals that may also be present at that time.

Experimental techniques are used to label genes and proteins in the developing nervous system

Because each cell subtype in the nervous system expresses a unique set of genes and proteins, researchers have developed several techniques to identify where and when these molecules are expressed during development and in adulthood. Among the techniques are those that use microscopy to identify the distribution of genes and proteins in tissues or individual cells. These approaches lead not only to understanding the cellular distribution of genes and proteins, but also provide a way to label or mark particular cells and track them over the course of development. This has been especially helpful, because the outward morphological appearance of embryonic neurons is often homogeneous, making it difficult, or impossible, to identify a cell with any certainty following any sort of experimental manipulation.

To visualize gene expression in neural tissues, scientist use ***in situ* hybridization**. With this technique, mRNA is visualized by incubating whole embryos, tissue sections, or cultured cells with probes made up of a DNA nucleotide sequence of interest. These probes are labeled with a radioactive or fluorescent marker so that when the DNA probe binds to the corresponding mRNA sequence, the cells expressing that mRNA can be visualized. Identifying gene expression patterns often provides insight into the putative function of that gene in a given cell, while also providing a labelling method to track changes in gene expression under normal and experimental conditions. Examples of experiments using *in situ* hybridization are found in Figures 3.24 and 4.20.

Scientists use **immunohistochemistry** or **immunocytochemistry** to label proteins in tissues or cells, respectively. These methods take advantage of the immune response in which an animal develops antibodies to a foreign substance. For example, one method of generating antibodies is to inject rabbits with a protein of interest. The animals develop antibodies to the protein that are then isolated from the blood serum. The resulting antibodies, called primary antibodies, are then added to tissues or cell cultures, where they bind to the target protein. These antibodies are visualized by either adding an enzymatic reporter molecule, such as horseradish peroxidase, or a fluorescent probe directly to the primary antibody, or by adding one of the labels to a secondary antibody that recognizes and binds the primary antibody. These methods often provide fine details on the distribution of a protein within a cellular region. Examples of such methods are shown in Figure 1.13, in which a neuron-specific protein is used to visualize the entire nervous system (Figure 1.13B), and another is used to identify a single presynaptic element known as the active zone protein (see Chapters 9 and 10), which is localized to presynaptic nerve terminals (Figure 1.13D). Another example of **immunolabelling** to identify the distribution of synaptic contacts on a single neuron is shown in Figure 10.1.

Altering development as a way to understand normal processes

One of the most common ways to assess normal developmental events is to alter some aspect of development and see what happens. This allows researchers to test whether a given tissue, cell, or protein is necessary for normal development to occur. Over the past century and a half, a number of methods have been used to alter development. Among the common approaches used today are techniques to manipulate tissues *in vivo* and *in vitro* and methods to evaluate naturally occurring and experimentally induced genetic mutations.

Tissue manipulations have been used since the earliest studies of developmental neurobiology. These methods typically involve surgically removing or rotating a particular region of the developing embryo or grafting extra tissue onto a region of the embryo. Several examples of these types of studies are highlighted in Chapters 2, 4, 7, and 8. Scientists can also observe effects of tissue manipulations in cell culture preparations. In these assays, tissues of interest are surgically dissected from an embryo at a given stage of development and placed into a cell culture dish. The dishes are often coated with substrate molecules that support the attachment and growth of the cells under investigation. The tissues are then covered in a nutrient-containing fluid (cell culture medium). To identify sources of signals that promote the survival, growth, or differentiation of a neural population, the tissues may be grown in the presence of other tissues. In some experiments, specific proteins may be added to test whether they have a direct effect on the developing cells. Examples using these approaches are discussed in Chapters 4, 7, and 8. Cell culture techniques to study neural development were introduced in the 1920s and remain a very popular method for analyzing the development of neural cells. An advantage of cell culture is the ability to test single reagents on a select population of cells. A limitation to the method is that the artificial environment removes other tissue-derived cues that may interact with and alter the effects of the reagent under investigation.

Scientists also observe the effects of additional or missing genes. Such **genetic manipulations** have been instrumental in understanding neural development in both invertebrate and vertebrate animal models. Studies of naturally occurring gene mutations in *Drosophila, C. elegans*, and mice have been documented for nearly a century. A number of these spontaneously occurring mutations have provided an extensive body of data on the development of the nervous system. As detailed in Chapter 6, scientists investigating *Drosophila* initially relied on naturally occurring mutations, but soon developed methods to experimentally mutate genes of interest. Methods for blocking, reducing, or increasing gene expression were also developed for many vertebrate animal models. Examples of mutations induced in frogs are found in Figures 2.10, 2.11, and 6.2, while examples from *Drosophila* are shown in Figures 7.20 and 7.22. A method to experimentally delete, or knock out, individual genes in mice was introduced in the 1980s. The development of a technique for generating **gene knockout mice** greatly advanced studies of mammalian neural development (**Box 1.2**).

Other methods to selectively interfere with gene expression use **short interfering RNAs (siRNAs)**. Segments of RNA consisting of 20–25 base pairs that are complementary to a gene sequence of interest are introduced to cells by **electroporation**, a method in which an electrical current is used to make cell membranes more permeable. siRNAs can be electroporated into specific regions of an embryo, where they degrade the target mRNA and prevent translation of the protein, thereby providing insight into the normal function of the protein *in vivo*. Examples of this approach are shown in Figure 4.10.

Researchers continue to refine techniques to selectively alter gene and protein expression in cells at specific stages of development, providing finer resolution of the molecular pathways involved in neural development. There are limits to these approaches, however, and researchers are aware that induced changes represent an artificial environment and that complementary studies are needed to test the role of the molecules during normal development. Despite the inherent limitations of these approaches, to date such tissue and genetic manipulation studies have provided considerable insight into mechanisms underlying normal neural development.

Box 1.2 Knockout mice

The term "knockout mouse" is now commonly used throughout the scientific literature. The technique has become so widely used and discussed that it may be difficult to imagine what a surprising and significant impact it had when it first emerged in the early 1980s. In fact, when Mario Cappecchi first proposed the technique to a funding agency, the proposal was turned down because reviewers believed the process could not work effectively. The technique relies on the process of **homologous recombination**—the ability of an inserted DNA sequence to line up in the correct orientation and location and replace a specific gene. Homologous recombination takes place naturally and frequently in bacteria, yeast, and viruses, but under normal conditions is rare in mammalian cells, except in germ-line and **embryonic stem (ES) cells**—that is, the cells that have the ability to give rise to all the cells in an organism. Mammalian cells are also capable of the process when foreign genes are intentionally inserted, such as occurs in the process of generating knockout mice.

Figure 1 outlines the steps used to generate mice lacking a gene of interest. In the first step, the target gene is removed from a segment of DNA and selector genes are inserted to create a targeting vector. Two commonly used selector genes are the *neomycin resistance* (neo^R) gene—a positive selector gene—and the *herpes thymidine kinase* (*tk*) gene—a negative selector gene. The neo^r gene is flanked by DNA present in the target gene, while the *tk* gene is located outside the targeted sequence.

In the second step, the target vector is introduced into mouse embryonic stem (ES) cells. Electroporation provides a small electrical charge that opens the cell membranes and permits entry of the DNA. The cells are grown in a culture medium that contains the drugs neomycin and glanciclivor. Cells that have inserted the neo^R gene in place of the targeted gene will survive in the medium containing the antibiotic neomycin. Glanciclivor will kill any cells that retain the *tk* gene; thus, the cells that have randomly inserted the targeting vector outside the gene sequence of interest will be eliminated. By using both positive and negative selector genes, all or nearly all of the cells surviving in the culture medium will be those with the targeted gene disrupted.

The remaining ES cells are then injected into blastocyst-stage mouse embryos from mice of a particular coat color (Figure 1, step 3). The blastocysts are implanted into a surrogate, or foster, female mouse of a different coat color to develop to term. The tissues of the pups from this first litter contain cells that arise from both mice. These chimeras, made up of genetic contributions from the blastocyst and surrogate mice, can be identified by a coat color that differs from that of the surrogate mother. The male chimeras are then mated with females of another coat color, such as white (step 4). The resulting pups are again selected by coat color to identify those that carry genes from the ES cells (for example, mice that are black). The DNA from these mice is then sequenced and those mice that are heterozygous—that is, the mice that contain one copy of the disrupted gene and one copy of the normal gene—are then mated with littermates that are also heterozygous for the mutation. One quarter of the resulting litter will contain mice that are homozygous for the mutation. These are the knockout mice that contain two copies of the mutated gene and are the ones to be examined for anatomical, physiological, or behavioral deficits. Other mice in the litter will be normal, or wild-type mice, which carry two copies of the normal gene, and the others will be heterozygous.

Depending on the particular gene disrupted, the resulting changes—the phenotype of the mice—can be mild, suggesting that another gene compensates for the loss of the targeted gene, or severe, sometimes even resulting in death of the embryo prior to birth. As scientists have found over the years, the heterozygous mice often have a milder phenotype than the homozygous mice, displaying a gene dosage effect—that is, those with one copy disrupted are impacted less than those with two copies disrupted. Embryonic lethal mutations can sometimes provide information on the role of the gene, depending on the stage when the embryos die. Such severe mutations are often of limited value, however, particularly if the embryo dies prior to the onset of gene function in the cell population of interest.

A refinement to the gene knockout technique was introduced in the late 1980s that allows for researchers to delete a gene in selected tissues or at specific stages of development and therefore overcome the limitations of deleting a gene in every cell of the body. The basic method used to create such **conditional knockout mice** involves inserting loxP (locus of X-over P1) in noncoding regions of the DNA sequence of interest using homologous recombination. These segments are said to be "floxed" (flanked by loxP). The floxed sites of the DNA are recognized by Cre recombinase that mediates the exchange of DNA. The conditional expression of the gene is regulated in one of two ways—namely, Cre recombinase can

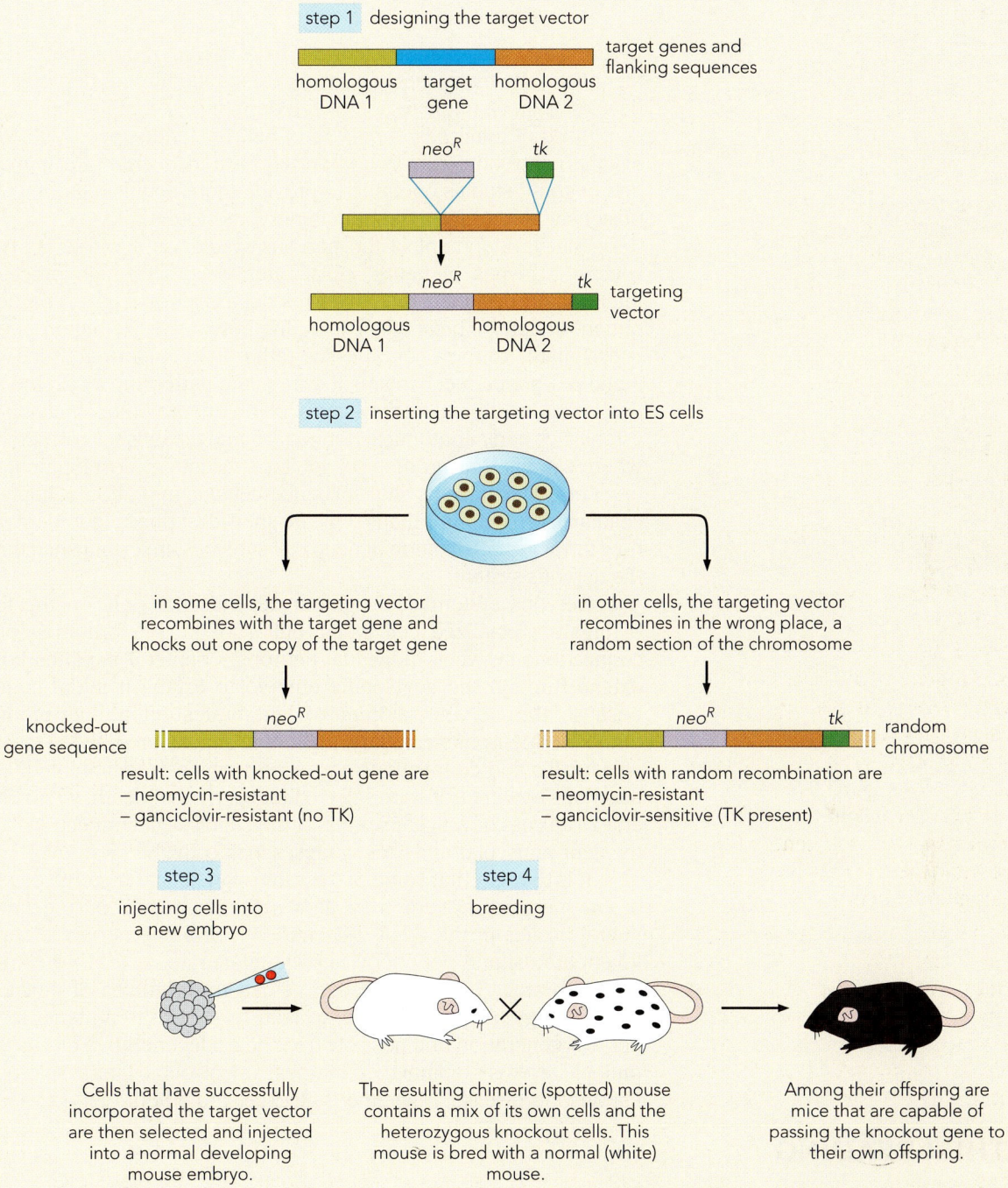

Figure 1 The creation of a knockout mouse. An outline of the steps used to generate mice with the targeted gene disruption.

be linked to tissue-specific promoters or to inducible proteins. Thus, depending on the method used, the gene of interest will only be altered in selected tissues or only at the developmental stages in which it is experimentally induced, thus allowing researchers to investigate the role of a gene in a selected cell population or at a specific time in development. A number of modifications to these initial knockout and conditional knockout methods have been made since they were first introduced over 25 years ago so that researchers now have the ability to track, delete, or overexpress multiple genes in a single animal.

SUMMARY

In the past 125 years, scientists have made remarkable progress in identifying the many mechanisms that lead the human nervous system to develop from a simple region of ectoderm into a complex nervous system comprised of hundreds of billions of neurons and supporting cells. These insights have come from several vertebrate and invertebrate animal models and ongoing studies continue to provide more details regarding the types of signals used to form the human nervous system.

The development of the nervous system as a whole, as well as of specific subtypes of neural cells, often occurs through processes that intentionally designate or actively restrict tissues or cells to a particular fate option. Chapter 2 begins by explaining how a specific region of ectoderm is designated to form neural tissue rather than epidermal tissue. Chapters 3 and 4 describe mechanisms that regulate patterning along the anterior–posterior and dorsal–ventral axes of the neural tube. Such patterning establishes early anatomical boundaries and establishes restricted regions where signals that promote or inhibit subsequent developmental events originate. Mechanisms that influence cellular migration (Chapter 5) and cellular fate options (Chapter 6) are then described, thus demonstrating how further specialization of neuronal subtypes are established throughout the nervous system.

Nervous system development depends not only on the formation of highly specialized cell types, but also on the formation of precise connections between individual neurons. Chapter 7 explains how axons extend through the developing embryo to establish initial connections. Because the nervous system of vertebrates often over-produces the number of neurons necessary, additional signals are needed to determine which cells and connections persist and which are eliminated. Chapter 8 describes how neurons are selected for survival or death, while Chapters 9 and 10 discuss how synapses are formed and select connections are lost at the neuromuscular and CNS synapses, respectively.

It is thought that some of the same signals that underlie development of the nervous system may also influence repair or regeneration of neural tissues in the adult. Regeneration of many nerves is common in invertebrates; and peripheral nerves regenerate, at least to some extent, in vertebrates. Thus, knowledge gained from studies of developmental neurobiology has the potential to lead to clinical treatments to enhance nerve regeneration and possibly treat neurodegenerative diseases such as multiple sclerosis, spinal cord injuries, amyotrophic lateral sclerosis (ALS), Parkinson's disease, Alzheimer's disease, and others.

FURTHER READING

Alberts B, Johnson A, Lewis J et al. (2015) Molecular Biology of the Cell, 6th ed. Garland Science.

Capecchi M (1994) Targeted gene replacement. *Sci Amer* 270(3)52–59.

Diaper DC and Hirth, F (2014) Immunostaining of the developing embryonic and larval *Drosophila* brain. In Brain Development Methods and Protocols (Sprecher SG ed), pp 3–17. Humans Press.

Friedel RH, Wurst W, Wefers B and Kuhn, R. (2012) Generating conditional knockout mice. In Transgenic Mouse Methods and Protocols, Methods in Molecular Biology. (Hofker MH & van Deursen JM eds), pp 205–231. Humana Press.

Gilbert SF (2014) Developmental Biology, 10th ed. Sinauer Associates.

Hamburger V & Hamilton HL (1951) A series of normal stages in the development of the chick embryo. *J Morph* 88(1):49–92.

Haretenstein V (1993) Atlas of *Drosophila* Development. Cold Spring Harbor Laboratory Press.

Kimmel CB, Ballard WW, Kimmel SR et al. (1995) Stages of embryonic development of the zebrafish. *Dev Dyn* 203:253–310.

Luo L (2016) Principles of Neurobiology. Garland Science.

Nieuwkoop PD & Faber J (1967) Normal Table of *Xenopus laevis*. Elsevier.

Oikonomou G & Shaham S (2011) The glia of *Caenorhabditis elegans*. *Glia* 59:1253–1263.

Sadler TW (2014) Langman's Medical Embryology, 13th ed. Wolters Kluwer Health.

Schoenwolf GC, Bleyl SB, Brauer PR & Francis-West PH (2014) Larsen's Human Embryology, 5th ed. Churchill Livingstone.

Stern CD & Holand PWH (1993) Essential Developmental Biology. IRL Press.

Stix G (1999) Profile (Mario Capecchi): Of survival and science. *Sci Amer* 281(2):26–27.

Vanderah TW & Gould DJ (2016) Nolte's The Human Brain: An introduction to its Functional Anatomy, 7th ed. Elsevier.

Westerfield M (1993) The Zebrafish Handbook, 2nd ed. University of Oregon Press.

ONLINE SOURCES FOR ATLASES OF EMBRYONIC DEVELOPMENT

The Endowment for Human Development: http://www.ehd.org/virtual-human-embryo/stage.php

eMouseAtlas

https://www.emouseatlas.org/emap/ema/theiler_stages/StageDefinition/stagedefinition.html

eChickAtlas

http://www.echickatlas.org/ecap/hamburger_hamilton_stages/hamburger_hamilton_stages.html

The Zebrafish Information Network

https://www.zfin.org/zf_info/zfbook/stages/stages.html

Xenbase

http://www.xenbase.org/anatomy/alldev.do

Atlas of *Drosophila* Development

http://www.sdbonline.org/sites/fly/atlas/00atlas.htm

C. elegans

Altun, Z. F. and Hall, D. H. 2005. Introduction to *C. elegans* anatomy. In *WormAtlas*. http://www.wormatlas.org/ver1/handbook/anatomyintro/anatomyintro.htm

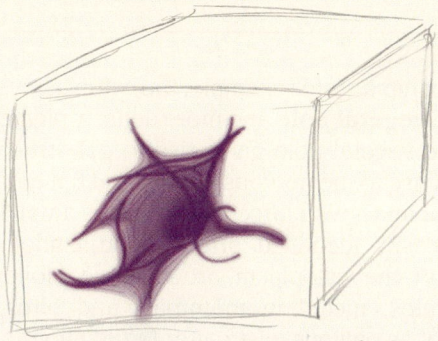

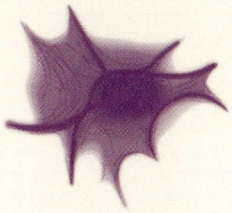

Neural Induction

2

The first step in neural development is **neural induction**, the process by which undifferentiated embryonic tissue is specified to develop as neural tissue. Once specified, neural regions continue to differentiate and specialize, eventually generating the many distinct regions of the nervous system. The steps involved in creating a functional nervous system are numerous and complex, and the process of neural differentiation begins very early in development. In fact, it is often within the egg cell itself that cellular regions competent to become neural tissue are first designated. Segregating neural regions at the earliest stages of development prepares cells to receive the correct signals at the right time so that normal development can occur. The following quote from Hans Spemann, a pioneer in the study of neural induction, emphasizes the importance of embryonic organization and segregation in the development of neural tissues:

> "We are standing and walking on parts of our body which we could have used for thinking if they had been developed in another position in the embryo."
>
> Hans Spemann, 1924

Neural induction has been the focus of research efforts for well over a century. As detailed below, these processes remain highly conserved across vertebrate species, though certain differences are noted and will be outlined in the examples that follow. Despite significant differences in nervous system morphology, the signaling molecules used to designate neural tissues are largely homologous in vertebrates and invertebrates, demonstrating the evolutionary conservation of this fundamental, early process of neural development.

THE ESTABLISHMENT OF NEURAL TISSUE DURING EMBRYOGENESIS

During the earliest stages of development, specific areas of the vertebrate embryo are designated to form neural tissues. Even prior to fertilization, an egg cell is compartmentalized as a first step in establishing neural tissue. In

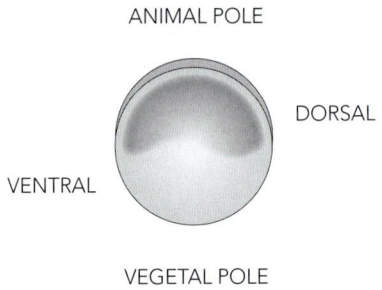

Figure 2.1 Regions competent to become neural tissue are established in the egg cell. Eggs from amphibians are divided into animal and vegetal poles that go on to establish areas that will give rise to specific tissues. The animal pole appears denser due to the accumulation of cytoplasmic lipids and granules. This pole will later give rise to epidermis and neural tissues. The lighter appearing vegetal pole will later give rise to structures associated with the gut.

egg cells from amphibians, for example, the cytoplasm is segregated into a lighter pigmented **vegetal pole** and more densely pigmented **animal pole** (**Figure 2.1**). The vegetal pole gives rise to gut structures, whereas the animal pole gives rise to the nervous system and the epidermis (the surface layer of skin). In animals with large eggs, such as amphibians, cytoplasmic differences are easily identified prior to fertilization, and subsequent rearrangements in the cytoplasm are observed following fertilization. In other vertebrates, including mammals and birds, the cytoplasmic differences are not as obvious, even after fertilization.

Following fertilization, the egg cell (**zygote**) divides into a number of cells called **blastomeres** that surround a central cavity known as a **blastocoel** (**Figure 2.2**). A group of blastomeres that aggregate above the cavity is called the **blastoderm**. This entire structure—that is, the cells and the hollow cavity they surround—is known by different names, depending on the species: the **blastula** in amphibians, the **blastocyst** in many mammals, and the **blastodisc** in birds, fish, and some mammals. The difference in terminology refers to the morphology of the cells. In birds, for example, the cells form a disc-like structure that is distinct from the cyst-like morphology observed in mice. In many cases, the term "blastula" is used as a generic term for all species.

The amphibian blastula-stage embryo (Figure 2.2A) is spherical, and like the egg itself, the embryo at this stage is characterized by animal and vegetal poles. In birds (Figure 2.2B), the blastodisc lies above the yolk, where it forms two sheets of cells designated the **epiblast** and **hypoblast**—regions that roughly correspond to the animal and vegetal poles, respectively. The epiblast and hypoblast are also found in mammals, however the size of the yolk, a source of nutrients for the embryo, is much greater in birds than in mammals. In fact, in most mammals the size of the yolk is negligible due to maternal sources of nutrients. Yet, despite species differences in blastula shape and yolk size, the next step in early development is comparable across vertebrate species.

Gastrulation creates new cell and tissue interactions that influence neural induction

Gastrulation is the next crucial stage in early embryogenesis—the stage during which cells first begin to reorganize into three germ cell layers to form specific tissues. These layers, the ectoderm, mesoderm, and endoderm, are called the **germ cell layers** or **primary cell layers**, because it is from these three layers that all tissues of the body arise. Gastrulation involves the migration, or invagination, of cells through an indentation on the outer

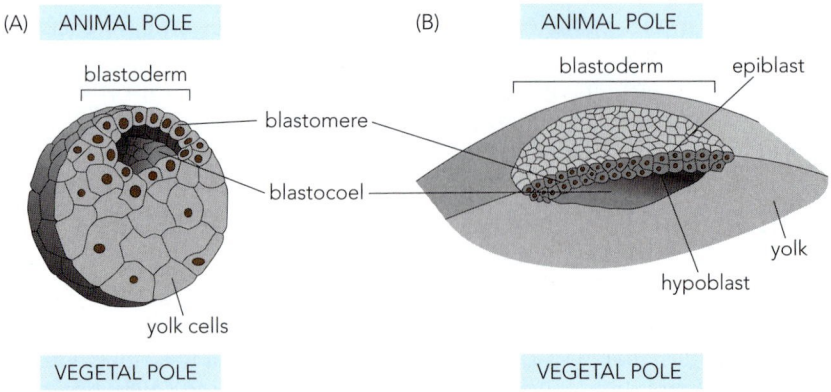

Figure 2.2 Comparison of blastula formation in amphibians and birds. Following fertilization, the egg cell divides repeatedly and the resulting new cells are called blastomeres that surround a hollow, fluid-filled cavity, the blastocoel. The consolidated group of cells above the blastocoel is called the blastoderm. In frogs (A), the blastula is a spherical structure, whereas in birds (B), the blastula, also called the blastodisc, appears as a flattened sheet of cells overlying the yolk, with the blastocoel cavity formed between. The two layers of cells in the blastodisc are termed the epiblast and hypoblast, regions that are roughly equivalent to the animal and vegetal poles. (Adapted from Patten BM [1958] Foundations of Embryology. McGraw-Hill.)

surface of the developing embryo. Here again, the terminology varies according to species. This site of indentation is called the **blastopore** in amphibians, the **embryonic shield** in zebrafish, and the **primitive streak** in birds and mammals.

As surface cells begin to migrate through the surface indentation (that is, the blastopore, embryonic shield, or primitive streak), new arrangements of the germ cell layers occur (**Figure 2.3**A, B). In all cases, as a result of gastrulation, the **endoderm** forms as the innermost germ cell layer and gives rise to the gut and organs associated with it. The **mesoderm** forms as the middle layer and gives rise to muscle, bone, connective tissues, and the cardiovascular and urogenital systems. The outermost layer, the **ectoderm**, remains at the embryo's surface (animal pole or epiblast) and gives rise to the epidermis and nervous system. In all of these animal models, during a limited period in the late gastrula-stage embryo, the ectodermal cells have the potential to become either epidermal cells or neural cells.

Figure 2.3C–E shows sketches of amphibian gastrulation modified from the original 1929 drawings by Walter Vogt. The sketches provide

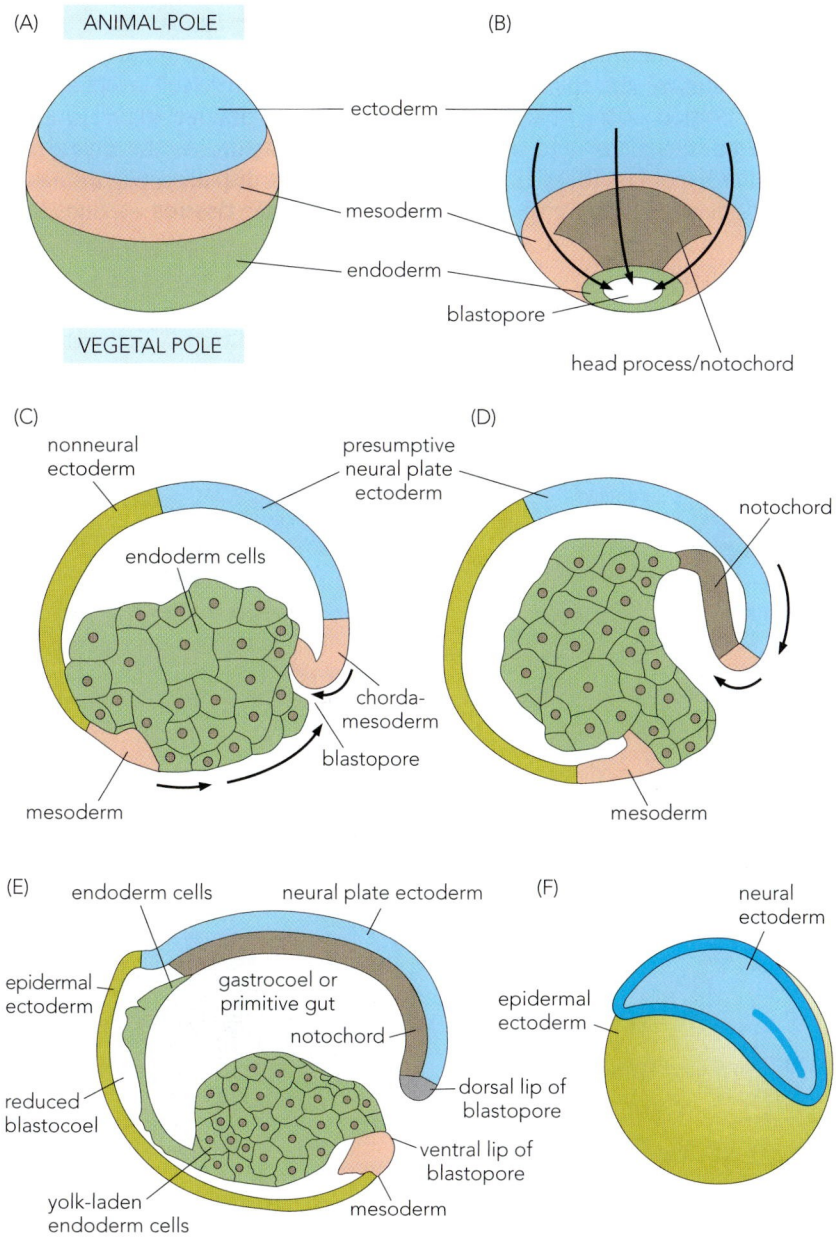

Figure 2.3 Transition from the blastula- to the gastrula-stage embryo in amphibians. (A) A surface view of an early stage amphibian blastula outlining the general areas of presumptive ectoderm, mesoderm, and endoderm. The animal pole will give rise to ectoderm (blue), the vegetal pole will give rise to endoderm (green), while the mesoderm (light red) will arise from the middle segment. (B) Schematic fate map shown from the dorsal surface of a blastula-stage amphibian embryo just prior to the onset of gastrulation. The sight of blastopore formation (white) indicates the area through which surface cells will migrate (arrows). The regions fated to become the three germ layers (ectoderm, mesoderm, or endoderm) and the head process/notochord (brown) are also shown. (C–E) show sagittal sections through the center of the blastula to reveal the inward movements of cells and the subsequent arrangement of tissue layers as gastrulation progresses. Dorsal is to the right in all panels. The arrows indicate the direction of cellular movements. As the cells of the presumptive endoderm (green) and mesoderm (light red) push inward through the blastopore (C, D) the primitive gut (gastrocoel) is established (E). At the dorsal surface (C), the chordamesoderm (axial mesoderm) also migrates inward, forming the notochord (brown, D) that extends below the forming neural plate ectoderm (blue, C–E). The cells of the ventral surface ectoderm will give rise to the nonneural, epidermal ectoderm (yellow). The mesoderm-derived dorsal and ventral blastopore lips at the margins of the blastopore are seen in panel E. Signals emanating from the dorsal blastopore lip (shown in gray) are critical for formation of neural ectoderm. (F) Surface view showing regions of neural (blue) and epidermal (yellow) ectoderm. (B, adapted from Nagy A, Gertsenstein M, Vinterstan K & Behringer R [2003] Manipulating the Mouse Embryo, 3rd ed. Cold Spring Harbor Press; C–E, adapted from Voght W (1929) Gestaltungsanalyse am amphibienkeim mit örtlicher vitalfarbung. II. Teil. Gastrulation und mesodermbildung bei oroden und anuren. *Wilhem Roux' Arch* 120: 385–706.)

remarkable details of the cellular rearrangements as they were viewed through the microscope at different stages. For example, the early migration of cells through the blastopore (Figure 2.3B) leads to the creation of the endoderm layer (Figure 2.3C, D) and the primitive gut (Figure 2.3E). In fact the word gastrulation comes from the Greek word *gaster* meaning stomach. Other cells then migrate through the blastopore and come to lie between the endoderm and surface ectoderm, thus forming the mesoderm. At the dorsal side, a specified group of these cells migrates inward to form a band of **chordamesoderm** (also called the axial mesoderm). This band of mesoderm gives rise to the notochord, a transient embryonic structure that lies below the surface ectoderm and is important for subsequent aspects of neural development (Figure 2.3D, E). At the opening of the blastopore, the margins form the mesoderm-derived dorsal and ventral blastopore lips. As described below, it is the **dorsal blastopore lip (DBL)** that is critical for designating neural ectoderm from epidermal ectoderm, the two cell types that remain at the outermost surface of the embryo (Figure 2.3E, F). Gastrulation in zebrafish would be similar to amphibians, because the embryonic shield more closely resembles the blastopore than the primitive streak.

As a comparison to amphibian gastrulation, **Figure 2.4** shows gastrulation in the chick embryo, a process that is similar in human embryos. The outer surface, the epiblast, gives rise to all of the future cell layers (Figure 2.4A). A portion of the cells migrates through the primitive streak to form the middle mesoderm and innermost endoderm layers. As the endoderm forms, the hypoblast that lies just above the yolk is displaced (Figure 2.4B). The hypoblast cells will give rise to **extraembryonic tissues**—tissues that

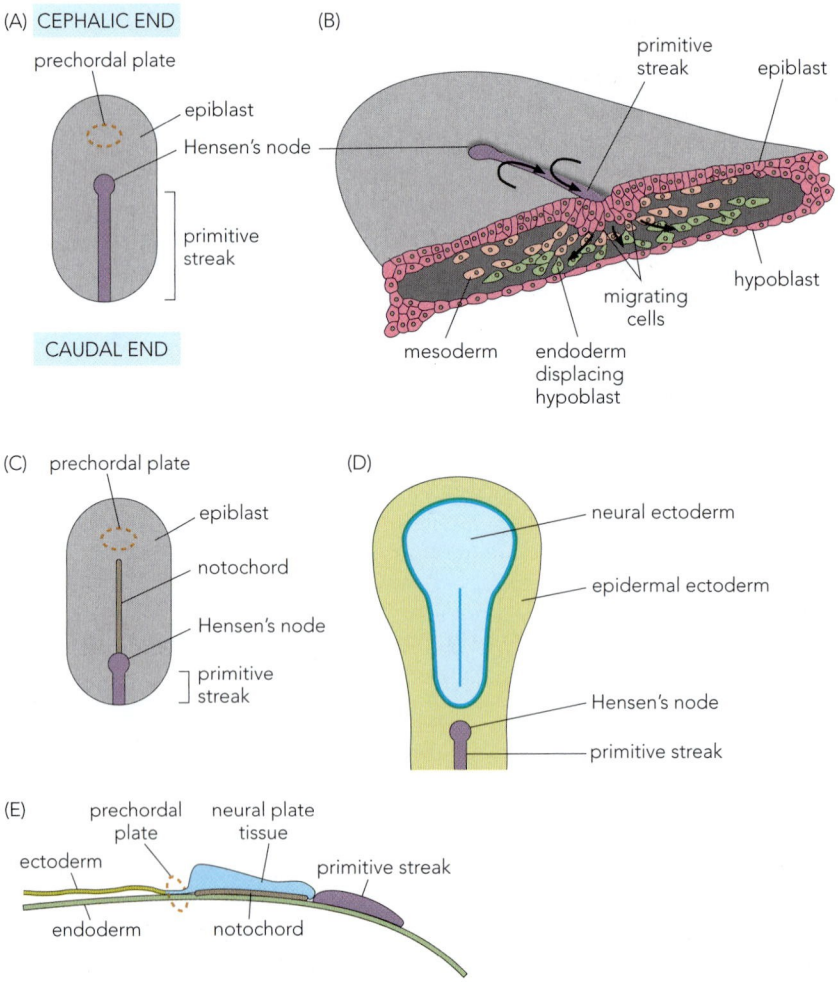

Figure 2.4 Gastrulation in chick embryos. (A) A surface view of a chick blastodisc at approximately 16 hours of development. The surface cells (epiblast) give rise to all of the future germ layers. The primitive streak, which forms the longitudinal axis of the embryo, establishes the site where cells migrate inward during gastrulation. Hensen's node, located at the anterior end of the primitive streak, is the equivalent of the amphibian dorsal blastopore lip. The area of the prechordal plate, where ectoderm and endoderm adhere, is outlined anterior to Hensen's node. (B) A cross section of a chick embryo demonstrates how the surface cells of the epiblast (gray) migrate medially into the opening of the primitive streak, then migrate laterally to establish the mesoderm and endoderm layers. The hypoblast, lying above the yolk, is displaced as the endoderm forms. The hypoblast later gives rise to extraembryonic tissues. (C) Cells migrating inward along the midline give rise to the notochord, which lies below the future neural plate (not shown). As more cells migrate inward, the primitive streak regresses. (D) The cells of the epiblast that remain at the surface form epidermal (yellow) and neural (blue) ectoderm, with the neural tissue forming anterior to the primitive streak. (E) A sagittal section through the embryo reveals the orientation of the neural tissue and primitive streak as well as the prechordal and notochord regions. The notochord lies beneath the neural plate tissue. The prechordal plate is formed where the surface ectoderm and underlying endoderm form a tight junction. The prechordal plate limits the anterior migration of notochord cells. Gastrulation in humans and other mammals that form a blastodisc is similar to that of chick. (A, adapted from Patten BM [1958] Foundations of Embryology. McGraw-Hill; C, E adapted from Balinsky BI [1975] An Introduction to Embryology, 4th ed. Saunders.)

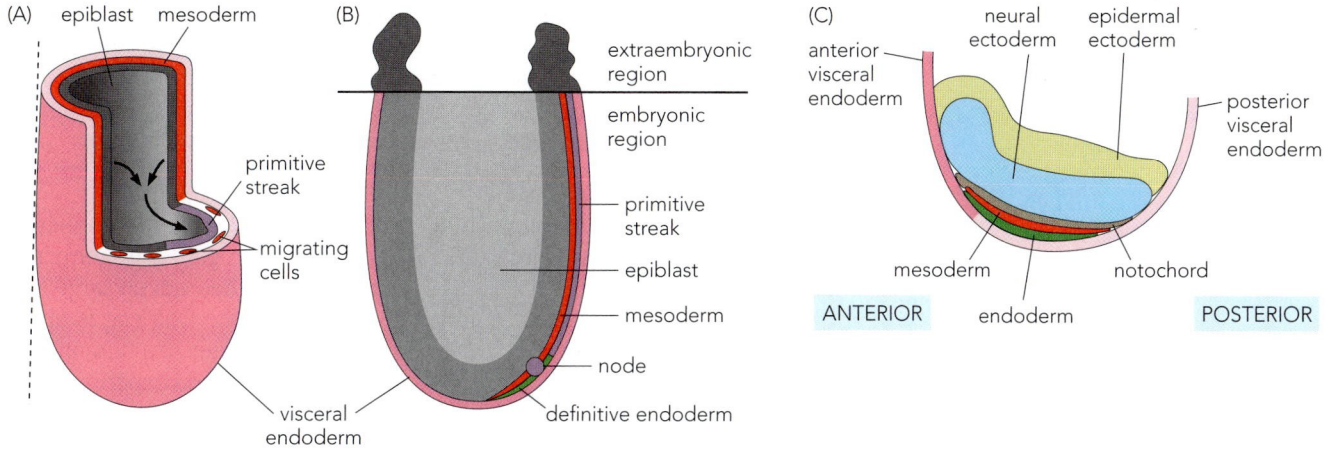

Figure 2.5 Gastrulation in mouse embryos is similar to that in the chick. If the blastocyst of a mouse embryo is flattened out, the movements of cells appear similar to those of the blastodisc. (A) The blastocyst with part of the posterior region cut away to reveal the orientation of the forming cell layers. If the blastocyst were cut open along the dashed lines and flattened out, the epiblast would be on the surface and the migration of cells (arrows) would be appear similar to the those of the chick blastodisc. (B) A sagittal section through the cup-shaped blastocyst of the mouse embryo reveals the epiblast in the inner region. The cells of the epiblast migrate through the primitive streak that begins at the junction of the extraembryonic and embryonic regions. The migrating epiblast cells begin to form germ layers between the epiblast and visceral endoderm. The visceral endoderm does not form part of the definitive endoderm of the embryo, but is an extraembryonic tissue that is replaced by cells that form the endoderm layer, similar to the hypoblast in the chick. The node is located at the anterior end of the primitive streak. (C) A section through the blastocyst at a later stage of development reveals the orientation of the neural ectoderm, epidermal ecdoderm, and notochord. The anterior visceral endoderm underlies the anterior portion of the neural tissue. (Adapted from Nagy A, Gertsenstein M, Vinterstan K & Behringer R [2003] Manipulating the Mouse Embryo, 3rd ed. Cold Spring Harbor Press.)

are not part of the embryo proper, but contribute to the maintenance of the embryo. Similar to amphibian embryo, a specified group of cells migrates inward along the midline to form the notochord. As the notochord elongates and the germ layers are established, the primitive streak regresses (Figure 2.4C, E). At the anterior region of the epiblast, the **prechordal plate** forms (Figure 2.4A, C, E). This is an area where the surface ectoderm adheres to the underlying endoderm. This tight association prevents cells of the notochord from migrating further anteriorly (Figure 2.4E). The epiblast cells remaining at the surface form neural ectoderm and epidermal ectoderm. The neural tissue forms as a sheet of cells that arises and extends from the anterior portion of the primitive streak. Like the dorsal blastopore lip, **Hensen's node**, a group of cells at the anterior (cephalic) end of the primitive streak, is necessary for development of neural ectoderm (Figure 2.4D, E), as is the **node**, the equivalent structure in mammals.

Gastrulation in mice is a bit more difficult to envision because the epiblast is found inside the blastocyst, a cuplike structure (**Figure 2.5**). If one imagines the blastocyst cut open (dashed line, Figure 2.5A) and flattened out, the similarities between gastrulation in a blastodisc and blastocyst are more apparent. The mouse blastocyst is surrounded by **visceral endoderm** (VE). Similar to cells of the blastodisc, the epiblast cells of the blastocyst migrate through the primitive streak to form the mesoderm and endoderm layers between the epiblast and VE (Figure 2.5A, B). The VE does not become part of the definitive endoderm and is similar to the hypoblast in that it becomes displaced by the forming endoderm germ layer. As described in Chapter 3, the **anterior visceral endoderm** (AVE) comes to lie below the anterior portion of the neural ectoderm and plays a role in defining regions of neural tissue (Figure 2.5C). Epiblast cells that remain at the inner surface of the blastocyst will form either the neural or epidermal ectoderm (Figure 2.5C). As noted above, it is during a specified period in the late gastrula-stage embryo that ectoderm can give rise to either neural or epidermal tissues. Precisely how different ectoderm cells are designated to become either neural or epidermal has been an active area of research for over a century.

EARLY DISCOVERIES IN THE STUDY OF NEURAL INDUCTION

By the early twentieth century, embryologists had established that dorsal ectoderm forms neural plate tissue, whereas ventral ectoderm forms epidermal tissues. What was not clear was how the ectoderm in these two regions became specialized as one tissue or the other. As scientists began to focus greater effort into identifying the origins of neural tissue, a groundbreaking discovery was made that indicated specific cells in one embryonic region can signal, or induce, dorsal ectoderm to become neural.

Amphibian models were used in early neuroembryology research and remain popular today

Many of the first studies of neural induction were conducted using amphibian embryos, particularly those from salamanders and frogs. Frogs remain a popular animal model today, in part because of the comparatively large size of their eggs and early-stage embryos. For example, the frog egg is about 2–3 mm in diameter compared to the 0.1–0.2 mm diameter of a human egg. This larger size makes cellular regions easier to identify and experimentally manipulate. Although such manipulations are more easily accomplished with the larger eggs and embryos of amphibians, early experiments still required considerable creativity (for example, designing tiny glass needles to use as surgical knives and using a baby's hair to separate blastomeres), tremendous patience, and very steady hands. The technical accomplishments of these first experiments, now almost a century old, are nearly as impressive as the scientific hypotheses that resulted from the studies.

A popular experimental method refined in the early twentieth century, and still used today, relies on grafting tissues from one embryo to another. This approach has proved critical to advancing our understanding of nervous system formation. Among the first studies conducted to evaluate neural specification of embryonic tissues were those in which scientists harvested tissue from one region of an early gastrula-stage amphibian embryo (the donor) and grafted it to another region of a second embryo at the same developmental stage (the host). The host embryo continued to develop and the effects of the donor and host tissues on subsequent developmental events were then evaluated. It was one such grafting experiment that first revealed the source of signals needed to induce neural tissue.

A region of the dorsal blastopore lip organizes the amphibian body axis and induces the formation of neural tissue

In 1924 Hans Spemann and Hilde Mangold published what has become one of the most cited studies in developmental neurobiology. In this study, the researchers completed grafting studies using two different species of newts. The pigmentation of the embryos from these two newts differed, allowing scientists to visualize the fate of donor and host tissues following surgical manipulations. Spemann and Mangold grafted the dorsal blastopore lip (DBL) of a pigmented, early gastrula-stage embryo to the ventral side of a nonpigmented embryo (the host) at the same stage of development (**Figure 2.6**). The ventral side of the embryo normally gives

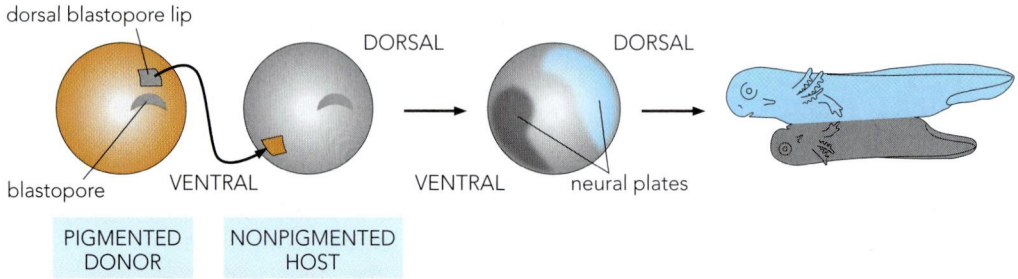

Figure 2.6 The classic grafting experiment performed by Hans Spemann and Hilde Mangold. The dorsal region of a blastopore lip from a pigmented donor embryo was grafted onto a nonpigmented host embryo's ventral surface, an area of ectoderm that does not give rise to neural tissue. As development proceeded, a neural plate and body axis from the host embryo developed along the dorsal surface as expected. In addition, a second neural plate and body axis formed from the ventral tissue of the host. This result indicated that the donor dorsal blastopore lip did not expand to form a new body axis, but instead produced signals that induced host ventral ectoderm to form an entirely new body axis complete with neural tissue.

rise to epidermal but not neural tissue. At this early stage of gastrulation, however, the regions of epidermal and neural ectoderm have not yet been specified, so the investigators were able to evaluate any inductive effects of the DBL. Spemann and Mangold discovered that the host embryos continued to develop and also formed a second body axis, complete with neural tissue. The key observation from this experiment was that the new body axis formed primarily from the nonpigmented, host embryo. This indicated that the ventral tissue of the host—ectoderm that normally becomes epidermal—was induced to form neural tissue when provided with a signal originating in the DBL. Thus, the second body axis did not result by a simple expansion of the pigmented donor dorsal lip tissue, as was presumed to be the case based on earlier grafting studies conducted by other investigators.

The band of DBL that induced the formation of new neural tissue came to be called **Spemann's organizer** because it organized the entire body axis. Spemann and his students, as well as investigators in many other labs, continued to explore the mechanisms that governed neural induction. Spemann ultimately received the Nobel Prize in Physiology or Medicine in 1935 for his work in this area. Tragically, Hilde Mangold died following a household explosion in 1924, when she was only 26 years old, and was unable to witness the continuing impact of her work. Her husband, Otto Mangold, also a student of Spemann's, went on to make additional contributions regarding the mechanisms underlying the processes of cell determination and induction of embryonic tissues.

The intriguing discovery of a neural inducer originating in the organizer—that is, the DBL of amphibians as well as its equivalent in other species, such as Hensen's node of the chick—led to numerous studies investigating the nature of this signal. Additional grafting experiments conducted in the 1930s noted that the signal arising from the organizer was not species-specific. For example, when grafts of organizer tissue from duck were transplanted to chick, neural tissues formed. Similarly, when grafts of chick organizer tissue were transplanted to either duck or rabbit, neural tissues were induced. Thus, the signal appeared to be universal in its ability to induce the formation of neural tissue from early-stage ectoderm. Other grafting experiments in the 1930s revealed that the medial or lateral portions of the amphibian DBL preferentially gave rise to head or tail structures, respectively. This suggested the possibility of multiple signals available to direct the formation of specific neural regions along the anterior–posterior body axis of the embryo (see Chapter 3).

The search for the organizer's neural inducer took decades of research

For almost 100 years, scientists have investigated the signals governing neural induction, using an ever-expanding array of experimental approaches. It is hard to capture in just a few paragraphs the extensive efforts made and the number of scientific debates and alternative hypotheses put forth during the search for putative neural inducers.

One serious obstacle to identifying neural inducers is the very small size of the organizer region. The small tissue size prevented scientists from isolating the signaling molecule directly; to overcome this problem, researchers screened other tissues for larger and more plentiful sources of inductive activity, tested other substances for inductive properties, and developed new ways to determine how the signal reached the ectoderm.

In the course of various studies, scientists determined that the neural inducer was diffusible and that, surprisingly, any number of biological and nonbiological substances induced competent undifferentiated ectoderm to form neural tissue. For example, tissues that were first boiled to denature active proteins still induced neural tissue formation, as did dyes and other chemicals known to be toxic to cells. For a number of years scientists struggled to distinguish true neural inducers from "artificial activators," as these nonspecific substances came to be called.

New tissue culture methods and cell-specific markers advanced the search for neural inducers

One of the experimental techniques first developed during the early era of neuroembryological research proved essential for identifying the true neural inducers. Scientists identified ways to isolate animal caps from blastula and early gastrula-stage amphibian embryos and maintain them in various tissue culture environments. As detailed below, this technique—first developed in the 1920s, and later refined in the 1960s —has been employed to assess interactions among cells from the animal pole, the vegetal pole, and the band of mesoderm that forms between the poles.

Additional technical advances in the latter part of the twentieth century provided the first opportunities to conclusively identify the specific cell types that arise following experimental manipulations. Because cells from early stages of development are not morphologically distinct from one another, identifying whether embryonic cells, particularly those in tissue culture, are endodermal, mesodermal, epidermal, or neural is often quite difficult. However, as techniques were developed for labeling sequences of DNA or RNA by ***in situ* hybridization** and specific proteins by **immunocytochemistry**, it became possible to identify cell types based on the expression of these **cell markers** (see Chapter 1). For example, **neural cell adhesion molecule (NCAM)** is found only in neural cells and provides scientists with a means of identifying these cells in a mixed cell population. The ability to identify different cell types also meant that scientists could now sort out whether a given experimental treatment caused direct or indirect induction of a cell or tissue type. Scientists recognized that any given treatment might induce the formation of neural cells directly, or might first induce the formation of organizer-like mesoderm, that would in turn induce neural cells.

Examples of some of the first discoveries made using the animal cap assay and labeling techniques are shown in **Figure 2.7**. When amphibian animal caps (that is, ectoderm) were cultured alone, epidermal tissues formed. However, when animal caps were cultured in the presence of mesoderm, the animal cap cells became neural (Figure 2.7A, B). As the DBL is formed from a band of cells that includes mesoderm tissue, these results were consistent with the notion that the organizer provides the signals needed to induce neural tissue in unspecified ectoderm.

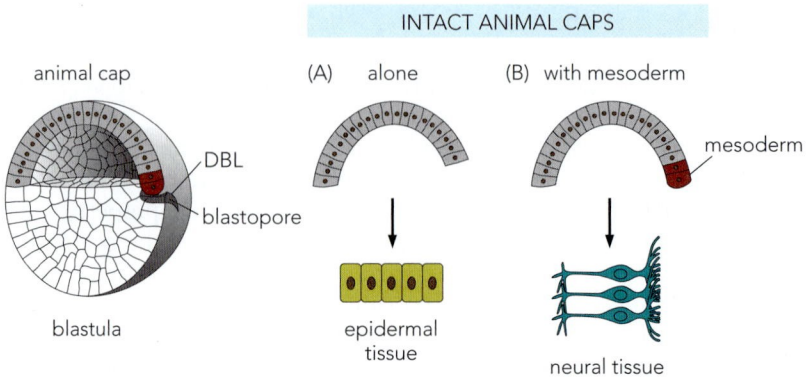

Figure 2.7 *In vitro* assays examining the effect of the mesoderm in determining the fate of cultured animal caps from *Xenopus* embryos. (A) When intact animal caps without any mesoderm tissue are removed and placed in culture, epidermal tissue forms. (B) When intact animal caps include mesoderm tissue, neural tissue forms. (DBL, dorsal blastopore lip.)

NEURAL INDUCTION: THE NEXT PHASE OF DISCOVERIES

Prior to the 1980s, it was generally assumed that neural inducers influenced ectoderm directly, instructing specific cells in the correct regions of that germ layer to become neural rather than epidermal tissue. However, more than 50 years after the first description of the organizer, several experiments carried out in different labs changed this fundamental assumption. The results of these experiments ultimately led to the realization that neural induction depended on signals that prevented the formation of epidermal ectoderm in presumptive neural regions.

Some of these advances were made possible by recognizing the limitations of previously used techniques. For example, scientists realized that animal cap cells from some amphibians are able to "auto-neuralize"—that is, become neural tissue in the absence of other tissues or added factors. This explained, in part, why so many different substances appeared to induce the formation of neural tissues; animal cap cells from these amphibians could form neural tissues independent of any added substances. It was also observed that animal cap cells of the *Xenopus laevis* embryo do not auto-neuralize to the same extent as those from other frogs. Therefore, researchers relied on these amphibians and designed further experiments to directly test for neural inducers.

The search for mesoderm inducers revealed that neural induction might involve removal of animal cap-derived signals

Among the key discoveries that led to the identification of neural inducers came from experiments originally designed to identify mesoderm inducers. As noted previously, mesoderm forms between the vegetal and animal poles. Scientists in the 1970s had already determined that mesoderm formation required interaction with one or both poles, and by the 1980s various groups were seeking to identify mesoderm-inducing signals emanating from either pole. Mesoderm formation was also of interest to scientists studying neural induction, because the organizer is a mesoderm-derived structure.

In the search for mesoderm inducers, growth factors from the **fibroblast growth factor (FGF)** and **transforming growth factor β (TGFβ)** families were added to cultures of animal cap cells. In these experiments, the isolated animal caps were treated with enzymes to break apart cell contacts. The resulting dissociated cells were added to cell cultures and treated with different growth factors. Normally, the dissociated animal cap cells reaggregated in the cell culture dishes and in the presence of the growth factors the reaggregated animal cap cells were induced to form mesoderm tissue (**Figure 2.8**A). During the course of these studies,

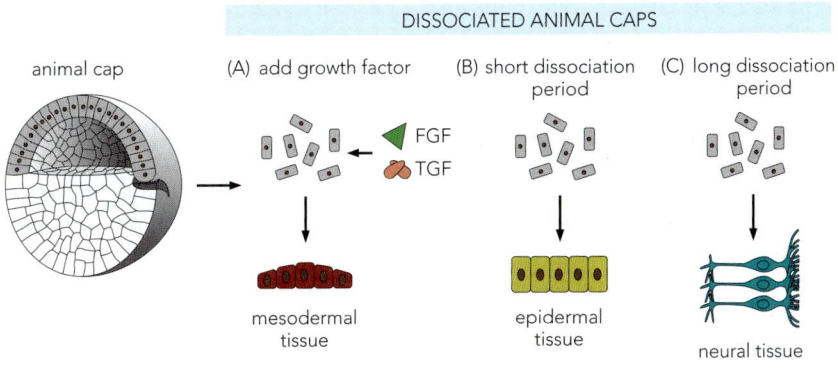

Figure 2.8 *In vitro* assays examining the effect of growth factors and dissociation time in determining the fate of dissociated animal cap cells from *Xenopus* embryos. (A) If animal caps without any mesoderm tissue are dissociated into single cells and then cultured in the presence of a growth factor, such as fibroblast growth factor (FGF) or transforming growth factor (TGF), the cells reaggregate to form mesodermal tissue. (B) If cells are kept dissociated for a brief period of time in the absence of additional growth factors, the cells reaggregate to form epidermal tissue. (C) If the animal cap cells are kept dissociated for longer periods (>3 h), the cells reaggregate to form neural tissue. This latter observation was one of the first clues to suggest that diluting or removing a signal present in animal cap cells would lead to the formation of neural rather than epidermal tissue.

scientists also observed that when untreated, control cultures (those lacking any additional growth factors) were kept dissociated for different lengths of time, the resulting reaggregated cells became either epidermal or neural cells. The scientists found that when cells were kept dissociated for only short periods prior to reaggregation, they became epidermal. However, if cells remained dissociated for longer periods of time (over 3 hours) prior to reaggregation, they became neural (Figure 2.8B, C). These observations led to the suggestion that during the longer dissociation period presumptive contact-mediated or secreted signals arising from the animal cap cells were no longer present at a sufficient concentration to prevent neural formation. Thus, without this signal, neural tissue formed instead of epidermal tissue. Other explanations were also offered, but the idea that neural tissue could form when a specific signal was absent received support from other studies conducted around the same period.

Mutation of the activin receptor prevents the formation of ectoderm and mesoderm but results in the formation of neural tissue

In another set of experiments designed to investigate the signaling pathways utilized in mesoderm formation, scientists focused on the role of the TGFβ signaling pathway. Several TGF-related proteins, including one called **activin**, had been identified as candidate mesoderm inducers. Activin, as well as its associated serine/threonine kinase receptors, were noted to be endogenous to blastula-stage animal cap regions. The necessity of activin signaling for mesoderm induction was therefore tested by interfering with receptor signaling.

To disrupt normal activin signaling pathways, investigators injected large amounts of synthetic RNA coding for a truncated (and thus inactive) version of the activin receptor into the two-cell stage animal pole of the frog *Xenopus*. The truncated receptors lacked only the kinase domain of the receptor subunits, but maintained the extracellular binding region. Therefore, activin could bind to the extracellular regions, but the ligand was unable to transduce intracellular signals. The truncated receptors were injected in excess so the mutated receptors would sequester the majority of activin or activin-like ligands and prevent signal transduction through any endogenous activin receptors. This is a type of **dominant negative** approach in which a mutation is great enough to interfere with a normal signaling event.

The *Xenopus* cells injected with a control RNA (**Figure 2.9**A) and those expressing the truncated activin receptors (Figure 2.9B) were allowed to develop to a blastula-stage embryo when the animal caps could be dissected and placed in culture. When control animal cap cells were placed in culture with increasing concentrations of activin, different mesoderm derivatives, including blood, muscle, notochord, and heart cells, formed (Figure 2.9C). These animal caps also gave rise to neural tissues, presumed to be induced by the newly formed mesoderm. In contrast, when intact animal caps expressing the truncated receptor were placed in tissue culture with activin, mesoderm formation did *not* occur. Thus, as hypothesized, interfering with activin signaling prevented mesoderm induction.

However, an additional finding was rather surprising and of considerable interest to those studying neural induction. The investigators noted that the intact animal caps with the truncated activin receptors expressed neural cell markers and formed only neural tissue (Figure 2.9D). This was unanticipated because in previous experiments, intact animal caps cultured in the absence of activin or mesoderm could form only epidermal tissue (see Figure 2.7A). This observation suggested that

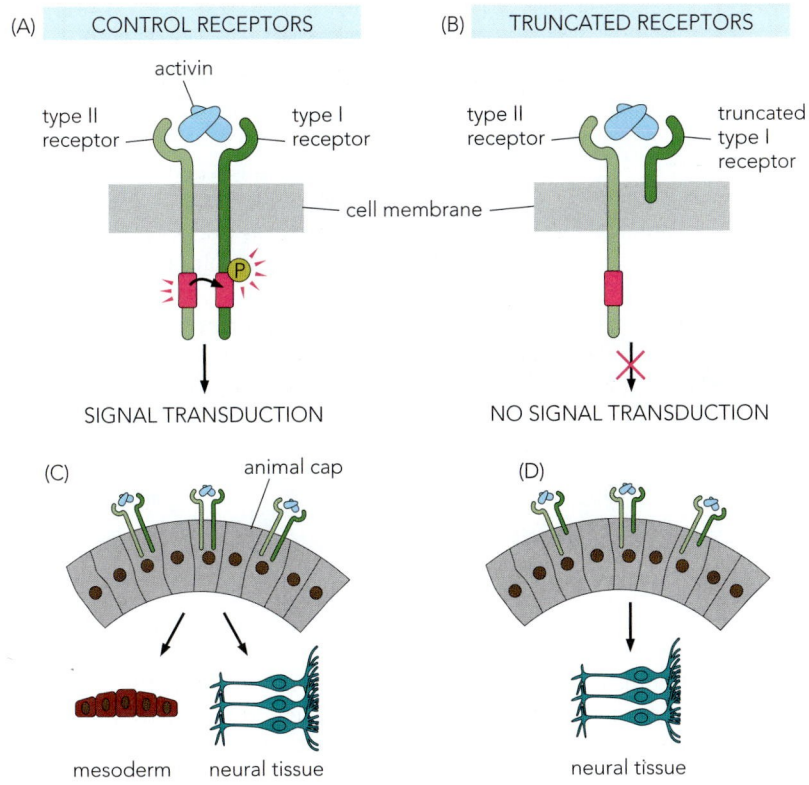

Figure 2.9 Blocking activin signaling disrupts mesoderm formation but leads to the formation of neural tissue. (A) Signal transduction through activin receptors is similar to other members of the TGF receptor family. The receptors are present as dimers that come into close association following ligand binding. As the type II and type I receptor subunits come together, the type II receptor is able to phosphorylate the intracellular region of the type I receptor, leading to signal transduction. (B) When the intracellular portion of the type I receptor is removed, ligand binding cannot initiate signal transduction because there is no site for phosphorylation to occur. Thus, these truncated receptors sequester the ligand but are nonfunctional (dominant negative approach). (C) When intact animal caps from control embryos (for example, injected with a control RNA) were cultured in the presence of activin, mesoderm formed, suggesting a role for activin signaling in mesoderm induction. Neural tissue also formed under these conditions, presumably in response to signals derived from the mesoderm. (D) When intact animal caps from embryos expressing the truncated activin receptor were cultured in the presence of activin, mesoderm was not induced, further supporting the hypothesis that activin signaling is needed for mesoderm formation. Surprisingly, however, neural tissue formed, rather than epidermal tissue, which normally arises from intact animal cap cells lacking mesoderm signals (see Figure 2.7A). Thus, the disruption of activin signaling led to neural tissue formation, suggesting neural tissue may be the default state, with epidermis requiring an inductive signal to form.

functional activin signaling was normally needed not only for mesoderm induction, but also for the formation of the epidermal regions of ectoderm. Because blocking activin signaling led to the formation of neural tissue instead of epidermal tissue, it appeared that neural tissue formed by default if epidermal formation was inhibited.

To explore this possibility further, additional studies were done in which truncated activin receptors were again injected into embryos at the two-cell animal cap stage, but this time the embryos were left to develop until the tail bud stage—the stage when the tail begins to emerge. These embryos had expanded regions of neural tissue, further suggesting that it was the inhibition of a signaling pathway that allowed the formation of neural tissues (**Figure 2.10**).

At first these results may appear to contradict earlier work. The tissue culture studies showed that neural tissue resulted when activin-mediated mesoderm induction was prevented, suggesting mesoderm was not important for neural induction. However, it was also well established that the mesodermal DBL induced neural tissue (see Figure 2.7B). Fortunately, at the same time these studies were being conducted, additional labs were investigating mesoderm and ectoderm formation. Together the results from these various labs helped revise the concept of how neural induction occurs and explained how mesoderm signals direct ectoderm to form neural tissue.

Modern molecular methods led to the identification of three novel neural inducers

One influential and elaborate set of experiments in *Xenopus* led to the first successful isolation of a neural inducer. This research utilized **ventralized embryos** that lack a nervous system and other dorsal structures and **hyperdorsalized embryos** that have a larger than normal nervous

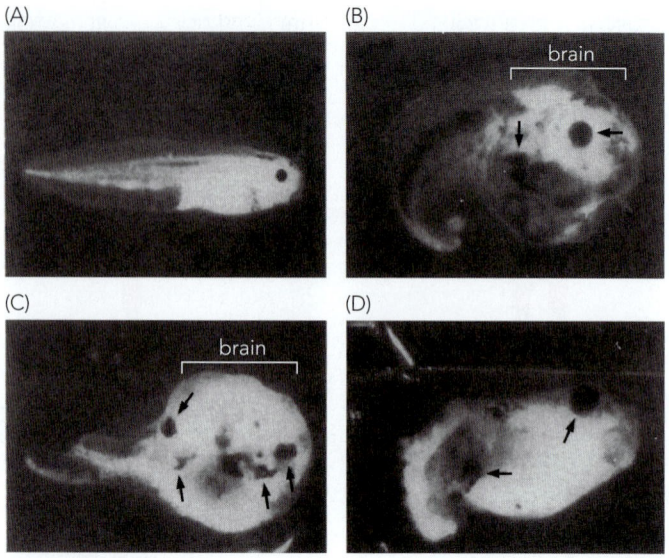

Figure 2.10 Disruption of activin signaling in whole embryos leads to increased neural tissue. *Xenopus* embryos at the two-cell stage were injected with mRNA for globin (control condition) or for the truncated activin receptor (experimental condition). (A) Embryos developed normally when injected with globin mRNA. (B–D) Embryos formed greater amounts of neural tissue, including extra eyes (arrows) and expanded brain (brackets, B, C) when activin signaling was disrupted. (Adapted from Hemmati-Brivanlou A & Melton DA [1994] *Cell* 77:273–281.)

system. Treatment of embryos with lithium chloride leads to the expansion of dorsal regions, including the nervous system. In contrast, ventralized embryos are created by exposure to ultraviolet (UV) light, which disrupts the process of cortical rotation, a step needed to establish the dorsal–ventral body axis in *Xenopus*. Notably, the ventralized embryos could be rescued to develop normally by grafting DBL tissue from a normal embryo. As a result, this preparation became a useful assay to identify molecules that induce the formation of neural structures.

Hypothesizing that the larger nervous systems found in hyperdorsalized embryos would provide a greater amount of neural-inducing signals than normal embryos, Richard Harland and William Smith used the hyperdorsalized embryos as a source of material to isolate candidate neural inducers. The scientists isolated fractions of mRNA from the hyperdorsalized embryos and tested them for the ability to rescue ventralized embryos. Smaller and smaller fractions of active mRNA extracts were tested until eventually only a single active molecule remained. When this final extract was injected into ventralized embryos, a normal embryo resulted (**Figure 2.11** A), similar to what occurred with grafts of DBL tissue. Through this process Harland, Smith, and colleagues ultimately identified a novel protein they called **noggin**.

Purified noggin protein was subsequently found to induce a normal body axis and neural tissue in UV treated embryos (Figure 2.11B). Furthermore, addition of noggin protein to the animal cap assay led to the formation of neural cells rather than epidermal cells, even in the absence of mesoderm tissue. This suggested that noggin was a direct inducer of neural tissue. Significantly, noggin expression was also found in the organizer regions, beginning in the late blastula stage. More than half a century after Spemann and Mangold revealed the importance of the organizer, noggin was the first novel neural inducer to be conclusively identified.

A second neural inducer was soon identified by Ali Hemmati-Brivanlou, Douglas Melton, and colleagues. In other developing tissues, it had already been shown that the reproductive hormone **follistatin** could bind to and inactivate activin. Subsequent studies noted that follistatin was expressed in the *Xenopus* gastrula-stage organizer region, suggesting a possible role in neural induction. Similar to noggin, follistatin induces expression of neural markers when injected into animal cap cells, yet mesoderm tissue does not

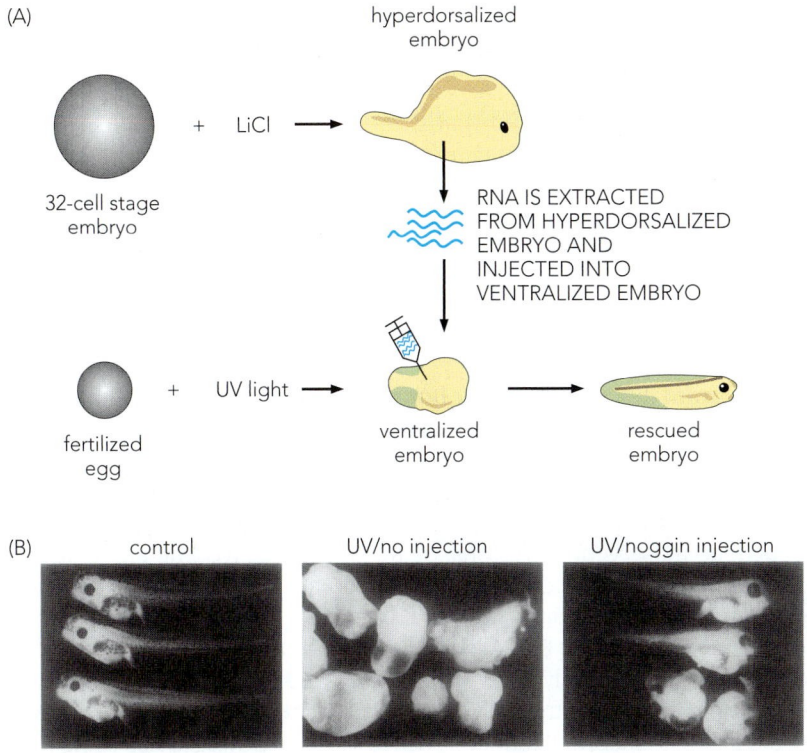

Figure 2.11 Discovery of the neural inducer noggin from studies using ventralized and hyperdorsalized embryos. (A) Experimental treatment of early stage amphibian embryos leads to expansion or loss of neural regions. (Top) When lithium chloride (LiCl) was added to the 32-cell stage embryo to inhibit dorsal–ventral axis specification, a hyperdorsalized embryo with expanded dorsal regions formed. The increased size of the neural regions served as a source of RNA for the screening of neural inducers. (Bottom) In embryos treated with UV light for 20–30 min following fertilization, dorsal structures failed to form because cortical rotation was disrupted. However, when RNA extracts from hyperdorsalized embryos were injected into the ventralized embryos, dorsal structures formed and the embryos were rescued. By testing smaller and smaller fractions of the RNA extracts, the protein noggin was ultimately identified. (B) Images from the 1992 study of Smith and Harland compare control embryos to those treated with UV light and those treated with UV light then rescued with injection of noggin protein. Injection of noggin was able to rescue many of the UV treated, ventralized embryos so that they developed dorsal structures comparable to controls. (Adapted from Smith WC & Harland RM [1992] *Cell* 70:833–840.)

form (**Figure 2.12**A). Thus, follistatin was proposed to be a direct neural inducer that could possibly act by inhibiting activin.

During this same period a third candidate neural inducer, **chordin**, was identified by Edward De Robertis and colleagues while screening for gene transcripts expressed specifically in the organizer region. When chordin mRNA was injected into the animal pole or when soluble chordin protein was added to animal cap cultures, neural tissue formed, but mesoderm did not (Figure 2.12A). When chordin was injected in the ventral regions of a frog embryo, it induced a second body axis reminiscent of the grafting experiment results observed by Spemann and Mangold (Figure 2.12B). Thus, within a span of about five years, three unrelated candidate neural inducers were identified using methods that were unavailable, and likely unimaginable, to earlier generations of neuroembryologists.

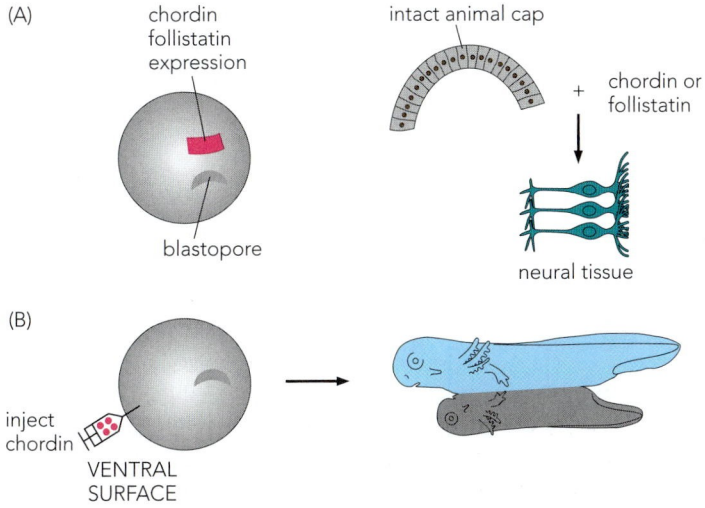

Figure 2.12 Identification of the neural inducers chordin and follistatin. (A) Both chordin and follistatin are expressed in the area of the dorsal blastopore lip (the organizer), and both molecules induce the cells of intact animal caps to form neural tissue rather than epidermal tissue. (B) Injection of chordin into the ventral (nonneural) surface of the embryo leads to the induction of a second body axis, similar to the results of the original grafting experiment by Spemann and Mangold.

NOGGIN, FOLLISTATIN, AND CHORDIN PREVENT EPIDERMAL INDUCTION

The previous data suggested that the newly discovered neural inducers acted directly, without the need for mesoderm tissue. Yet, studies also indicated that blocking activin, a molecule used in mesoderm formation, induced neural tissues. Research efforts in multiple areas resolved this apparent contradiction.

Studies of epidermal induction revealed the mechanism for neural induction

One important discovery was that **bone morphogenetic proteins (BMPs)** induced epidermal markers while suppressing neural markers. The BMPs are a large family of molecules and are part of the larger TGFβ superfamily. BMPs play a role in a number of developmental events; despite their name, BMP functions are not limited to bone formation. BMP2 and BMP4 are expressed throughout most ventral regions of the frog embryo, but are ultimately restricted to regions of ectoderm associated with epidermal, but not neural, tissue. In cultures of animal cap cells dissociated for three hours or more, treatment with BMP2 or BMP4 led to the formation of epidermal rather than neural cells, as previously observed (**Figure 2.13**A, B; see also Figure 2.8C). Further, animal cap cells treated with BMPs did not express any neural markers, but did express epidermal markers such as keratin. Other studies using intact animal cap assays revealed that BMP4, at a high enough concentration, could inhibit the ability of chordin, noggin, or follistatin to form neural cells, instead inducing the formation of epidermal tissue (Figure 2.13D, E). In contrast, the experiments that blocked BMP signaling, including those using truncated activin receptors, resulted in the formation of neural rather than epidermal tissue. Thus, the activation of signaling pathways for BMP4, BMP2, and possibly BMP7 appeared necessary for the formation of epidermal tissues, whereas blocking BMP signaling allowed neural tissue to form.

The discovery of neural inducers in the fruit fly *Drosophila* contributed to a new model for epidermal and neural induction

Emerging data from experiments in the fruit fly species, *Drosophila melanogaster*, helped link the roles of neural inducers and BMPs in the formation of epidermal and neural tissues (**Box 2.1**). Scientists discovered that similar to their *Drosophila* homologs, chordin, as well as noggin and follistatin, interacted with BMP molecules and blocked BMP signaling. Experiments with noggin or chordin revealed specific binding between each of these inducers and BMP2, BMP4, and BMP7 (**Figure 2.14**A). Together, the accumulated data pointed to a mechanism by which

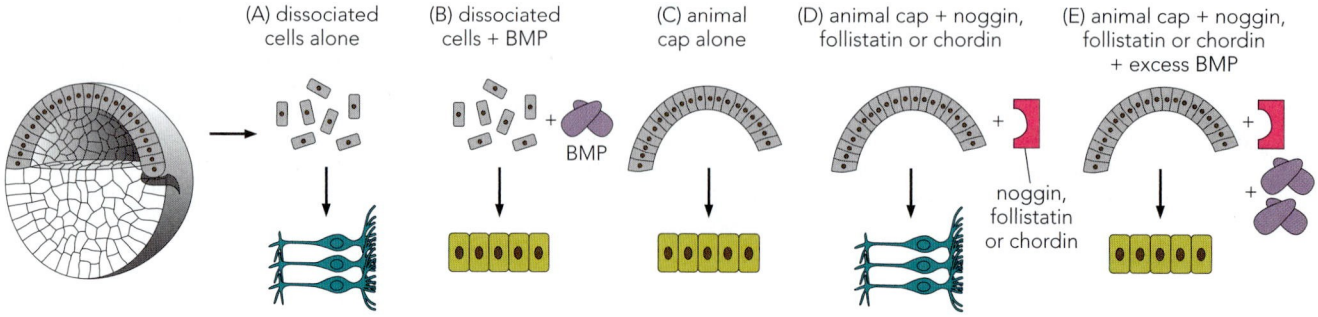

Figure 2.13 Animal cap assays demonstrate that bone morphogenetic proteins (BMPs) induce epidermal tissue. (A) Animal cap cells that remain dissociated for >3 h typically form neural tissue, as previously seen (Figure 2.8C). (B) Adding BMP to dissociated animal cap cells leads to the formation of epidermal tissues. (C) Intact animal caps in cell culture form epidermal tissue (see also Figure 2.7). (D) The addition of noggin, chordin, or follistatin to intact animal cap cultures results in the formation of neural tissue. (E) Adding an excess concentration of BMP to these cultures leads to the formation of epidermal tissue. In these experimental conditions the concentration of BMP is sufficient to prevent the neural inducers noggin, chordin, or follistatin from interfering with BMP signaling. Together these experiments show that BMPs actively induce the formation of epidermal tissue from the animal cap ectoderm.

Box 2.1 Neural induction: Lessons from invertebrates

Major advances in understanding how changes in bone morphogenetic protein (BMP) signaling established epidermal versus neural fate in vertebrate animal models stemmed from studies conducted in the fruit fly *Drosophila melanogaster*. Despite major structural and developmental differences between *Drosophila* and the vertebrate animal models, there is a remarkable conservation of signaling pathways across species. Studies in *Drosophila* first identified genes expressed on the opposite sides of the embryo that could differentially induce epidermal or neural tissues. The identification of these genes led scientists to evaluate the role of their homologs, BMPs and chordin, in vertebrate neural induction.

Fruit fly embryogenesis is quite different from that of vertebrates. Following fertilization of the ovoid-shaped *Drosophila* egg, nuclei divide rapidly within the cytoplasm prior to migrating to the cortex of the cell, where they form a **syncytial blastoderm**. The nuclei are then surrounded by cell membrane, thus creating the individual cells of the **cellular blastoderm** (**Figure 1**).

Beginning during the syncytial blastoderm stage, gradients of a protein called Dorsal are established such that the highest concentrations are located ventrally and the lowest concentrations are found dorsally. At the dorsal surface, the lower concentration of Dorsal protein induces expression of the gene *decapentaplegic* (*dpp*). Intermediate levels of Dorsal protein lead to the expression of the *short of gastrulation* (*sog*) genes in ventrolateral regions, while the highest concentrations of Dorsal protein in the ventral most region induces *twist*, a gene important in mesoderm induction (**Figure 2**).

Similar to their vertebrate homologs, the protein products of *dpp* and *sog* induce the formation of either epidermal or neural ectoderm, respectively. Unlike vertebrate animal models, however, *Drosophila* epidermis is formed on the dorsal side of the embryo, whereas neural tissue is formed on the ventral side (Figure 2; see also Chapter 1). Despite this inverted body axis, the homologous invertebrate and vertebrate signaling molecules function in a similar manner.

Studies in *Drosophila* found that the gradients of Dorsal protein induced gradients of secreted proteins, Dpp and Sog, with higher concentrations of Dpp located dorsally and higher concentrations of Sog ventrolaterally. The high concentration of Dpp on the dorsal side of the embryo induced extraembryonic (amnioserosa) and epidermal tissue, whereas high concentration of Sog ventrolaterally induced neural tissue. Sog was also found to antagonize Dpp. Thus, on the ventrolateral sides of the embryo, the high levels of Sog are able to block Dpp signaling, prevent the formation of epidermis, and induce neural tissue. In contrast, the lower levels of Sog on the dorsal side of the embryo are insufficient to inhibit Dpp signaling and therefore lead to the induction of epidermis. Thus, in *Drosophila*, Sog blocks Dpp signaling to retain neural fate, just as chordin blocks BMP activity to maintain neural fate in vertebrates.

The importance of Dpp and Sog interactions in defining epidermal and neural regions was seen in multiple experiments. For example, if *dpp* expression was eliminated, ectopic neuroectoderm formed in the presumptive epidermal regions on the dorsal side of the fly embryo. Conversely, experimental overexpression of *dpp* converted regions of presumptive neuroectoderm to epidermis.

Evidence for cross-species conservation of these signaling pathways was demonstrated in experiments in which *Drosophila sog* mRNA was injected into *Xenopus* embryos, thus leading to the formation of a

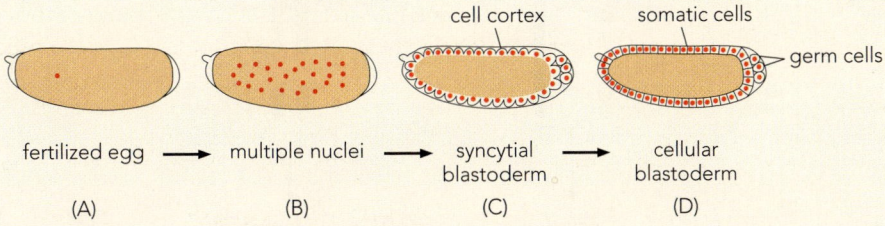

Figure 1 Formation of the blastoderm in *Drosophila*. After the ovoid-shaped egg cell of a *Drosophila* is fertilized (A), nuclei begin to divide rapidly within the egg cytoplasm (B). The nuclei then migrate to the cell surface (cortex), forming the syncytial blastoderm (C). The nuclei become surrounded by cell membranes and form the cellular blastoderm that contains both the somatic and germ cells (D).

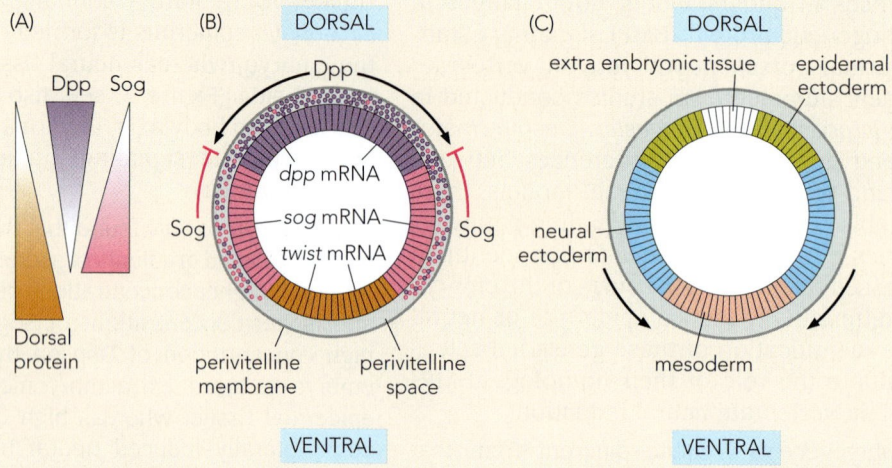

Figure 2 Signals direct the formation of epidermal tissue at the dorsal surface of the *Drosophila* embryo. (A) The protein Dorsal is produced in a gradient, with the highest levels produced ventrally and the lowest levels produced dorsally. The concentration of Dorsal differentially induces expression of the genes including *decapentaplegic (dpp)* that induces epidermal ectoderm and *short of gastrulation (sog)* that induces neural ectoderm. (B-C) Cross sections through the *Drosophila* embryo reveal the origins of the signals that induce epidermal and neural tissues. Note that the epidermis arises on the dorsal surface of the embryo, whereas in vertebrates, neural tissue forms at the dorsal surface of the embryo. The lower concentrations of Dorsal protein in the dorsal-most region of the embryo induce expression of the mRNA for *dpp*. Intermediate levels of Dorsal protein lead to the expression of *sog* mRNA in ventrolateral regions, while the highest concentrations of Dorsal protein in the ventral-most region induces expression of mRNA for Twist required for mesoderm induction. The protein products that result from *dpp* and *sog* expression are set up as gradients within the perivitelline space that surrounds the embryo such that Dpp is found in the highest concentration in dorsal regions of the embryo, whereas Sog is found in higher concentrations at progressively more ventral regions. Similar to the homologous vertebrate neural inducer chordin, Sog inhibits the activity of Dpp—the *Drosophila* homolog of BMP—to block the induction epidermal ectoderm in the regions in the ventrolateral regions that are to give rise to neural ectoderm. (C) The high concentrations of Dpp in the dorsal region induce epidermal ectoderm and extraembryonic tissue. Higher concentrations of Sog in the more ventral regions are sufficient to block the activity of Dpp and prevent the induction of epidermal ectoderm in the regions that give rise to neural ectoderm. Mesoderm is located to the ventral-most region prior to gastrulation, but will later invaginate to form the middle germ cell layer, while the neural ectoderm will extend more ventrally (arrows).

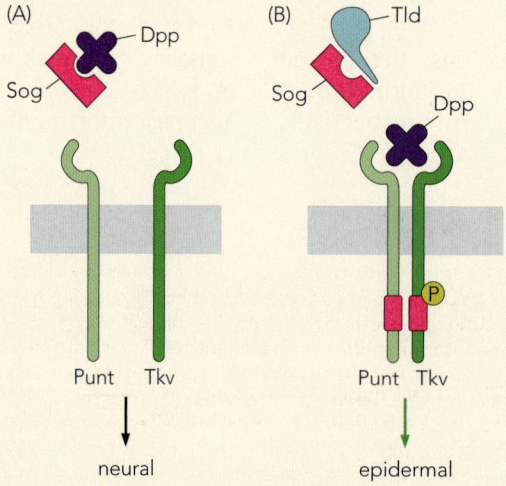

Figure 3 Sog and Dpp interactions regulate formation of neural and epidermal tissue. (A) Similar to vertebrate neural inducers, the *Drosophila* neural inducer Sog binds to Dpp to prevent it from activating the serine/threonine kinase receptor subunits Punt and Thick veins (Tkv). Under normal conditions, Sog protein binds to Dpp and diffuses dorsally to transport the bound Dpp back to its region of origin, and thereby further increase the level of Dpp in the dorsal region. (B) Once in the dorsal region, local concentrations of the protein Tolloid (Tld) cleave Sog away from the Dpp so that Dpp is now able to bind to the Punt and Tkv receptor subunits and activate the signaling pathways that lead to expression of the genes needed for epidermal induction.

second body axis, much like that observed in the early transplantation studies of Spemann and Mangold. Thus, studies in the fruit fly not only provided clues to the ways in which neural inducers interacted with BMPs, but also revealed the evolutionary conservation of signaling mechanisms involved in the crucial early steps needed for the formation of the nervous system.

The details of how Sog and Dpp interact to differentially induce neural and epidermal structures continue to be explored. Like BMPs, homodimers of Dpp proteins bind to type II and type I serine/threonine kinase receptors named Punt and Thick veins (Tkv), respectively. Sog functions as a binding protein to block Dpp from interacting with the receptor subunits. In addition to the gradients of Dpp and Sog proteins established through the activity of Dorsal protein, investigations also suggest that Sog itself increases the Dpp concentration by diffusing dorsally and transporting bound Dpp back to the dorsal region. In the dorsal region, the Sog is cleaved from Dpp by the protein Tolloid (Tld), thus releasing Dpp so it can bind to local receptors (**Figure 3**). Other proteins related to Dpp and Sog are also differentially expressed in the fly embryo, where they contribute to the signaling pathways that induce the formation of neural, epidermal, and extra embryonic tissues. The mechanisms by which these proteins interact, and how homologs of the proteins function in vertebrates, continue to be explored.

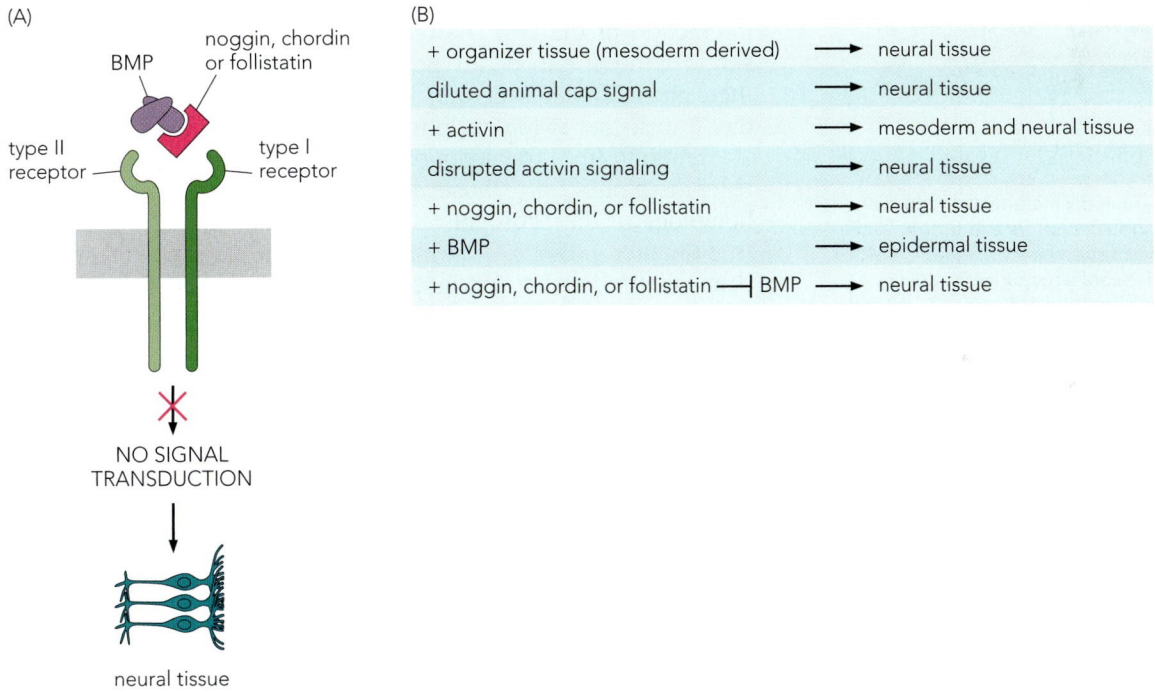

Figure 2.14 Blocking the BMP signal transduction pathway directs neural tissue formation. (A) The neural inducers, (noggin, chordin, and follistatin) bind to BMP and prevent it from activating the type II and type I BMP receptors. Because this signal transduction pathway is blocked, neural tissue forms by default. (B) Numerous studies conducted over several decades determined that neural tissue forms by default and actively prevents the neural regions from adopting epidermal characteristics. Among the key discoveries that led to this realization were: (1) The organizer, a mesoderm-derived tissue, induced neural tissue. (2) Diluting a signal present in animal cap cells induced neural tissue. This suggested that under normal conditions, a signal was present in the animal cap region that prevented neural formation. (3) The protein activin, a member of the TGFβ superfamily, induced mesoderm and neural tissues. (4) Disrupting activin signaling also induced neural tissue. (5) Three neural inducers were identified in the organizer region. (6) BMP proteins, also members of the TGFβ superfamily, induced epidermal tissue. BMPs are present throughout most of the undifferentiated regions of ectoderm in the embryo. (7) The neural inducers bind to BMP to block BMP signal transduction and prevent epidermal formation. Neural inducers are present in the organizer and diffuse into the dorsal regions of the embryo to inhibit endogenous BMP signals. In contrast, the ventral ectoderm is not exposed to sufficient levels of inducers, so BMP is able to bind to the endogenous BMP receptors and induce ventral ectoderm to form epidermal tissue.

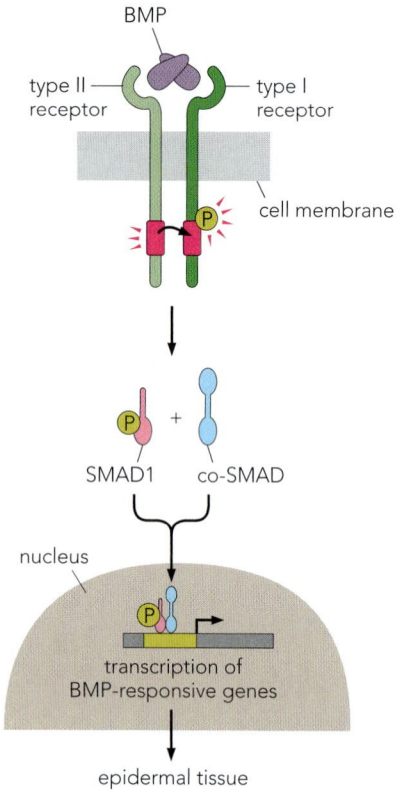

Figure 2.15 A BMP signal transduction pathway induces epidermal tissue. Similar to other TGFβ superfamily members, BMP binds to heterodimeric serine/threonine kinase receptors. The BMP binds to the type II receptor that associates with the type I receptor, leading to phosphorylation. The now-activated type I receptor can phosphorylate C-terminus serines on SMAD1, one of the receptor-regulated SMAD proteins (R-SMADs). SMAD1 then associates with co-SMAD and both are transported to the nucleus to activate BMP-responsive genes that lead to the formation of epidermal tissue.

ectoderm becomes epidermis when BMP is able to activate corresponding BMP receptors localized to the ectoderm. However, if this BMP signal is blocked by one of the candidate neural inducers, neural tissue forms. As this model indicates, the elusive neural inducers sought by scientists for so many years were actually the inhibitors of epidermis-inducing signals. After nearly a century of work and the contributions of dozens of labs, the basic mechanisms for neural induction were finally beginning to be understood (Figure 2.14B).

Interestingly, activin itself is not part of the neural induction pathway. It was later determined that the truncated activin receptor used in the original experiments was not specific for activin, but recognized many ligands of the TGFβ family, including the BMPs. The fortuitous use of this nonspecific receptor contributed greatly to our understanding of both epidermal and neural inducers.

BMP signaling pathways are regulated by SMADs

More recent studies have revealed ways in which the neural inducers block the intracellular BMP signaling pathways in order to promote the formation of neural tissue. BMPs, like other members of the TGFβ superfamily, such as activin, bind to heterodimeric serine/threonine kinase receptors. The interaction of the two receptor subunits leads to phosphorylation of a serine or threonine on the type I subunit. The phosphorylated type I subunit then phosphorylates certain transcription factors within the cytoplasm that translocate to the nucleus to activate target genes (**Figure 2.15**).

Among the transcription factors phosphorylated following BMP receptor binding are those of the SMAD family. The *Smad* genes are homologs of the *Drosophila Mad* and *Caenorhabditis elegans sma* genes (see Chapter 4, Box 4.1). In mammals, eight SMAD family members have been identified and are designated as SMAD1, -2, -3, -4, -5, -6, -7, and -8/9. SMAD8/9 is the same protein, but has been referred to as either SMAD8 or SMAD9 in earlier studies.

SMAD signaling is transduced by a trimer comprised of two receptor-regulated SMADs (R-SMADs) and one common-mediator SMAD (co-SMAD). R-SMADs include SMAD1, -2, -3, -5, and -8/9. SMAD4 functions as the co-SMAD, while SMAD6 and SMAD7 function as inhibitory SMADs (I-SMADS; see Chapter 4). SMAD1 is the R-SMAD important for epidermal induction. SMAD1 is phosphorylated at its C-terminus serine residues, and once phosphorylated, it joins the co-SMAD and travels to the nucleus to influence the transcription of BMP-responsive genes that lead to the formation of epidermal tissue (Figure 2.15). Because chordin, noggin, and follistatin interfere with BMP signaling, SMAD1 is never phosphorylated and therefore the transcription of BMP-responsive genes does not occur and the ectoderm remains neural.

Additional signaling pathways may influence neural induction in some contexts

As the default pathway for neural tissue formation was established and the general mechanisms by which BMPs induced epidermal tissue were identified, it appeared that the process of neural induction was at last understood. However, ongoing studies continue to indicate that the signaling pathways that regulate formation of neural and epidermal ectoderm are even more complex. In addition, several findings now suggest that the neural inducers described above are used mainly to designate anterior brain structures (see also Chapter 3) and that some of the associated signaling pathways may be most relevant to amphibians and fish, rather than all vertebrates.

Species differences may determine which additional pathways are needed for neural induction

Research on the process of neural induction in other vertebrates has revealed a number of interesting differences and discrepancies. For example, the expression of chordin, noggin, and follistatin in the node of the chick does not entirely coincide with the timing of neural induction. Furthermore, chordin alone is unable to induce neural tissue in chick, but if the tissue is transiently exposed to a graft of node tissue, neural induction occurs, suggesting that other signals are present in the node at earlier stages. Further, in mice genetically altered so as not to produce one or more of the neural inducers, neural tissue was still detected, though not always in its entirety. Thus, while the organizer clearly plays a role in the induction of neural tissue, multiple signals, including those that are expressed prior to gastrulation, may be needed for neural induction in some species.

Among the signaling molecules currently proposed to be involved at various stages of neural induction are FGFs and Wnts (homologs of the *wingless* genes of *Drosophila*). FGFs have multiple roles and act at several stages during neural induction. For example, in at least some species, FGFs in the hypoblast have been shown to regulate which cells remain at the epiblast surface during gastrulation. At later stages, FGFs expressed in the organizer may influence which cells become neural. There are 22 FGF ligands in vertebrates that bind with high affinity to specific FGF receptors. The four vertebrate FGF receptors are tyrosine kinase receptors that dimerize and cross-phosphorylate upon ligand binding (**Figure 2.16**). FGF dimers and their receptors contain a heparan sulfate proteoglycan (HSPG) binding region that allows HSPGs in the extracellular space to facilitate ligand binding and signal transduction. The phosphorylation of the FGF receptors leads to the activation of various intracellular pathways, including the Ras/Erk cascade detailed in Chapter 3.

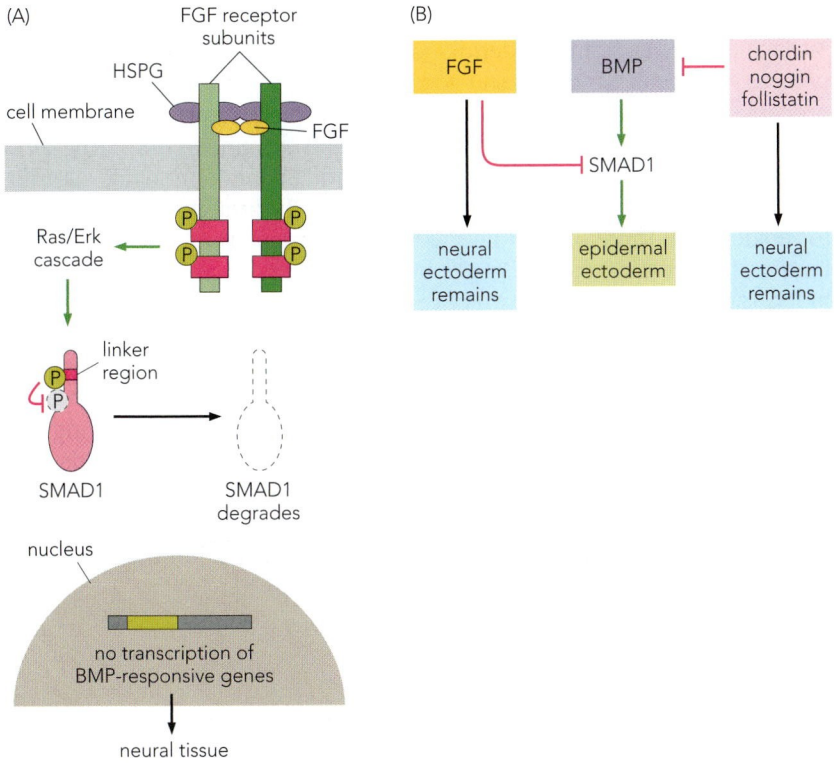

Figure 2.16 Fibroblast growth factors interfere with SMAD1 phosphorylation to block epidermal formation. (A) Fibroblast growth factors (FGFs) form dimers that bind to tyrosine kinase receptors. Ligand binding causes the receptor subunits to dimerize and cross-phosphorylate one another. Binding to the receptors is enhanced by heparan sulfate proteoglycan (HSPG) localized to extracellular spaces. Both the FGF ligands and receptors contain HSPG binding regions. The phosphorylation of the FGF receptors activates various intracellular pathways, including the Ras/Erk cascade (see Chapter 3). Activation of the Ras/Erk pathway leads to phosphorylation of the linker region of SMAD1. This inhibits the BMP-induced phosphorylation on the C-terminus serines and causes degradation of SMAD1 in the cytoplasm. Without SMAD1, BMP-responsive genes cannot be transcribed, so the default neural tissue remains. (B) Formation of epidermal ectoderm requires activation of intracellular signaling pathways (green arrows). Signals that interfere with BMP signal transduction (red bars) permit the default neural ectoderm to remain (black arrows). FGFs stop the BMP-induced phosphorylation of SMAD1, thus inhibiting the intracellular signaling pathways that would lead to epidermal formation. Neural inducers (chordin, noggin, and follistatin) associate with BMPs to prevent the ligands from binding to their receptors and thus prevent the intracellular pathways that are necessary to induce epidermal tissue.

FGF is now known to inhibit BMP signaling by phosphorylating a specific region of the SMAD1 transcription factor, called the linker region. Phosphorylation of the linker region stops the phosphorylation of the SMAD1 C-terminus serines that occurs following BMP receptor binding (Figure 2.15). Phosphorylation at the linker region leads to degradation of SMAD1, ending the BMP-induced gene transcription pathway needed for epidermal formation and instead promoting the formation of neural tissue (Figure 2.16). Thus, BMP signaling can be inhibited by neural inducers that prevent receptor binding, or by FGFs that interfere with SMAD1 activity (Figure 2.16)

The role of Wnt family members in neural induction is also an active area of research at this time. Studies in chick and *Xenopus* have reported that members of the Wnt family up-regulate BMP signaling and thus lead to the formation of epidermal tissue, whereas a lack of Wnt signaling leads to neural induction. In the absence of Wnt, glycogen synthase kinase-3β (GSK-3β), like FGF, is able to phosphorylate the linker region of SMAD1, thus terminating the activity of SMAD1 and preventing the transcription of BMP-responsive genes. Wnt promotes BMP signaling by inhibiting GSK-3β, thereby retaining the phosphorylation of SMAD1 C-terminus serines and inducing ectoderm tissue. However, several contradictory results have been reported, and the role of Wnt proteins in the induction of neural tissues remains controversial. Researchers continue to investigate how Wnt pathways may regulate neural induction in different species. Some of the difficulty in establishing the precise role of Wnts may be due to species differences, subtle variations in the way the experiments were conducted, and whether the canonical or noncanonical Wnt pathways are activated (see Chapter 4 and 10). Thus, while the basic mechanisms of neural induction are now well defined, details regarding how different signaling pathways interact to influence neural induction across species continues to be investigated.

SUMMARY

Surface ectoderm forms epidermis (skin) through an instructive or active process, whereas neural tissue forms through a permissive or passive process. The ectoderm itself secretes the signals necessary to induce epidermal formation (for example, BMPs) and the underlying mesoderm tissue (the organizer) secretes molecules (for example, noggin, chordin, and follistatin) to block epidermal induction. By inhibiting the pathway for epidermal formation, the neural tissues remain. Thus, the spatial and temporal expression pattern of the neural inducers (that is, the epidermal blockers) ensures that the correct regions of ectoderm receive the signal that prevents epidermis formation and permits the ectoderm to maintain a neural identity. The regions of ectoderm not exposed to the signals from the organizer remain uninhibited and thus transduce the BMP signals to generate epidermal tissue. Although Spemann had no way of knowing that neural inducers functioned in this way, his quote from 1924 still holds true. If the ectoderm that formed the epidermis on the soles of our feet had been localized to another region of the embryo, it may have been exposed to an inhibitor of BMP signaling and been induced to form neural tissue instead.

FURTHER READING

Appel B (2000) Zebrafish neural induction and patterning. *Dev Dyn* 219(2):155–168.

Attisano L & Lee-Hoeflich ST (2001) The Smads. *Genome Biol* 2(8).

Dorey K & Amaya E (2010) FGF signalling: diverse roles during early vertebrate embryogenesis. *Development* 137(22):3731–3742.

Eivers E, Fuentealba LC & De Robertis EM (2008) Integrating positional information at the level of Smad1/5/8. *Curr Opin Genet Dev* 18(4):304–310.

Hemmati-Brivanlou A & Melton DA (1994) Inhibition of activin receptor signaling promotes neuralization in *Xenopus*. *Cell* 77(2):273–281.

Hemmati-Brivanlou A, Kelly OG & Melton DA (1994) Follistatin, an antagonist of activin, is expressed in the Spemann organizer and displays direct neuralizing activity. *Cell* 77(2):283–295.

Holley SA, Jackson PD, Sasai Y et al. (1995) A conserved system for dorsal-ventral patterning in insects and vertebrates involving sog and chordin. *Nature* 376(6537):249–253.

Huang CT, Tao Y, Lu J et al. (2016) Time-course gene expression profiling reveals a novel role of non-canonical WNT signaling during neural induction. *Sci Rep* 6:32600.

Imbard A, Benoist JF & Blom HJ (2013) Neural tube defects, folic acid and methylation. *Int J Environ Res Public Health* 10(9):4352–4389.

Kaneda T & Motoki JY (2012) Gastrulation and pre-gastrulation morphogenesis, inductions, and gene expression: similarities and dissimilarities between urodelean and anuran embryos. *Dev Biol* 369(1):1–18.

Lamb TM, Knecht AK, Smith WC et al. (1993) Neural induction by the secreted polypeptide noggin. *Science* 262(5134):713–718.

Mulligan KA & Cheyette BN (2012) Wnt signaling in vertebrate neural development and function. *J Neuroimmune Pharmacol* 7(4):774–787.

Nieto MA (1999) Reorganizing the organizer 75 years on. *Cell* 98(4):417–425.

Sasai Y, Lu B, Steinbeisser H et al. (1994) *Xenopus* chordin: a novel dorsalizing factor activated by organizer-specific homeobox genes. *Cell* 79(5):779–790.

Shimmi O, Umulis D, Othmer H & O'Connor MB (2005) Facilitated transport of a Dpp/Scw heterodimer by Sog/Tsg leads to robust patterning of the *Drosophila* blastoderm embryo. *Cell* 120(6):873–886.

Smith WC & Harland RM (1992) Expression cloning of noggin, a new dorsalizing factor localized to the Spemann organizer in *Xenopus* embryos. *Cell* 70(5):829–840.

Smith WC, Knecht AK, Wu M & Harland RM (1993) Secreted noggin protein mimics the Spemann organizer in dorsalizing *Xenopus* mesoderm. *Nature* 361(6412):547–549.

Spemann H & Mangold H (2001) Historical article: Induction of embryonic primordia by implantation of organizers from a different species. 1924. *Int J Dev Biol* 45(1):13–38.

Stern CD (2005) Neural induction: old problem, new findings, yet more questions. *Development* 132(9):2007–2021.

Stern CD & Downs KM (2012) The hypoblast (visceral) endoderm): an evo-devo perspective. *Development* 139:1059–1069.

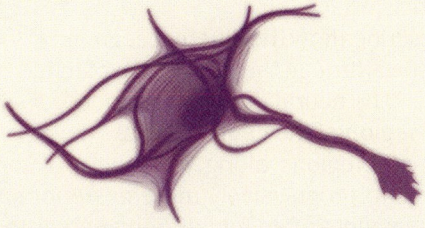

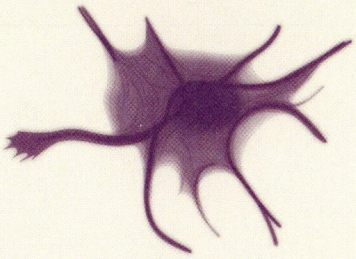

Segmentation of the Anterior–Posterior Axis

3

In vertebrates, starting shortly after the induction of neural ectoderm, the formation of the neural plate begins. The **neural plate** forms as the flattened structure along the surface ectoderm that then undergoes a series of steps to form the **neural tube**. Subsequent expansions and constrictions at various sites along the length of the neural tube serve to designate the early boundaries for future anatomical regions of the central nervous system (CNS). A number of signaling molecules have been identified that regulate the closure and anatomical organization of the vertebrate neural tube along its anterior–posterior (A/P) axis. These signaling molecules interact with each other to ensure that only the signals necessary to pattern a given region are active at the correct time in development.

Many of the advances in understanding how the A/P axis is segmented occurred in the later part of the twentieth century as a result of the advances in techniques that allowed scientists to investigate the genetic and molecular aspects of development in greater detail. As methods of identifying genes and proteins continued to improve throughout the twentieth century, scientists were able to evaluate the molecular signals governing cellular development. In addition, methods for generating transgenic mice were introduced in the 1980s. This approach had an enormous impact on studies of A/P axis formation. As described in Chapter 1, transgenic mice are those that have an altered gene incorporated into a segment of DNA. Using transgenic mice, scientists are able to evaluate the effects of the altered gene on development, thereby providing insight into the normal function of the gene in neural development. There are limitations to this approach, some of which are especially important in developmental biology. In some cases, other related genes compensate for the missing gene. In other instances, the mutation is lethal to the embryo and the animal dies before the gene function can be determined. Despite these limitations, however, the use of transgenic mice has proved to be invaluable for understanding aspects of A/P segmentation along the vertebrate neural tube. These methods, and those developed for manipulating gene expression in chick embryos and *Drosophila melanogaster*, combined with classic methods of tissue grafting and cell culture, have elucidated many of the complex and often shared mechanisms used to regulate segmentation of the A/P axis in multiple animal models.

This chapter describes how the different segments along the vertebrate A/P axis are established through the interaction of inductive, transforming, and inhibitory signals. The neural tissue is first induced with characteristics of the anterior-most region of the neural tube. As development progresses, signals arise at specific regions along the length of the neural tube to transform the tissue into progressively more posterior structures. Signals arising in adjacent segments often inhibit one another to limit the actions of each signal to the proper region of the neural tube. The importance of these signals is highlighted by examples of clinical disorders that arise when these signaling pathways are disrupted during embryogenesis.

NEURAL TUBE FORMATION

One of the first morphological changes noted in the developing neural plate is a central indentation along its length called the **neural groove**. The edges of the neural plate then begin to curl upward to create the **neural folds** (**Figure 3.1**). In most vertebrates, the folds are pushed upward by cellular contractions and forces within the neural plate, as well as by the lateral epidermal ectoderm migrating medially and pushing the folds together. The lateral edges of the folds continue to wrap around and

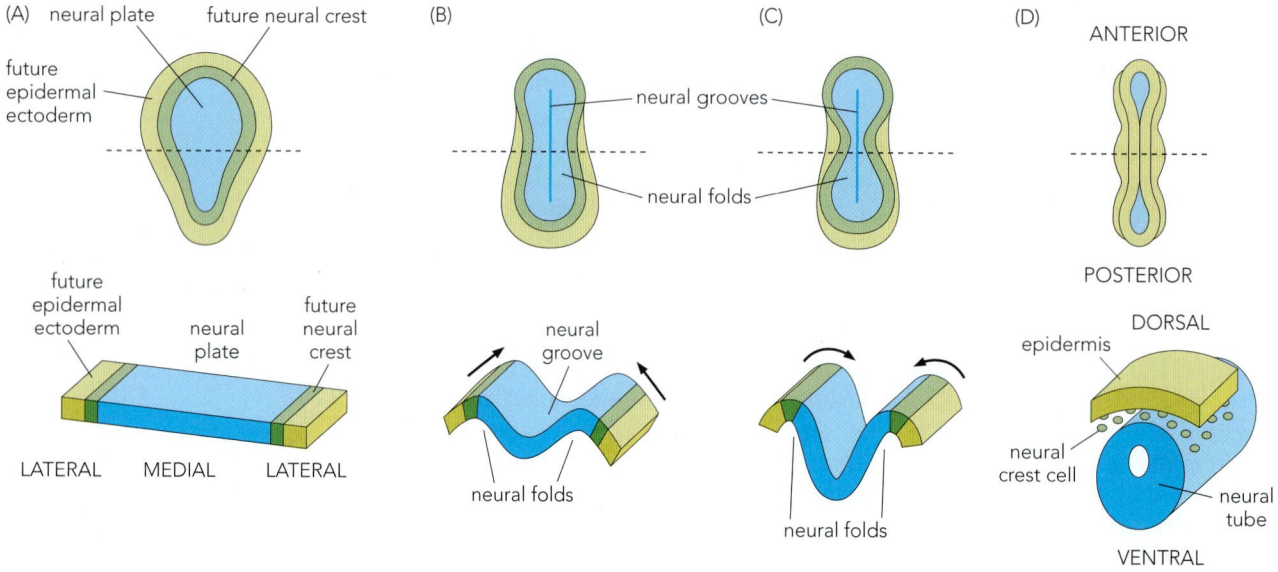

Figure 3.1 The neural tube forms from the neural plate. The general process of neural tube formation is similar across vertebrates. The stages shown in this figure occur between approximately embryonic days 7.5–10.5 in mice, gestation days 18–28 in humans, and 23–33 hours post fertilization in the chick. In A–D, the top panels reveal surface views of the neural plate and forming neural tube. Dashed lines indicate the plane of section shown in the lower panels. (A) The specified area of ectoderm that forms the neural plate (blue) indents to form the neural groove (B) while the edges curve upward to form neural folds (B, C). As the lateral margins of the neural plate continue to curve over (arrows in B and C), the neural tube closes, forming the dorsal surface of the neural tube where the neural crest cells arise (D, lower panel). The remaining ectoderm originating at the lateral margins of the neural plate form the epidermis overlying the neural tube (D). (E) The neural tube is elevated above the surface ectoderm. In mammals, the neural tube first closes in the mid-anterior regions then progresses in more anterior and posterior regions as indicated by the arrows. The somites and primitive streak are also labeled.

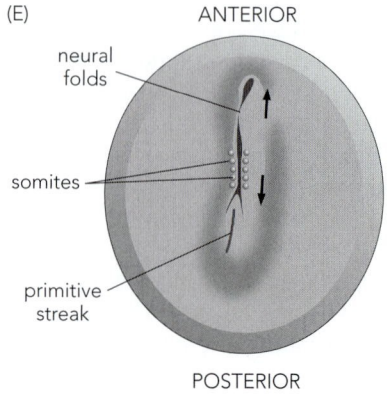

eventually connect to one another, forming the dorsal surface of the neural tube. Later, the neural crest cells that give rise to much of the peripheral nervous system will emerge as a separate cell population from the dorsal surface (see Chapters 4 and 5). Thus, upon closure of the neural tube, the neural plate's lateral and medial regions become the neural tube's dorsal and ventral regions, respectively (Figure 3.1).

The location where neural tube closure begins along the A/P axis varies slightly in different vertebrate animal models. In frogs, the neural tube closes at about the same time at all A/P levels. In chicks, the anterior neural tube closes first, followed by progressively more posterior regions. In mammals, the neural tube first closes at the mid-anterior region, then progresses to the more anterior (rostral) and posterior (caudal) regions. Failure of the neural tube to close results in human birth defects that can have mild to fatal outcomes, depending on which region along the A/P axis is affected. Thus, the process of neural tube formation and closure is studied to reveal the basic cellular and molecular mechanisms that govern these events and to understand how **neural tube defects (NTDs)** can be prevented in children (**Box 3.1**).

Box 3.1 Adequate intake of folic acid can reduce incidence of neural tube defects

In the United States, about one pregnancy per 2000 results in a neural tube defect (NTD), with anencephaly and spina bifida the most prevalent. When the cranial region fails to close, the result is **anencephaly**, a condition in which the anterior portions of the brain and back of the skull fail to develop and close properly (**Figure 1**). While pregnancy can continue through full term, the condition is fatal. In contrast, **spina bifida** arises from incomplete closure of the posterior (caudal) region of the neural tube, leaving the spinal cord exposed through the opened vertebral column (**Figure 2**). Spina bifida occurs in many forms that result in mild to severe disabilities.

The incidence of NTDs can be dramatically decreased with proper diet. **Folic acid** (folate, vitamin B_9) was first discovered in 1941; by the 1960s, it was recognized to have potentially important nutritional effects during pregnancy. In the 1980s, folic acid was specifically noted to decrease the occurrence of NTDs. Studies found that women who included folic acid in their daily diets during the first four weeks of pregnancy decreased the risk of neural tube defects by 50–70%. In 1996, the United States Preventive Services Task Forces recommended that women capable of conception take folic acid supplements, and the United States Food and Drug Administration published regulations to require the addition of folic acid to enriched breads, cereals, flours, and other grains. It is now recommended that women of child-bearing age take 400 micrograms of folic acid daily.

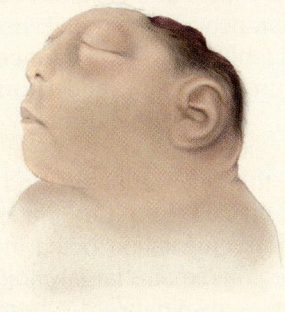

Figure 1 An infant with anencephaly. Anencephaly occurs when the neural tube fails to close in the anterior-most regions, leaving the skull open and the remaining brain tissue exposed. While pregnancy may continue full term, the condition is fatal.

The neural tube closes between 21 and 28 days of gestation, so women are unlikely to know they are pregnant at the time of neural tube closure. The average U.S. diet is estimated to include only 200 micrograms of folic acid. Therefore, to ensure adequate intake, women are encouraged to add folic acid supplements to their diets and eat foods rich in folate, such as spinach, orange juice, and fortified cereals, to reduce the risk of NTDs.

The precise mechanisms by which folic acid prevents NTDs are still not entirely clear, but several clues have been identified in recent years. In some mothers whose children were born with NTDs, autoantibodies that target the folate receptor were detected, revealing one pathway by which folate uptake can be limited. In several animal models, reduced nucleotide synthesis has been associated with NTDs. Because folic acid is involved in the synthesis of nucleotides, it may prevent NTDs by improving otherwise reduced rates of synthesis. Folic acid is also involved in the methylation pathway, an important epigenetic modification that can influence

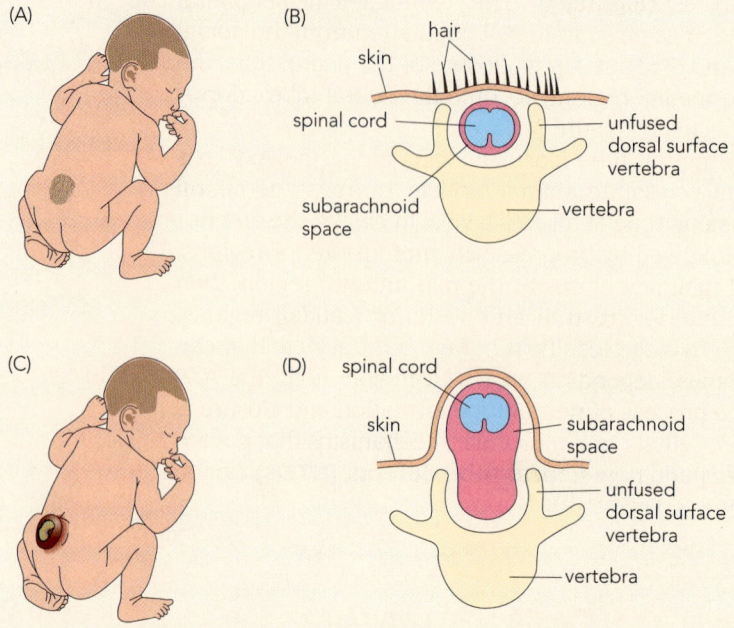

Figure 2 Spina bifida. (A) An infant with a mild form of spina bifida in which the vertebrae fail to fuse. (B) The spinal cord develops normally but is only protected by a thin patch of skin that is often covered by a patch of hair. (C) A more severe form of spina bifida results when the spinal cord bulges outward through the opened vertebral column (D).

gene expression. In mice in which the methylation cycle was inhibited, NTDs developed. Further, hypomethylated DNA was found in the brains of mice with NTDs. Thus, folic acid can influence a number of cellular events and multiple genetic and environmental factors may be involved in determining the likelihood of a NTD occurring. Whatever the precise mechanisms involved, taking the recommended amount of folic acid daily can substantially reduce the risk of this common birth defect.

Early segmentation in the neural tube helps establish subsequent neural anatomical organization

As cells continue to proliferate during neural tube formation, the length of the neural tube extends at the posterior end. In addition to lengthening, the neural tube constricts, expands, and bends at specific sites along much of its length to establish early anatomical and cellular boundaries for the future CNS. Such changes in appearance are first noted shortly after neural tube closure and are soon readily identified as the **prosencephalon** (future forebrain), **mesencephalon** (future midbrain), and **rhombencephalon** (future hindbrain; **Figure 3.2**A). These three primary brain vesicles become the five secondary brain vesicles as development continues. The prosencephalon is divided into the telencephalon and diencephalon, while the rhombencephalon is divided into the metencephalon and myelencephalon. In contrast to the segmentation noted in the developing brain, the developing spinal cord region remains largely uniform in outward appearance (Figure 3.2B). Differences in spinal cord levels are more easily observed along the dorsal-ventral axis (see Chapter 4).

The expansions that form along the A/P axis during the early stages of neural tube development are often described as metameric segments, or **neuromeres**, a series of enlargements that provides regionally restricted areas for further development. In the developing forebrain region, these

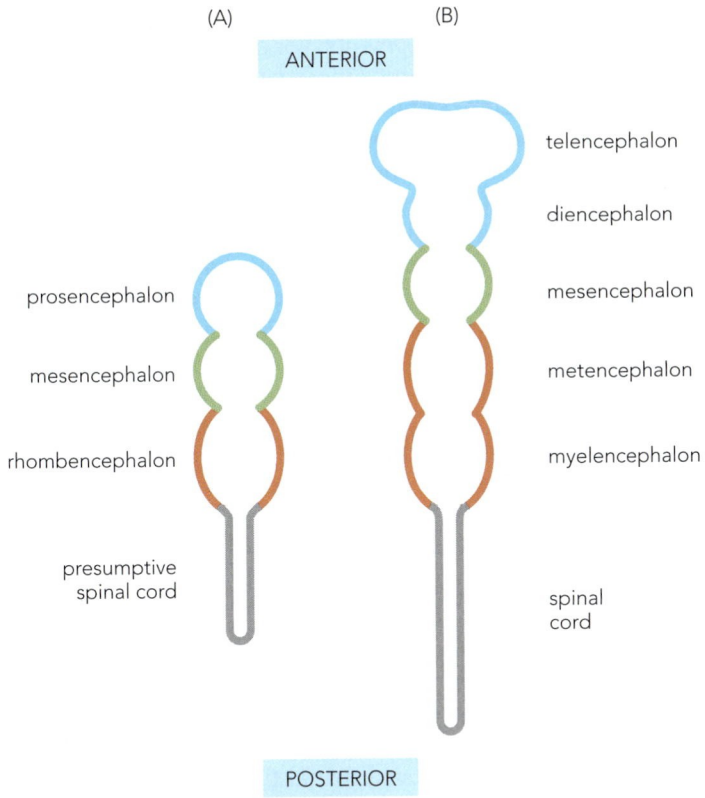

Figure 3.2 The neural tube becomes segmented into brain vesicles. (A) A dorsal view of the neural tube reveals that shortly after neural tube closure, three primary brain vesicles are visible. These are the prosencephalon (future forebrain), mesencephalon (future midbrain), and rhombencephalon (future hindbrain). (B) As development progresses, the five secondary brain vesicles are formed as the prosencephalon is divided into the telencephalon and diencephalon and the rhombencephalon is divided into the metencephalon and myelencephalon.

expansions are called **prosomeres**, and in the hindbrain, **rhombomeres** (**Figure 3.3**A). Although the neuromeres correspond to general anatomical areas found in the adult nervous system, the purpose of segmentation at such early stages is not to designate rigid boundaries for adult structures, but rather to establish initial patterning of the nervous system. The transient boundaries between segments appear to serve many functions, such as limiting the movement of cells until they have been exposed to critical signals that impact their subsequent development.

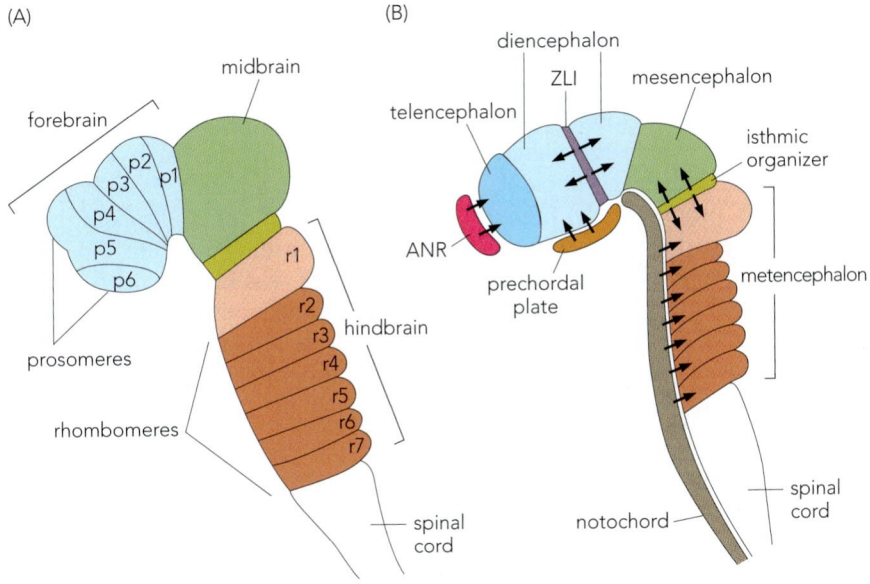

Figure 3.3 Early boundary formation along the anterior–posterior axis. (A) Areas of expansions called neuromeres form along the anterior–posterior (A/P) axis. In the forebrain, these expansions are called prosomeres and are numbered from 1-6 (p1-p6) beginning at the junction of the forebrain (prosencephalon) and midbrain. In the hindbrain, the expansions form rhombomeres that are numbered sequentially beginning at the midbrain–hindbrain junction (r1-r6). The neuromeres are transient structures that are thought to provide regionally restricted segments that produce unique signals required for early cell development. (B) Signaling centers that provide local development cues (arrows) are located along the A/P axis. The anterior neural ridge (ANR) lies at the anterior-most end of the neural tube (telencephalon) to provide signals that influence development of the forebrain. Signals influencing forebrain development are also generated by the prechordal plate and the zona limitans intrathalamica (ZLI) in the diencephalon. The isthmic organizer is found at the boundary between the mesencephalon (midbrain) and metencephalon (hindbrain), where signals are generated to influence development of these regions of the developing nervous system. The notochord extends from the mesencephalon to the spinal cord and provides signals that influence the development of cells in these more posterior regions of the neural tube.

Signaling centers that help pattern the A/P axis of the vertebrate nervous system are located adjacent to segments. For example, the **anterior neural ridge** (**ANR**) forms the anterior-most signaling center to influence development of the **telencephalon** portion of the prosencephalon. More posteriorly, the **zona limitans intrathalamica** (**ZLI**), located at the junction of the prechordal and notochordal regions, is thought to provide signals to regulate development of the **diencephalon** portion of the prosencephalon. The **midbrain–hindbrain border** (**MHB**) or **isthmus** is located between the mesencephalon and metencephalon and is the site of a signaling center called the **isthmic organizer** (**IsO**) that influences development of the future midbrain, pons, and cerebellum. In addition, the **notochord** extends along much of the neuroaxis and is an important source of signals for patterning hindbrain and spinal cord regions. The notochord extends as far anteriorly as the mesencephalon, but is prevented from extending further by the prechordal plate, the area where ectoderm and mesoderm are tightly adherent to one another (Figure 3.3B; see also Figure 2.4). Specific molecules identified in the various signaling centers are described below.

Temporal–spatial changes in the signals required to induce head and tail structures

Soon after the discovery that the dorsal blastopore lip (DBL) of the gastrula stage amphibian embryo (Spemann's organizer) was a source of signals that induced neural tissue, scientists made other discoveries that indicated additional signals were used to direct neural development along the A/P axis. For example, grafting studies conducted independently by Hans Spemann and Otto Mangold demonstrated that the age of the organizer tissue determined whether anterior or posterior regions of the body axis formed (**Figure 3.4**). Younger organizer tissues induced anterior structures, whereas slightly older tissues induced posterior regions. If the DBL from an early stage gastrula or a section of the anterior (younger) region of chordamesoderm were grafted into a host embryo, then extra ectopic head structures formed. Conversely, if slightly older DBL or posterior (older) chordamesoderm were grafted in the host embryo, then extra tail structures formed. Together, these results suggested that the organizer generated different signals at different times in development to direct the formation of these areas along the A/P axis.

Over the years, many scientists concluded that the anterior (forebrain-like) region of the neural tube is a "default" state for neural tissue, with additional signals required to transform neural tissues into more posterior regions. In the 1950s, work by Pieter Nieuwkoop suggested a model by which the signal for neural induction, the "activator," was needed to induce competent ectoderm to form anterior brain regions, but that progressively higher concentrations of other "transforming" signals were needed to convert early neural tissue into hindbrain and spinal cord. Although identification of such molecules was elusive at that time, the general mechanisms first proposed by this "activation–transformation" model of Nieuwkoop have been largely supported by more recent studies. Additional signals that restrict the activity of the transforming signals have also been discovered.

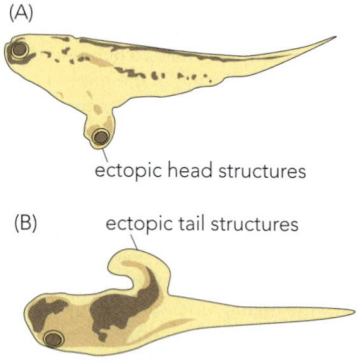

Figure 3.4 The age and region of organizer tissue regulates which regions of the embryo are induced to form. Grafting studies by Otto Mangold and others demonstrated that when younger (anterior) organizer tissue was grafted into a host embryo, the grafted organizer induced extra ectopic head structures (A). In contrast, grafts of older (posterior) organizer tissue induced ectopic tail-like structures in the host embryos (B). (Adapted from Stern CD [2001] *Nat Rev Neurosci* 2:92–98. With permission from Macmillan Publishers Ltd.)

Activating, transforming, and inhibitory signals interact to pattern the A/P axis

The current model to explain A/P patterning suggests that a complex and elegant balance of activating, transforming, and antagonizing signals is required to establish forebrain, midbrain, and hindbrain regions. It now appears that neural inducers first promote formation of primarily anterior-like tissue in the neural plate, then region-specific signals emerge to refine each area along the A/P axis of the developing neural tube. Some of the signals transform previously induced areas into new regions, while other signals inhibit (antagonize) those signals to prevent transformations (**Figure 3.5**). The signals often overlap and interact through concentration gradients established by **morphogens**—diffusible molecules that act on cells at a distance. As will be seen below, cells within different regions along the A/P axis respond best to specific concentrations of each signal and it is the interactions of these signals, in combination with other region-specific cues, that establish anatomical specializations along the A/P axis of the nervous system.

Although anterior regions are generally specified prior to more posterior regions, the signals used to pattern each A/P level are not simply produced in a wave from anterior to posterior. Instead, signals are produced at several locations and at various times in order to pattern nearby areas while also restricting the influence of molecules from adjacent territories. For example, signals that specify midbrain structures must be prevented from extending to more anterior or more posterior regions. Additionally, midbrain structures must not be converted by signals originating in the forebrain or hindbrain regions. As will be seen, the expression of regionally specific signals and the repression of adjacent signals are equally important in patterning the A/P axis.

Many of the genes and signal transduction pathways for A/P axis patterning are conserved across invertebrate and vertebrate species, although some differences in signaling mechanisms have been noted in the various animal models that may reflect, in part, the differences in brain anatomy. Further, different members of a given gene family may govern patterning in a particular A/P region in the different species. The examples that follow focus primarily on the gene families involved in patterning the specific regions of the vertebrate neural axis, rather than species-specific family members.

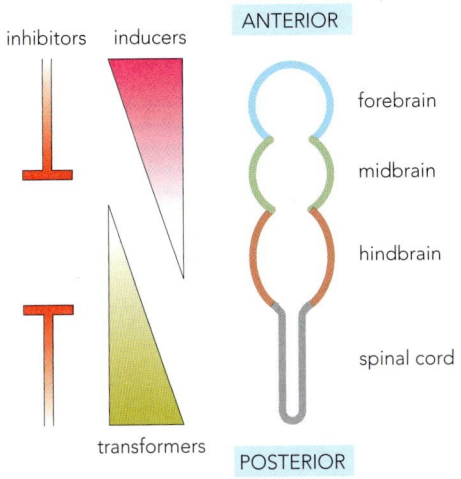

Figure 3.5 Schematic representation of the interplay between inducers, transformers, and inhibitors in establishing gradients of signals. Cells along the A/P axis respond best to a given concentration of a morphogen. An inducer may be secreted such that those cells closest to the source of the inducer receive the highest concentration, while those further away receive lower concentrations. Inducers are thought to form primarily forebrain-like regions, while transforming signals modify the effects of the inducers in order to influence development of more posterior-like regions. Various inhibitory signals limit the spread of inducers and transformers to help maintain the proper boundaries of the morphogen signals. Multiple signals help pattern the neural tube along the A/P axis and gradients of specific inducers, transformers, and inhibitors may operate over long or short distances to pattern different segments.

SPECIFICATION OF FOREBRAIN REGIONS

Blocking bone morphogenetic protein (BMP) signaling by chordin, follistatin, or noggin leads to the induction of neural ectoderm, with anterior-like brain regions forming initially (see Chapter 2). However, additional signaling molecules are necessary to delineate the telencephalon and diencephalon regions of the forebrain.

Signals from extraembryonic tissues pattern forebrain areas

The **anterior visceral endoderm** (**AVE**) in the mouse has been long recognized as a signaling center important for head and forebrain development (**Figure 3.6**; see also Chapter 2). Unlike the organizer region,

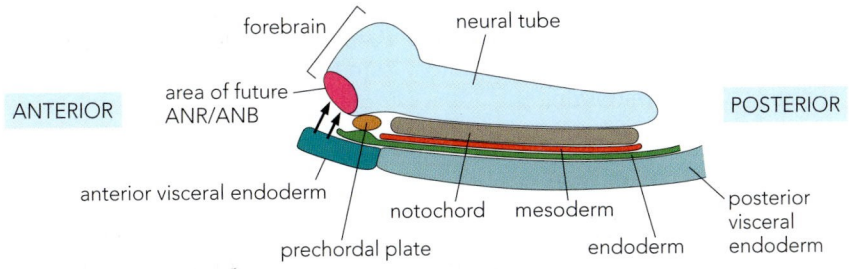

Figure 3.6 The anterior visceral endoderm (AVE) influences forebrain development. The AVE is the anterior-most region of the visceral endoderm that underlies the developing embryo. The AVE is a population of extraembryonic cells found in mice that is similar to the hypoblast of chick embryos. The AVE serves as a signaling center to influence head and forebrain development and produces signals such as fibroblast growth factors that specify the cells of the anterior neural ridge (ANR; also called the anterior neural border, ANB, in zebrafish). The ANR/ANB is a signaling center located at the anterior region of the neural tube that further influences gene expression patterns and development of the forebrain. The relative locations of signaling centers that are important at later stages of development (the prechordal plate and notochord) are also shown.

the AVE—the population of extraembryonic cells that comes to lie beneath the anterior neural plate—does not induce neural tissue. However, the AVE appears necessary to pattern the anterior regions of the neural tube. For example, in mice lacking the AVE, anterior brain regions fail to form. The hypoblast in the chick serves a similar function.

The AVE and hypoblast have been proposed to influence forebrain patterning in a number of ways. For example, the cells in these regions produce signals that specify cells located at the most anterior limit of the neural ectoderm, called the anterior neural ridge (ANR, mouse and chick) or the **anterior neural border** (**ANB**, fish). In chick, hypoblast-derived fibroblast growth factor (FGF) has been associated with specifying the ANR. The ANR/ANB in turn serves as a signaling center that regulates forebrain development and the expression of forebrain-specific genes. In zebrafish, the ANB cells are important at the mid-late gastrula stages to establish the forebrain/telencephalon region and initiate A/P polarity. When ANB cells were transplanted to posterior regions of the embryo, genes normally restricted to the telencephalon were now expressed in these regions. The AVE may regulate anterior development by inhibiting posteriorizing signals originating in the developing midbrain.

Forebrain segments are characterized by different patterns of gene expression

The developing forebrain is often described as being segmented into divisions called prosomeres. Six different prosomeres have been identified beginning with prosomere 1 (p1) adjacent to the mesencephalon and progressing anteriorly to prosomere 6 (p6) in the telencephalon. Regional expression of genes, including members of the *Otx* (orthodenticle homeobox), *Pax* (paired homeobox), and *Six* (sineoculus homeobox) families, have been described within boundaries of these forebrain segments (**Figure 3.7**). These molecules play direct roles in forebrain patterning and are homologs of *Drosophila* genes that often have similar patterning functions during fly development. For example, a number of experiments have demonstrated the necessity of *Otx* genes in forebrain development across species. In vertebrates, *Otx* genes are expressed widely in the regions anterior to the isthmus/MHB. In mice lacking *Otx* genes, head structures do not form. However, if the *Drosophila* homolog, the *Otd* (orthodenticle) gene, replaces the missing *Otx* genes, the head structures are rescued. Conversely, human *Otx* genes can rescue *Drosophila Otd* mutants, further demonstrating the highly conserved nature of these genes and their functions in patterning forebrain and head structures.

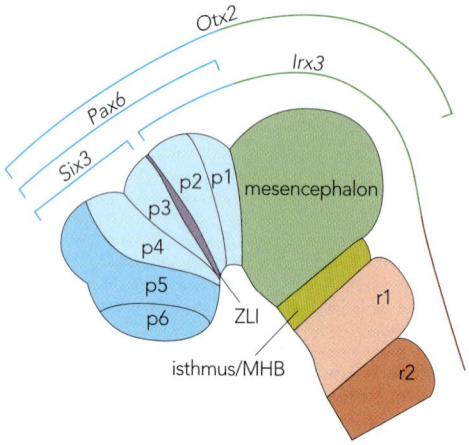

Figure 3.7 Restricted patterns of gene expression regulate development of forebrain regions. The prosomeres are the six neuromere segments located in the forebrain region anterior to the mesencephalon (p1-p6). Gene expression is limited within many of these prosomeres to establish boundaries for further neural development. *Otx* genes such as *Otx2* expressed in p1-p6 regulate development of the forebrain, as well as the mesencephalon and other head structures. *Pax6* is expressed in p1-p6 and is needed for normal forebrain, eye, and ear formation. *Six3*, which is localized to p3-p6, opposes expression of *Irx3* in p1, p2, and more posterior regions.

Genes of the *Pax* family are also expressed widely in the developing neural tube and *Pax6* is required for forebrain development. *Pax6* is expressed in the telencephalon through diencephalon regions, where it is required for normal eye and ear formation. Loss of *Pax* genes leads to small or absent eyes in mice and humans.

Beginning at the neural plate stage, the ZLI is bordered anteriorly by *Six3* and posteriorly by *Irx3*, a member of the Iroquois homeobox gene family. *Six3*, like *Pax6*, is needed to direct development of forebrain regions (Figure 3.7). The effects of *Six3* and *Pax6* are restricted to these more anterior regions in part by inhibitory signals such as Irx3 and Wnt that originate in the midbrain.

Signals prevent Wnt activity in forebrain regions

Although the role of Wnt proteins in neural induction remains unclear (see Chapter 2), the role of these proteins in patterning regions along the

A/P axis is well established. The interaction of neural inducers and Wnt in patterning forebrain versus midbrain regions has been shown in several experiments. For example, when Wnt was overexpressed in the presence of a neural inducer (for example, chordin, follistatin, or noggin), genes and structures associated with anterior brain regions were suppressed. In contrast, if endogenous Wnt expression was inhibited, anterior brain regions expanded into areas that would normally comprise midbrain regions.

A number of Wnt-inhibiting molecules, including Cerberus, soluble members of the Frizzled family, and Dickkopf, have been identified. These molecules are initially expressed in the organizer and later found in the AVE, two signaling centers found to be important for patterning the anterior regions of the neuraxis.

A Wnt receptor complex is used in some Wnt signaling pathways. This receptor complex is comprised of a transmembrane protein of the Frizzled family of receptors (Frz) and a low-density lipoprotein receptor (LRP5 or LRP6), both of which interact with Wnt (**Figure 3.8**A). **Cerberus** is a secreted protein that binds to Wnt and BMP and thus blocks the activity of both (Figure 3.8B). When Cerberus mRNA was injected into early-stage *Xenopus* embryos, one or more incomplete ectopic heads formed. Based on this finding, the molecule was named Cerberus after the mythical three-headed dog that guards the gates of Hades. Although the *Xenopus* embryos injected with the excess mRNA did not develop three complete heads, they clearly had extra head structures.

A second group of proteins associated with anterior regionalization is the Frizzled family. In addition to the Frizzled (Frz) transmembrane proteins that serve as receptors for Wnt, there are additional small, soluble molecules of the Frizzled family called **secreted Frizzled-related proteins (sFRPs)**. The sFRPs sequester Wnt and thus block signaling through Frz receptors (Figure 3.8B). Injection of such sFRPs into *Xenopus* embryos resulted in formation of larger heads, and sFRP overexpression in zebrafish resulted in expansion of the telencephalon.

Another secreted molecule discovered shortly after Cerberus was **Dickkopf (Dkk)**, which binds directly to subunits of the Wnt receptor

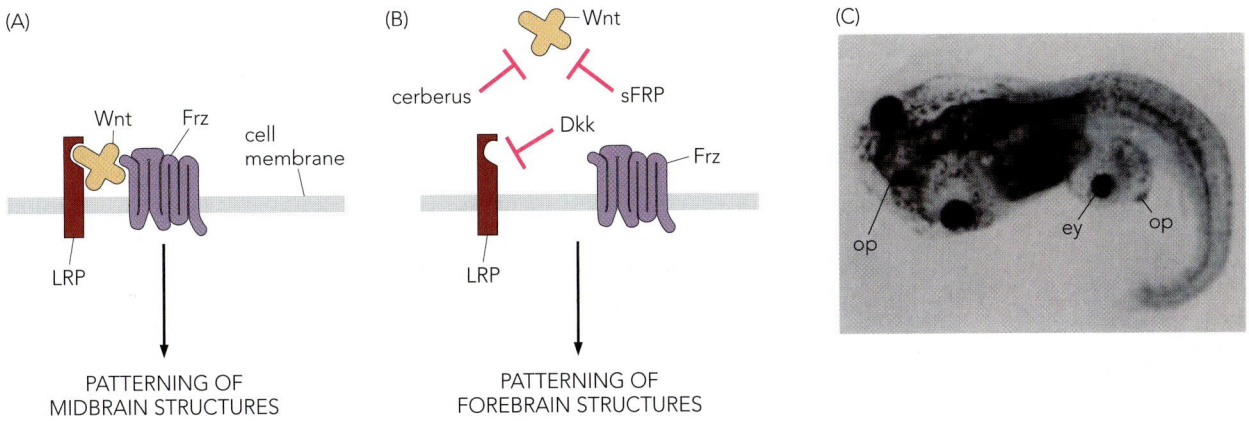

Figure 3.8 Wnt signaling is inhibited in forebrain regions. In order to develop forebrain structures, signals such as Wnt that are used to pattern the midbrain must be prevented from extending too far anteriorly and converting forebrain tissue into more posterior-like regions. (A) Wnt can signal through a receptor complex formed by a transmembrane protein of the Frizzled family (Frz) and a low-density lipoprotein receptor (LRP5 or LRP6). (B) Several forebrain-derived signals inhibit the activity of Wnt. Cerberus and secreted Frizzled-related proteins (sFRPs) prevent Wnt from binding to the receptor complex. Dickkopf binds to subunits of LRP5/LRP6 to inhibit Wnt signaling. (C) *Xenopus* embryos injected with excess Dickkopf that blocked endogenous Wnt signaling developed larger head and brain regions that extended into the midbrain regions. (op, olfactory placode; ey, eye. C, from Bouwmeester T, Kim S, Lu B & DeRobertis EM [1996] *Nature* 382:595–601. With permission from Macmillan Publishers Ltd.)

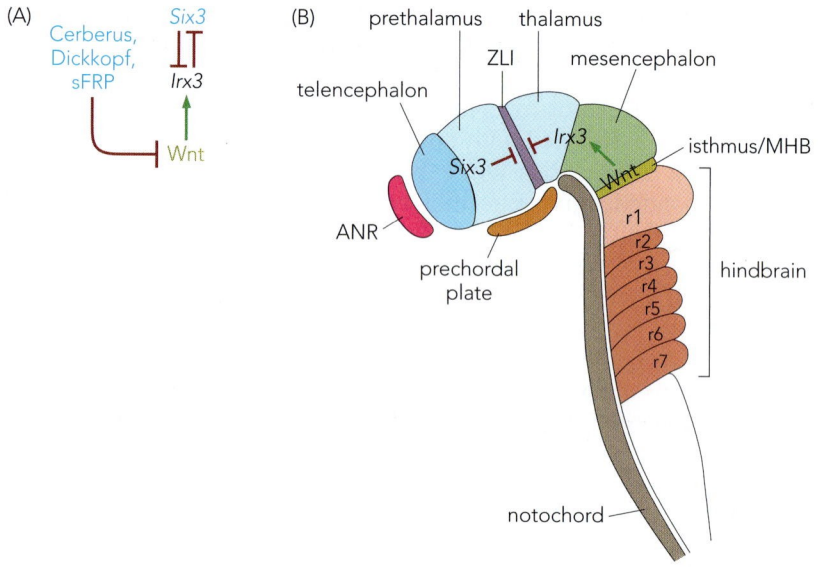

Figure 3.9 Restricted expression of signaling molecules along the A/P axis influences boundary formation by inducing or blocking expression of adjacent genes. A number of examples of signaling molecules interacting at different segment boundaries have been identified. Cerberus, Dickkopf, and secreted Frizzled-related proteins (sFRPs), initially expressed in the organizer, inhibit Wnt signals to prevent posteriorization of anterior regions. *Six3* in the forebrain and *Irx3* in the posterior diencephalon/midbrain region antagonize the expression of each other, thus limiting expression of each gene to a certain region. Wnt located at the isthmus (midbrain–hindbrain border, MHB) regulates expression of Irx3 levels to ensure sufficient Irx3 is present to limit expansion of the forebrain into these more posterior regions.

complex. By binding to the LRP5 or LRP6 subunits, Dkk interferes with Wnt signaling and prevents formation of midbrain structures in more anterior regions. The importance of Dkk signaling in formation of anterior regions of the nervous system is seen in experimental manipulations of *Xenopus* embryos. Truncated (inactive) BMP receptors had previously been shown to induce neural tissue (see Chapter 2). When these truncated receptors and Dkk were co-injected into *Xenopus* embryos, a larger head and brain region formed (Figure 3.8C). The resulting phenotype of an expanded head and brain seen in the Dkk-treated embryos led to the name of the gene. Dickkopf is German for "big head" or "stubborn." Conversely, if antibodies raised against Dkk were also injected, the Dkk-induced expansion of head structures was inhibited and actually led to a smaller head and incomplete brain development (microcephaly). Similarly, mice lacking the *Dkk1* gene developed with a truncated head and reduced brain size. Thus, multiple signals are present in the developing forebrain regions to restrict the activity of midbrain-associated Wnt.

Other signals important for limiting Wnt activity in anterior regions of the neural tube include members of the Six family of homeodomain transcription factors that are expressed in the developing forebrain. Balancing the role of Six3 in the normal embryo is the Irx family of transcription factors expressed in the posterior diencephalon and midbrain (see Figure 3.7). The restricted expression domains of the Six and Irx family members are maintained because these two proteins suppress one another. *Irx3* expression is also influenced by Wnt. Loss of Wnt causes a loss of *Irx3* expression, whereas increasing Wnt levels leads to expansion of *Irx3* expression into more anterior regions (**Figure 3.9**). Thus, coordinated expression of Wnt and Irx are needed to pattern posterior diencephalon regions and delineate them from more anterior *Six3*-expressing regions of the forebrain.

REGIONALIZATION OF THE MESENCEPHALON AND METENCEPHALON REGIONS

Posterior to the diencephalon is the mesencephalon—the future midbrain region of the vertebrate brain that forms the dorsal tectal and ventral tegmental structures. The mesencephalon forms a border with the metencephalon, the region that gives rise to the pons and cerebellum.

The metenecephalon is typically described as extending posteriorly to rhombomeres 1 and 2 (r1 and r2), the anterior-most rhombomeres in the hindbrain.

Intrinsic signals pattern the midbrain–anterior hindbrain

As noted above, the mesencephalon and metencephalon regions are separated by a narrow constriction called the midbrain–hindbrain border (MHB) or isthmus. This anatomical constriction also serves as one of the intrinsic signaling centers in the developing nervous system known as the isthmic organizer (IsO; Figure 3.3). The IsO produces multiple molecules that pattern the midbrain and anterior hindbrain regions, as well as signals that prevent the spread of other signals originating in the forebrain or posterior hindbrain.

Signaling centers intrinsic to the mesencephalon and metencephalon regions have been identified using a number of experimental approaches. For example, in the late 1980s, transplantation studies using quail and chick embryos revealed the presence of cues intrinsic to the mesencephalon and metencephalon. The quail–chick transplantation method was utilized because the cells originating from the quail donor could be distinguished from those of the chick host with the Feulgen nucleolar stain, or with specific antibodies that label the condensed heterochromatin characteristic of quail cells (**Figure 3.10**A). When quail mesencephalon or metencephalon was grafted to corresponding regions of chick embryo, the chick embryos developed the correct midbrain or hindbrain structures, respectively. These results indicated that signals were present in the donor tissue to induce formation of these brain regions. The results also demonstrated that the mesencephalon- and metencephalon-derived signals were *not* species specific, an observation that proved helpful in determining the cellular origins of different signals in other experiments, including transplant experiments between mouse and chick.

The mesencephalon also produced signals needed to form the cerebellum, as well as signals that induced adjacent midbrain tissue. For example, Salvador Martinez and colleagues found that when quail metencephalon was rotated

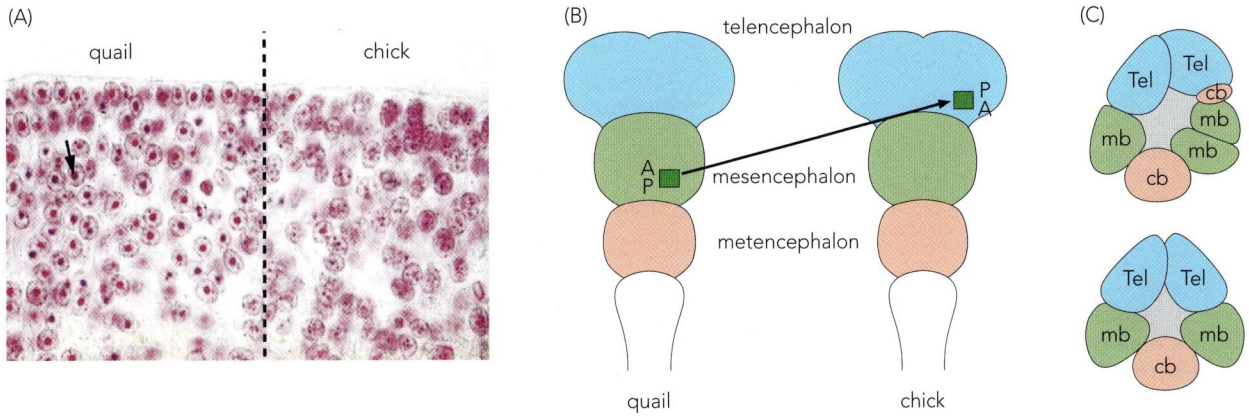

Figure 3.10 Midbrain tissues can induce the formation of ectopic midbrain and anterior hindbrain structures in forebrain tissues. Transplantation studies in which tissue from quail is grafted into chick embryos at similar stages of development reveal intrinsic signaling centers in the midbrain. (A) Cells originating from a quail donor embryo (left side of panel) can be distinguished from those of the chick host (right side of panel) with the Feulgen nucleolar stain that reveals the condensed heterochromatin (arrow) that is characteristic of quail but not chick cells. (B) Grafts of quail mesencephalon tissue were rotated 180 degrees and placed in the forebrain (telencephalon) region of chick embryos. (C) The graft of mesencephalon tissue led to the formation of an extra ectopic midbrain and induced the formation of an ectopic cerebellum. When viewed from the dorsal surface, the new structures were observed to form in an opposite, or mirror image, orientation (top panel) compared to the normal cerebellar/midbrain region (bottom panel). Tel, telencephalon; cb, cerebellum; mb, midbrain. (A, from Le Douarin NM [2004] *Mech Dev* 121:1089–1102. With permission from Elsevier Inc.)

180 degrees and grafted into certain areas of the developing chick forebrain, an ectopic cerebellum formed and adjacent midbrain structures were induced in an orientation consistent with the rotated graft (Figure 3.10B, C). However, a new cerebellum and midbrain were only found if the transplant occurred posterior to the ZLI. The expression of *Six3* anterior to the ZLI may have prevented the induction of midbrain structures in this region.

Multiple signals interact to pattern structures anterior and posterior to the isthmus

The signals originating in the isthmus/MHB are now known to extend anteriorly and posteriorly to influence signals arising in adjacent segments of the A/P axis. Similar to what was described in forebrain regions, some of these signals act primarily to maintain gene expression patterns, while others repress the activity of molecules in adjacent structures. Many of the genes for these signals are set up during gastrulation or early neural plate stages. For example, beginning in gastrula-stage embryos and continuing through A/P patterning, a division between anterior expression of *Otx* in the mesencephalon and posterior expression of *Gbx* in the metencephalon is observed (**Figure 3.11**).

Interactions between Otx and Gbx have been demonstrated in a number of studies. In mice lacking *Otx* genes, the areas anterior to the isthmus that normally express Otx are respecified and take on more posterior-like characteristics. Similarly, if *Gbx* is missing, midbrain areas extend more posteriorly leading to midbrain-like characteristics in regions posterior to the isthmus. Thus, with the loss of either gene, the adjacent area is able to dominate and expand to re-pattern neighboring regions of the neural tube. Later in development, additional signals are employed to restrict Otx and Gbx to regions anterior and posterior to the isthmus, respectively.

Two well-characterized gene products that participate in midbrain–anterior hindbrain patterning are FGF8 and Wnt. There are 22 known members of the FGF family. FGF8, FGF17, and FGF18 are expressed in the isthmus region, but the role of FGF8 is the most fully characterized to date. Wnt is initially expressed throughout much of the midbrain but eventually becomes restricted to a narrow band in the isthmus, immediately anterior to FGF8 expression. Numerous experimental findings suggest a direct role for FGF8 in cerebellar induction and an indirect role in the patterning of adjacent regions of the nervous system. In contrast, Wnt does not directly induce midbrain–hindbrain structures, but instead interacts with other signaling molecules, such as Otx and Gbx to regulate development in these areas.

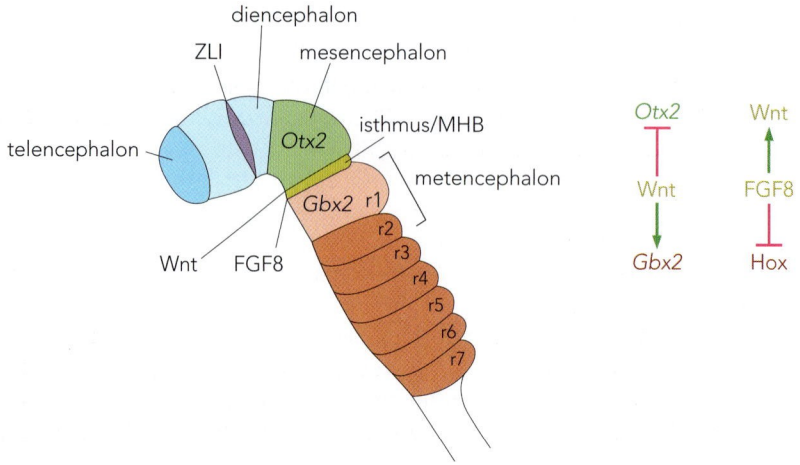

Figure 3.11 The isthmus or midbrain–anterior hindbrain border (MHB) is a source of FGF8 and Wnt. A narrow constriction at the MHB, the isthmus, expresses FGF8. Wnt is initially expressed throughout much of the midbrain but becomes restricted to a band just anterior to FGF8. These signals influence the expression of genes that further pattern these regions. For example, Wnt inhibits activity of *Otx2* anteriorly and activates *Gbx2* posteriorly, whereas FGF8 activates Wnt expression in the region of the isthmus, but inhibits expression of *Hox* genes in rhombomere 1 (r1).

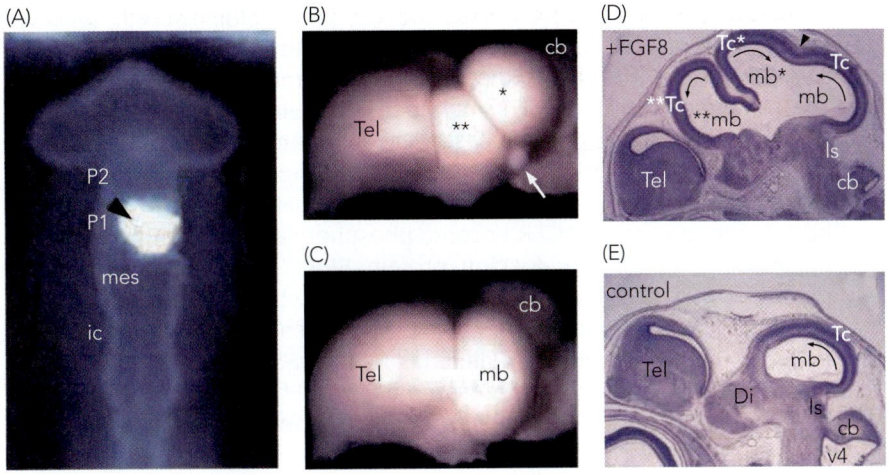

Figure 3.12 Exogenous FGF8 induces ectopic midbrain and anterior hindbrain structures at the juncture of the midbrain and forebrain. When beads coated with FGF8 were implanted into the posterior forebrain of the mesencephalon–prosomere 1 boundary (arrowhead, A), ectopic midbrain and cerebellar tissues were induced to form. Side views of the brain show extra midbrain regions (asterisks, B) and an ectopic cerebellar-like outgrowth (arrow, B) in the FGF8-treated embryos compared to untreated controls (C). Sagittal sections through FGF8-treated (D) and FGF8-untreated (E) brains show that the two ectopic midbrain regions (asterisks, D) formed in an opposite orientation (arrows, D). These experiments demonstrated a role for FGF8 in patterning midbrain and cerebellar regions. Double asterisks indicate ectopic regions that are anterior to those marked by a single asterisk. Tel, telencephalon; Cb, cerebellum; Mb, midbrain; Tc, tectum; Di, diencephalon; Is, Isthmus; ic, isthmic constriction; mes, mesencephalon; v4, fourth ventricle. (From Martinez S, Crossley PH, Cobos I et al. [1999] *Development* 126:1189–1200. With permission from The Company of Biologists Ltd.)

FGF is required for development of the cerebellum

A role for FGF8 in cerebellar induction was demonstrated in studies in which experimental beads were soaked in FGF8 protein. FGF8 remained attached to the beads that were then implanted into embryonic chick brains at the three primary vesicle stage. When beads were implanted into the posterior segment of the chick forebrain (near the prosomere 1 and mesencephalon border; **Figure 3.12**A), the FGF8-coated beads induced ectopic cerebellar and midbrain structures in these more anterior regions (Figure 3.12B). In several of the host embryos, the orientation of the cerebellum and midbrain were opposite of that found in control conditions, similar to the results obtained when rotated metencephalon tissue was transplanted into the forebrain (Figure 3.12D).

In the posterior regions of the forebrain, the FGF8 beads also altered gene expression patterns. The cells adjacent to the bead now failed to express the forebrain gene *Otx2*, but instead expressed *Engrailed 2 (En2)*, *FGF8*, and *Wnt*, genes that are normally found in midbrain–anterior hindbrain regions. Additionally, like the transplant studies, beads placed in more anterior forebrain regions failed to induce midbrain structures, suggesting signals present in the anterior forebrain inhibited midbrain formation.

FGF isoforms and intracellular signaling pathways influence cerebellar and midbrain development

Studies of mice with decreased levels of FGF8 revealed defects in midbrain and cerebellum formation, further demonstrating the requirement for FGF8 in patterning this region of the developing nervous system. Other experiments suggest that the concentration of FGF8 influences which regions form, with lower concentrations impacting the development of the midbrain and higher concentrations primarily influencing the cerebellum. Different isoforms of FGF8 may also affect development in this region. For example, FGF8a appears to be the isoform important for midbrain induction, whereas FGF8b, the isoform used in the bead implantation experiments, appears necessary for cerebellar induction.

It has also been proposed that the different effects of these two isoforms result from differences in receptor binding strength. The four FGF receptors are receptor tyrosine kinases, and FGF8 binds with highest affinity to FGFR3 and FGFR4. The addition of 11 amino acids in FGF8b may allow this isoform to bind to the FGF receptors more strongly than the FGF8a isoform. Such enhanced binding is thought to increase the strength of the FGF signal, leading to the induction of cerebellar tissue.

The importance of FGF signaling in cerebellar formation is also seen when intracellular signaling cascades are manipulated. Under normal conditions, the Ras–Erk pathway is one of the signaling cascades initiated when FGF binds to its receptor. This pathway is activated when the tyrosine kinase domains of the FGFR become phosphorylated, causing subsequent phosphorylation of the docking protein FRS2 (FGFR substrate 2) and recruitment of the adaptor protein Grb2 (growth factor receptor-bound protein 2). Grb2 then interacts with SOS (son of sevenless) to activate Ras (rat sarcoma). Activation of Ras leads to the sequential phosphorylation of Raf (rapidly accelerated fibrosarcoma kinase), MEK (MAP kinase/Erk kinase) and Erk (extracellular signal-regulated kinase). Phosphorylated Erk then enters the nucleus, where it regulates transcription of FGF-responsive genes (**Figure 3.13**). Experiments in zebrafish revealed that when this pathway was inhibited, the expression of *gbx2* was repressed in the metencephalon while *otx2* expression was induced in its place. This led to the formation of more anterior structures in areas that normally give rise to the cerebellum and demonstrated that FGF8 normally induces *gbx2* expression via the Ras–Erk pathway.

FGF also provides a negative regulator of the Ras–Erk pathway under normal circumstances to prevent expansion of metencephalon structures too far anteriorly. This occurs because FGF also induces expression of the genes *sprouty*, *sef-b*, and *MKp3* in the isthmus, all of which inhibit the Ras–Erk pathway at specific sites (Figure 3.13). By inducing expression of these genes, FGF is able to regulate its own activity.

FGF and Wnt interact to pattern the A/P axis

As noted in the bead implantation experiments, FGF8 not only directly induced formation of cerebellum, but also influenced expression of additional genes used to pattern the midbrain region. Further, in addition to FGF8 maintaining the expression of Wnt and En in the regions

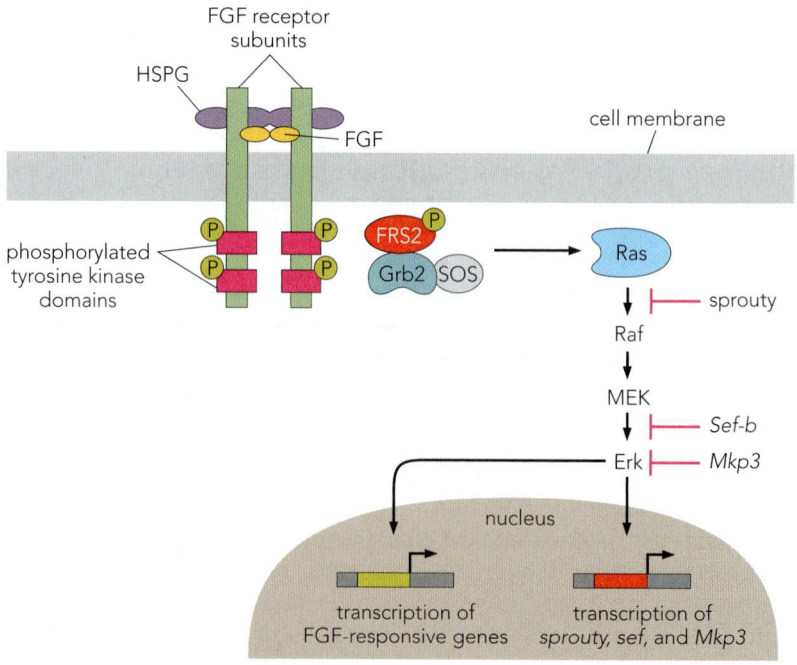

Figure 3.13 FGF regulates transcription of genes important for cerebellar development. Fibroblast growth factors (FGFs) form dimers that bind to tyrosine kinase receptors. Ligand binding causes the receptor subunits to dimerize and cross-phosphorylate one another. Binding to the receptors is enhanced by heparan sulfate proteoglycan (HSPG) localized to extracellular spaces. Both the FGF ligands and receptors contain HSPG binding regions. The phosphorylation of the FGF receptors leads to subsequent phosphorylation of the docking protein FRS2 (FGFR substrate 2) and recruitment of the adaptor protein Grb 2 (growth factor receptor-bound protein 2). Grb2 then interacts with SOS (son of sevenless) to activate Ras (rat sarcoma). Activation of Ras leads to the sequential phosphorylation of Raf (rapidly accelerated fibrosarcoma kinase), MEK (MAP kinase/Erk kinase), and Erk (extracellular signal-regulated kinase). Phosphorylated Erk then enters the nucleus, where it regulates transcription of FGF-responsive genes. FGF also provides a negative regulator of the Ras–Erk pathway under normal circumstances by inducing expression of the genes *sprouty*, *sef-b*, and *MKp3*, each of which inhibits the Ras–Erk pathway at a specific site. These genes are also expressed in the isthmus region, thus providing a means of regulating the activity of FGF signals and preventing FGF from extending too far anteriorly.

immediately anterior to FGF8 expression, FGF8 also inhibits the expression of *Hox* genes in the first rhombomere (**Figure 3.14**). As detailed below, *Hox* genes are necessary for the formation of the more posterior rhombomere segments of the hindbrain. The suppression of *Hox* genes in rhombomere 1 (r1) permits the formation of cerebellar tissues from this segment, rather than rhombomere-like structures that would arise following the activation of *Hox* genes. Thus, FGF8 is a critical signaling molecule for cerebellar formation and the regulation of genes that pattern midbrain regions.

Wnt also plays a vital role in midbrain and cerebellar development, as is demonstrated by the absence of midbrain and cerebellar regions in mice lacking Wnt. However, unlike FGF8, Wnt does not induce midbrain–anterior hindbrain regions directly. Instead, the coordinated expression of Wnt, FGF8, and En1/En2 enables the formation and maintenance of boundaries within the midbrain–anterior hindbrain region (Figure 3.14). Loss of any one of these molecules not only causes defects, but also impacts the expression of the other two signals. For example, in mice lacking the transcription factors En1 and En2, the midbrain and cerebellar regions are dramatically reduced in size. Because Wnt is necessary to maintain expression of En, a loss of Wnt likewise disrupts development of the midbrain and anterior hindbrain. Wnt, FGF8, and En2 also work together to repress expression of genes associated with the formation of forebrain-like regions, such as *Pax6* and *Otx2* and thereby prevent anteriorization of midbrain areas.

The various signaling molecules in the midbrain–anterior hindbrain region likely serve multiple purposes in segregating cell types, limiting migration of cells into adjacent regions, and maintaining the expression of genes and the transduction of signals that impact final cell fate. Scientists continue to make discoveries that help define the interplay between signaling centers and boundary formation in this region of the developing nervous system.

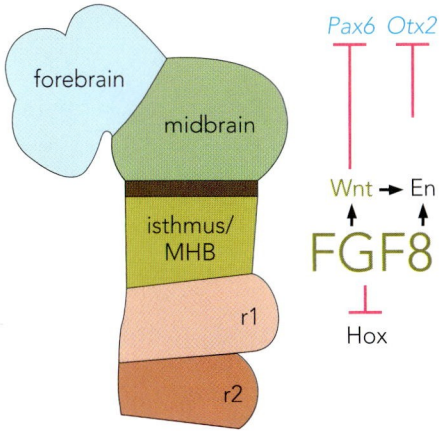

Figure 3.14 FGF8 influences expression of genes in the midbrain and hindbrain. FGF8 maintains expression of *Wnt* and *En* in the regions immediately anterior to FGF8 expression. Wnt and En then inhibit expression of forebrain-associated genes including *Pax6* and *Otx2*. FGF8 also inhibits expression of *Hox* genes in the first rhombomere (r1) so that the cerebellum, rather than Hox-dependent hindbrain structures, can form in this region.

RHOMBOMERES: SEGMENTS OF THE HINDBRAIN

In terms of A/P segmentation of the developing neural tube, the rhombomeres of the hindbrain are perhaps the most studied and best characterized to date. The rhombomeres are numbered in an anterior-to-posterior sequence beginning with rhombomere 1 (r1) in the anterior hindbrain and extending to the junction of the spinal cord (**Figure 3.15**). The exact number of rhombomeres varies from seven to nine, depending on the species and the criteria used for designating the segments.

Each rhombomere expresses a unique set of proteins that impacts proliferation, differentiation, and axonal growth of the developing hindbrain cells. For example, the formation of several of the cranial nerves is impacted by early rhombomere boundaries. Figure 3.15 provides an example from the mouse of the rhombomere origin of various cranial motor neurons in r2–r7. As will be seen below, defects in rhombomere formation lead to abnormal cranial nerve development.

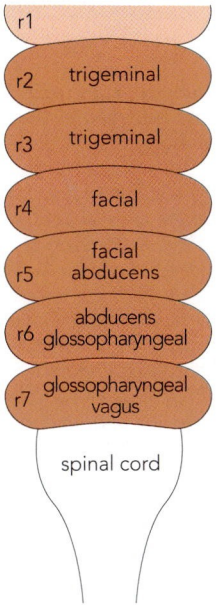

Figure 3.15 Segmentation of the hindbrain. Rhombomeres form segments that extend from the junction with the midbrain to the junction with the spinal cord. Rhombomeres are transient swellings that provide restricted regions for early neural development along the hindbrain. A number of cell types, such as cranial motor neurons associated with the trigeminal, facial, abducens, vagus, and glossopharyngeal nerves, originate in specific rhombomeres as shown in rhombomere 2 (r2) through rhombomere 7 (r7). Unlike the other rhombomeres, r1, the largest rhombomere, contributes to formation of the cerebellum.

Cells usually do not migrate between adjacent rhombomeres

Rhombomeres provide transient segregated regions for early and region-specific developmental cues, ensuring local cells receive the signals they require. One mechanism used to restrict cells to specific rhombomeres is to limit the migration of cells between adjacent rhombomeres. It is estimated that under normal conditions, fewer than 10% of the cells are able to migrate across rhombomere boundaries during development. Limiting cellular migration into adjacent rhombomeres at early stages ensures that cells are exposed to rhombomere-specific developmental cues required to influence subsequent cell fate. At later stages, when the cells need to migrate to another region of the nervous system, the transient rhombomere boundaries are no longer present.

Cell surface molecules appear important in establishing proper cellular groupings in the hindbrain. In particular, differences in adhesive properties and boundary restrictions have been found in the cells of even- versus odd-numbered rhombomeres. For example, a series of experiments by Andrew Lumsden and colleagues revealed that cells from even- or odd-numbered rhombomeres preferentially adhere to cells from other even- or odd-numbered rhombomeres, respectively. In one *in vitro* study, dissociated cells from odd-numbered rhombomeres reaggregated with cells from the same rhombomere or those from another odd-numbered rhombomere, but not with cells from even-numbered rhombomeres. Similarly, cells of even-numbered rhombomeres preferentially reaggregated with cells from other even-numbered rhombomeres. Moreover, grafting studies in chick embryos confirmed that boundaries are normally established between adjacent rhombomeres. In one series of experiments, the scientists removed rhombomere segments from one side of both donor and host embryos of the same developmental stage (**Figure 3.16**A). The donor segments were then grafted into the region of missing segments of the host embryo and allowed to develop for another 1.5–2.5 days. When segments of odd- and even-numbered rhombomeres were grafted adjacent to one another in the host embryo, a rhombomere boundary remained and the migration of cells between the two segments was restricted, similar to the untreated side of the embryo (Figure 3.16B). However, surgically recombining two even-numbered segments allowed intermixing of cells from these different segments (Figure 3.16C). Together, *in vitro* and *in vivo* studies suggested

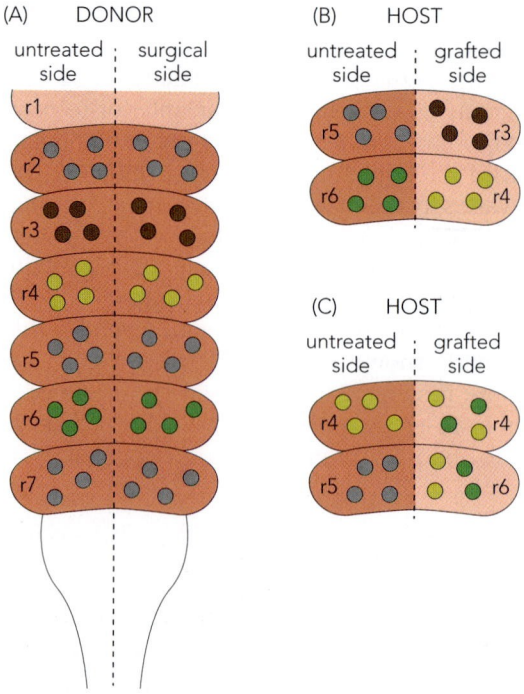

Figure 3.16 Inhibitory cues restrict migration of cells between even- and odd-numbered rhombomeres. (A) Under normal conditions, cells from adjacent rhombomeres do not intermingle. In this example, the cells in each rhombomere are shown as a single color to indicate the segregation of cells within rhombomere boundaries. Grafting experiments in chick embryos tested whether the normal boundaries between adjacent rhombomeres inhibited the intermixing of cells from even- and odd-numbered rhombomeres. Rhombomere segments were removed from one side of both donor (surgical side, A) and host (grafted side, B) embryos at the same stage of development. (B) When grafts of rhombomeres from adjacent odd and even rhombomeres (r3 and r4) were transplanted into a different level in the host embryo, cell migration was still inhibited across rhombomere boundaries, similar to control conditions. (C) Intermingling across rhombomere boundaries was observed when two even-numbered rhombomere segments (r4 and r6) were grafted adjacent to one another in the host embryo. In these grafts, the cells intermingled, unlike those on the untreated side in which the cells of r4 and r5 remained segregated. Similar results were obtained if two odd-numbered segments were grafted next to one another. Together, these experiments suggested that under normal conditions there are inhibitory cues present in odd- and even-numbered rhombomeres to limit the migration of cells between adjacent rhombomeres.

that adhesive cues group cells together within a rhombomere segment, while inhibitory cues limit migration to adjacent segments.

Some of the signals responsible for establishing and maintaining hindbrain segments have been identified

Scientists subsequently determined that the migration of cells between even- and odd-numbered rhombomeres is inhibited by proteins of the Eph family of tyrosine kinase receptors (Ephs) and their associated cell surface ligands (ephrins). Several Ephs and ephrins reveal different expression levels in r2 through r7. The receptors and ligands are often present in alternating patterns along the hindbrain. Evidence that some of these receptor–ligand pairs play a role in rhomobmere patterning comes from studies of r2–r6. In r3 and r5, EphA4, EphB2, and EphB3 receptors are highly expressed. One or more of the corresponding ligands (ephrin A2, ephrin B1, and ephrin B3) are highly expressed in r2, r4, and r6. The interaction of membrane-bound **ephrin** ligands with their corresponding **Eph** receptors leads to bidirectional signaling. In other words, a signal is transduced through both the receptor and the ligand (**Figure 3.17**A). In this group of rhombomeres (r2–r6), the Eph/ephrin signaling is repulsive, thus providing a mechanism to prevent the mixing of cells between adjacent rhombomeres.

Other studies revealed mechanisms by which EphA4 receptor expression is regulated in r4. The Krox20 transcription factor, expressed in r3 and r5, had previously been found to play a role in hindbrain segmentation (Figure 3.17B). In mice lacking *Krox20*, rhombomeres r3 and r5 show initial

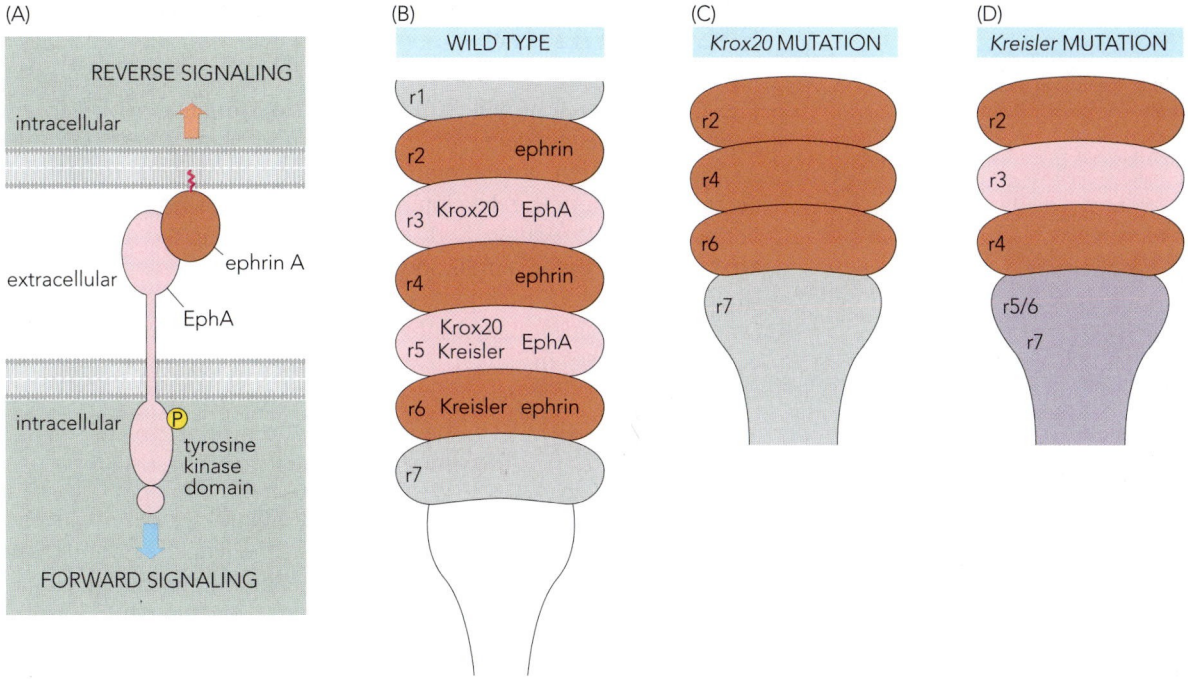

Figure 3.17 Multiple signals interact to regulate the formation of rhombomere boundaries. (A) EphA receptors and ephrin A ligands initiate forward and reverse signaling pathways in cells located in adjacent rhombomeres. An ephrin A ligand binds to an EphA receptor to initiate tyrosine phosphorylation (P) and forward signaling in an adjacent cell. The EphA receptor also initiates signal transduction and reverse signaling through the membrane-attached ephrin A ligand. The ephrin A ligand interacts with co-receptors (not shown) to initiate signal transduction in the ligand-bearing cell. (B) EphA receptors are present on cells in odd-numbered rhombomeres (r3 and r5), while the corresponding ephrin ligands are present on cells of even-numbered rhombomeres (r2 and r4). The resulting bidirectional signaling limits migration between adjacent rhombomeres. The transcription factor Krox 20 is also present in r3 and r5 and regulates expression of these Eph receptors. Another transcription factor, kreisler, is localized to r5 and r6. Both Krox 20 and kreisler are needed to maintain rhombomere boundaries. (C) In mice lacking Krox 20, r3 and r5 are not maintained and the remaining even-numbered rhombomeres, r2–r6, are joined together, creating a shortened hindbrain. (D) When *Kreisler* is absent, r5 and r6 fuse, with no further segmentation observed posteriorly.

signs of formation, but they fail to develop and are ultimately lost. These results show that *Krox20* is not needed for the initial specification of r3 and r5, but is clearly important for their maintenance. The even-numbered rhombomeres that would normally lie adjacent to r3 and r5 are then brought together resulting in a shortened hindbrain (Figure 3.17C). The neural structures associated with r3 and r5 are also lost and the growth of axons from cranial nerve neurons originating in the r2, r4, and r6—those associated with the trigeminal, facial, and glossopharyngeal nerves, respectively—are rerouted within the shortened hindbrain structure.

Another transcription factor important in hindbrain development was identified in the Kreisler mouse, a strain of mice carrying a *Kreisler 1* (*Krml1/MafB*) gene mutation. Kreisler is a member of the Maf (musculo-aponeurotic fibrosarcoma) transcription factor family, a large group of transcription factors named for the origin of the first identified member. In the Kreisler mutant mice, r5 and r6 fail to form, producing an unsegmented hindbrain posterior to r4 (Figure 3.17D). The loss of r5 and r6 results in several abnormalities of the associated hindbrain structures, including loss of the abducens and glossopharyngeal cranial nerves and malformations of the inner ear that normally develops adjacent to r5.

Both Krox20 and Kreisler/MafB transcription factors also regulate expression of other genes required for normal rhombomere development (for example, *Hox* genes, which are discussed below). Thus, the expression patterns of several different molecules interact to establish and maintain the rhombomere boundaries that serve as a first step in establishing future anatomical and cellular specializations of the nervous system.

GENES THAT REGULATE HINDBRAIN SEGMENTATION

The rhombomeres provide an excellent experimental model to elucidate how region-specific developmental cues contribute to the early stages of A/P segmentation. Segmentation is not unique to the rhombomeres; it occurs throughout the developing body plan across species. In fact, many of the genes that regulate body segmentation along the A/P axis of insects are highly conserved across species and are used in hindbrain patterning. Understanding how such genes are organized and regulated in the fruit fly *Drosophila* provides insight into how segmentation genes function in the vertebrate hindbrain.

The body plan of *Drosophila* is a good model for studying the roles specific genes play in segmentation

The fruit fly has proven an exceptionally helpful model for investigating genes that regulate segmentation of both the main body axis and the nervous system. Scientists use X-ray or chemical exposure to mutate single genes and observe how those genes impact normal development. Some of these mutations give rise to altered body plans, such as flies with missing or misplaced body parts or, in extreme cases, bodies with no observable body segmentation. The wild type *Drosophila* body plan consists of three easily identifiable segments—the head, thorax, and abdomen—with each segment contributing to the formation of specific body parts or regions (**Figure 3.18**A). For example, under normal conditions the head segment gives rise to antennae, whereas the thoracic segments give rise to legs and wings. As scientific techniques advanced, investigators were able to identify many of the genes that caused the observed developmental changes in this body plan organization.

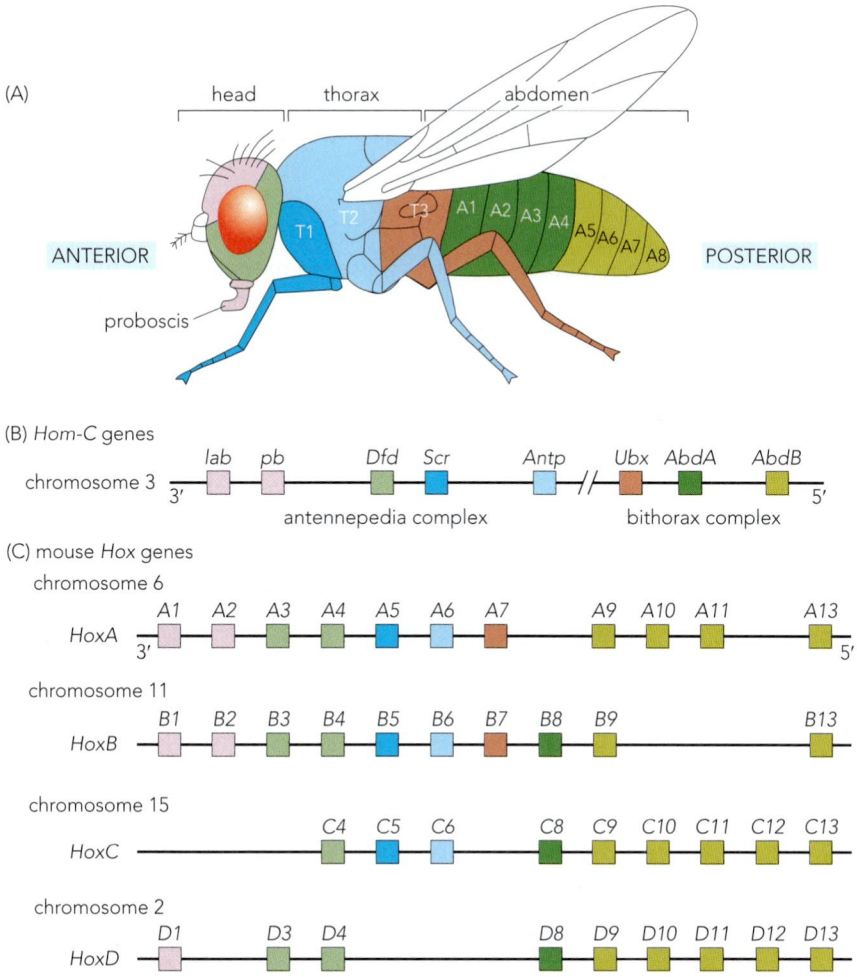

Figure 3.18 Homeotic genes are conserved across species. (A) The fruit fly body plan is organized into distinct segments for the head, thorax, and abdomen. (B) The HOM-C genes (also called *Hox* genes) that regulate this body segmentation are clustered in the antennepedia and bithorax complexes found on chromosome 3. These genes are arranged in order from the 3′ end to the 5′ end of the chromosome so that anterior segments develop in response to the genes expressed closer to the 3′ end, while progressively more posterior regions develop in response to the genes expressed closer to the 5′ end. This is known as the principal of co-linearity, where the relative position of a gene along the chromosome corresponds to the relative position along the A/P axis. (lab, *labial*; pb, *proboscipedia*; Dfd, *deformed*; Antp, *antennepedia*; Ubx, *ultrabithorax*; AbdA, *abdominal A*; AbdB, *abdominal B*; Scr, *sex combs reduced*.) (C) Segmentation in the mouse body plan is regulated by homologs of the HOM-C genes, the *Hox* genes. The *Hox* genes are found in four clusters (A–D) on four different chromosomes (chromosomes 6, 11, 15, and 2). Each cluster contains a subset of the 13 *Hox* gene subfamilies. Similar to *Drosophila*, the relative position of a *Hox* gene from the 3′ to 5′ end of the cluster corresponds to the relative position along the A/P body axis. (Adapted from Alberts B, Johnson A, Lewis J et al. [2015] Molecular Biology of the Cell, 6th ed. Garland Science.)

Among the genes identified were the **segmentation genes**. Segmentation genes include those of the gap, pair-rule, and segment polarity classes. Each class of genes works in sequence to divide the body into smaller and smaller segments along the A/P axis. The **gap genes** are the first class to be active and establish the larger boundaries of the head, thorax, and abdomen. Many of these genes, including *caudal*, *hunchback*, *Krüppel*, and *orthodenticle*, contribute to multiple aspects of neural development. Combinations of gap genes then control the expression of the **pair-rule genes** that divide the three segments into smaller units. Pair-rule genes include *paired*, *even-skipped*, *hairy*, and *odd-paired*. These genes in turn regulate expression of the **segment polarity genes** that establish smaller boundaries within each existing segment along the A/P axis (**Figure 3.19**). Segment polarity genes also play important roles in establishing characteristics of the cells that are restricted to a given segment. Examples of segment polarity genes include *hedgehog*, *wingless*, and *engrailed*. Similar to the gap genes, the pair-rule and segment polarity genes play additional roles at other stages of neural development. Mammalian homologs of some of the common segmentation genes important in neural development are listed in **Table 3.1**.

The homeotic genes that are active in establishing segment identity are conserved across species

The gap and pair-rule genes also regulate expression of a group of homeotic genes, originally called the *homeotic complex (HOM-C)* genes that code

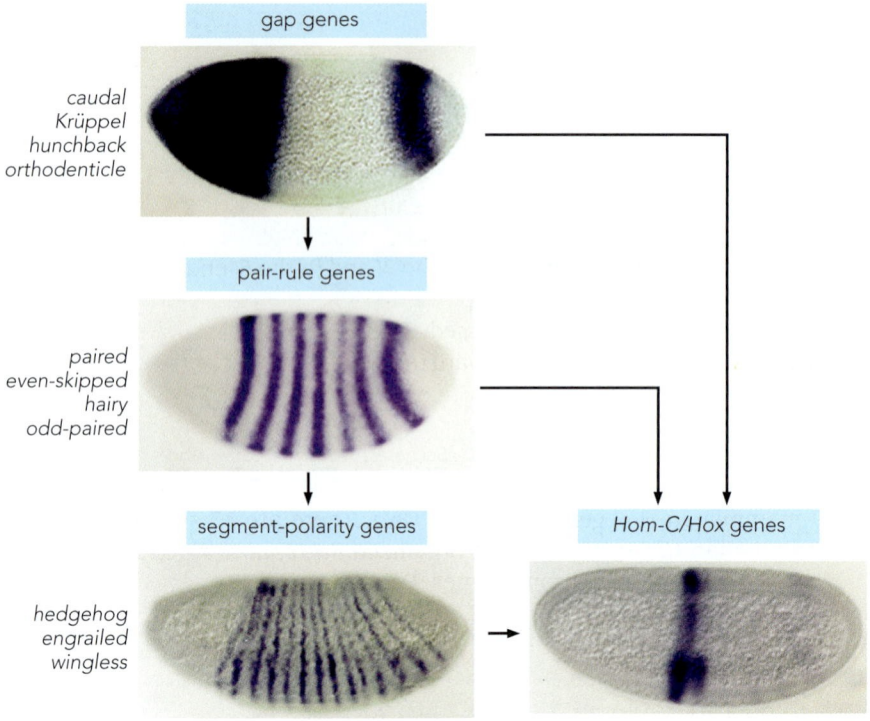

Figure 3.19 Segmentation genes work in sequence to pattern progressively smaller segments along the fruit fly body axis. (A) Gap genes are the first group of segmentation genes to be active and are needed to establish the boundaries between the head, thorax, and abdomen. Gap genes include *caudal*, *hunchback*, *Krüppel*, and *orthodenticle*. (B) The three main body segments are then divided into smaller segments by the activity of pair-rule genes whose expression is regulated by the gap genes. Pair-rule genes include *paired*, *even-skipped*, *hairy*, and *odd-paired*. (C) Segment polarity genes are then expressed in response to the pair-rule genes, leading to further segmentation of the existing boundaries along the A/P axis. Segment polarity genes include *hedgehog*, *wingless*, and *engrailed*. (D) Gap and pair-rule genes also regulate the expression of the homeotic complex (*HOM-C*)/*Hox* (homeobox) genes that code for a group of helix-turn-helix transcription factors critical for further patterning of regions along the A/P axis. (Adapted from Alberts B, Johnson A, Lewis J et al. [2015] Molecular Biology of the Cell, 6th ed. Garland Science.)

for a group of helix-turn-helix transcription factors. The *HOM-C* genes are the homologs of vertebrate *Hox* (Homeobox) cluster genes that function by activating or repressing downstream genes. These transcription factors contain a 60-amino-acid DNA binding domain (the **homeodomain**) that is encoded by a specific 180-base-pair region of DNA termed the **homeobox**. The prefix "homeo" refers to similarity or sameness; mutations in homeotic genes caused one segment of the fruit fly body to become similar to another.

Although originally called *HOM-C* genes in *Drosophila*, the genes are now typically referred to as *Hox* genes because they all arise from the same ancestral gene complex. The remarkable conservation across species is highlighted in studies in which experimental substitution of a mouse *Hox*

Table 3.1 Examples of segmentation genes

	Drosophila	Mammalian homolog
Gap genes	*caudal*	*Cdx* (caudal type homeobox)
	hunchback	*Ikaros*
	Krüppel	*Krüppel-like*
	orthodenticle	*Otx* (orthodenticle homeobox)
Pair-rule genes	*even-skipped*	*Evx-1* (even-skipped homeobox-1)
	hairy	*Hes1* (hairy and enhancer of split 1)
	odd-paired	*Zic* (zinc finger of the cerebellum)
	paired	*Pax* (paired box)
Segment polarity genes	*wingless*	*Wnt* (wingless-related integration)
	hedgehog	*Shh* (Sonic hedgehog)
	engrailed	*En* (Engrailed)

gene restored function in a fly mutant lacking the homologous *Drosophila* gene.

In *Drosophila*, the *Hox* genes are comprised of the antennepedia and bithorax complexes located on chromosome 3 (Figure 3.18B). The antennepedia complex includes the *labial*, *proboscipedia*, *Deformed*, *Sex combs reduced*, and *Anntennepedia* genes. The bithorax complex includes the *Ultrabithorax*, *Abdominal A*, and *Abdominal B* genes.

Hox genes in both *Drosophila* and mammals are arranged on chromosomes in a linear fashion from the 3′ end to the 5′ end. Each gene's relative position in its cluster corresponds to the relative position along the A/P axis where that gene is expressed. This is known as the principle of **co-linearity**. *Hox* genes located toward the 3′ end of a cluster are expressed earliest and at the anterior end of the embryo, while those located toward the cluster's 5′ end are expressed later and at the posterior regions.

The mammalian *Hox* genes are not identical to *Drosophila* genes, however. Over the course of evolution, gene duplications have resulted in 39 mammalian *Hox* genes (Figure 3.18C). There are 13 subfamilies of the mammalian *Hox* genes that are organized into four *Hox* gene clusters located on four different chromosomes. Each *Hox* cluster contains 9–11 of the 13 *Hox* gene subfamilies. In mice, the *Hox* genes are designated *Hoxa–d* with the corresponding gene number (for example, *Hoxa1*), whereas the human *Hox* genes are designated *HOXA–D* with the corresponding number (for example, *HOXA1*). **Paralogous groups** of genes, which share homology as a result of gene duplication, are located at the same relative position in the cluster. For example, the mammalian genes *Hoxa1*, *Hoxb1*, and *Hoxd1* are paralogs (Figure 3.18B).

Just as mutations in the *Drosophila* homeotic genes lead to altered body segmentation, a loss of *Hox* genes in mice can lead to altered formation and patterning of rhombomere segments. Some *Hox* gene mutations result in an absence of rhombomere boundaries, causing adjacent rhombomeres to merge and the development of neural precursors associated with those rhombomeres to take on characteristics of the neurons found in the adjacent rhombomeres. Although some *Hox* mutations in mice are embryonic lethal—that is, the embryos do not develop to term—in many cases the severity of a single *Hox* gene mutation is generally not as dramatic as seen with a single homeotic gene mutation in *Drosophila*. The decreased severity in *Hox* mutations is thought to be due in part to the overlap of *Hox* gene expression within a given rhombomere. Overlapping *Hox* gene expression would allow remaining *Hox* genes to partially compensate for the loss of a single *Hox* gene. The ability of *Hox* gene **enhancers** to act over long distances may also protect against deficits arising from a single gene mutation (**Box 3.2**).

A unique set of expressed *Hox* genes defines the patterning and cell development in each rhombomere

Among the most important signals necessary for segmentation of the hindbrain region are the homeodomain-containing transcription factors encoded by the *Hox* genes. In the hindbrain, *Hox* genes are expressed in distinct, yet often overlapping patterns within each rhombomere. To ensure normal hindbrain development, the expression of individual *Hox* genes must be carefully regulated so that the right *Hox* genes are activated at the right time and in the right segment of the forming nervous system. The particular combination of *Hox* genes expressed in each hindbrain segment varies. The combination of genes expressed in any given segment has traditionally been referred to as the ***Hox* code**, which is thought to

Box 3.2 Developing neuroscientists: Additional insights into *Hox* gene regulation

Hillary Mullan graduated from Oberlin College in 2014 where she majored in Neuroscience and Biology. Following graduation, Hillary was a research associate in a neurology lab prior to beginning medical school in 2016. During the summer of 2012 she worked in the laboratory of Robb Krumlauf at the Stowers Institute for Medical Research in Kansas City, Missouri. Below is her description of the project she worked on during that period.

As described in this chapter, *Hox* genes are a group of genes that are critical for normal embryogenesis in many species. These genes are activated during gastrulation and produce transcription factors that bind to enhancer regions on DNA and, in turn, initiate a number of downstream cascades. Within the mammalian genome, the 39 *Hox* genes are organized into four clusters (A, B, C, and D). An interesting characteristic of these genes lies in the relationship between their structural organization along the chromosome and the order in which they are spatially and temporally activated during development. Genes located closer to the 3′ end of the cluster are expressed earlier and more anteriorly along the body axis than those located toward the 5′ end of the cluster. *Hox* genes at the 3′ end are activated by lower levels of retinoic acid (RA) than those at the 5′ end, which require a higher concentration of RA to be activated. As also noted in this chapter, RA binds to the retinoic acid response element (RARE) in the promoter region of the gene to regulate *Hox* expression. These RAREs are associated with sequences adjacent to the corresponding *Hox* gene. However, the deletion or mutation of specific RAREs does not result in the same morphological deficits observed with loss of the *Hox* gene itself. This suggested that other enhancers may be available to regulate *Hox* gene expression and compensate for the loss of specific RAREs. Other experimental observations also suggested additional enhancers are available to regulate RA-dependent *Hox* gene expression. In the spinal cord, a subset of *Hox* genes exhibits an expression pattern known as rostral expansion. The expression of these *Hox* genes expands anteriorly from their initial site in the spinal cord to include sites in the hindbrain. For example, in mice expression of *Hoxb5*, *Hoxb6*, and *Hoxb8* is first detected in the spinal cord at embryonic day 9.5 (E9.5), but by E11.5 is found in the hindbrain as well. Although this shift in expression has been associated with the presence of retinoic acid, there are no RAREs near these 5′ *Hoxb* genes. However, RAREs are located in the 3′ region of the *Hoxb* genes. This suggested the possibility that 3′ RAREs function as enhancers to influence transcription of 5′ *Hox* genes and influence their rostral expansion.

Candidates for this long-distance regulation of 5′ *Hox* gene transcription include the RA-dependent enhancers early neural enhancer (ENE) and distal enhancer (DE) located near the 3′ end of the *Hox* cluster, where they activate expression of *Hoxb4* and *Hoxb5* in the hindbrain. ENE is located 3′ to *Hoxb4* and DE is located 3′ to *Hoxb5*.

A set of experiments conducted by Ahn, Mullan and Krumlauf tested whether these 3′ enhancers regulate the rostral expansion of 5′ *Hoxb* genes. The researchers produced mice with mutations in either one or both of these RAREs and observed the effect of these mutations on the expression of different *Hoxb genes* genes. Gene expression patterns were visualized by genetically engineering the mice so that each 5′ *Hoxb* gene was labeled with a different fluorescent or protein tag.

The experiments demonstrated that while both RA-dependent enhancers are needed to regulate the rostral expansion of the *Hoxb* genes, the ENE-RARE primarily influence the nearby *Hoxb4* gene while the DE-RARE influenced all the 5′ members of the *Hoxb*

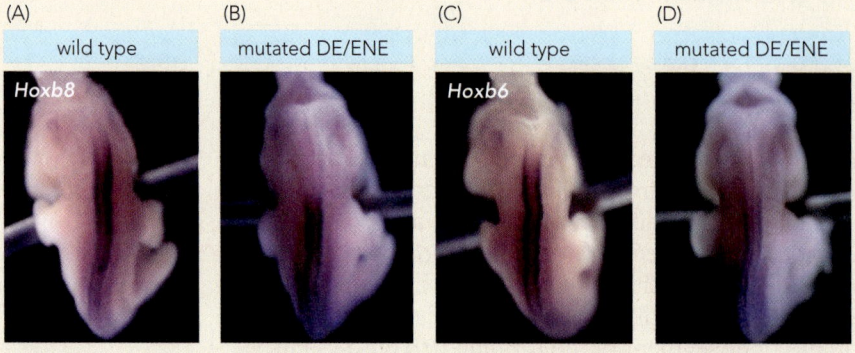

Figure 1 The normal rostral expansions of *Hoxb8* and *Hoxb6* are lost when 3′ RAREs are mutated. *Hoxb8* (A) and *Hoxb6* (C) are seen in the hindbrain regions of wild type mice at embryonic day 11.5. This expression was lost when both the early neural enhancer (ENE) and distal enhancer (DE) were mutated (B, D). (Adapted from Ahn Y, Mullan H, & Krumlauf R (2014) *Dev Biol* 388:134-144.)

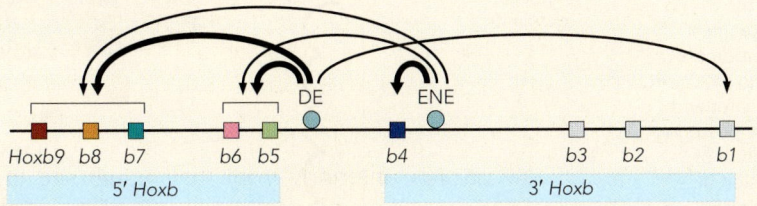

Figure 2 Schematic representation of the shared long-range retinoic acid response elements (RAREs). The RAREs DE and ENE are both located close to 3′ *Hoxb* genes (*Hoxb1–Hoxb4*). However, as indicated by the arrows, the enhancers can also act over a longer distance to influence rostral expansion of the 5′ *Hoxb* genes (*Hoxb5–Hoxb9*).

cluster, *Hoxb5–Hoxb9* (**Figure 1**). Thus, DE-RARE functions over a long distance to influence multiple *Hox* genes and regulate the rostral expansion of 5′ *Hoxb* genes (**Figure 2**). The research also supports previous studies suggesting enhancers that work at a distance and influence multiple *Hox* genes provide a means of compensating for a single genetic mutation. *Hox* genes may have remained clustered over the course of evolution in order to ensure the different *Hox* genes are within reach of the various enhancers.

impact the patterning and development of cells that arise in each axial segment (**Figure 3.20**).

Hox expression patterns are not fixed, however; temporal and spatial changes in *Hox* gene expression patterns are noted during hindbrain development, and these changes lead to further specialization of the cells that originate in each segment. For example, *Hoxa1* is expressed in the presumptive r4–r7 regions of the mouse embryo, but the expression is lost by the time the rhombomeres are fully formed. Other *Hox* genes, such as *Hoxb5*, *Hoxb6*, and *Hoxb8*, are initially expressed in spinal cord regions, but extend their expression anteriorly to rhombomeres as development progresses (see Box 3.2). Thus, *Hox* gene expression patterns shift as development progresses.

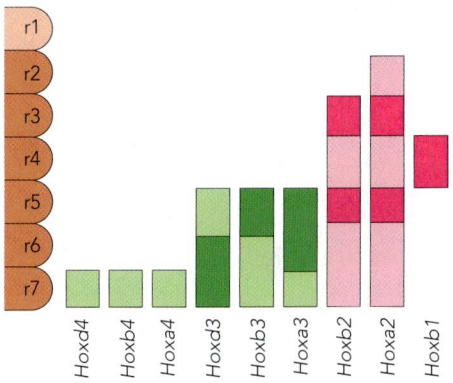

Figure 3.20 The *Hox* code specifies different rhombomere segments. Each rhombomere expresses specific *Hox* genes at relatively higher (darker shading) or lower levels (lighter shading). The expression of *Hox* genes in each segment is known as the *Hox* code. The combination of genes expressed in a rhombomere influences further development of cells arising in that segment. The expression of *Hox* genes can vary slightly between species and at different times in development. The example shown here is from an embryonic day 9.5 mouse. (Adapted from Alexander T, Nolte C & Krumlauf R [2009] *Annu Rev Cell Dev Biol* 25:431–456. With permission from Annual Reviews.)

Retinoic acid regulates *Hox* gene expression

Similar to the organization of *Homeotic* genes in *Drosophila*, *Hox* genes are arranged in a linear pattern along the chromosomes. As in *Drosophila*, the *Hox* genes at the 3′ end (for example, paralog groups 1–4) are expressed earliest and influence development of the more anterior structures, in this case the rhombomeres 2–7 in the hindbrain (**Figure 3.21**A), while those at the 5′ end (for example, paralog groups 5–13) are expressed later and influence development of the more posterior structures (for example, the spinal cord).

During normal development, *Hox* genes are not expressed in r1 or regions anterior to it. However, if r1 is transplanted to a more caudal region of the hindbrain, the grafted r1 will begin to express the *Hox* genes associated with that region. Similarly, misexpression of *Hox* genes in r1 leads to formation of neural cells that are characteristic of those normally found in the corresponding *Hox*-expressing hindbrain region. Thus, under normal developmental conditions, *Hox* gene expression is established within a segment of the hindbrain and the resulting gene expression patterns establish which structures will form at that A/P level.

The expression of the different *Hox* genes is regulated by the coordinated efforts of multiple interacting signaling pathways. **Retinoic acid (RA)**, a vitamin A derivative, is one of the critical modulators of *Hox*

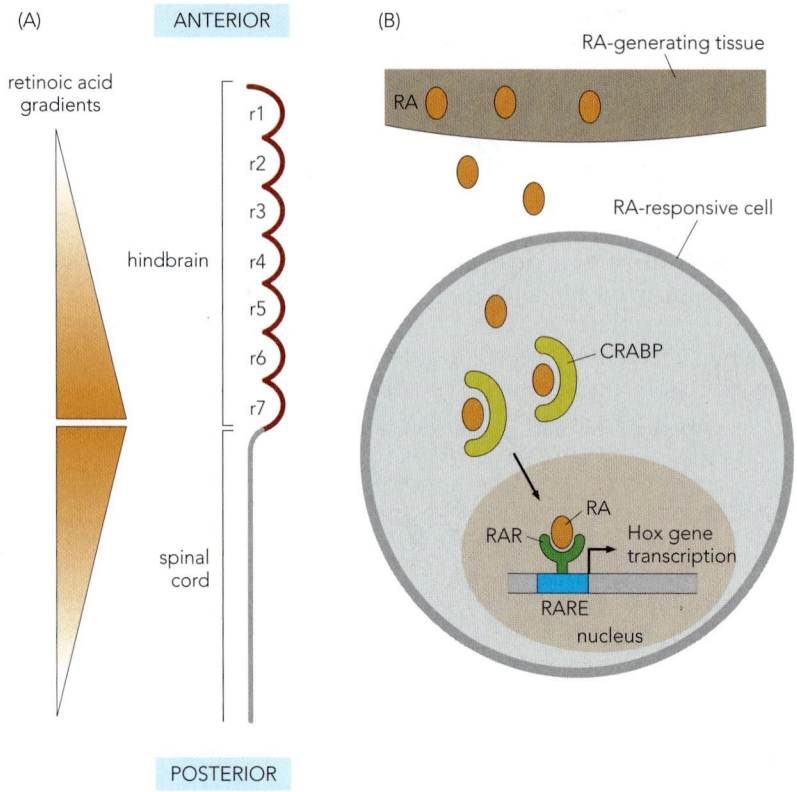

Figure 3.21 ***Hox*** **genes are arranged in a linear pattern on the chromosome and are differentially affected by retinoic acid.** *Hox* genes located closer to the 3′ end of the chromosome are expressed earliest and regulate development of rhombomere segments r2–r7. *Hox* genes closer to the 5′ end of the chromosome are expressed later and influence development of the more posterior, spinal cord regions. The concentration of retinoic acid (RA) regulates expression of the *Hox* genes. The RA concentration is lowest in the anterior hindbrain, where it influences expression of the 3′ *Hox* genes, and increases as it approaches the junction with the spinal cord, where it influences *Hox* genes required for development of the posterior rhombomeres. RA is also found in an opposing concentration gradient in the spinal cord, where it influences development cells along the dorsal–ventral axis (see Chapter 4). (B) Retinoic acid (RA) binds to a receptor to act as a transcription factor for *Hox* gene transcription. RA released from RA-generating tissues adjacent to the neural tube enters cells of the hindbrain, where it is then translocated to the nucleus. Translocation may be aided by association with cellular RA-binding proteins (CRABPs). Once in the nucleus, RA binds to a retinoic acid receptor (RAR), which subsequently associates with the RA-response element (RARE) of a *Hox* gene's promoter to activate gene transcription.

gene expression in the hindbrain. RA is able to directly activate *Hox* gene expression by binding to receptors located in the nuclei of responsive cells. RA is taken up by the cell and translocated to the nucleus, often aided by cellular RA-binding proteins (CRABPs). The RA then associates with one of the retinoic acid receptors (RARs) to act as a transcription factor that binds to a retinoic acid response element (RARE) in the promoter region of the target *Hox* gene. In this way, RA is able to regulate expression of the individual *Hox* genes (Figure 3.21B).

In the developing chick, mouse, and zebrafish, RA is expressed beginning in the gastrula stage embryo and is later secreted by the mesoderm adjacent to the neural tube. RA is synthesized by the enzyme retinaldehyde dehydrogenase 2 (Raldh2, also called aldehyde dehydrogenase 1A2) that is present in the paraxial mesoderm and becomes localized adjacent to the posterior hindbrain. RA is found in two opposing gradients anterior and posterior to the junction of the presumptive hindbrain and spinal cord. The hindbrain RA gradient increases as it approaches the junction with the spinal cord (Figure 3.21A). The spinal cord gradient decreases as it approaches the posterior-most regions. These different RA gradients contribute to cellular specializations along the A/P as well as the dorsal-ventral (D/V) axis (see Chapter 4). The focus of this chapter is on the role of the hindbrain RA gradient in patterning A/P segmentation.

In the hindbrain, each *Hox* gene was observed to respond preferentially to a given RA concentration. *Hox* genes located at the 3′ end of the chromosome were found to be activated by lower concentrations of RA than the *Hox* genes located nearer the 5′ end. Thus, because *Hox* gene location on the chromosome corresponds to the development of structures along the A/P axis, the structures that form in the more anterior regions of the hindbrain develop in response to lower RA concentrations, whereas the more posterior structures develop following exposure to higher concentrations of RA (Figure 3.21).

Early investigations of RA–*Hox* gene interactions in the hindbrain noted that experimentally increasing the RA concentration to which embryos were exposed led to the expansion of more posterior structures, while limiting the formation of anterior structures. Additionally, when RA was added to embryonic stem cells, low concentrations led to expression of *Hox* genes that are active in anterior hindbrain regions, whereas higher RA concentrations led the embryonic stem cells to express *Hox* genes that act in progressively more posterior hindbrain segments. Thus, several lines of evidence supported the role of RA concentration differences in regulating *Hox* gene expression and influencing A/P regionalization.

Although there is a clear link between RA concentration and *Hox* gene expression in the hindbrain, ongoing research efforts indicate that the regulation of these interactions is more complex than originally envisioned. As explained below, the levels of RA are regulated by multiple signals and the concentration requirements at specific A/P regions appear to change as development progresses, at least in some animal models. Additionally, RA levels regulate the expression of other genes, which together with the *Hox* genes pattern the hindbrain segments and prevent segmentation of the spinal cord.

The RA-degrading enzyme Cyp26 helps regulate Hox gene activity in the hindbrain

Just as *Hox* gene expression must be carefully regulated, so too must RA levels. The Cyp26 family of proteins is a group of enzymes that degrade RA. Cyp26 family members are thus a primary modulator of RA levels and one of the essential signals required for hindbrain development. There are three or four members of the Cyp26 family detected in most vertebrate species. Specific Cyp26 enzymes exhibit overlapping and unique functions in different animal models. The following description is a general summary of the role of Cyp26 enzymes as a group rather than the individual role of each enzyme in the different vertebrate animal models.

Cyp26 enzymes are expressed in a concentration gradient along the hindbrain and spinal cord. In the hindbrain, higher concentrations of Cyp26 are found in the more anterior regions of the hindbrain and are therefore able to degrade more RA. This results in the lower concentrations of RA needed for activation of *Hox* genes in the anterior hindbrain (**Figure 3.22**A). In contrast, lower concentrations of Cyp26 are found at more posterior rhombomere segments leading to less RA degradation and thus the higher RA concentrations needed for activation of *Hox* genes associated with the development of more posterior rhombomeres.

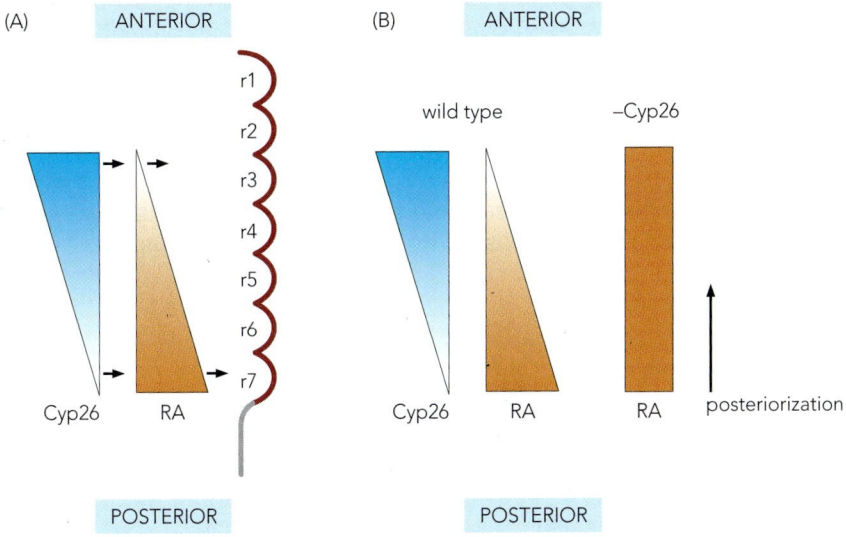

Figure 3.22 The RA concentration gradient in the hindbrain is regulated by gradients of Cyp26 enzymes. (A) Higher concentrations of Cyp26 family members, the enzymes that degrade RA, are found in the anterior hindbrain segments, where lower concentrations of RA are needed to activate 3′ *Hox* genes. Progressively lower concentrations of Cyp26 are found approaching the posterior hindbrain, where higher RA concentrations are needed to induce *Hox* genes. (B) Cyp26 enzymes are required for maintaining the RA gradient during normal developmental conditions (wild type). When Cyp26 was experimentally inhibited in mice or zebrafish (–Cyp26), RA levels increased throughout the hindbrain, leading to the posteriorization of anterior hindbrain segments.

If Cyp26 enzymes fail to degrade sufficient RA in the anterior regions, RA levels increase, leading to expression of the *Hox* genes that normally regulate the development of posterior regions. The importance of tightly regulating Cyp26 was noted in studies of mice and zebrafish, where loss of Cyp26 led to increases in anterior RA levels and posteriorization of otherwise anterior hindbrain segments (Figure 3.22B). Thus, Cyp26 enzymes may be the primary signal utilized to generate the RA gradient necessary for patterning the anterior hindbrain.

RA can also influence Cyp26 levels to some extent. RA levels naturally fluctuate in response to levels of vitamin A available to the embryo. Experiments in zebrafish found that as RA synthesis fluctuates in the future hindbrain regions, the Cyp26 levels also change. Further, when RA synthesis was experimentally increased, Cyp26 activity increased. Conversely, when RA synthesis was decreased, Cyp26 activity also decreased. The coordinated activity of these two molecules allowed for a consistent level of RA to be available to each region of the hindbrain, which would be necessary during embryogenesis to maintain proper activation of hindbrain-associated genes as development progressed.

RA and FGF signaling interactions differentially pattern posterior rhombomeres and spinal cord

In addition to the roles of FGFs in forebrain and midbrain patterning, members of the FGF family also influence patterning of the posterior regions of the neural tube. In the spinal cord, FGF is present in a gradient that increases at progressively more posterior regions, where it regulates transcription of posterior *Hox* genes and influences the levels of RA synthesis. The mechanisms by which FGF regulates RA levels are not entirely clear. There appear to be variations in signaling mechanisms utilized by the different animal models, as well as differences in mechanisms used in different organ systems of the same species. Several studies have demonstrated that in various contexts FGF can influence the synthesis of RA either directly or indirectly via Wnt signaling. Increases in RA levels, in turn, inhibit FGF signaling in the hindbrain (**Figure 3.23**A). Thus, by increasing RA synthesis, FGF restricts its effects to the more posterior regions.

FGF also appears to influence the shape of the Cyp26 gradient in the hindbrain. Studies reveal that a loss of FGF leads to broader expression of Cyp26 (Figure 3.23B). In zebrafish lacking FGF, the addition of RA restored the Cyp26 gradient, suggesting that FGF may normally have an indirect effect on Cyp26 by regulating RA synthesis, which in turn refines the existing Cyp26 gradient (Figure 3.23). Precisely how FGF and RA interact and how they influence Cyp26 expression patterns in different animal models remains under investigation.

Cdx transcription factors are needed to regulate *Hox* gene expression in the spinal cord

FGF also patterns the posterior neural tube by regulating *Hox* gene transcription. This is accomplished by FGF interacting with the Cdx family

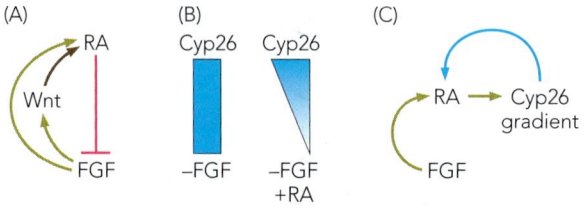

Figure 3.23 FGF interacts with RA to pattern the posterior hindbrain and spinal cord. (A) Under different conditions, members of the fibroblast growth factor (FGF) family appear to influence levels of RA synthesis either directly or via Wnt. RA, in turn, inhibits FGF signaling in the hindbrain. Thus, by regulating RA levels, FGF is able to restrict its activity to the spinal cord. (B) In the absence of FGF, the normal Cyp26 gradient is lost. Addition of RA restores this gradient, suggesting that FGF regulates RA levels and RA then modifies the existing Cyp26 gradient, which in turn influences RA levels (C).

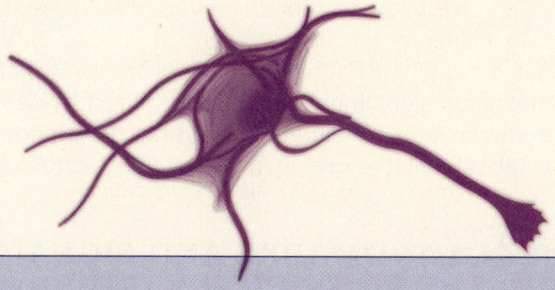

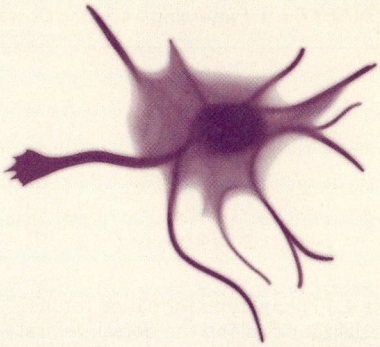

Patterning along the Dorsal–Ventral Axis

4

Looking at the vertebrate nervous system, even at early embryonic stages, it is relatively easy to follow the morphological changes along the anterior–posterior (A/P) axis. As development proceeds, the segments, curves, folds, and expansions along the A/P axis of the neural tube become progressively more apparent (see Chapter 3). In contrast, differences along the **dorsal–ventral (D/V) axis** of the neural tube are not as readily observed at the macroscopic level. However, like A/P polarity, D/V polarity in the early neural tube is crucial for segmenting cell types within the developing nervous system. While segmentation of cell types occurs along the D/V axis at all levels of the nervous system, the mechanisms that underlie this patterning at posterior (caudal) levels are the best characterized to date and are therefore the focus of this chapter.

Both invertebrate and vertebrate animal models share many of the same basic mechanisms for patterning along the D/V axis. In general, the signals that pattern neuronal types along this axis are concentrated in nonneural tissues located at or near the dorsal and ventral halves of the neural tube. These signals regulate the expression of specific transcription factors in progenitor cells located at various distances from the site of the signal. **Neuronal progenitors** or **precursors** are cells that have the capacity to develop into a restricted number of neuron types but have not yet taken on all of the characteristics of a specialized cell type. Each neuronal type that develops along the D/V axis is exposed to a particular concentration of signals that helps establish the unique **transcription factor code** in a given cell type. The expression of specific transcription factors regulates, in turn, the expression or repression of the genes that determine the morphological and behavioral characteristics of each neuronal type. Much more is discussed about cell fate options and cell differentiation in Chapter 6. Here in Chapter 4, the focus is on the initial patterning of progenitor cells along the D/V axis and how scientists identify the cell populations that give rise to different neuronal subtypes. Therefore, exploring the mechanisms that govern D/V axis organization is also an opportunity to begin to examine how gene expression patterns regulate **neuronal specificity**, the processes by which precursor cells adopt specific neural characteristics.

This chapter begins with an overview of the anatomical features found along the D/V axis of the vertebrate nervous system. Next, sources of signals known to influence the development of specific D/V cell types and the associated signaling pathways that govern D/V patterning are described. Homologous signaling mechanisms utilized by *Drosophila* are also discussed. The chapter concludes with a brief description of what is currently known about D/V patterning in the anterior neural tube.

ANATOMICAL LANDMARKS AND SIGNALING CENTERS IN THE POSTERIOR VERTEBRATE NEURAL TUBE

Perhaps the most distinctive feature of the D/V axis of the vertebrate neural tube is the general segregation of sensory and motor structures. The separation of areas that will give rise to the distinct neuronal cell types that become part of sensory or motor pathways is first apparent during the early stages of neural tube formation. As described in Chapter 3 and reviewed in **Figure 4.1**, the neural tube forms from the neural plate. The lateral edges of the neural plate curve upward, forming the neural folds. The neural folds then come together to form the dorsal surface of the neural tube (Figure 4.1A–D). At this dorsal junction the neural tube forms a wedge-shaped region of specialized glial and neuroepithelial cells called the **roof plate**. As the neural tube closes, the former medial region of the neural plate becomes the ventral surface of the neural tube and is marked by another region of glia cells called the **floor plate** (Figure 4.1E). The floor plate lies just above the **notochord**, the mesodermal structure that runs the length of the main body axis. Along the lumen of the neural tube new neurons are generated in the ventricular zone (VZ; see Chapter 5). The lumen of the neural tube later forms the ventricles of the adult nervous system.

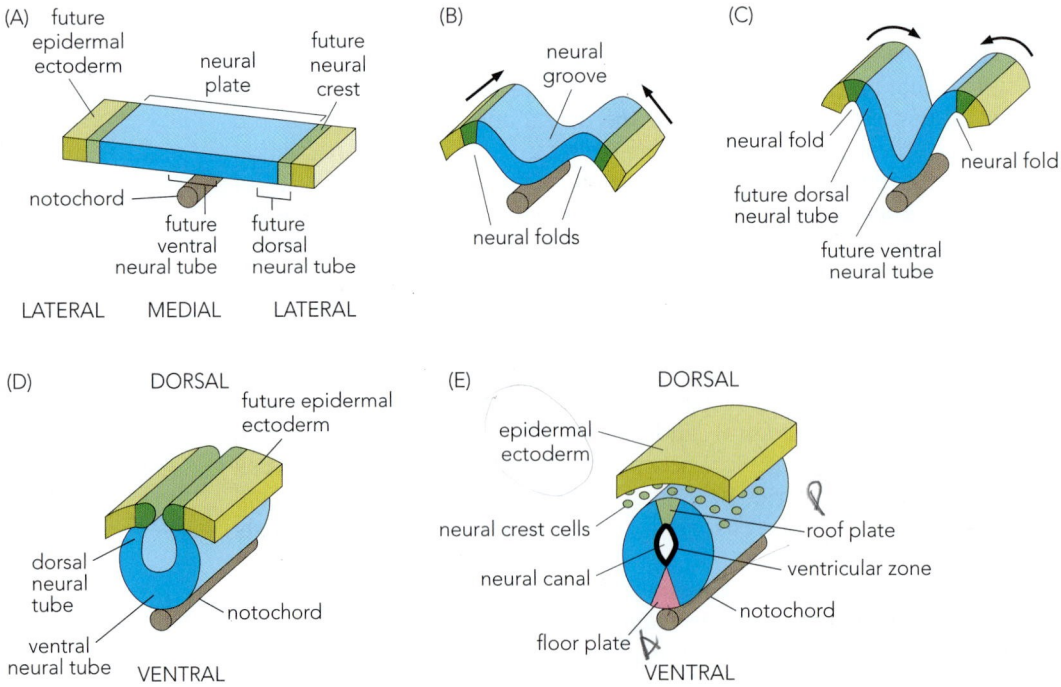

Figure 4.1 Neural tube formation results in specializations along the dorsal–ventral axis. (A) At the neural plate stage, the lateral edges mark where structures associated with the dorsal surface of the neural tube will arise. The lateral-most edges will ultimately form epidermal ectoderm (yellow), while the adjacent regions will form neural crest cells (green) and the dorsal region of the neural tube (blue). Future ventral neural tube structures will arise in the medial region of the neural plate above the notochord—the mesodermal structure running along the length of the caudal neural tube. (B) The lateral edges of the neural plate begin to curve upward, creating the neural folds. The bending of the neural plate also creates the neural groove that lies above the notochord. (C, D) The lateral edges continue to curve upward and approach one another as the neural tube (blue) begins to take shape, thereby creating distinct dorsal and ventral neural tube regions. (E) As the edges of the neural tube come together dorsally, a strip of specialized cells—the roof plate—is identified. The epidermal ectoderm is now separated from the neural tube, and neural crest cells begin to migrate away from the dorsal surface. At the ventral surface, another strip of specialized cells—the floor plate—is seen above the notochord. The closure of the neural tube also results in the formation of the centrally located neural canal.

The sulcus limitans is an anatomical landmark that separates sensory and motor regions

At the early neural tube stage, a longitudinal groove along the inner surface of the lateral walls provides an anatomical landmark that provides a visual cue to where sensory and motor structures will arise. This groove is called the **sulcus limitans** and is most apparent in the spinal cord and hindbrain regions, but also extends through the midbrain (**Figure 4.2**A). In the spinal cord region of the developing neural tube, the alar plate is detected between the sulcus limitans and the roof plate, while the basal plate is found between the sulcus limitans and the floor plate. The alar plate contains sensory interneuron precursor cells, whereas the basal plate contains the precursors to motor neurons and motor interneurons.

In the adult spinal cord, the division of the sensory and motor neurons in the gray matter remains. Sensory neurons are located in the dorsal horn, whereas motor neurons are located in the ventral horn. An intermediate zone containing a mixture of sensory and motor neurons is also visible at some spinal cord levels. The former neural canal persists as the much smaller central canal, and the former ventricular zone is replaced by a lining of **ependymal cells**, another type of specialized glial cell (Figure 4.2B). The functional significance of the sulcus limitans is currently not known, but it provides a visible landmark for developmental neurobiologists to easily distinguish early patterning along the D/V axis of the nervous system. As the nervous system matures, the sulcus limitans is no longer visible in most regions of the nervous system as it becomes obscured by the expanding neuronal populations.

The roof plate and floor plate influence gene expression patterns to delineate cell groupings in the dorsal and ventral neural tube

Signals originating in the epidermal ectoderm and roof plate help pattern the dorsal half of the neural tube (**Figure 4.3**). Initially, the cells of the roof

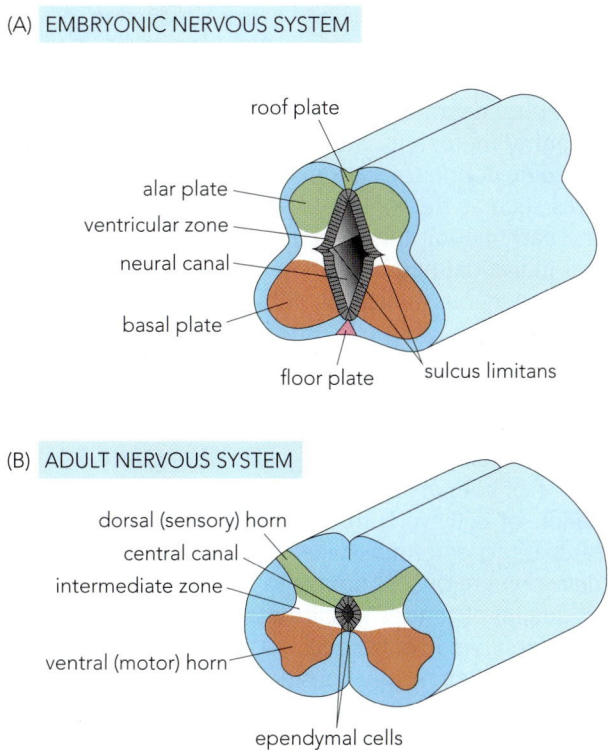

Figure 4.2 The sulcus limitans marks the boundary between sensory and motor neurons of the developing nervous system. (A) In the posterior (caudal) regions of the embryonic nervous system, a longitudinal groove—the sulcus limitans—is readily observed along the inner surface of neural tube. This anatomical landmark separates the alar plate from the basal plate. The alar plate gives rise to sensory interneurons and is located between the sulcus limitans and the roof plate. The basal plate gives rise to motor neurons and associated interneurons. The basal plate is located beween the sulcus limitans and the floor plate. Newly formed neurons arise in the ventricular zone (VZ) adjacent to the neural canal. (B) In the mature spinal cord, the division between sensory and motor neurons remains with sensory interneurons located in the dorsal horn and motor neurons and associated interneurons located in the ventral horn. An intermediate zone containing a mixture of sensory and motor interneurons is also noted at some spinal cord levels. As neuronal populations expand over the course of development, the sulcus limitans becomes obscured in most regions of the nervous system. Additionally, the former neural canal becomes the much smaller central canal lined with ependymal cells.

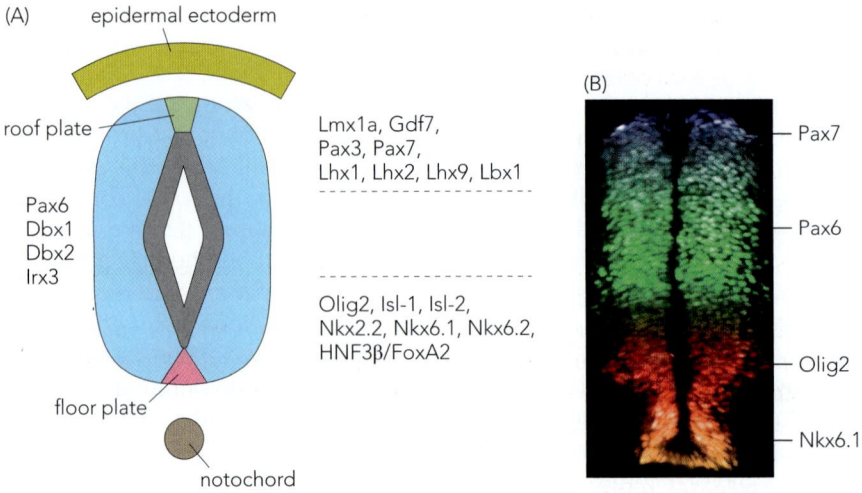

Figure 4.3 The distribution of selected molecular markers in the dorsal and ventral segments of the neural tube. A number of transcription factors have been localized to cells of either the dorsal or ventral half of the neural tube. (A) Transcription factors that serve as markers for dorsal progenitor cells and mature interneurons include Lmx1a, Gdf7, Pax3, Pax7, Lhx1, Lhx2, and Lhx9. Ventral markers include Olig2, Isl-1, Isl-2, Nkx2.2, Nkx6.1, Nkx6.2, and HNF3β/FoxA2. Some transcription factors span portions of both dorsal and ventral regions (for example, Pax6, Dbx1, Dbx2, and Irx3). (B) Expression of select cell markers in the mouse neural tube provides an example of where different progenitor cell populations arise along the D/V axis. Scientists study the spatial and temporal expression patterns of these transcription factors and their genes to gain insight into their roles in neural development. Researchers also use them as molecular markers to identify cell populations following a given experimental treatment. (B, from Briscoe J [2009] *EMBO J* 28:457–465. With permission from EMBO.)

plate extend to contact the surface epidermal ectoderm. The epidermal ectoderm will later separate completely from the neural tube, as will a population of cells that originated in the neural folds—the neural crest cells (see Figure 4.1). Individual neural crest cells take defined migratory routes throughout the developing embryo to form specific peripheral neurons, including the sensory neurons found in the dorsal root ganglia that come to lie along the dorsal region of the spinal cord (see Chapter 5). To pattern cell types in the dorsal region of the neural tube, the dorsal epidermal ectoderm provides signals that influence the roof plate, which in response produces signals for patterning the dorsal horn sensory interneurons, a group of neurons that receive and process sensory information from the body and connect with motor neurons in other regions of the nervous system.

The floor plate located at the ventral surface of the neural tube is influenced by signals originating in the notochord (Figure 4.3). The notochord is ultimately incorporated into the vertebral column in most vertebrates, but prior to that time it serves multiple functions important for neural development in the embryo, including the production of proteins that induce formation of the floor plate. Similar to the roof plate in the dorsal region, the floor plate then produces signals that regulate the specialization of cells in the ventral neural tube. Along the D/V axis, therefore, both dorsal sensory and ventral motor specializations are induced partly in response to signals provided by nonneural cells in the roof plate and floor plate.

Structures such as the roof plate and floor plate are relatively easy to identify in the early neural tube (Figure 4.3). However, the morphological appearance of individual neurons along the D/V axis of the neural tube is remarkably homogeneous, making it difficult to identify individual cell types. Fortunately, by the late-twentieth century, methods had been developed that allowed scientists to identify and localize mRNA and proteins that were differentially expressed in different cell types using such methods as *in situ* hybridization and immunocytochemistry, respectively (see Chapter 1). With these techniques, scientists can now readily distinguish dorsal and ventral precursor cell types both *in vivo* and *in vitro*. These methods not only allow the identification of cell types throughout normal development, but also permit researchers to determine whether a given experimental treatment directs the development of cells unique to either region.

Some examples of gene products localized to progenitors or postmitotic neurons in the dorsal and ventral regions of the developing spinal cord are shown in Figure 4.3. Genes now considered markers of dorsal neural tube progenitor cells include *Lxm1a*, *GDF7*, *Pax3*, and *Pax7*. Markers of ventral progenitor cell types include *Olig2*, *Nkx2.2*, *Nkx6.1*, *Nkx6.2*, and *HNF3β* (*FoxA2*). Other identified progenitor cell markers are located at intermediate regions and span ventral and dorsal segments. For example, *Pax6*, *Dbx1*, *Dbx2*, and *Irx3* mark cells in portions of both ventral and dorsal segments. Mature neurons are identified in dorsal regions by the markers *Lbx1*, *Lhx1*, *Lhx2*, and *Lhx9*. *Isl-1* and *Isl-2* identify mature neurons specific to ventral regions.

The discovery of these genes occurred over the course of many years, so some of the experiments described below took place before relevant genes were identified. It is important to emphasize that while the expression of these different neuronal markers is very helpful as an experimental tool, the expression of the markers during development is required to establish the ultimate function of the cell. Thus, the dorsal and ventral signaling centers influence the expression of specific transcription factors in each cell type, and the resulting transcription factor code determines which genes are expressed or repressed to establish the final characteristics of each neuronal cell type along the D/V axis. The importance of the transcription factor code in D/V patterning is highlighted in experiments described below in which expression levels were experimentally manipulated and corresponding changes in cell development were observed.

VENTRAL SIGNALS AND MOTOR NEURON PATTERNING IN THE POSTERIOR NEURAL TUBE

It is now known that the D/V patterning signals originating in dorsal epidermis and roof plate coordinate with signals originating in the notochord and floor plate. In this section, patterning of the ventral segment of the neural tube is described prior to patterning of the dorsal segment, because many of the mechanisms governing the development of ventral structures were discovered first. Further, because the patterning mechanisms in the ventral and dorsal regions often oppose each other, understanding how patterns are established in the ventral region helps clarify the processes involved in dorsal patterning as well.

The notochord is required to specify ventral structures

The ventral spinal cord is characterized by the spinal motor neurons lying in the ventral horn on either side of the midline. Embryonic motor neurons arise in response to signals released by the floor plate, which is first induced by signals produced by the underlying notochord (**Figure 4.4**A). The importance of the notochord for floor plate and subsequent motor neuron induction was demonstrated over several decades using a variety of experimental approaches. In 1939, for example, the classic work of Johannes Holtfreter in amphibians demonstrated that surgically removing the notochord resulted in an absence of ventral structures, including the floor plate (Figure 4.4B). Subsequent studies suggested that surgically inhibiting or disrupting notochord formation led the cells of the ventral neural tube to appear similar to those of the dorsal neural tube. This was confirmed once markers for different neuronal types were identified. In the absence of notochord, the expression of genes associated with the dorsal neural tube, such as *Pax3* or *Pax7*, expanded into ventral regions.

In another series of experiments in amphibians, it was noted that grafting extra (supernumerary) notochord tissue at the lateral plate regions of the developing neural tube induced floor plate tissue. A number of subsequent studies in chick, mice, and zebrafish confirmed the necessity of the

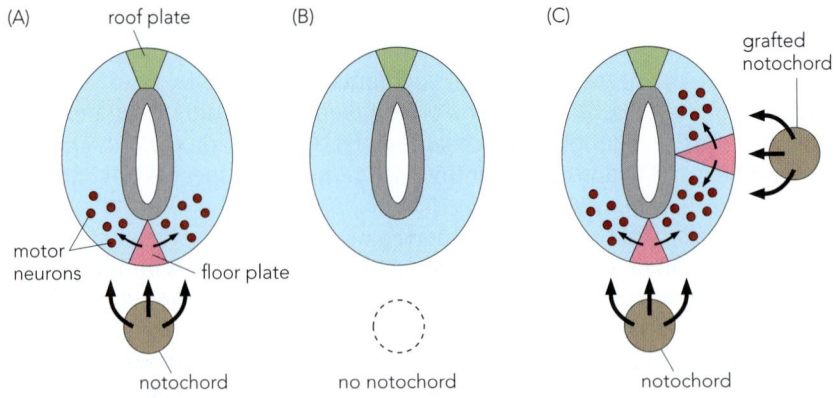

Figure 4.4 The notochord induces ventral specializations, including the floor plate and motor neurons. (A) Under normal developmental conditions, the notochord provides signals (large arrows) that can induce the formation of the floor plate and adjacent motor neuron. Once formed, the floor plate becomes the primary source of signals (small arrows) to induce motor neurons. (B) Following surgical removal of the notochord, neither floor plate nor motor neurons form. (C) When notochord tissue is grafted to lateral regions of the neural tube, the signals that arise from a notochord graft induce the formation of an ectopic floor plate with motor neurons on either side.

notochord in providing signals to induce ventral structures of the neural tube. For example, **ectopic** motor neurons formed on each side of these misplaced floor plate regions (Figure 4.4C). Thus, lateral plate regions of the neural tube that normally do not give rise to floor plate or motor neurons were capable of forming ventral specializations when provided with a signal from the notochord.

In other experiments it was observed that if the size of two grafted notochords differed, the larger graft induced a larger floor plate and more motor neurons than the smaller graft. Thus, the amount of signal produced by the notochord determined the size of the induced tissues. Other studies suggested that contact with the notochord was necessary for formation of the floor plate. When another tissue, such as mesenchyme, was placed between a grafted notochord and the neural tube, the neural tube floor plate tissue was no longer induced, presumably because the notochord signal was unable to penetrate the mesenchymal tissue.

Cell culture assays further demonstrated the role of the notochord in inducing ventral patterning. When segments of neural tube were cultured in contact with a segment of notochord, ventral markers such as the floor plate marker *HNF3β* (*FoxA2*) and the motor neuron markers *Isl-1* and *Isl-2* were identified in the neural tube region adjacent to the notochord.

Thus, several studies from multiple species conducted over several decades indicated that the notochord is essential for ventral neural tube patterning. In the absence of notochord tissue, dorsal-like patterning occurred instead, suggesting that the dorsal characteristics are the default state for the neural tube.

Sonic hedgehog (*Shh*) is necessary for floor plate and motor neuron induction

Based on the strong evidence that the notochord was a source of inducing signals for ventral specializations in the neural tube, several research groups searched for the signal arising from the notochord. Similar to the pace of progress in other areas of developmental neurobiology, it required several decades of investigation to identify the signal originating from the notochord. The following is just a short overview of some of the many findings that led to the identification of a notochord-derived signal and its mechanism of action.

The first studies to identify a specific notochord-derived signal were published in the early and mid-1990s. During this period, several labs published studies suggesting roles for ***Sonic hedgehog (Shh)*** in the development of vertebrate limb and neural tissues. *Shh* is one of three vertebrate genes related to the *Drosophila hedgehog* (*hh*) gene (**Box 4.1**). Studies in

Box 4.1 How genes are named and why they are sometimes renamed

You will no doubt notice some odd and funny names for genes. In some cases it is easy to figure out why a gene was given a particular name. In other cases, the origin of the name is not as readily apparent to someone outside the lab that first discovered and named the gene. Many gene names discussed in this book originated in *Drosophila* and were often intended to help scientists remember the names of the numerous genes being rapidly discovered.

Several *Drosophila* genes are named based on the appearance or behavior of the flies. An example of a gene name that ties to the mutation is the *sevenless* gene, described in Chapter 6. In flies with a mutation in this gene, the R7 photoreceptor cell is missing in the fly eye. Thus, without the gene there is no R7 cell and the fly is therefore "sevenless." Here the gene name describes what happens when the gene is missing and the origin of the name is easy to follow.

Other gene names are not as intuitive. For example, the vertebrate *Smad* genes described in this chapter were named by combining the names of the homologous genes first discovered in *C. elegans* and *Drosophila*. In *C. elegans*, the reduced size of worms lacking the gene led to the gene name small (*sma*), the first part of the vertebrate *Smad* gene name. In *Drosophila*, the gene was named *Mad*, the second half of the vertebrate gene name. *Mad* stands for "mothers against decapentaplegic." A mutation in *Mad* represses the gene *decapentaplegic*, which is required for the formation of the 15 (decapenta) imaginal discs of the fly used to form a number of tissues including limbs. "Mothers against decapentaplegic" was derived from the campaign in the United States and Canada called Mothers Against Drunk Driving (MADD). So, while the name *Smad* may be easy to remember and its origin makes sense when explained, it is not a gene name one can easily relate to its function.

The long tradition of creative gene naming has recently come under scrutiny by a number of groups. When a gene first discovered in *Drosophila* has a human homolog that is later found to be associated with a particular disease, the clever name may create unexpected problems. In recent years a number of articles have discussed the problem of genes names and scientific panels have been created to review and rename genes if needed.

In some cases, the name of a gene can seem insensitive to families. For example, families may feel that the severity of a condition is not being recognized when they are told the condition is caused by "Sonic Hedgehog." The origin of the gene name *Sonic hedgehog* (*Shh*) was based on the original *Drosophila* gene *hedgehog* (*hh*), which was named after the physical appearance of the flies. Flies lacking the gene were covered with small, spiky projections that reminded the scientists of a hedgehog. Later three vertebrate homologs were discovered. *Indian hedgehog* (*Ihh*) and *Desert hedgehog* (*Dhh*) were named after types of hedgehogs and the third was named *Sonic hedgehog* (*Shh*) after the video game character. None of these names seemed offensive until it was discovered that some human birth defects, including holoprosencephaly (HPE), arise from disruption in the *Shh* signaling pathway. Some parents felt doctors and researchers were not taking the condition seriously when they found the gene that causes this tragic defect was named after a cartoon video character.

In some cases, genes have been renamed after threats of lawsuits. For example, in 2005 a mammalian cancer proto-oncogene was discovered and originally named *Pokémon* after characters in a popular game. *Pokémon* stood for POK erythroid myeloid ontogenic. While the gene name may be easy to remember, the company that makes Pokémon did not appreciate having a popular product associated with a cancer-causing gene. The scientists renamed that gene *ZBTB7*, which is a member of the POK (POZ and *Krüppel*)/ZBTB (zinc finger and BTB) protein family. One can see how the original name *Pokémon* was easier to remember, but one can also understand why the company was concerned about the negative associations that could be made with its product.

How to name genes, and when and if to rename them, will continue to be an issue for scientists. Today scientists find they must balance creativity and humor with sensitivity when naming newly discovered genes. For further exploration of the topic, some suggested readings are as follows:

Simonite T (2005) Pokémon blocks gene name. *Nature* 438:897.

Schwartz J (2006) Rename That Gene "Sonic Hedgehog" Sounded Funny, at First

The New York Times (http://www.nytimes.com/ 2006/ 11/12/weekinreview/12schwartz.html).

Krulwich, R. (2009) Fruit fly scientists swatted down over "Cheap Date." National Public Radio, All Things Considered,(http://www.npr.org/templates/story/story.php?storyId=100468532).

the developing neural tube revealed that the *Shh* gene was first expressed in the notochord then subsequently expressed in the floor plate. The location and timing of *Shh* expression in the notochord and floor plate made it a plausible candidate for regulating ventral neural tube specializations, and several subsequent studies provided evidence to support the role of *Shh* in mediating ventral neural tube development.

Figure 4.5 outlines some of the *in vitro* experiments conducted by Thomas Jessell and colleagues that revealed a critical role for *Shh* in ventral patterning. These experiments employed the immortalized COS cell line to express Shh protein. An immortalized cell line is a group of cells that carries a mutation that permits continuous proliferation. The COS cell line was named based on the fact that it originates from cells called CV-1 that have been immortalized with the SV40 virus (CV-1 in Origin SV40). COS are often used experimentally to express a protein of interest to test the putative functions of that protein. Normally, COS cells alone have no influence on neural tube development. However, COS cells transfected with the *Shh* gene produce the Shh protein and are able to induce floor plate and motor neuron formation in cultures of neural tube segments (Figure 4.5A). These studies also noted that different modes of Shh delivery resulted in the induction of different ventral cell types. When Shh-expressing COS cells were grown in contact with the neural tube cultures (Figure 4.5A), floor plate tissue formed first, as indicated by the expression of the transcription factor HNF3β. The formation of the floor plate was followed by the induction of motor neurons. In contrast, if the cultured neural tube tissues were physically separated from the COS cells, only motor neurons formed, as revealed by *Isl-1* and *Isl-2* expression. The neural tube segments and COS cells were separated either by placing a filter membrane between the COS cells and the neural tube segment or by placing the two tissues on opposite sides of the cell culture dish (Figure 4.5B, C). These results indicated that Shh was a diffusible signal because it was able to act through a filter membrane and at a distance. The results also suggested that different Shh concentrations induced different ventral specializations. These experiments suggested that floor plate formed in response to the higher Shh concentrations that resulted from direct contact with the COS cells, whereas lower concentrations provided by the diffusible Shh signal influenced formation of the motor neurons. Other studies soon supported the hypothesis that notochord-derived Shh was necessary for floor plate induction while floor-plate-derived Shh contributed to motor neuron development.

In another set of preparations, antibodies to Shh protein were added to the cultures of neural tube segments and Shh-expressing COS cells. In the presence of these antibodies, ventral structures failed to form in the neural tube segments. These experiments confirmed that COS cells alone were not having an influence on ventral neural tube development and demonstrated the specificity of Shh signaling in inducing floor plate and motor neurons in the ventral portion of the neural tube (Figure 4.5D).

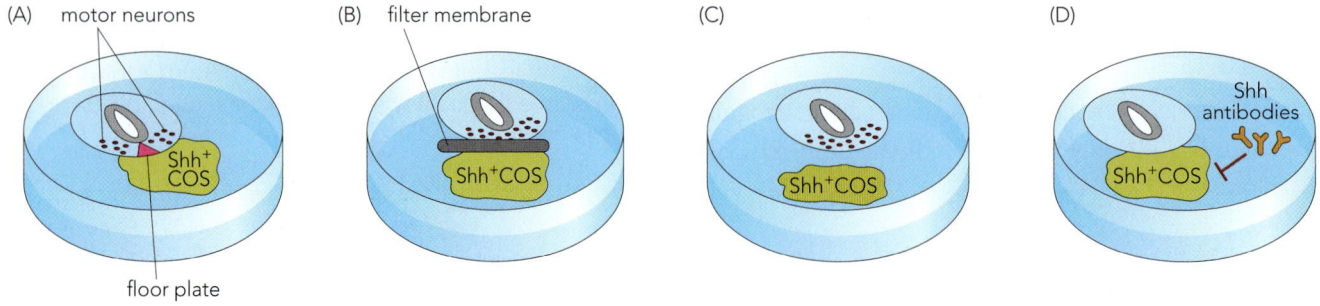

Figure 4.5 ***In vitro*** **assays revealed that Sonic Hedgehog (Shh) mimics the activity of notochord tissues.** Scientists used the COS cell line to express Shh protein and test the effects of Shh on ventral neural tube differentiation in different assay conditions. (A) When COS cells that express Shh protein were cultured in contact with neural tube explants, both floor plate and motor neurons formed. (B, C) Culture conditions that prevented contact between the neural tube explants and Shh-producing COS cells led to the induction of only motor neurons. The ability of COS-derived Shh to induce motor neurons under these conditions indicated Shh functions as a diffusible factor. Contact was prevented either by placing a filter membrane between the explant and COS cells (B) or by placing the tissues apart from each other in the culture dish (C). (D) The specificity of Shh was confirmed when antibodies to Shh were added with cultures of Shh-expressing COS cells. Under these conditions, neither floor plate nor motor neurons formed.

A number of *in vivo* experiments further confirmed the importance of Shh in inducing ventral structures. For example, mice lacking the *Shh* gene failed to develop several ventral cell types, including the floor plate, motor neurons, and the ventral-most interneurons. In fact, when *Shh* was eliminated, the dorsal interneurons and the intermediate cell types located in the lateral plate expanded into the ventral region. Additional support for the role of Shh was seen in transgenic mice in which *Shh* was misexpressed in the dorsolateral regions of the neural tube. Under these conditions, ectopic floor plate and motor neurons arose, similar to what was observed when supernumerary notochords were transplanted in these regions. Thus, the presence of Shh promoted ventral neural tube differentiation, whereas blocking production of Shh *in vitro* or *in vivo* inhibited the formation of ventral cell types in the neural tube and led instead to the expansion of dorsal interneuron types.

Shh concentration differences regulate induction of ventral neuron subtypes

Several studies support the idea that varying concentrations of Shh are necessary and sufficient to instruct cellular identity for most ventral cell types of the neural tube. As suggested by the cell culture experiments described above, the highest concentrations of Shh induce floor plate and lower Shh concentrations generate motor neurons. It is now also known that progressively lower concentrations of Shh induce the different ventral interneuron subtypes that modify the activity of the motor neurons. Ventral interneurons include the V3, V2, V1, and V0 populations. These interneurons are located at specific locations in the ventral half of the neural tube and arise from different precursor cell populations (**Figure 4.6**).

In vivo the concentration differences needed to induce the floor plate and the different populations of ventral neurons arise as a result of the

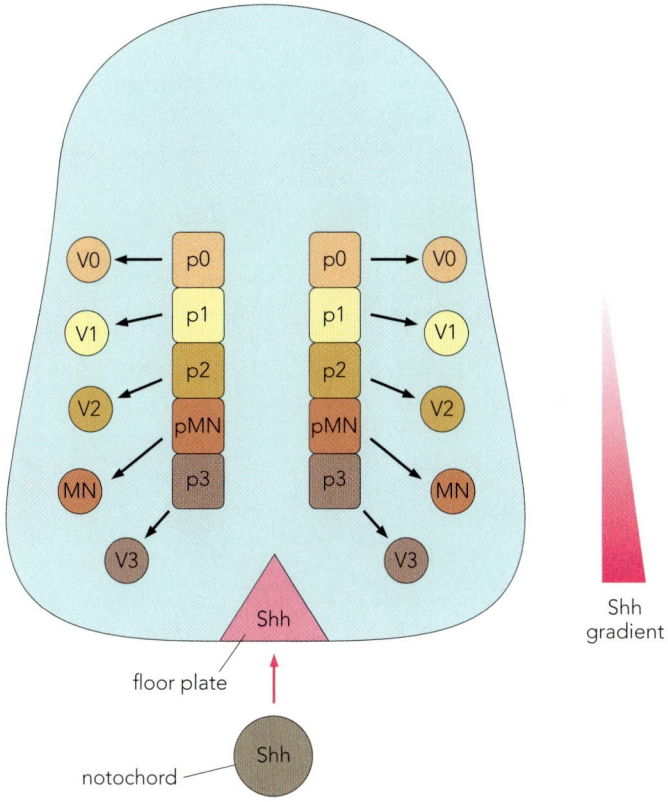

Figure 4.6 Neurons in the ventral neural tube arise in response to graded concentrations of Shh. In the ventral segment, the floor plate is located at the midline, above the notochord. Motor neurons (MN) and interneurons (V3, V2, V1, and V0) lie on either side of the midline. These neurons arise from specific progenitor cell groups located in a ventral-to-dorsal progression (p3–p0). The concentration of Shh is greatest near the ventral midline, where it is released by notochord and floor plate. As Shh diffuses from these regions, the concentration gradually decreases toward the dorsal segment of the neural tube.

cleavage of the Shh protein into amino (Shh-N) and carboxyl (Shh-C) fragments. Shh uses autoproteolysis to generate these two fragments and only the Shh-N fragment is used for ventral cell patterning. Shh-N is further modified by the addition of a cholesterol molecule that allows the Shh to be tethered to the surface of notochord cells. Thus, much of the Shh-N remains at the cell surface, providing the higher concentration needed for induction of the floor plate. The remaining Shh-N is able to diffuse away from the sites of production, first from the notochord and later from the floor plate, to provide the progressively lower concentrations needed for induction of the various subtypes of ventral neurons (Figure 4.6). The graded concentrations of Shh therefore lead to the generation of specific neuronal subtypes within the ventral neural tube, with those located most ventrally (the V3 interneurons and motor neurons) requiring higher concentrations than the interneurons located closer to the intermediate zone (V2, V1, and V0). Thus, in the ventral neural tube, Shh functions as both an inducer to floor plate tissue and as a morphogen signaling gradient to cells along the D/V axis.

Genes are activated or repressed by the Shh gradient

The concentration of Shh activates or represses genes expressed by the motor neuron and interneuron progenitor cells in the ventral segment of the neural tube (**Figure 4.7**). Among these genes are those that code for class I and class II transcription factors that include homeodomain proteins of the Dbx, Nkx, Irx, and Pax families, and the basic helix-loop-helix (bHLH)

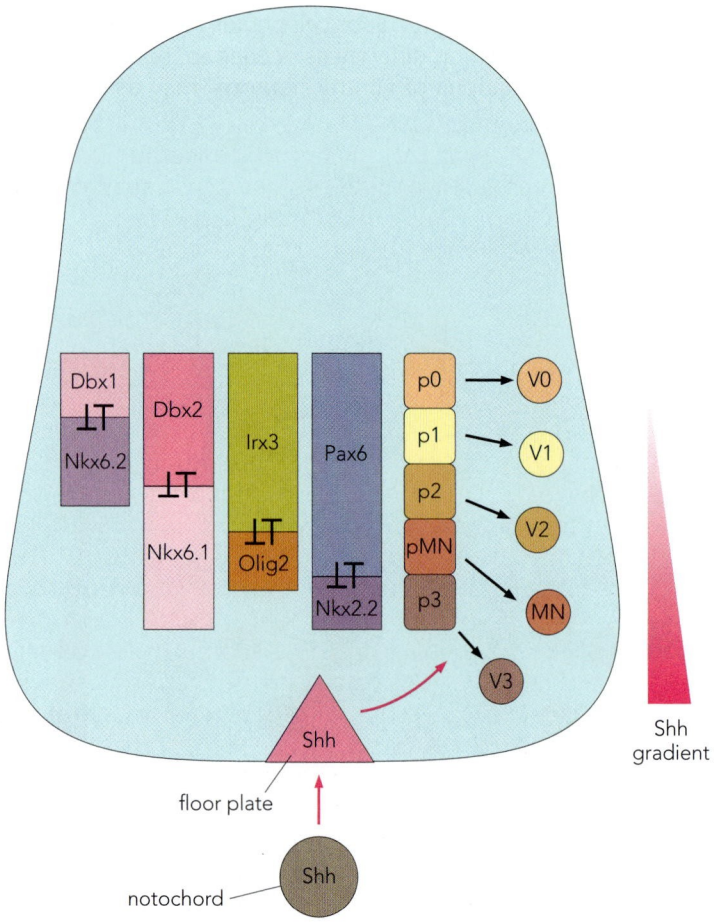

Figure 4.7 Shh levels differentially regulate class I and class II transcription factors. Shh either activates or represses specific transcription factors to specify the cell types in the ventral neural tube. Class II transcription factors (for example, Nkx6.2, Nkx6.1, Olig2, and Nkx2.2) are activated by Shh, whereas class I transcription factors (for example, Dbx1, Dbx2, Irx3, and Pax6) are repressed by Shh. These classes of transcription factors cross-repress each other (black bars) so that only one is activated in a given cell at any time. The gradient of Shh helps determine which transcription factor will predominate at a given site along the ventral neural tube. (Adapted from Dessaud E, McMahon AP & Briscoe J [2008] *Development* 135:2489–2503. With permission from The Company of Biologists.)

protein Olig2. The class I transcription factors, such as Dbx1, Dbx2, Irx3, and Pax6, are expressed in portions of the ventral segment, but are later repressed by Shh in a gradient-dependent manner. For example, in the region of V1 and V0 interneurons, the dorsal-most of these interneurons, low levels of Shh are sufficient to repress these genes. In contrast, at the progressively more ventral regions closer to the floor plate, higher Shh concentrations are required to repress class I transcription factors (Figure 4.7). Unlike the class I transcription factors, the class II transcription factors are activated by the higher concentrations of Shh located in the ventral region of the neural tube. Nkx2.2, Nkx6.1, Nkx6.2, and Olig2 are examples of the class II transcription factors that are activated by the higher Shh concentrations found closest to the floor plate (Figure 4.7).

Specialization of the different ventral precursors occurs because class I and class II transcription factors cannot be active in the same cell at the same time. To maintain the boundaries between progenitor cell types in the ventral neural tube, cross-repression of pairs of class I and class II transcription factors occurs in response to the graded concentrations of Shh (Figure 4.7). For example, consider the paired transcription factors Nkx2.2 and Pax6. As the Shh concentration increases above a certain threshold, *Nkx2.2* in the ventral-most region becomes activated, while *Pax6* expression in that region is repressed. Experiments revealed that in the presence of antibodies to inhibit Shh, *Nkx2.2* expression was absent and neural tube explants expressed only *Pax6*. Conversely, in mice lacking *Pax6*, *Nkx2.2* expression extended further dorsally so that the area devoted to V3 interneurons expanded while the region for motor neurons decreased.

Similar cross-repression appears to take place between the other pairs of class I and class II transcription factors. For example, Olig2 pairs with Irx3, while Nkx6.1 pairs with Dbx2, and Nkx6.2 pairs with Dbx1 (Figure 4.7). Thus, because different concentrations of Shh regulate the repression or activation of specific class I or class II transcription factors, the resulting neural subtypes (V3, MN, V2, V1, or V0) are patterned accordingly in response to the Shh gradient. Under normal conditions, the levels of Shh generated throughout the ventral neural tube are sufficient to regulate the expression of the distinct transcription factors that induce the formation of the specific ventral progenitor cell types.

Shh binds to and regulates Patched receptor expression

The signaling pathways for Hh and Shh have been investigated in multiple systems in developing *Drosophila* and vertebrates, respectively. The discussion here addresses Shh, though the basic mechanisms are similar for both proteins. In both cases, intracellular signaling is differentially regulated depending on whether the Hh/Shh ligand is bound to the Patched (Ptc) receptor. In the absence of ligand binding, an intracellular signaling complex represses the transcription of target genes, whereas with the binding of the ligand to Ptc, the intracellular complex activates the transcription of target genes.

Among the proteins that make up the vertebrate Shh signaling complex is Suppressor of Fused (SuFu), a cytoplasmic protein that interacts with the GLI (glioma-associated) zinc finger transcription factors that regulate gene transcription. In *Drosophila*, SuFu interacts with ci (cubitus interruptus) proteins—homologs of vertebrate GLI transcription factors.

In the absence of Shh binding, the Shh signaling complex is associated with microtubules. Further, Ptc that is not bound by Shh inhibits the activity of a transmembrane protein called Smoothened (Smo), likely by sequestering it in intracellular vesicles. However, the mechanism by

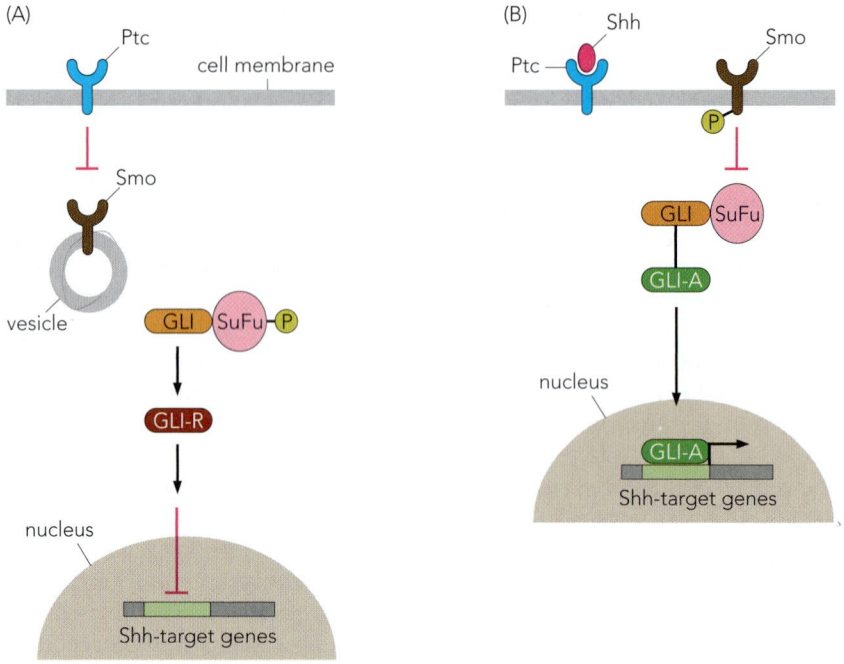

Figure 4.8 Shh signal transduction pathway. The binding state of Patched (Ptc) receptors on the cell surface determines whether target genes are repressed or activated. (A) In the unbound state, Ptc inhibits the activity of the transmembrane protein Smoothened (Smo) by keeping it sequestered in intracellular vesicles. With Smo sequested, intracelluar kinase activity activates the microtubule-associated Shh signaling complex comprised of suppressor of fused (SuFu) and GLI proteins. Phosphorylation (P) of SuFu leads to cleavage of GLI into the transcriptional repressor form (GLI-R) that travels to the nucleus to inhibit Shh-target genes. (B) Binding of Shh to Ptc stops the inhibition of Smo, allowing Smo to translocate to the cell membrane, where it is phosphorylated (P). Smo undergoes a conformational change and inhibits SuFu. With SuFu inactivated, GLI is not cleaved to the repressor form, so the activator form of GLI (GLI-A) travels to the nucleus to initiate transcription of Shh-target genes.

which Ptc suppresses Smo activity is still not fully understood. What is known is that when Smo activity is blocked, intracellular protein kinase activity leads to proteolytic cleavage of GLI proteins associated with SuFu. This cleavage results in the processing of the transcriptional repressor form of GLI (GLI-R) that then travels to the nucleus to block the transcription of target genes. Thus, in the absence of Shh binding, Shh-target genes are repressed (**Figure 4.8**).

In contrast, when Shh binds to Ptc, Ptc is internalized and no longer inhibits Smo. Smo then translocates and fuses to the cell membrane, where it is phosphorylated by protein kinase A (PKA) and other kinases. Phosphorylation leads to a conformational change in the Smo protein that allows it to interact with and inhibit SuFu. Inactive SuFu is unable to cleave GLI into repressor forms, so the activator forms of the GLI (GLI-A) transcription factors remain. The activator proteins travel to the nucleus, where they initiate transcription of Shh-responsive genes, including the floor plate marker *HNF3β* and the receptor *Ptc*. Thus, the intracellular pathway initiated with activated Smo leads to the induction of genes that typify ventral neuron subtypes (Figure 4.8B). Vertebrates have three known GLI proteins (GLI1, GLI2, and GLI3). In the absence of Shh binding, GLI3 predominates and functions as the repressor form. Some species, including mammals, appear to rely primarily on GLI2 for transcriptional activation, whereas other vertebrates, such as zebrafish, rely on GLI1. With the binding of Shh, the activator forms of GLI predominate and override the effects of the repressor forms.

The necessity of balancing the activator and repressor forms of GLI proteins is seen in mice lacking genes for both Shh and GLI3. *Shh*$^{-/-}$ mice have a loss of ventral cell types, particularly those closest to the floor plate. However, if the *GLI3* gene is also absent, then the repressor activity of GLI3 is lost and some of the ventral neurons remain, particularly those furthest from the floor plate. These results indicate that under normal conditions, Shh is not needed to specify all ventral neuronal precursors. Instead, the gradient of Shh influences the gradient of GLI and establishes a balance between activator and repressor forms of the protein. The pattern of ventral subtypes along the D/V axis is thus influenced by the levels of the different GLI proteins available.

The importance of Shh signaling to patterning ventral cell types was also seen when Smo activity was experimentally manipulated. For example, a constitutively active form of the *Smo* gene was introduced into mice or chick embryos. The transgenic mice were bred to carry the activated form of the *Smo* gene and chick embryos had the mutated *Smo* gene introduced into neural tube cells by electroporation. In both experimental models the regions of ventral neuronal cell types were expanded so that areas of the neural tube that normally give rise to dorsal precursor cell types now produced ventral cell types.

Among its many roles, Shh regulates *Ptc* expression through a feedback loop, so the relative concentrations of Shh and Ptc must be carefully balanced to ensure the proper levels of Shh signaling in a given cell. Under normal conditions, *Ptc* is up-regulated in cells that respond to Shh. For example, *Ptc* expression is greatest near the floor plate and gradually decreases dorsally, in the same pattern as the Shh gradient. A direct link between *Shh* and *Ptc* expression is seen in mice that lack *Shh*; these mice also have lower levels of *Ptc* expression.

The effects of altering the Shh/Ptc ratio can be seen under experimental conditions. Studies of both transgenic mice bred to produce excess Ptc receptors and chick embryos electroporated to express excess Ptc receptors in new locations confirmed the importance of this signaling pathway. In both types of studies, the overexpression of *Ptc* led to a repression of Shh activity in the ventral neural tube. This occurs because an increase in Ptc receptors means that available Shh has a greater chance of binding to a Ptc receptor that is not interacting with Smo. Such receptors act as a "sink" for Shh, thus sequestering Shh and preventing it from binding to any Ptc receptors associated with Smo. The unbound Smo-associated Ptc receptors continue to inhibit Smo, preventing activation of genes associated with the ventralization of the neural tube (**Figure 4.9**).

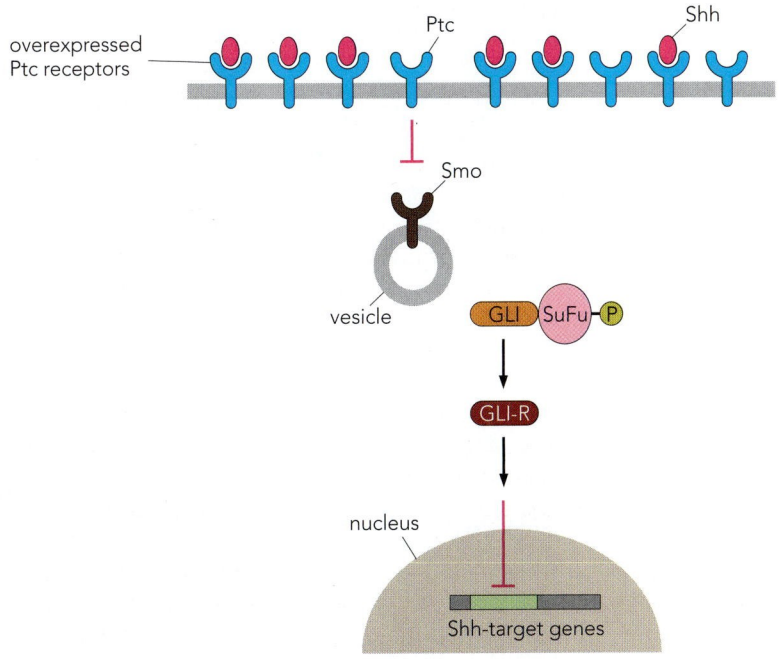

Figure 4.9 Overexpression of Ptc prevents activation of Shh-responsive genes.
(A) When Ptc receptors are overexpressed, the excess number of Ptc receptors that are not interacting with Smo act as a "sink" to sequester available Shh. This results in fewer Smo-associated Ptc receptors binding to the available Shh and therefore the repressor form of GLI (GLI-R) is processed and Shh-target genes, including the gene encoding the Ptc receptor, are repressed.

In another series of experiments, a mutated form of the Ptc receptor was developed that lacked one of the binding sites for Hh/Shh. The receptor was unable to bind the ligand, even when excess concentrations of Shh were present and therefore Smo remained inhibited. Thus, the mutated receptor acted as type of dominant inhibitor because too few normal receptors were available to transduce the Shh signal. The effects of the mutated receptor were examined in a number of contexts, including *Drosophila* wing development and chick neural tube development. In the chick experiments, the mutated receptor was electroporated into the neural tube of chick embryos (**Figure 4.10**A). The cells in the ventral portion of the neural tube that now expressed the mutated receptor revealed a ventral-to-dorsal shift in cell identity. Those cells that would normally respond to high concentrations of Shh and express markers of ventral cell types instead expressed cell markers such as *Pax7* and *Pax6* (Figure 4.10B). Recall that under normal conditions *Pax6* is repressed in the ventral-most areas by low levels of Shh. The mutated receptors interfered with normal Shh signaling, thereby preventing the Shh-mediated repression of these transcription factors and allowing the expansion of dorsal cell types into ventral regions of the neural tube. Only those cells that incorporated the mutated gene shifted cell fate, indicating that the receptor functioned in a **cell-autonomous** manner, the process in which only the cells that carry the gene mutation have an altered phenotype. This supported the hypothesis that Shh acts directly on the cells of the ventral neural tube.

Many of the specific steps in the processing of the transcriptional repressors and activators mediated by Shh remain unclear. However, scientists continue to identify the mechanisms by which differences in the concentration and timing of Shh exposure impact the development of cells along the D/V axis.

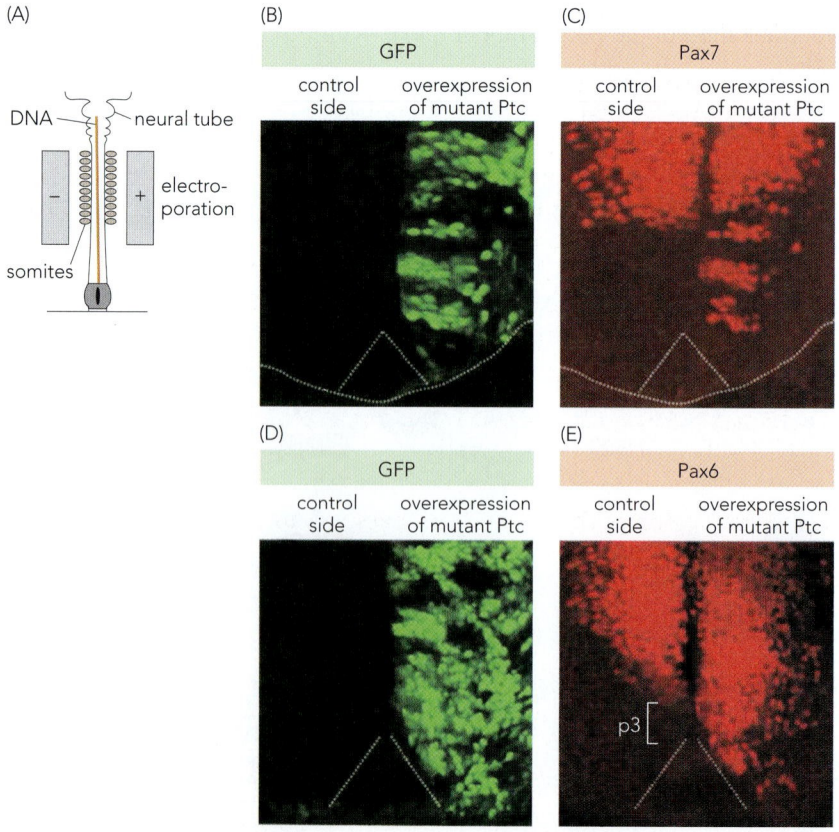

Figure 4.10 Mutated Ptc receptors interfere with Shh signaling. (A) DNA for a mutated Ptc receptor lacking a Shh binding site was introduced to one side of a chick neural tube by electroporation (+). The shaded region indicates the area of the neural tube that received the mutated DNA. The mutated receptor was overexpressed so that a greater concentration of mutated receptors was present compared to endogenous receptors. Thus, the mutated receptors functioned as a dominant inhibitor because too few normal receptors were available to transduce the Shh signal. (B, D) Cells overexpressing the mutant Ptc receptors are labeled with GFP (green). (C, E) The expression of Pax7 and Pax6 (red) remains restricted to the dorsal regions of the control side of the embryo but expression expands into ventral regions on the side expressing the mutant Ptc receptors. Thus, without sufficient Shh signaling on the treated side of the embryo, the repression of *Pax6* and *Pax7* in ventral regions was no longer possible. The bracket in E indicates the region where p3 ventral progenitor cells are found. The triangle outlines the region of the floor plate. GFP, green fluorescent protein. (Adapted from Briscoe J, Chen Y, Jessell TM & Struhl G [2001] *Mol Cell* 7:1279–1291. With permission from Elsevier.)

RA and FGF signals are also used in ventral patterning

Not surprisingly, Shh is not the only signal used in ventral patterning of the posterior neural tube. Evidence has indicated that in zebrafish, floor plate cells can still form in the absence of notochord tissue, and other studies suggest that only lateral floor plate cells require Shh, whereas the medial floor plate cells also rely on nodal, a member of the transforming growth factor β (TGFβ) superfamily. Other studies suggest both Shh and nodal cooperate to regulate early and late stages of floor plate induction in both chick and zebrafish.

Additionally, signals associated with A/P patterning, such as retinoic acid (RA) and fibroblast growth factor (FGF) may also refine cell specializations in the D/V axis. For example, a gradient of RA produced by the paraxial mesoderm (see Chapter 3) helps regulate expression of transcription factors along the D/V axis. RA appears to be important for patterning the ventral cell types by promoting expression of class I transcription factors associated with motor neurons and V3–V0 interneurons, including Olig2, Pax6, Dbx1, Dbx2, and Irx3. For example, in one *in vitro* study of lateral regions of spinal cord segments, the addition of RA induced expression of genes such as *Dbx1*, *Dbx2*, and *Irx3* while also repressing the expression of the dorsal marker *Pax7*. Conversely, disruption of RA production or signaling leads to a loss of expression of class I transcription factors. In mice lacking the RA-synthesizing enzyme Radlh2 (*Radlh2*$^{-/-}$), ventral markers such as *Pax6*, *Irx3*, *Nkx6.2*, and *Olig2* are down-regulated. Similar results are observed in vitamin-A-deficient quails that lack RA. Further, when chick embryos were electroporated with a dominant negative form of a retinoic acid receptor (RAR), there was also a decrease in the expression of the ventral markers *Pax6*, *Irx3*, *Dbx1*, and *Dbx2*. Thus, a number of studies in different animal models demonstrate the need for RA to help pattern the ventral neural tube. This effect may be due to a direct action of RA on the promoter regions of the transcription factors associated with ventral cell types. RA may also repress the actions of Shh at progressively more dorsal regions of the ventral segment so that transcription factors such as Dbx1 and Dbx 2 can be expressed.

FGFs have been associated with suppressing expression of transcription factors in progenitor cell populations to ensure the cells remain in the progenitor pool long enough to produce a sufficient number cells. Once FGF signals cease, progenitors begin to express class I and class II transcription factors. The activity of these classes of the transcription factors is subsequently regulated by the concentration of Shh they encounter (**Figure 4.11**). Thus, opposing gradients of RA and FGFs are utilized in both A/P and D/V patterning to specify cellular differences along the neuroaxis.

DORSAL PATTERNING IN THE POSTERIOR NEURAL TUBE

The dorsal region of the neural tube gives rise to various neuronal types of the peripheral and central nervous systems. The central neurons in the

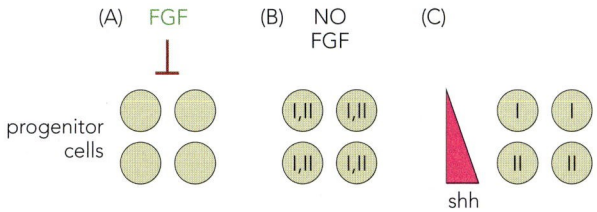

Figure 4.11 FGF inhibits expression of class I and II transcription factors in ventral progenitor cells. (A) The presence of FGF prevents progenitor cells in the ventral region of the neural tube from expressing class I or class II transcription factors. (B) When FGF production stops, the progenitor cells are able to express both class I and class II transcription factors. (C) The gradient of Shh then selectively activates either class II or class I transcription factors, with class II transcription factors activated by the higher levels of Shh.

dorsal horn of the spinal cord are sensory interneurons that process and relay sensory information entering the spinal cord from the periphery. The signals that regulate the specification of dorsal sensory structures are distinct from those that regulate ventral motor specializations. However, many dorsal patterning signals act in an antagonistic manner to the signals used for ventral patterning. Similar to the ventral neural tube region, the signals that pattern the dorsal interneurons do so by regulating the expression of transcription factors that lead to the development of different interneuron populations with unique characteristics. Within the dorsal horn, six distinct regions of progenitor cells are identified based on the transcription factors they express.

TGFβ-related molecules help pattern the dorsal neural tube

As noted earlier, the dorsal neural tube appears to be the default fate for the neural tube, and many of the markers that are later associated with dorsal regions are first expressed throughout the neural plate until the time of neural tube closure. For example, Pax3 and Pax7 are initially widely distributed, but as the neural tube begins to close, the floor-plate-derived signals help limit their expression to the dorsal segment. However, ventral signals cannot convert all neural tube regions to ventral structures. Although notochord grafted to lateral regions of the neural tube led to ectopic formation of a floor plate and motor neurons (see Figure 4.4), grafts to the dorsal neural tube did not have the same effect. In experiments in which the notochord was transplanted to the dorsal surface, the roof plate cells persisted (**Figure 4.12**A), although the expression of some dorsal genes was inhibited. Similarly, if a transplanted neural tube was rotated 180° so that the host notochord was now adjacent to the neural tube's dorsal surface (Figure 4.12B), a roof plate still formed above the notochord (Figure 4.12C). Thus, the roof plate could not be converted to a floor plate by notochord signals. Together these studies indicated the developmental potential of floor plate and roof plate cells was restricted and that formation of the roof plate was independent of signals that originate from the notochord. Scientists therefore began to search for signals that induce the roof plate and the specification of dorsal cell types.

By the late 1990s, scientists investigating patterning mechanisms in the dorsal neural tube had identified members of the transforming growth factor β (TGFβ) superfamily as candidate molecules for dorsal specializations. Recall that during neural induction, bone morphogenetic protein (BMP) family members are needed to form the nonneural, epidermal ectoderm by inhibiting formation of neural tissue (see Chapter 2). At this later stage of development, however, BMPs are used to specify distinct neuronal

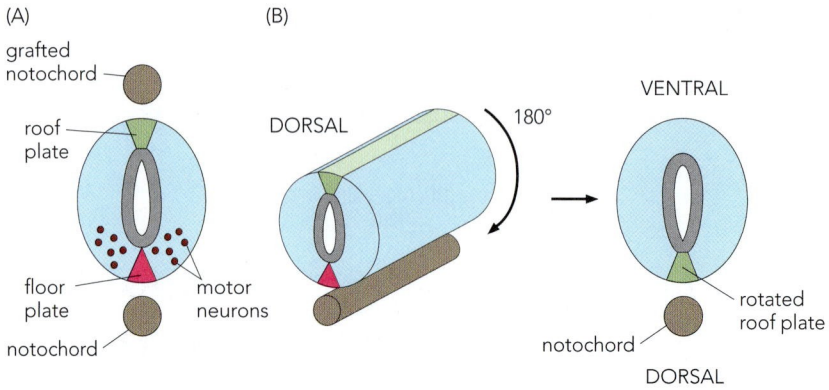

Figure 4.12 Notochord does not induce floor plate in the dorsal neural tube. As previously seen, grafting notochord to lateral regions of the neural tube induced ventral tissues, including a floor plate and motor neurons (see Figure 4.4C). (A) Notochord grafted to dorsal regions, however, did not induce floor plate or motor neurons and was unable to convert the roof plate to a floor plate. Rotation of the neural tube 180° to align the dorsal surface with the notochord (B) also failed to induce floor plate tissue in the dorsal region. The rotated dorsal surface, now located above the notochord, continued to produce the roof plate. Together, these experiments show the roof plate forms in response to signals other than those associated with the notochord.

cell types, including the neural crest cells, the roof plate cells, and the six subtypes of dorsal interneurons.

Experiments soon revealed that expression of BMP-4 and BMP-7 in the epidermal ectoderm overlying the neural tube induced formation of the roof plate, much like the notochord induces the floor plate (**Figure 4.13**). The epidermal BMPs induce the expression of the LIM homeodomain transcription factor Lmx1a in the roof plate cells, so Lmx1a is used as a marker for the roof plate. Similar to the floor plate in the ventral neural tube, the roof plate appears to serve as a secondary, local source of signals to pattern cell types unique to the dorsal segment of the neural tube. Investigators found that the roof plate becomes a secondary source of BMPs as well as other TGFβ-related molecules, including dorsalin-1 (Dsl-1), growth differentiation factor 7 (GDF7), and activin (Figure 4.13).

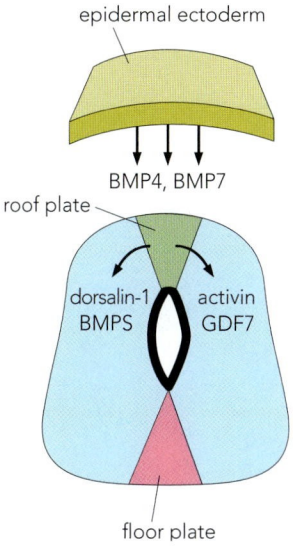

Figure 4.13 TGFβ-related molecules help pattern the dorsal neural tube. Epidermal ectoderm overlying the dorsal surface of the neural tube secretes BMP4 and BMP7. The roof plate becomes a secondary source of signals, releasing dorsalin-1, BMPs, activin, and growth differentiation factor 7 (GDF7), all members of the TGFβ superfamily.

Roof plate signals pattern a subset of dorsal interneurons

In vitro evidence supported the importance of roof plate signals for inducing dorsal interneuron cell types. In cell culture experiments, roof plate tissue induced formation of interneurons in the dorsal horn of the spinal cord (**Figure 4.14**A). Similarly, BMPs, Dsl-1, and activin induced formation of dorsal interneurons (Figure 4.14B). Further evidence to support the role of TGFβ-like molecules was seen with the addition of BMP inhibitors such as noggin or follistatin. As hypothesized, inhibiting BMP signaling blocked the expression of dorsal transcription factors and therefore the formation of the dorsal horn interneurons (Figure 4.14C).

In vivo evidence for the necessity of the roof plate in patterning dorsal interneurons was found in a naturally occurring mouse mutant known as the *Dreher* mutant. Among the developmental abnormalities detected in these mice are cerebellar defects, an absence of roof plate tissue, and a lack of spinal cord dorsal interneurons. Mice with this mutation have several neurological deficits, including an ataxic gait that was found to be due to alterations in the axonal projections from the dorsal spinal cord to the cerebellum.

BMP-related signals pattern class A interneurons

Based on studies of roof-plate-derived signals, scientists ultimately distinguished two classes of progenitor cells (class A and class B) that give rise to the six subtypes of dorsal interneurons (**Figure 4.15**). The class A progenitors (dp1–dp3) depend on signals from the roof plate and are the most studied to date. These populations go on to form the dorsal interneurons 1–3 (dI1–dI3). In literature prior to 2002, dorsal interneurons 1–3 were most often designated as D1, D3A, and D2, respectively. In contrast to class A neurons, class B progenitor populations (dp4–dp6) are independent of roof-plate-derived signals and appear to rely on non-BMP-related signaling molecules for their specialization. Class B progenitors give rise to dorsal interneurons 4–6 (dI4–dI6; formerly D3–D5).

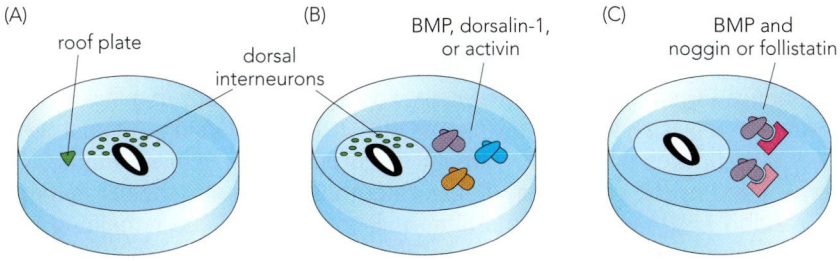

Figure 4.14 ***In vitro*** **studies confirmed the importance of roof-plate-derived signals for dorsal neural tube patterning.** (A) When roof plate tissue was cultured with explants of neural tube, dorsal interneurons formed (green). (B) In the absence of roof plate tissue, the addition of BMP, dorsalin-1, or activin was sufficient to induce dorsal interneurons. (C) Adding the BMP inhibitors noggin or follistatin to the BMP culture media prevented formation of dorsal cell types, thus demonstrating the specificity of BMP signals for developing dorsal interneurons.

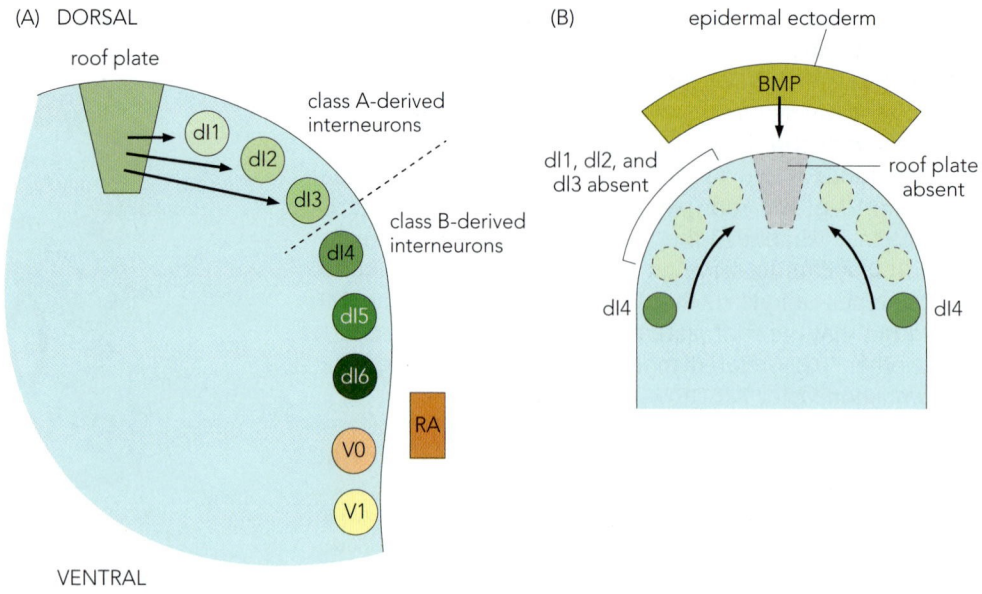

Figure 4.15 Class A and class B progenitor cells in the dorsal neural tube respond differently to roof-plate-derived signals. Sensory interneuron progenitor cell populations are located in a dorsal-to-ventral progression within the dorsal segment of the neural tube. (A) Progenitor populations (class A) for interneurons dI1, dI2, and dI3 depend on roof-plate-derived signals for development. In contrast, progenitor populations (class B) for interneurons dI4, dI5, and dI6, which are located between the class A interneurons and the dorsal-most ventral interneurons (V0 and V1), do not require roof-plate-derived signals for differentiation. However, retinoic acid (RA) located in paraxial mesoderm appears to contribute to the development of class B dorsal interneurons and the V0 and V1 ventral interneurons. (B) The roof plate fails to form in mice genetically engineered to express the diptheria toxin gene in place of *GDF7*, a roof-plate-specific gene. With this method, the toxin causes neuronal degeneration specifically in roof plate cells. The loss of roof plate does not interfere with other BMP signals arising from the epidermal ectoderm. In the absence of the roof plate, the dI1, dI2, and dI3 interneurons fail to form. The dI4 interneurons are not influenced by roof plate signals and therefore develop normally and expand into the more dorsal regions normally occupied by dI1-dI3 interneurons (arrows).

The effects of BMP on patterning the class A dorsal interneurons are most apparent during the early stage of development when dorsal interneuron progenitor populations are present. If BMP levels are decreased at these stages of development, the resulting dI1–dI3 interneurons are decreased in number in the dorsal-most regions of the neural tube. The class B (dI4–dI6) interneurons then expand to occupy the regions typically associated with the dI1–dI3 interneurons.

Studies of the dI1 and dI2 interneurons indicate that specification of these populations appears to be regulated in part by concentration differences and the timing of exposure to BMP-related molecules. In experiments manipulating available levels of BMP, higher concentrations of BMP led primarily to dI1 induction, while lower concentrations induced only dI2 interneurons. In contrast, the dI5 and dI6 subtypes were unaffected by changes in available BMP levels.

The *Dreher* mouse mutants also reveal the importance of roof-plate-derived signals in mediating the expression of the transcription factors needed to specify dI1 neurons. Investigations of these mutant mice ultimately revealed that the phenotype arises due to a mutation in the LIM homoeodomain transcription factor *Lmx1a* that is expressed in roof plate cells of mice. BMP from the epidermis normally signals to the roof plate to induce expression of this transcription factor and specify the roof plate cells. However, due to the mutation in the *Lmx1a* gene, the BMP signal is unable to induce formation of roof plate cells. Loss of *Lmx1a* leads to a lack of roof plate as well as an accompanying loss of *BMP* expression in the spinal cord. Abnormal development in the dI1 interneurons in the *Dreher* mutant mice was noted by the generation of fewer neurons expressing the dI1 markers *Lhx2* and *Lhx9*, as well as the abnormal migration of the remaining Lhx positive neurons.

GDF7 is another BMP-related signal that is present in roof plate cells, but not other regions of the spinal cord or in the overlying epidermis. Mice lacking *GDF7* also revealed a loss of dI1 interneurons. In these mutant mice, no other subpopulations of dorsal interneurons were affected, suggesting roof-plate-derived GDF7 was a signal specific for dI1 neurons. The necessity of roof plate signals in the development of dI1–dI3 interneurons was also seen in transgenic mice that had roof plate cells selectively destroyed by the diphtheria toxin gene. This was accomplished by inserting the gene for the diphtheria toxin in place of the *GDF7* gene that is normally expressed in the roof plate. The diphtheria toxin led to selective degeneration of the cells expressing the gene, so all roof plate cells were lost in these mice. The selective removal of roof plate cells allowed the investigators to examine the importance of roof-plate-derived signals on dorsal patterning without altering BMP-related signals arising from the epidermal tissues. In these mice, dorsal interneurons dI1–dI3 failed to form, whereas formation of dI4 interneurons was not inhibited. With the loss of the dorsal-most interneurons (dI1–dI3), the dI4 cells were able to expand dorsally into former dI1–dI3 and roof plate areas similar to what was observed in experiments in which BMP signals were decreased. Thus, the roof-plate-derived BMP-like signals are required for normal development of dI1-dI3 interneurons, whereas GDF7 appears specific for dI1 interneurons (Figure 4.15B).

Activin, another TGFβ-related molecule, has been shown to specifically influence development of the dI3 subpopulation. When chick embryos were electroporated with a constitutively active from of an activin receptor, the dI3 population was induced independent of BMP signaling. Whereas the level of BMP signaling influenced which cell populations developed, the level of activin receptor activation only affected dI3 interneurons. This result suggested that activin and BMP signaling pathways function independently of one another in the dorsal neural tube.

BMP-like signaling pathways are regulated by SMADS

As described in Chapter 2, BMPs, like other members of the TGFβ superfamily, bind to heterodimeric serine/threonine kinase receptors. BMP binds to the type II subunit of the receptor, which then associates with the type I subunit. The interaction of the two subunits leads to phosphorylation of a serine or threonine on the type I subunit. The phosphorylated type I subunit then phosphorylates certain SMAD proteins within the cytoplasm that translocate to the nucleus to activate specific genes (**Figure 4.16**).

The *Smad* genes are homologs of the *Drosophila Mad* and *C. elegans sma* genes (see Box 4.1). In mammals, eight SMAD family members have been identified and are designated as SMAD1, -2, -3, -4, -5, -6, -7, and -8/9. SMAD signaling is transduced by a trimer comprised of two receptor- regulated SMADs (R-SMADs) and one common-mediator SMAD (co-SMAD). R-SMADs include SMAD1, -2, -3, -5, and -8/9. SMAD4 is the co-SMAD. When R-SMADs are phosphorylated by the BMP receptor, the R-SMADs join the co-SMAD and travel to the nucleus to activate target genes, such as those associated with patterning the dorsal neural tube.

SMADs not only activate gene transcription, but two SMAD proteins (SMAD6 and SMAD7) function as inhibitory SMADs (I-SMADS). I-SMADs block the BMP signaling cascade by inhibiting phosphorylation of R-SMADs, thus preventing translocation to the nucleus and the activation of BMP-responsive genes (Figure 4.16).

The importance of BMP signal transduction can be explored by manipulating parts of the SMAD pathway. For example, chick embryos were electroporated with **short interfering RNA (siRNA)** against the co-SMAD, SMAD4. siRNAs are synthetic double-stranded RNAs that transiently block

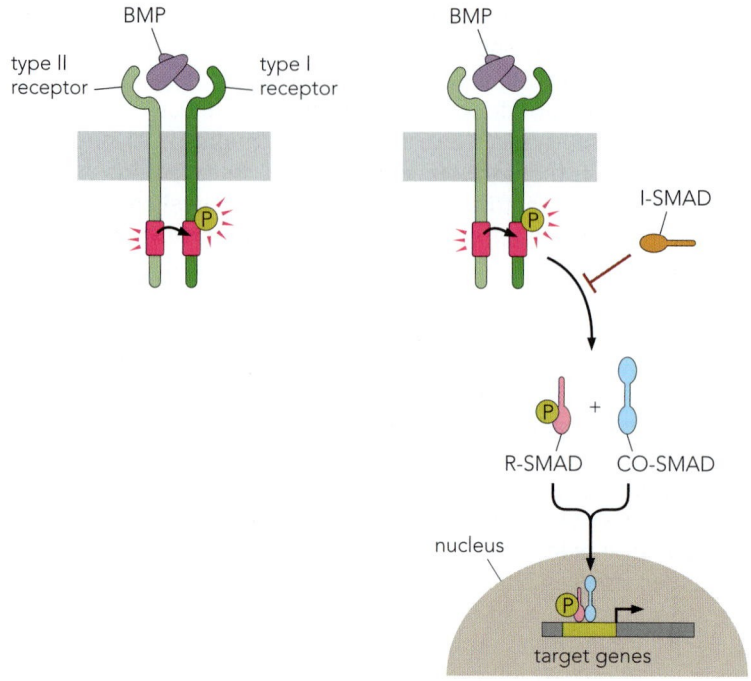

Figure 4.16 BMP signal transduction pathway. BMP and other TGFβ superfamily members bind to heterodimeric serine/threonine kinase receptors. The type II receptor binds BMP, then associates with the type I receptor, leading to phosphorylation. The now-activated type I receptor can phosphorylate receptor-regulated SMAD proteins (R-SMADs), which then associate with co-SMAD and are transported to the nucleus to activate or inhibit BMP-responsive genes. However, inhibitory SMAD proteins (I-SMADs) can block this gene activation pathway by preventing the phosphorylation of R-SMADs.

gene transcription by binding to complementary mRNA sequences, thus causing the cleavage and degradation of the targeted region. When siRNAs against SMAD4 were selectively introduced into dorsal neural tube cells outside of the roof plate, a specific loss of dI1–dI3 interneurons was noted. This indicated a cell-autonomous role for SMAD signaling in specifying the dorsal-most interneurons. In the areas normally occupied by dI1–dI3 interneurons, markers for dI4–dI6 interneurons were now detected. These results were similar to previous studies showing a need for BMP-related, roof-plate-derived signals in regulating the expression of the transcription factors expressed in dI1–dI3 interneurons. Together, the investigations indicate that roof-plate-derived signals utilize the SMAD pathway to selectively express the genes associated with the dI1–dI3 interneurons.

Although several molecules have been identified that impact class A progenitors, the signals used to induce specification of class B progenitors are still under investigation. RA is one plausible candidate, because it also impacts development of the ventral interneuron populations V0 and V1 that lie closest to the class B interneurons in the dorsal segment of the neural tube (see Figure 4.15A).

Wnt signaling through the β-catenin pathway influences development in the dorsal neural tube

Other signals are likely to be involved in the patterning of dorsal regions of the neural tube, and several candidates have been identified. For example, Wnt family members have also been implicated as roof-plate-derived signals that complement the actions of TGFβ superfamily members. In Chapter 3, discussion of Wnt (also called Wnt1) focused on its role in patterning midbrain regions and preventing the expansion of forebrain regions into more posterior territories. Along the D/V axis, Wnt signaling appears to act downstream of BMP to regulate cell cycle control or specification of interneuron subtypes. This hypothesis is supported by studies that show a loss of BMP or BMP-related signals in the dorsal neural tube leads to changes in Wnt expression patterns. BMPs may function to activate Wnt ligands and receptors in dorsal regions and inhibit Wnt ligands and receptors in ventral regions through signaling mechanisms that intersect with Shh pathways originating in the ventral neural tube.

The Wnt family of ligands includes up to 19 members, depending on the species. Among the Wnt ligands important for D/V patterning in the nervous system of mice and chick are Wnt1 and Wnt3a. The specific role of these ligands in neural patterning along this axis is still unclear, but it appears that Wnts in the dorsal regions of the neural tube influence the proliferation of cells as well as modify the effects of BMP and Shh signaling on the gene expression patterns typical of various dorsal interneuron populations.

The most common intracellular signaling cascade initiated by the binding of a Wnt ligand to its receptor complex is the β-catenin or **canonical pathway** introduced in Chapter 3. This pathway involves Wnt binding to the receptor complex comprised of Frizzled and Lrp4 or Lrp6. This leads, in turn, to recruitment of a scaffolding protein called Dishevelled (Dvl), phosphorylation of Lrp 4/6 by the protein kinases GSK3 (glycogen synthase kinase 3) and CKI (casein kinase 1), and the recruitment of the scaffolding protein axin and the protein APC (adenomatous polyposis coli). The association of these proteins with the Wnt-bound receptor complex allows cytoplasmic β-catenin to travel to the nucleus, where it displaces a co-repressor protein called Groucho and interacts with LEF1/TCF (lymphoid enhancer factor 1/T-cell factor) to initiate transcription of Wnt-target genes (**Figure 4.17**A). In the absence of Wnt binding, β-catenin is degraded in the cytoplasm by a degradation complex comprised of axin, APC, GSK3, and CK1. Active GSK3 and CK1 phosphorylate β-catenin, leading to its degradation. Thus, β-catenin is unavailable to translocate to the nucleus and initiate transcription of Wnt-responsive genes (Figure 4.17B). Instead, the Wnt-target genes are repressed by the binding of the co-repressor Groucho to LEF1/TCF. Wnt binding and the stabilization of β-catenin therefore changes the actions of LEF1/TCF from interacting with Groucho and acting as a transcriptional repressor to interacting with β-catenin and functioning as a transcriptional activator.

Wnt–β-catenin signaling appears to regulate the proliferation of neuronal precursor cells by up-regulating the specific enzymes needed to drive cellular proliferation (cyclin-dependent kinases, CDKs; see Chapter 5).

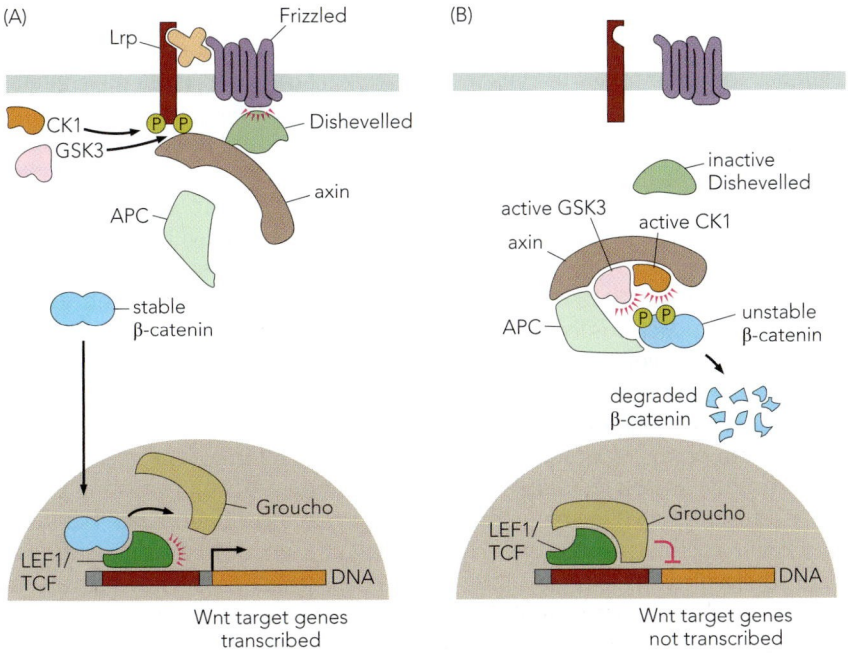

Figure 4.17 The Wnt–β-catenin pathway determines activation of Wnt-target genes. The canonical Wnt–β-catenin pathway is the most common Wnt signaling pathway. (A) When Wnt binds to a receptor complex comprised of Frizzled receptors and Lrp4 or Lrp6, a series of events occurs that allows β-catenin to travel to the nucleus and activate Wnt-target genes. Binding of Wnt leads to the phosphorylation of the Lrp subunit by protein kinases CK1 (casein kinase 1) and GSK3 (glycogen kinase 3) and the recruitment of a complex that includes the scaffolding proteins Dishevelled and axin, as well as the protein APC (adenomatous polyposis coli). With the recruitment of these to the Wnt receptor complex, the β-catenin is free to travel to the nucleus, where it displaces Groucho to interact with LEF1/TCF (lymphoid enhancer factor 1/T-cell factor) and initiate transcription of target genes. (B) In the absence of Wnt binding, the scaffolding protein Dishevelled is inactive and the other proteins associate as a degradation complex, which is able to phosphorylate β-catenin. This causes β-catenin to become unstable and degraded in the cytoplasm. Because β-catenin is unavailable to travel to the nucleus, Groucho remains associated with LEF1/TCF and together they function as a transcriptional repressor to prevent Wnt-target gene transcription.

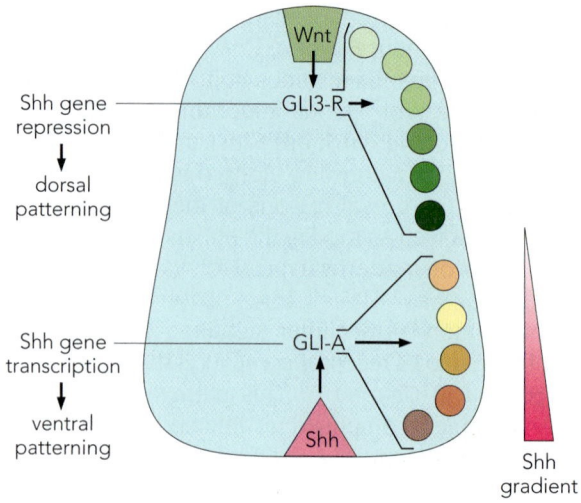

Figure 4.18 Wnt and Shh pathways interact with GLI proteins to pattern dorsal and ventral regions of the neural tube. Wnt1 and Wnt3a in the dorsal neural tube bind to the enhancer region of GLI3, the predominant repressor form of the GLI proteins (GLI-R). The Shh concentration in the dorsal region is too low to activate Ptc-induced GLI-A activity, so the repressor GLI dominates and prevents Shh gene transcription in dorsal interneurons. In the ventral neural tube, Shh is present in a high enough concentration to bind to Ptc receptors and initiate the activator form of GLI (GLI-A) that promotes the Shh gene transcription that leads to ventral cell patterning.

Due to this effect on proliferation, it was thought that Wnt may influence D/V patterning indirectly by causing an increase in cell number and expansion of the dorsal neural tube regions, rather than directly influencing the transcription factor code and gene expression patterns of cell types along the D/V axis. However, several studies also suggest that the Wnt–β-catenin pathway intersects with the Shh–GLI pathway to directly influence patterning. Wnt1 and Wnt3a appear to induce activity of GLI3—the predominant repressor form of GLI—in the dorsal regions of the neural tube (**Figure 4.18**). When sufficient levels of Wnt signaling are present, the stablized β-catenin is able to bind to enhancer regions of the *GLI3* gene. Because the Shh concentration in the dorsal neural tube is too low to activate sufficient Ptc receptor signaling and induce transcription of the activator forms of GLI, the Wnt-induced GLI3-R form represses *Shh* gene transcription. Thus, Wnt signaling increases GLI3-R activity and prevents *Shh* gene transcription in the dorsal neural tube, thus allowing for the expression of transcription factors associated with dorsal interneuron cell fates.

Gradients of BMP and Shh antagonize each other to form D/V regions of the neural tube

Mutually antagonistic gradients of BMP and Shh help establish the boundaries between dorsal and ventral structures in the vertebrate neural tube, similar to homologous signals in *Drosophila* (**Box 4.2**). BMPs up-regulate genes in the dorsal neural tube, such as *Pax3* and *Pax7* (see Figure 4.3). In contrast, these same genes are repressed in the ventral neural tube by Shh. Experimentally manipulating the concentration of one signal relative to the other helped demonstrate how BMP and Shh interact to specify whether dorsal or ventral cell types form. For example, when levels of BMP were increased relative to Shh, ventral structures were absent. Thus, higher concentrations of BMP inhibited the ventralizing capabilities of Shh so that only dorsal structures formed. In contrast, when BMP antagonists were also included, only ventral structures formed, even when Shh was present at low concentrations that would not usually induce ventral structures. Thus, the concentrations of BMP and Shh normally produced *in vivo* appear to interact in opposing gradients to ensure specialization of the various neuronal subtypes needed along the D/V axis (**Figure 4.19**). Opposing gradients of BMP and Shh may similarly act to maintain boundaries between forebrain regions along the A/P axis, such as proposed in the zona limitans intrathalamica (ZLI) region (see Chapter 3). These observations further demonstrate that patterning of the A/P and D/V axes do not require two entirely separate and distinct groups of signaling molecules.

Box 4.2 Signaling mechanisms that regulate D/V patterning in the *Drosophila* embryo

Similar to the role of signaling molecules in vertebrates, signals originating in dorsal and ventral regions of the *Drosophila* embryo differentially regulate the expression and repression of genes characteristic of different types of neural cells found along the dorsal–ventral (D/V) axis of the developing nervous system. As described in Box 2.1, the *Drosophila* nervous system arises from the ventrolateral region of the ectoderm and ultimately forms on the ventral side of the embryo. The epidermal ectoderm arises from the ectoderm located on the dorsal side of the embryo, opposite the orientation in the vertebrate embryo (**Figure 1**A). The ventrolateral neuroectoderm extends to the ventral side as the mesoderm begins to invaginate (Figure 1B, C). The cells of the mesoderm and neuroectoderm are ultimately surrounded by the epidermis (Figure 1D). Initially, the neural region is comprised of three rows of undifferentiated neural cells (neuroblasts) that form as individual cells delaminate from the neuroectoderm region. The cells in each row are identified by the expression of different homeodomain transcription factor genes: *msh*, muscle segment homeobox; *ind*, intermediate neuroblasts defective; and *vnd*, ventral nervous system defective. Theses genes are homologs of the vertebrate genes *Msx1/2*, muscle segment homeobox; *Gsh*, genomic screen homeobox; and *Nkx2.2*, NK homeobox 2.2, respectively.

The specific neural genes expressed in the three primary rows of neuroblasts are determined by the actions of the opposing gradients originating from the dorsal and ventral signaling centers. Similar to vertebrate BMP signals, the *Drosophila* Dpp signals in the dorsal region first play a role in repressing the expression of neural genes destined to become epidermal cells (see Box 2.1) and later regulate expression of different neural genes through a dosage-dependent manner established by the gradients of Dpp. Neuronal cell types are further influenced by the gradient of Dorsal protein originating on the ventral side of the blastoderm stage embryo. In the fly, the mechanisms that govern neural gene expression along the D/V axis are said to be ventral dominant—that is, the signals originating from the ventral region of the embryo are the primary signals that first shape neuronal gene expression patterns. The gene expression patterns are then modified by a threshold-dependent repression of neural genes by signals from the dorsal region of the embryo. The expression of these genes begins in the neuroectoderm regions of the blastoderm. In addition, the genes expressed more ventrally are able to repress the expression of the other genes. Thus, *vnd* represses *ind* and *msh*, while *ind* represses *msh* (**Figure 2**).

The cells that form from the neuroectoderm closest to the ventral region express *vnd*, those in the central layer express *ind*, and those located closest to the dorsal side express *msh*. Ventrally derived Dorsal protein influences the expression of these genes in a gradient-dependent manner. The higher concentration of Dorsal at the ventral region induces expression of *vnd*, while progressively lower concentrations induce expression of *ind* and *msh*. The expression of *vnd* in the ventral-most region also represses *ind* and *msh* expression in the ventral region, further contributing to the boundaries of gene expression found in the future neuroblasts. The higher concentration of Dorsal protein in the ventral region is thought to be particularly important for establishing the border between *vnd*- and *ind*-expressing cells (Figure 2A).

The Dpp originating on the dorsal side further shapes the D/V axis by repressing the expression of neural genes.

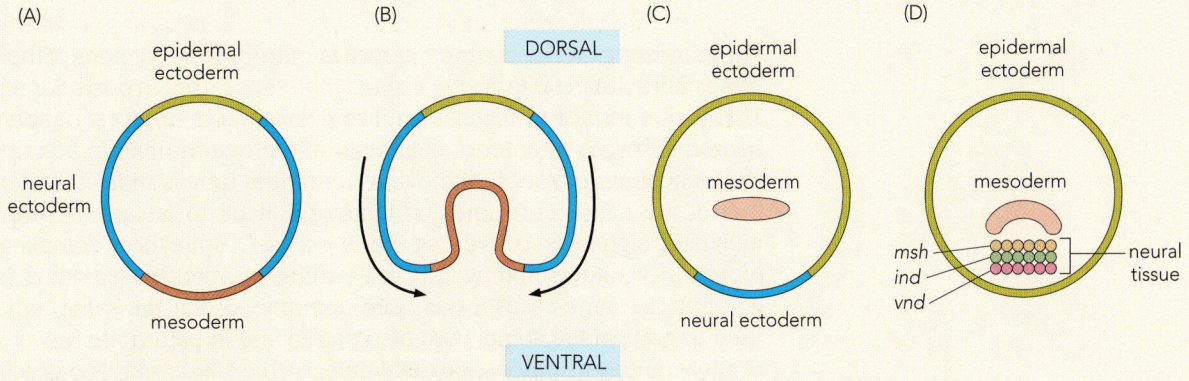

Figure 1 Neural ectoderm arises from the ventrolateral regions of the *Drosophila* embryo. (A) Neural ectoderm is found along the lateral sides of the *Drosophila* embryo. (B) As mesoderm at the ventral side invaginates, the neuroectoderm extends further ventrally (arrows) until it lies on the ventral side of the embryo (C). Individual cells of the neuroectoderm delaminate to form three rows of neuroblasts (D). Each row expresses a different transcription factor gene. The ventral-most cells express *vnd*, the cells in the middle row express *ind*, and those of the dorsal-most row express *msh*.

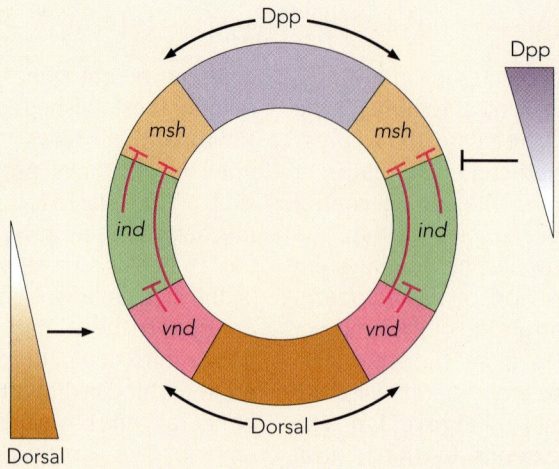

Figure 2 Signals arising from dorsal and ventral regions organize the transcription factor expression in neuroblasts along the D/V axis. The protein Dorsal on the ventral side of the embryo regulates the expression of each transcription factor in a concentration-dependent manner. Higher levels of Dorsal are able to induce expression of *vnd*, whereas progressively lower concentrations induce expression of *ind* and *msh*. The boundaries between these cell regions are further specified by the repression of adjacent genes. *vnd* is able to repress expression of *ind* in the ventral-most regions to maintain a boundary between the *vnd*- and *ind*-expressing cells. The gradient of Dpp that originates on the dorsal side of the embryo represses neural gene expresssion. The higher concentration at the dorsal-most region is insufficient to influence *msh* expression. However, the lower concentration at the mid-regions of the neuroectoderm is able to repress *ind* expression. With *ind* repressed, the inhibitory effect of *ind* on *msh* is lost and therefore *msh* is expressed in the most dorsal regions of the neuroectoderm. In this way Dpp helps establish the boundary between *ind* and *msh*. (Adapted from Mizutani CM, Meyer N, Roelink H & Bier E [2006] *PLoS Biol* 4(10):e313. With permission from PLoS.)

Dpp can repress *ind* expression to a greater extent than it can repress *msh* expression. Thus, the lower concentration of Dpp found near the cells originating in the middle region of neuroectoderm is sufficient to repress *ind* expression. However, the higher concentration at the more dorsal side is still insufficient to repress *msh* expression. When Dpp blocks expression of *ind*, it is no longer available to repress the more dorsally located *msh*, so this gene is able to be expressed in the dorsal-most region of neuroectoderm (Figure 2B). The boundary between *msh*- and *ind*-expressing cells is believed to be influenced primarily by the gradient of Dpp originating at the dorsal side of the embryo.

Thus, in the *Drosophila* embryo, gradients of Dpp and Dorsal are coordinated so that expression of the different genes is tightly regulated in the three areas of neuroectoderm that give rise to the neuroblasts oriented along D/V axis.

D/V PATTERNING IN THE ANTERIOR NEURAL TUBE

The patterning mechanisms employed in more anterior regions of the vertebrate neural tube are less well defined than those used in posterior regions. This is due in part to fewer identified cell markers for dorsal and ventral neuron subtypes in anterior segments of the neural tube. In addition, the anatomical organization of the anterior neural tube is more complex than that of the spinal cord, making it more difficult to isolate the impact of individual signals on cells along the D/V axis. Despite these complications, progress in understanding D/V patterning in anterior segments is progressing. In some cases it has been determined that the same molecules used to pattern the spinal cord also participate in patterning the D/V axis at more anterior regions. For example, in the hindbrain, the gradient of Shh appears to induce different motor neuron populations. Studies have shown that the concentrations of Shh required to induce the more ventrally located visceral motor neurons are higher than the Shh concentrations required to induce the more dorsally located somatic motor neurons. Thus, levels of available Shh regulate the types of motor neurons that develop in

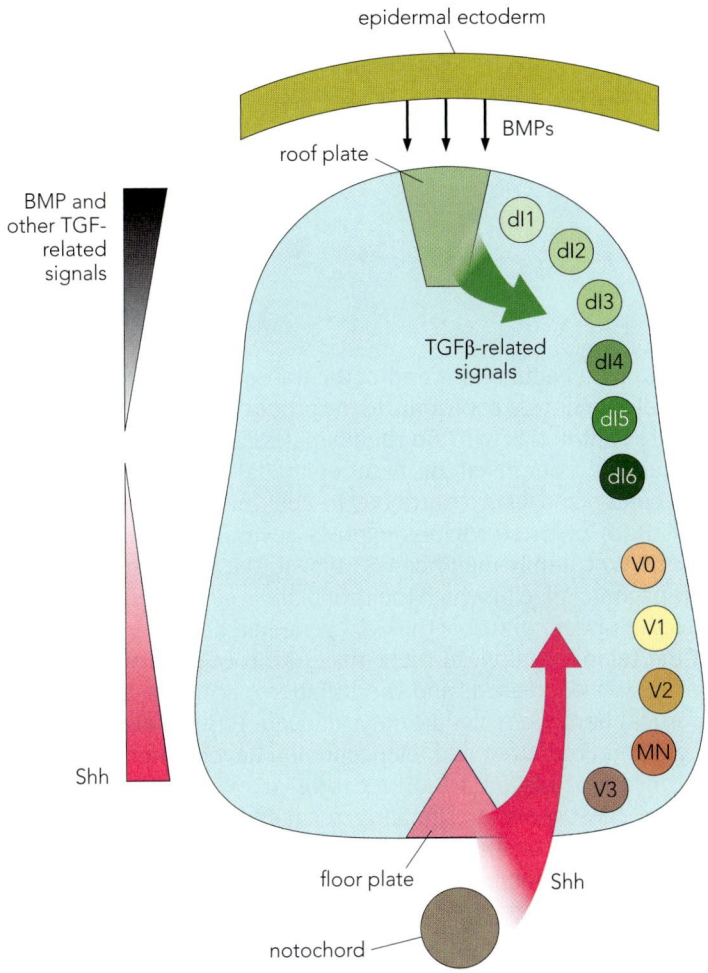

Figure 4.19 Signals from the dorsal and ventral regions interact to organize dorsal–ventral patterning in the posterior neural tube. BMP and Shh both create concentration gradients that interact to induce specific cell types along the dorsal–ventral axis. At the dorsal segment, the epidermal ectoderm releases BMPs to induce formation of the roof plate. The roof plate then releases BMPs and other TGFβ-related molecules to induce formation of sensory interneurons (dI1–dI3). Both the timing and concentration of BMP affect which cell types form from the progenitor cells. The roof plate signals are not required for the induction of dI4–dI6. At the ventral surface the notochord releases Shh to induce the floor plate, which then also releases Shh. Together these sources of Shh induce motor neurons (MN) and interneurons (V3, V2, V1, and V0) in the ventral segment in a concentration-dependent manner. D/V patterning is further established because BMP and Shh antagonize each other, causing genes up-regulated by one to be repressed by the other.

ventral regions of the hindbrain, similar to the process that occurs in the spinal cord.

Roof plate signals pattern the anterior D/V axis by interacting with the Shh signaling pathway

As noted in Chapter 3, the cerebellum arises from the dorsal region of rhombomere 1 (r1) of the hindbrain. The roof plate in this region of the neural tube expresses BMP6, BMP7, and GDF7, similar to the spinal cord roof plate. When the roof plate adjacent to the region of cerebellar formation is reduced in size, such as occurs in *Dreher* mutant mice, the cerebellum that forms is much smaller in size and lacks a vermis, the central region of the cerebellum. The granule cells of the cerebellum (see Chapter 5) are the dorsal-most cells to arise in the cerebellum and they appear to rely on BMP signals to develop. When BMP is added to cultures of rhombomere 1 tissues, granule cells are formed, thus supporting the hypothesis that BMPs are used in dorsal regions of the anterior hindbrain as well as in the spinal cord.

Roof plate signals are also important in the diencephalon portion of the neural tube. In various transgenic mice in which roof plate cells have been depleted, dorsal markers in this region such as *Pax6* are down-regulated. In addition, dorsal structures such as the pineal gland and the posterior commissure, a major axonal bundle in this region of the nervous system, are lost.

The roof plate in the telencephalon is morphologically quite different than that of more posterior regions of the neural tube. The roof plate is located between the forming cerebral hemispheres and later differentiates

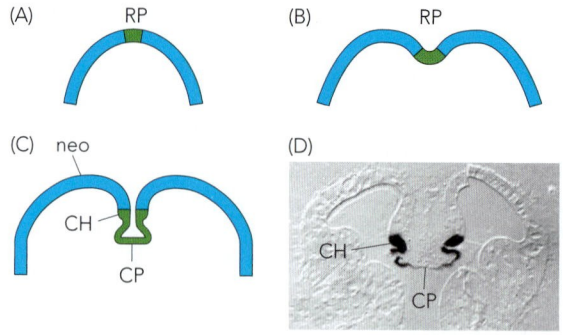

Figure 4.20 Roof plate formation in the telencephalon. (A) In the anterior-most region of the neural tube, the roof plate (RP) forms along the dorsal surface similar to that observed at more posterior regions. As the telencephalon begins to invaginate (B), the roof plate comes to lie between the forming cerebral hemispheres of the neocortex (neo). (C). The roof plate cells give rise to choroid plexus epithelum (CP), which produces cerebral spinal fluid in the ventricles. The cortical hem cells (CH) later contribute cells to the outermost layer of the cerebral cortex. During development, these cells function as a signaling center to influence the development of dorsal regions of the telencephalon. (D) Image of a mouse embryo at embryonic day 12.5. The expression of *Lmx1a* (dark staining), a roof plate marker, is revealed by *in situ* hybridization. (Adapted from Chizhikov VV & Millen KJ [2005] *Dev Biol* 27:287–295. With permission from Elsevier.)

into the choroid plexus epithelium and cortical hem. The choroid plexus consists of epithelial cells that contribute to the production of cerebral spinal fluid (CSF) and are continuous with the ependymal cells lining the ventricles. The cortical hem is comprised of neuroepithelial cells that function as a signaling center and later contribute to cells of the cerebral cortex (**Figure 4.20**). Both of these regions express several BMP-related molecules as well as Wnt family members. Similar to the experiments in the spinal cord, transgenic mice in which the roof plate is selectively destroyed by diphtheria toxin under control of the *Gdf7* gene had defects in the dorsal structures of the telencephalon. In these mice the roof plate was absent, *Lhx2* expression was decreased, and the number of cells in the choroid plexus and cortical hem were significantly reduced. Further evidence that BMP signaling is required for dorsal telencephalon development was seen in transgenic mice that lacked a BMP receptor or overexpressed a BMP receptor. Loss of BMP receptor signaling resulted in a loss of choroid plexus cells, whereas a constitutively activated form of the receptor led to the entire alar plate region becoming choroid plexus. In these mice, the ventral region was not altered, even though the receptor was expressed throughout both dorsal and ventral regions of the telencephalon.

Wnt expression is also widely distributed in the dorsal telencephalon, where it induces expression of dorsal markers such as *Pax6*. The Wnt signals also block the ability of ventral signals to convert dorsal structures to a ventral fate, likely by interacting with Shh signaling pathways similar to the mechanisms used in the dorsal spinal cord.

Zic mediates D/V axis specification by integrating dorsal and ventral signaling pathways

The integration of dorsal and ventral signaling pathways in the midbrain and hindbrain is also influenced by the expression of various *Zic* (zinc finger of the cerebellum) genes, the vertebrate homologs of the *Drosophila odd-paired* genes (see Chapter 3). Vertebrates have five *Zic* genes that are expressed in different patterns throughout the developing embryo and regulate a variety of functions, including those related to neural and limb development. In the nervous system, *Zic* genes are expressed during different stages of neural development and have important roles during neural induction, neural crest development, and D/V patterning. *Zic* expression is regulated by a number of other signals including BMPs, Wnts, and Shh. Elucidating the mechanisms by which these signals interact to pattern the D/V axis is currently an active area of research. During neural induction, BMPs appear to repress *Zic* genes, thereby preventing a neural fate in BMP-responsive cells that become ectoderm. In contrast, the neural inducers chordin, noggin, and follistatin interfere with BMP signaling and allow *Zic* gene expression. However, later in development when *Zic* genes are expressed in dorsal regions of the neural tube, BMP signals promote *Zic* expression. Wnt signals also promote expression of

Zic genes in dorsal areas of the neural tube. Zic proteins subsequently promote the expression of genes such as *Pax3* and *Wnt* that are needed for normal development of dorsal structures, particularly in the midbrain and hindbrain. Further patterning along the D/V axis occurs because Shh from ventral regions prevents expansion of *Zic* gene expression to the ventral areas of the neural tube.

The exact mechanisms that regulate the different effects of signaling on *Zic* gene expression patterns at different stages of development are still unclear. However, the ability of multiple signals to regulate *Zic* expression appears to occur because the signals intersect at specific DNA binding domains to influence the transcription of various genes associated with dorsal or ventral regions. For example, Zic proteins associated with the development of dorsal cells and GLI-A proteins activated downstream of Shh in ventral cells may inhibit the actions of each other by binding to each other's zinc finger domains (**Figure 4.21**). In this way, the different levels of Zic and GLI proteins available along the D/V axis ensure that Shh signaling occurs only in ventral regions.

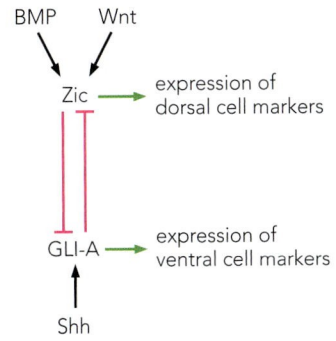

Figure 4.21 Zic proteins integrate dorsal and ventral signaling pathways along the D/V axis. The expression of *Zic* (zinc finger of the cerebellum) genes is regulated by the BMP and Wnt signals arising in dorsal regions of the neural tube. The Zic proteins interact with GLI-A proteins activated by Shh in the ventral neural tube. The Zic and GLI proteins inhibit the zinc finger domains of one another. This interaction contributes to the repression of Shh signaling in the dorsal neural tube and the repression of BMP and Wnt signaling in the ventral neural tube.

The location of cells along the A/P axis influences their response to ventral Shh signals

An important point to keep in mind when thinking about structures differentiating along the neural tube is that A/P and D/V patterning often occur at the same time, through a process described as analogous to a Cartesian grid. In this way cells at each point along the two axes process multiple signals that combine to regulate differentiation of specific cell types at each segment of the nervous system. Thus, the location of signals along the A/P axis often influences the response of cells along the D/V axis. For example, the types of neurons that develop in response to Shh depend, in part, on the location of the cells along the A/P axis. In the hindbrain, Shh from the notochord and floor plate regulate the development of serotonergic neurons, whereas in the midbrain Shh regulates the development of dopaminergic neurons. This was also observed *in vitro*. Addition of Shh induced dopaminergic neurons in cultures of midbrain explants, but addition of Shh could *not* induce dopaminergic neurons in cultures of hindbrain explants (**Figure 4.22**). This provides one example of how location along the A/P axis impacts the patterning of neuronal populations along the D/V axis.

Shh also plays a role in patterning more anterior regions. Forebrain patterning involves the formation of two symmetrical cerebral hemispheres and the segregation of the eye fields. Shh and nodal released by the prechordal plate appear to influence this patterning. In experiments in which the prechordal plate was removed, a fused cerebral hemisphere and a single eye were observed, indicating the necessity of prechordal-plate-derived signals such as Shh in patterning these forebrain structures. Thus, Shh regulates

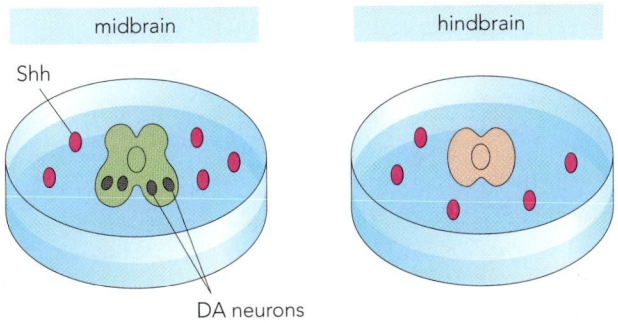

Figure 4.22 Shh regulates patterning of ventral cell types in the anterior regions of the neural tube. At more anterior regions of the neural tube, the notochord and floor plate release Shh to induce formation of specific neuronal types in the ventral regions of the hindbrain and midbrain. In the midbrain, Shh induces dopaminergic neurons. (A) The additon of Shh to cultures of midbrain neural tube segments induces dopaminergic neurons (DA). However, when Shh is added to cultures of hindbrain segments (B), no DA neurons form. These studies confirmed that Shh differentially regulates formation of neurons based on the location along the A/P axis.

many aspects of D/V patterning and the specific function of Shh depends on its location along the A/P axis.

Analysis of birth defects reveals roles that D/V patterning molecules play in normal development

The failure of the two hemispheres to separate is called **holoprosencephaly (HPE)**, a birth defect that typically impacts both brain and craniofacial development. There are several subtypes of HPE that produce neurological deficits that range from mild to severe. One severe form of HPE occurs when the forebrain regions fail to separate. This leads to a single undivided hemisphere and often a single midline eye (cyclopia). This is a relatively common birth defect in humans, occurring in 1 out of 250 conceptions and over 1 in 10,000 live births, though in the most severe cases the infant dies shortly after birth. HPE also occurs in farm animals, and studies of those animals led to our understanding of the importance of Shh signaling in preventing this form of HPE.

Cyclopia in farm animals has been observed since the 1950s. It was found that when pregnant sheep or goats grazed on corn lily plants (*Veratrum californicum*) during the first two weeks of gestation, up to one quarter of the offspring were born with cyclopia—the observable manifestation of this form of HPE (**Figure 4.23**A, B). Later, scientists realized that the plants contain a naturally occurring alkaloid, now termed cyclopamine. It was subsequently determined that cyclopamine inhibits the action of Shh. This inhibitory effect was confirmed by *in vitro* experiments in which neural tube explants were cultured in the presence of Shh and cyclopamine. In contrast to control cultures that only contained Shh, the cyclopamine-treated explants failed to form either floor plate or motor neurons (Figure 4.23C). Thus, a naturally occurring birth defect was confirmed to result from disruption of the Shh signaling pathway.

Another typically less severe form of HPE, the middle interhemispheric (MIH) form, is caused by disruption in BMP signaling from the

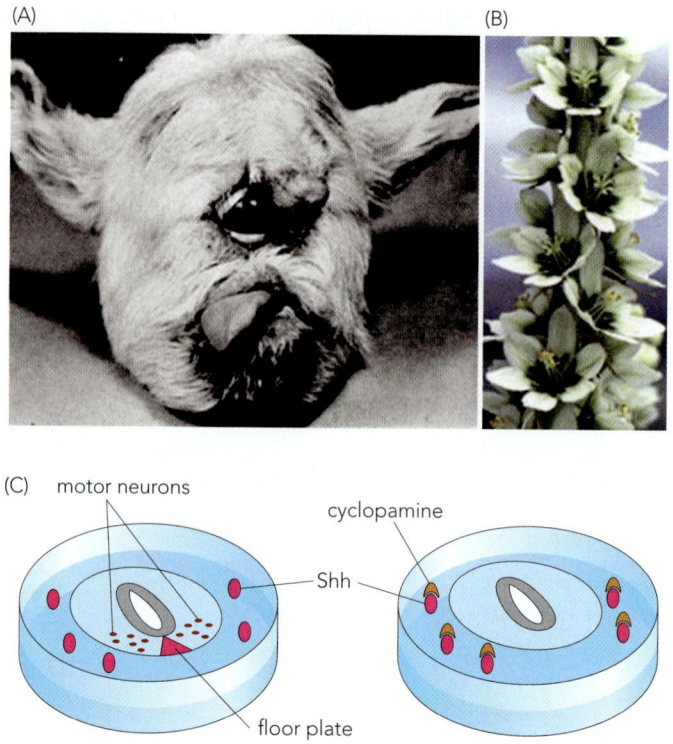

Figure 4.23 Inhibition of Shh by cyclopamine leads to congenital malformations in the forebrain. (A) Farmers discovered that sheep grazing on the corn lily plant often give birth to offspring with a single eye (cyclopia). Cyclopia is one characteristic of holoprosencephaly (HPE), a birth defect that is characterized by fused cerebral hemispheres and craniofacial defects. It was later determined that this birth defect occurs because the plant (B) contains a naturally occurring alkaloid, cyclopamine, that inhibits Shh signaling. (C) The effects of cyclopamine on Shh signaling were confirmed *in vitro*. When explants of neural tube were cultured in the presence of Shh, the floor plate and motor neurons formed. When cyclopamine was added to the Shh treated cultures, no floor plate or motor neurons were induced. (A and B courtesy of Philip Beachy/HHMI)

roof plate. Disruption of the roof plate signals in the dorsal region of the anterior neural tube causes only mild craniofacial deformities in most cases. However, disruption of this BMP signaling leads to defects in the posterior frontal and the parietal lobes. These developmental changes typically result in mild-to-moderate developmental delays, learning problems, and cerebral palsy.

The above descriptions of congenital malformations in farm animals and humans clearly illustrate that disruption of ventral and dorsal patterning molecules can lead to congenital birth defects. Thus, by focusing on the basic science underlying the cellular and molecular pathways utilized in neural tube patterning, scientists discover important information relevant to veterinary and human medicine.

SUMMARY

The mechanisms that pattern neural tube structures along the D/V axis demonstrate that developmental processes often utilize opposing gradients of signaling molecules to activate specific downstream regulators of cellular specialization. The gradients influence the particular transcription factors in a given cell population and these, in turn, establish gene expression patterns that result in cellular specialization along the D/V axis. It is also important to keep in mind that differentiation of both the A/P and D/V axes often occurs at the same time. Thus, cells must recognize and respond to multiple signals simultaneously. Scientists have described this process in terms of a Cartesian grid in which cells at each point along the two axes process the various signals that lead to differentiation of specific cell types at each segment of the developing nervous system. Scientists continue to identify new gene expression patterns along both axes and to unravel how signals are integrated in each cell population. The intersection of the various signaling pathways provides one means for cells to integrate information at each point along the A/P and D/V axes. It also appears that for many cells, fluctuations in signaling molecule concentrations are integrated over time, and the sum of the effect leads to activation or repression of different genes. As information continues to accumulate, our understanding of the early stages of neural tube segmentation and boundary formation will also expand.

FURTHER READING

Attisano L & Lee-Hoeflich ST (2001) The Smads. *Genome Biol* 2(8):1–8.

Briscoe J & Small S (2015) Morphogen rules: design principles of gradient-mediated embryo patterning. *Development* 142(23):3996–4009.

Briscoe J & Therond PP (2013) The mechanisms of Hedgehog signalling and its roles in development and disease. *Nat Rev Mol Cell Biol* 14(7):416–429.

Chang C & Hemmati-Brivanlou A (1998) Cell fate determination in embryonic ectoderm. *J Neurobiol* 36(2): 128–151.

Chesnutt C, Burrus LW, Brown AM & Niswander L (2004) Coordinate regulation of neural tube patterning and proliferation by TGFβ and WNT activity. *Dev Biol* 274(2): 334–347.

Dessaud E, McMahon AP & Briscoe J (2008) Pattern formation in the vertebrate neural tube: a sonic hedgehog morphogen-regulated transcriptional network. *Development* 135(15): 2489–2503.

Goodrich LV, Jung D, Higgins KM & Scott MP (1999) Overexpression of ptc1 inhibits induction of Shh target genes and prevents normal patterning in the neural tube. *Dev Biol* 211(2):323–334.

Gross MK, Dottori M & Goulding M (2002) Lbx1 specifies somatosensory association interneurons in the dorsal spinal cord. *Neuron* 34(4):535–549.

Hynes M, Ye W, Wang K et al. (2000) The seven-transmembrane receptor smoothened cell-autonomously induces multiple ventral cell types. *Nat Neurosci* 3(1):41–46.

Ille F, Atanasoski S, Falk S et al. (2007) Wnt/BMP signal integration regulates the balance between proliferation and differentiation of neuroepithelial cells in the dorsal spinal cord. *Dev Biol* 304(1):394–408.

Kim JJ, Gill PS, Rotin L et al. (2011) Suppressor of fused controls mid-hindbrain patterning and cerebellar morphogenesis via GLI3 repressor. *J Neurosci* 31(5):1825–1836.

Le Dreau G & Marti E (2012) Dorsal-ventral patterning of the neural tube: a tale of three signals. *Dev Neurobiol* 72(12):1471–1481 (doi: 10.1002/dneu.22015).

Lee KJ, Dietrich P & Jessell TM (2000) Genetic ablation reveals that the roof plate is essential for dorsal interneuron specification. *Nature* 403(6771):734–740.

Liem KF Jr, Tremml G & Jessell TM (1997) A role for the roof plate and its resident TGFβ-related proteins in neuronal patterning in the dorsal spinal cord. *Cell* 91(1):127–138.

Litingtung Y & Chiang C (2000) Control of Shh activity and signaling in the neural tube. *Dev Dyn* 219(2):143–154.

Lumsden A & Graham A (1995) Neural patterning: a forward role for hedgehog. *Curr Biol* 5(12):1347–1350.

Maden M (2006) Retinoids and spinal cord development. *J Neurobiol* 66(7):726–738.

Megason SG & McMahon AP (2002) A mitogen gradient of dorsal midline Wnts organizes growth in the CNS. *Development* 129(9):2087–2098.

Millen KJ, Millonig JH & Hatten ME (2004) Roof plate and dorsal spinal cord dI1 interneuron development in the dreher mutant mouse. *Dev Biol* 270(2):382–392.

Mizutani CM, Meyer N, Roelink H & Bier E (2006) Threshold-dependent BMP-mediated repression: a model for a conserved mechanism that patterns the neuroectoderm. *PLoS Biol* 4(10):e313.

Monsoro-Burq AH, Bontoux M, Vincent C & Le Douarin NM (1995) The developmental relationships of the neural tube and the notochord: short and long term effects of the notochord on the dorsal spinal cord. *Mech Dev* 53(2):157–170.

Pachikara A, Dolson DK, Martinu L et al. (2007) Activation of Class I transcription factors by low level Sonic hedgehog signaling is mediated by Gli2-dependent and independent mechanisms. *Dev Biol* 305(1):52–62.

Patten I, Kulesa P, Shen MM et al. (2003) Distinct modes of floor plate induction in the chick embryo. *Development* 130(20):4809–4821.

Persson M, Stamataki D, te Welscher P et al. (2002) Dorsal-ventral patterning of the spinal cord requires Gli3 transcriptional repressor activity. *Genes Dev* 16(22): 2865–2878.

Roelink H, Porter JA, Chiang C et al. (1995) Floor plate and motor neuron induction by different concentrations of the amino-terminal cleavage product of sonic hedgehog autoproteolysis. *Cell* 81(3):445–455.

Timmer J, Chesnutt C & Niswander L (2005) The activin signaling pathway promotes differentiation of dI3 interneurons in the spinal neural tube. *Dev Biol* 285(1):1–10.

Tozer S, Le Dreau G, Marti E & Briscoe J (2013) Temporal control of BMP signalling determines neuronal subtype identity in the dorsal neural tube. *Development* 140(7): 1467–1474.

Van Kampen KR & Ellis LC (1972) Prolonged gestation in ewes ingesting *Veratrum californicum*: morphological changes and steroid biosynthesis in the endocrine organs of cyclopic lambs. *J Endocrinol* 52(3):549–560.

Wilson L & Maden M (2005) The mechanisms of dorsoventral patterning in the vertebrate neural tube. *Dev Biol* 282(1): 1–13.

Wilson SW & Houart C (2004) Early steps in the development of the forebrain. *Dev Cell* 6(2):167–181.

Yamada T, Placzek M, Tanaka H et al. (1991) Control of cell pattern in the developing nervous system: polarizing activity of the floor plate and notochord. *Cell* 64(3):635–647.

Proliferation and Migration of Neurons

5

As the neural tube forms and begins to segregate into distinct regions along the anterior-posterior (A/P) and dorsal-ventral (D/V) axes, neuroepithelial cells are dividing rapidly to produce the cells that later form the neurons and glia of the central nervous system (CNS). The proliferating cells originate along the ventricular surface that lies adjacent to the neural canal. Beginning in early stages of development, selected cells in the neuroepithelium begin to express proneural genes that will set them on the path to become neurons. Other neuroepithelial cells will express genes that lead them to a glial cell fate. As development continues various intrinsic and extrinsic signals lead to the activation or repression of other genes that will determine the specific type of neuron or glial cell that will form.

As seen in previous chapters, the location along the A/P and D/V axes determines which extrinsic signals are available to a particular group of cells at any given stage of development. Further, as cells migrate away from the ventricular surface to reach a new destination, other signals are encountered. Thus, newly formed CNS precursor cells are exposed to a variety of spatially and temporally regulated signals. Similarly, neurons and glia of the peripheral nervous system (PNS) that originate in the neural crest are influenced by numerous spatially and temporally regulated signals that regulate gene expression patterns and final cell fate.

This chapter explores how vertebrate neuroepithelial cells begin to form different cell types and how the migration of cells from the cerebral cortex, cerebellum, and neural crest influences subsequent cell fate options. In Chapter 6, examples of cell determination and differentiation in these and other CNS and PNS populations from vertebrate and invertebrate animal models are described. Even though aspects of segmentation, neurogenesis, migration, and cell fate are described separately in this textbook, many of these events take place concurrently within the developing nervous system.

NEUROGENESIS AND GLIOGENESIS

Early in nervous system development, neuroepithelial cells divide rapidly to generate the precursor cells that later give rise to new neurons (**neurogenesis**) and glial cells (**gliogenesis**). The earliest precursor, or progenitor, cells have the ability to form neurons or glia. As development continues, precursors stop dividing and the cells begin to express genes that limit cellular potential to a neural or glial fate. Once fate is established, further development or experimental manipulation can no longer alter the type of cell that forms. This is the stage of **cell fate determination**. Subsequent gene activation or repression initiates the gradual process of **cellular differentiation** that leads to the unique cellular characteristics associated with the various neuronal and glial subtypes.

The mechanisms that determine whether a precursor cell becomes a neuron or a glial cell have been actively studied for more than one hundred years. Scientists in the nineteenth century debated whether neurons and glia arose from the same population of precursor cells or from two separate populations. More recent investigations have focused on intracellular signaling mechanisms that regulate the formation of specific neuronal or glial cell subpopulations.

Scientists debated whether neurons and glia arise from two separate cell populations

Published reports by early neurobiologists such as Wilhelm His (1886, 1889) and Santiago Ramón y Cajal (1894) provided detailed descriptions of the cellular composition of the vertebrate embryonic nervous system. Both produced elaborate drawings of cells arranged in layers near the lumen of the neural tube (that is, the ventricular surface). His and Cajal correctly identified these cell layers to be the precursor, or germinal, cells of the nervous system, though they interpreted the organization differently.

Based on the appearance of the cells under the microscope, His hypothesized that neural and glial cells form from two different precursor populations (**Figure 5.1** A). He proposed that the glial cells arose from the "spongioblasts," a layer that remained along the ventricular surface as a **syncytium**, a connected network of cells. The idea that cells formed this sort of continuous network was popular at the time and was described in many other tissues as well.

The concept of a spongioblast network was consistent with the appearance of the cells as they were prepared by His. Cajal's own observations

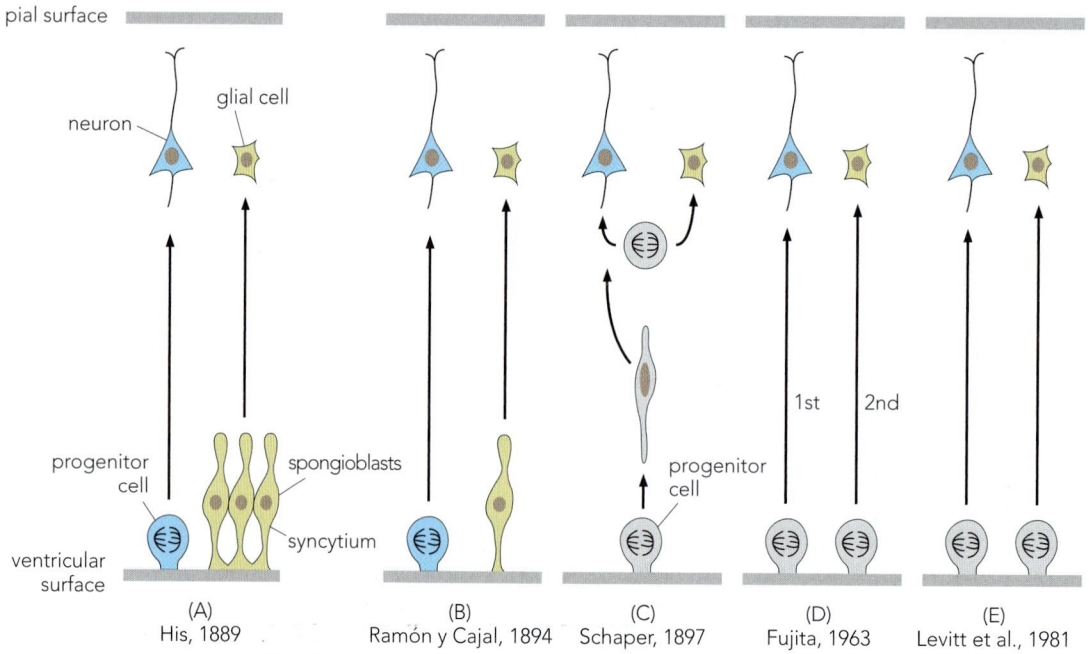

Figure 5.1 Evolving views on the origin of neurons and glia. (A) In 1889, Wilhelm His proposed that neurons and glia arise from two different populations located at the ventricular surface of the neural tube. Based on the specimens examined by His, he proposed that neurons (blue) arose from one cell population and that glia (yellow) arose from another cell population he called spongioblasts. The spongioblasts were thought to form an interconnected network of cells. (B) In 1894, Ramón y Cajal agreed with much of His's hypothesis, but suggested that neurons and glia each arose from individual cells and that the glia were not connected by a syncytium. (C) Alfred Schaper proposed in 1897 that neurons and glia arise from the same population of cells (gray). He described the cells as forming at the ventricular surface and migrating toward the pial surface, where they would become neurons, or would divide again before giving rise to a neuron (blue) and glial cell (yellow). (D) Setsuya Fujita also proposed that neurons and glia arise from the same cell population. In his 1963 model, the cells at the ventricular surface (gray) first divide to produce neurons (blue). Once all neurons are generated, the cells then produce glia (yellow). (E) The 1981 model by Pat Levitt and colleagues also shows that neurons (blue) and glia (yellow) arise from the same populations at the ventricular surface (gray); in this model, however, neurogenesis is *not* required prior to gliogenesis. (Adapted from Levitt P, Cooper ML & Rakic P [1981] *J Neurosci* 1:27–39.)

confirmed much of what His stated, but based on his own preparations, Cajal described the glia as individual cells that were *not* connected as a syncytial network. Subsequent reports from other scientists ultimately supported Cajal's conclusion that the ventricular surface was lined with individual cells and that glial cells did not form a connected network (Figure 5.1B).

This period also marked the start of a long-standing debate over whether the adult nervous system functioned as a reticular network or by the action of individual cells. Later, it was shown that, with the exception of the electrical synapses formed by gap junctions between adjacent neural cells, nervous system functions are carried out by individual cells rather than by a network. Thus, as histological evidence accumulated and microscopic techniques advanced to the level of electron microscopy, the idea of an interconnected syncytial network present during development or in the adult was dismissed.

While many scientists in the late nineteenth century thought that neurons and glia arose from two separate precursor populations, other scientists, including Alfred Schaper (1897), believed that precursor cells originating at the ventricular surface could form neurons or glia. Schaper proposed that the cells along the ventricular surface were a single population of uncommitted precursor cells that migrated out of the ventricular zone to become neurons or divided again outside the ventricular zone to form neurons and glia (Figure 5.1C). However, this idea contrasted with reports of two cell populations described by Cajal and His and therefore Schaper's idea remained unpopular at the time. Schaper's interpretation was ultimately shown to be correct, however, as studies in the 1960s (Figure 5.1D) and 1980s (Figure 5.1E) confirmed that a single population of cells gives rise to both neurons and glia. Further, neuroepithelial cells give rise to the radial glial (RG) cells that serve as **multipotent** progenitor cells and provide a scaffold for new neurons to extend upon. The RG cells first generate more neurons then give rise to glia. Thus, both neuroepithelial cells and RG cells are considered neural **stem cells**.

Precursor cell nuclei travel between the apical and basal surfaces

In the neural tube, cell bodies are observed at different distances from the ventricular to pial surfaces. Many scientists in the late nineteenth century believed that these cells represented different populations of neuroblasts settled in different layers. In contrast, Schaper reported that the cell bodies observed at the ventricular (apical) surface and the cell bodies seen closer to the pial (basal) surface are the same cells viewed at different phases of the cell cycle (**Figure 5.2**A). However, it was not until 1935 that experimental evidence first supported Schaper's proposal.

The morphological changes that occur during neural tube formation explain the origins of the terminology used to describe the apical and basal regions. These surfaces are first identified in the neural plate stage (Figure 5.2B). As the neural plate curves over to form the neural tube, the former apical surface is located surrounding the lumen (the future ventricles), while the former basal surface becomes the outer surface of the neural tube located just below the pial membrane.

In a series of experiments conducted in multiple species, Fred C. Sauer reported that cells located near the ventricular surface extend thin processes that contact both the apical and basal surfaces. The nuclei of these cells then travel through these processes as the cell cycle progresses. In order to monitor the cell cycle at different locations from the apical to basal surfaces, Sauer measured the size of the nucleus—a method that provides an indirect measure of the amount of DNA in a cell. From these

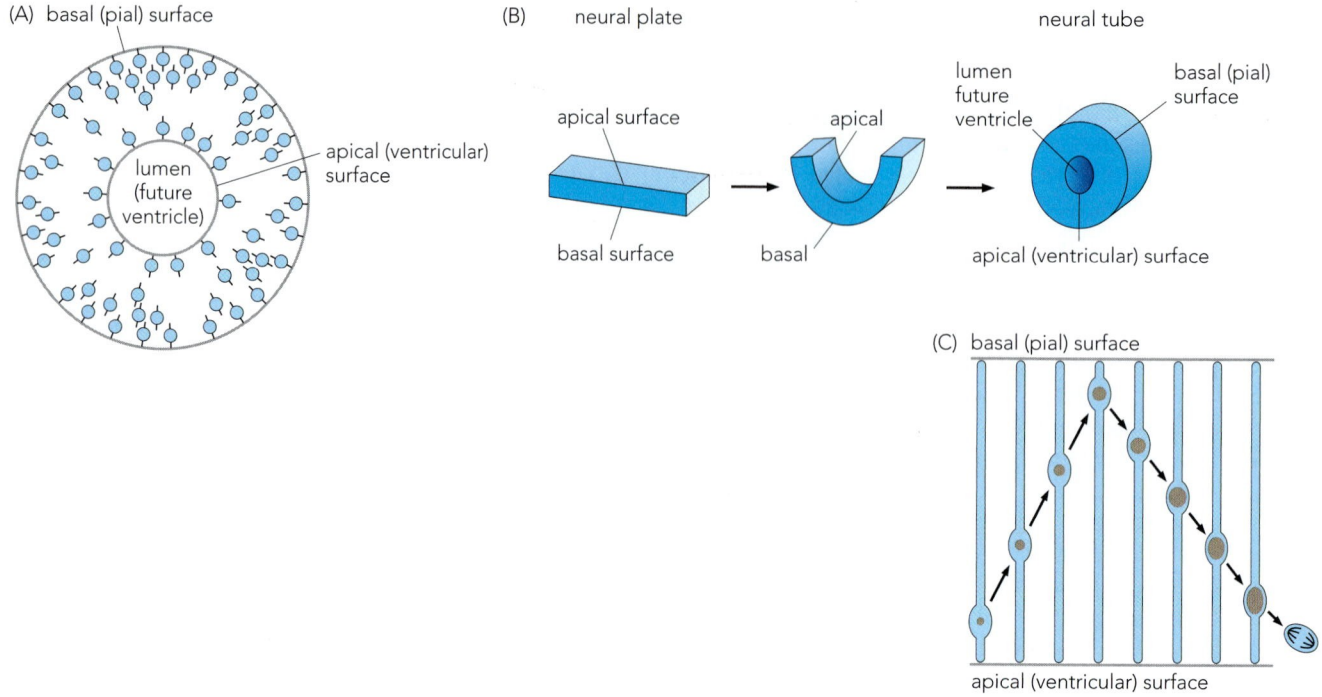

Figure 5.2 During the cell cycle, cells undergo interkinetic nuclear migration as they migrate from the apical to basal surface. (A) In the neural tube, cell bodies (blue) are located at various locations from the apical (ventricular) to the basal (pial) surface in the neural tube, appearing to form different cell layers. (B) Morphological changes account for the terminology used to describe the location of cells relative to the ventricular and pial surfaces of the neural tube. As the neural plate curves over to form the neural tube, the apical surface of the neural plate becomes the ventricular surface surrounding the lumen of the neural tube and the basal surface becomes the outer (pial) surface. (C) In 1935, F. C. Sauer was the first to report that the cells along the ventricular surface extend cytoplasmic processes that contact both the pial and ventricular surface. The nucleus of the cell then migrates through the cytoplasmic process using a mechanism called interkinetic nuclear migration. The nucleus first travels to the basal (pial) surface, then returns to the apical (ventricular) surface, where it undergoes mitosis (cell division). Sauer further noted that that the size of the nucleus (shown in brown), which provides an indirect measure of the amount of DNA, first increased in size as the cell approached the basal surface, then continued to increase as it approached the apical surface. (C, adapted from Sauer FC [1935] *J Comp Neurol* 62:377–405.)

measurements, he determined that the nucleus increases in size as it approaches the pial (basal) surface and further doubles in size again as it moves back to the ventricular (apical) surface. From these observations, Sauer suggested that the nucleus moves through the cytoplasm of the processes to reach the basal surface, and then returns to the apical surface to divide (Figure 5.2C). The movement of the nucleus through the cytoplasmic process was called intermitotic migration of nuclei or, as is commonly called today, **interkinetic nuclear migration**.

Interkinetic movements are linked to stages of the cell cycle

F.C. Sauer's work provided the first evidence of interkinetic nuclear migration, though the interpretation of his results remained controversial for some time. After his death in 1936, his wife, Mary Elmore Sauer, followed up on his work. In the 1950s she and colleagues conducted a number of experiments using nuclear stains and the newly developed tritiated thymidine method that confirmed nuclear migration was linked to phases of the cell cycle.

The proliferative **cell cycle** begins with the **G1 (gap 1)** stage, during which cells prepare for DNA synthesis. Upon exiting G1, the cells enter **S phase**, the DNA synthesis phase of the cell cycle. Once chromosomes are copied, the cells progress to the **G2 (gap 2)** stage, preparing for mitosis. The final stage of the cycle is **M phase**, when mitosis occurs and two daughter cells are generated from the precursor cell (**Figure 5.3**A).

At the 1959 meeting of the American Association of Anatomists, M. E. Sauer reported the pattern of nuclear migration in neural precursor cells in the neural tube of the chick. This report was one of the first neurobiology experiments to use the tritiated thymidine labeling technique developed in 1957. With this method a radioactive label (tritium) is incorporated into DNA during the synthesis phase of the cell cycle. Sauer's lab found cells undergoing DNA synthesis were concentrated at the pial surface (Figure 5.3B), as first suggested by the nuclear measurements made by F. C. Sauer in 1935. Subsequent studies confirmed that the nuclei of neurons in the developing neural tube that undergo interkinetic nuclear migration

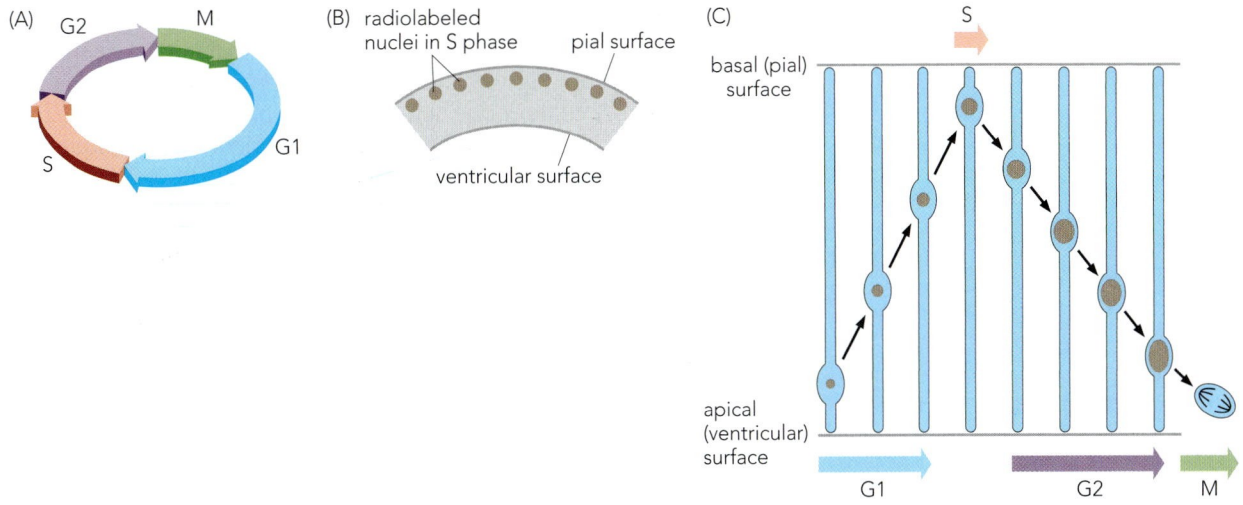

Figure 5.3 Interkinetic nuclear migration is linked to phases of the cell cycle. (A) The cell division cycle begins with gap 1 (G1) phase in which the cell prepares for DNA synthesis. The cell then enters the S phase, the phase in which DNA is synthesized. The cell progresses to the gap 2 (G2) phase, which prepares the cell for mitosis. The final stage is the M phase (mitosis) in which the cell divides to produce two daughter cells. (B) In 1959 M. E. Sauer demonstrated that cells that incorporated a radiolabel during the S phase of the cell cycle were located near the pial (basal) surface of the neural tube. (C) Together, the work of F. C. Sauer and M. E. Sauer confirmed that the translocation of the nucleus corresponds to different stages of the cell cycle. The cell enters the gap 1 (G1) phase of the cell cycle as its nucleus travels toward the pial (basal) surface. At the pial surface the cell enters the DNA synthesis (S) phase, then continues to the gap 2 (G2) phase as it travels back toward the ventricular surface. Once the nucleus reaches the ventricular (apical) surface, it enters the mitotic (M) phase in which the cell divides.

do so in a sequence linked to the stages of the cell cycle. Thus, nuclei of cells at the G1 phase of the cell cycle are located at the ventricular zone. When the migrating nucleus reaches the pial surface, the cell is in the S phase of the cell cycle, after which the nucleus travels back to the ventricular surface, where the cell enters the M phase and divides into two daughter cells (Figure 5.3C). The resulting daughter cells will either migrate away from the ventricular surface, or extend a process to the pial surface and continue through another series of interkinetic movements in order to divide again. As development progresses and more precursor cells are generated, the wall of the neural tube thickens so that the cell nuclei must travel increasingly greater distances to the pial surface. Once sufficient precursor cells are generated, the interkinetic movements cease.

The purpose of interkinetic movements is not fully understood. It is thought that the nucleus may be exposed to growth factors or other signals in the cytoplasm as it travels through the different regions of the cell process during the various stages of the cell cycle. It has also been suggested that the movement of nuclei into different layers provides more space for proliferation to occur in the very small region of the early neural tube. Although the exact purpose is not yet known, interkinetic movements in the developing neural tube are a highly orchestrated and precise mechanism that likely provides essential cues for further development and specification of cellular characteristics.

These unique and very precisely ordered patterns of interkinetic movement are seen only in regions of pseudostratified epithelium and the ventricular zone of the neural tube. Other sites of neurogenesis in the nervous system, including the subventricular zone of the cerebrum and external granule cell layer of the cerebellum discussed below, do not undergo nuclear translocation prior to cell division. Thus, the intriguing process of interkinetic movements is restricted to specific regions of the developing embryo.

The plane of cell division and patterns of protein distribution determine whether a cell proliferates or migrates

Following interkinetic translocation of the nucleus and its return to the ventricular surface, the precursor cell divides. The cleavage plane of cell division is critical in determining whether a cell continues to divide or migrates away from the ventricular zone. Cell division occurs along a vertical cleavage plane (perpendicular to the ventricular surface) or either a horizontal (parallel to the ventricular surface) or an oblique cleavage plane (at an angle to the ventricular surface).

With vertical cleavage, a cell divides symmetrically to generate two identical daughter cells that remain at the ventricular surface and undergo further proliferation (**Figure 5.4**A). The basal process is either split between the two daughter cells, or one cell inherits the basal process while the other produces a new process. In contrast, with a horizontal or oblique cleavage plane, a cell divides asymmetrically to generate two nonidentical daughter cells. The basal daughter cell (the cell closer to the pial surface) does not inherit the basal process and also loses attachment to the ventricular surface. This daughter cell then migrates away from the ventricular surface as an immature neuron. During this early stage of neurogenesis, the basal cells begin to express **proneural genes**, the genes that provide a cell with the potential to become a neuron. Later in development, additional genes will be expressed that establish the type of neuron that will form.

Meanwhile, the apical daughter cell (the cell closer to the ventricular surface) retains the basal process and remains attached at the ventricular surface, where it continues to proliferate (Figure 5.4B). In contrast to the basal daughter cell, proneural genes are repressed in the apical cells so that they continue to divide. Cells will continue to proliferate until enough neuronal precursors are generated. Later the RG stem cells will shift from producing neurons to producing the precursors of oligodendrocytes, followed by ependymal cells, and finally astrocytes.

Distinct proteins are concentrated at the apical and basal poles of progenitor cells

At the ventricular surface, the plane of cell division in a progenitor cell is regulated by many factors. Numerous proteins are segregated at the apical and basal poles of cells that function to mediate the attachment of cellular processes to the apical and basal surfaces, direct the orientation of the mitotic spindles, and regulate the expression of genes that allow for either continued proliferation or neural differentiation and migration. For example, adherens junctions are detected in apical cells at the ventricular surface. **Adherens junctions** contain the calcium-dependent adhesion molecule cadherin and are among the cellular elements required to maintain proliferation. Adherens junctions are not found in migrating basal daughter

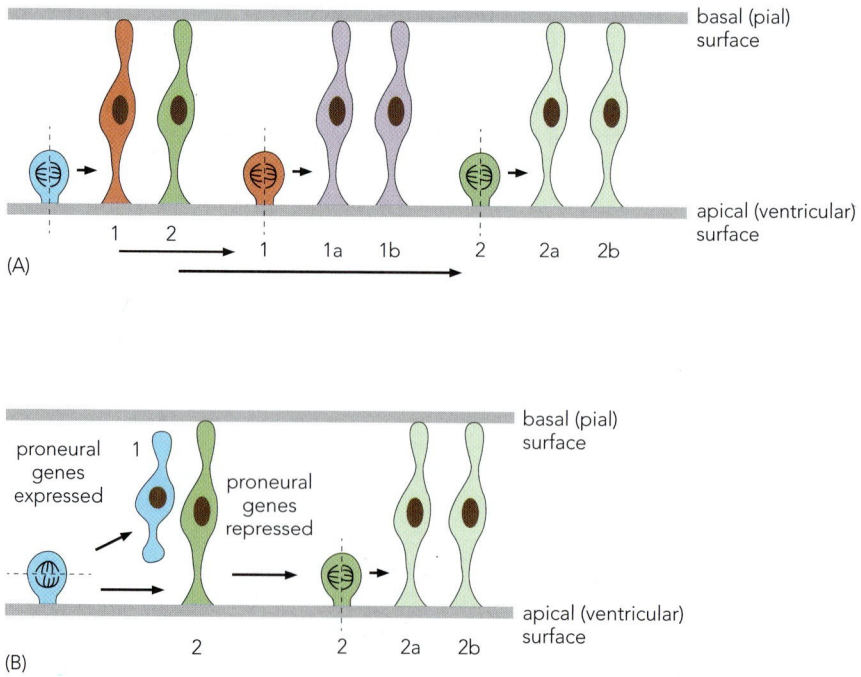

Figure 5.4 The plane of cell division determines whether a cell continues to proliferate or migrates out of the ventricular zone. (A) Vertical cleavage (dashed lines) occurs when the plane of cell division is perpendicular to the ventricular surface. In this case, the cell divides symmetrically, producing two daughter cells (cell 1 and cell 2) that remain at the ventricular (apical) surface of the neural tube. These cells continue to proliferate, producing new daughter cells (cells 1a, 1b, and cells 2a, 2b). (B) Horizontal cleavage (dashed lines) occurs when the plane of cell division is parallel to the ventricular surface. In such cases, the cell divides asymmetrically, producing a basal daughter cell that will migrate away from the ventricular region toward the pial (basal) surface (cell 1) and an apical daughter cell (cell 2) that will continue to proliferate at the ventricular (apical) surface (cells 2a and 2b). Similar events occur with an oblique cleavage plane. In the basal daughter, proneural genes begin to be expressed, whereas in the apical daughter cell, proneural genes are repressed.

cells that arise during asymmetrical cleavage. During asymmetrical cleavage, the basal cell down-regulates the expression of cadherins. The loss of this adhesion molecule leads to the detachment of the apical process and allows the basal cell to migrate away from the ventricular surface (**Figure 5.5**A, B).

The orientation of the mitotic spindle determines the position of the cleavage furrow and therefore establishes whether a cell undergoes symmetrical or asymmetrical division. In cells undergoing symmetric division, multiple proteins ensure that the mitotic spindles adhere to the lateral cell membrane. When the mitotic spindle is oriented for symmetrical cleavage, basal and apical polarity proteins are equally distributed in both daughter cells, so these are able to continue to proliferate (Figure 5.5A).

When the mitotic spindle is oriented in a horizontal or oblique plane, an unequal segregation of proteins into apical and basal poles is observed in the daughter cells. Initially, it appeared that horizontal cleavage was necessary for segregating apical and basal proteins, but it is now known that even a slight shift in spindle orientation away from the vertical plane provides enough of a difference in protein distribution to initiate asymmetrical division and to produce two unique daughter cells (Figure 5.5B).

Much of what is known about the distribution of apical and basal proteins in vertebrate cells was first discovered in *Drosophila*, where homologous proteins are found to regulate the formation of neuronal and nonneuronal cells (see Chapter 6). Among the proteins involved during vertebrate neurogenesis are the Notch receptors and their corresponding membrane-bound ligands. The receptors and ligands are expressed in all precursor cells, but at differing levels. The distribution of the apical and basal proteins within a daughter cell helps to establish the level of Notch receptor activity, and therefore determines whether a cell proliferates or migrates. Proteins unique to the apical pole of vertebrate CNS neurons

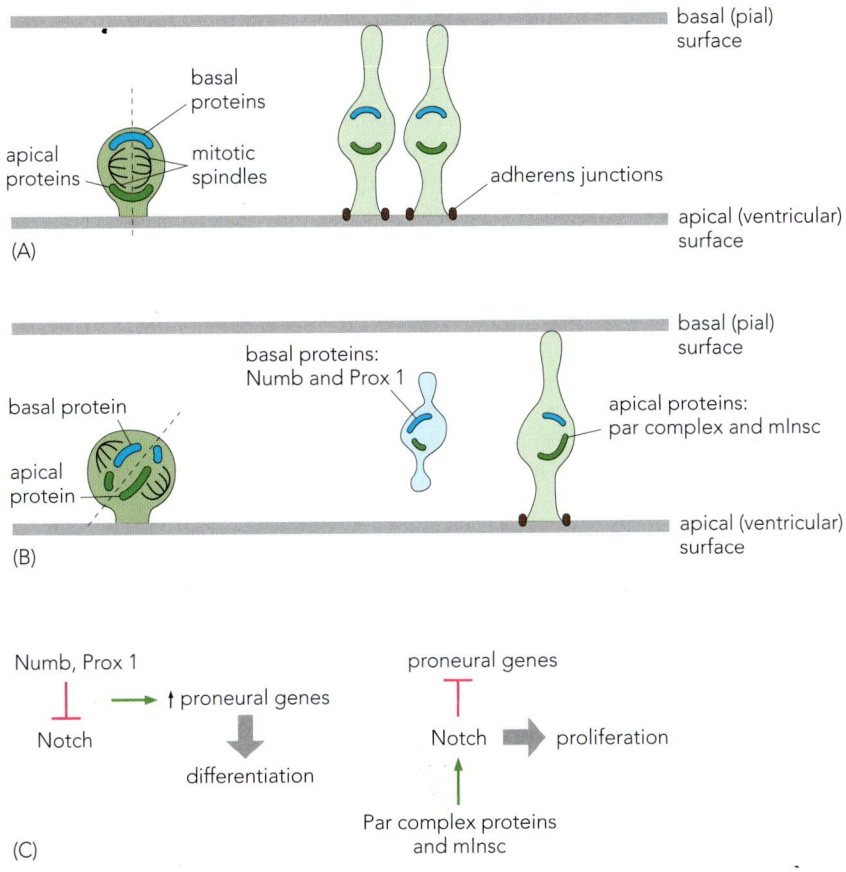

Figure 5.5 The distribution of proteins helps determine if a cell will continue to proliferate or migrate away. (A) In cells undergoing symmetrical (vertical) cleavage, the mitotic spindles are attached at the lateral walls of the cell and the proteins concentrated at the basal and apical poles of the cell become equally distributed in the resulting daughter cells. Both cells are attached at the ventricular surface by adherens junctions that are needed to retain a proliferative state. (B) In cells that undergo asymmetrical (horizontal or oblique) cleavage, the mitotic spindles are oriented differently and the basal and apical proteins end up unequally distributed in the daughter cells. The migrating basal cell will inherit a higher concentration of the basal proteins Numb and Prox 1, while the apical daughter cell will inherit a higher concentration of apical proteins, including those of the Par complex and mInsc (mammalian Inscuteable). (C) These differences in protein level lead to differences in the activity of the Notch receptor that is required for continued cell proliferation. In the basal daughter cell, Numb and Prox 1 are able to decrease the activity and expression of Notch, respectively. Decreased Notch activity in the basal daughter cell allows for the expression of proneural genes and cell differentiation. In the apical cell, the Par complex and mInsc allow the Notch receptor to have a higher level of activity. Notch receptor activity also leads to a suppression of proneural genes to help maintain the undifferentiated, proliferative state in that cell. Dashed lines indicate the plane of cleavage in A and B.

include those of the Par protein complex. Par proteins recruit additional proteins, including mammalian Inscuteable (mInsc) and AGS3 (activator of G-protein signaling 3). Together these proteins appear to influence the orientation of the mitotic spindle and help keep basal proteins, such as Numb and Prox-1 (prospero homeobox1), concentrated at the opposite pole of the cell.

With symmetric cell division, both daughter cells maintain similar levels of Notch receptor activity, so both cells continue to proliferate. With asymmetric division, however, the higher concentration of basal proteins in the basal cell interferes with Notch signaling in that cell. Numb, for example, binds to the intracellular region of the Notch receptor to inhibit its activity and prevent proliferation of the basal daughter cell. As the basal cell begins to migrate out of the ventricular region, Notch gene expression is down-regulated. Further, as cells begin to exit the cell cycle, *Prox 1* expression increases. *Prox 1*, like the *Drosophila* homolog *Prospero*, inhibits the proliferation of precursor cells by binding to the Notch promoter region, preventing Notch expression. Reduced Notch receptor activity or gene expression ends proliferation and allows for the expression of proneural genes in the basal daughter. In this way, the basal daughter cell is established as a neuronal precursor cell and begins the differentiation process (Figure 5.5C).

In the apical daughter cells that arise following asymmetric cleavage, the level of Numb protein is insufficient to interfere with Notch receptor activity. Therefore, Notch activity in that cell is able to suppress proneural gene expression and promote cell proliferation (Figure 5.5C). Proneural genes are needed to induce *Prox 1* expression, so Notch receptor activity also prevents *Prox 1* expression. Thus, Notch and *Prox 1* are able to cross-inhibit one another to carefully balance the number of proliferating and migrating precursor neurons to ensure the correct number of cells is produced at each phase of neurogenesis.

Further details on Notch signaling and the regulation of gene cascades that establish cell fate are described in Chapter 6. What is significant for this stage of development is that the distribution of specific proteins into apical and basal poles during cell division establishes the first step in determining whether a neural progenitor remains in the proliferative state or migrates away from the ventricular surface to begin the differentiation process.

The rate of proliferation and the length of the cell cycle change over time

The balance of symmetric and asymmetric cleavage changes over the course of development. Initially, symmetric cleavage predominates so that enough cells are produced. Such proliferation is particularly evident in the neural plate and early neural tube stages. Over time, as a sufficient number of neuronal precursors are generated, the need for production of new cells lessens. As this happens, more and more cells reach their final cell division (terminal mitosis) and migrate out of the ventricular zone to occupy specific regions of the developing neural tube.

Several studies have documented how the cell proliferation rate changes as development progresses. One means of determining the proliferation rate in a given tissue is to monitor the length of the cell cycle. Several methods have been developed for measuring cell cycle length, including tritiated thymidine, BrdU (bromodeoxyuridine, an analog of thymidine), and retroviral labels.

As cell cycle time lengthens, fewer cycles per day are possible, and fewer new cells are produced. In the developing mouse cortex, studies revealed that cells undergo about 2.5 to 3 cell cycles per day on day one of proliferation, but by day six, toward the end of this region's proliferative phase, the cells complete only one cycle per day. The length of the cell cycle, which was observed to last about 8 hours on the first day of proliferation, increased to 18 hours by the end of the proliferative phase. Similarly, in the developing chick optic tectum, the cell cycle was found to lengthen from about 8 to 15 hours over a three-day period of proliferation. In most regions of the CNS, it is the length of the G1 phase that increases over time, with the length of the other stages remaining fairly constant. In some regions of the CNS, however, it appears that the time spent in both G1 and S phase is lengthened. The purpose of the lengthening of cell cycle phases is not entirely clear. One hypothesis is that the longer G1 phase allows for exposure to growth factors that influence subsequent developmental events (**Box 5.1**). Many growth factor proteins, including members of the

Box 5.1 Cyclins and Cdks regulate progression through cell cycle stages

Specific cell cycle regulators act at each stage of the cell cycle to stimulate or inhibit progression to the next stage. Much of the current understanding of cell cycle regulation originated in work with model systems such as yeast and sea urchins. Scientists soon discovered that knowledge gained from these studies is broadly applicable to most vertebrate neurons as well.

Cell cycle regulation is primarily controlled by two families of interacting proteins: cyclins and cyclin-dependent kinases. Both families contain multiple members, including some that are not associated with the cell cycle.

The **cyclins** are a group of proteins that were initially identified on the basis of an unusual expression pattern: Levels of various cyclins oscillate throughout the different stages of the cell cycle. Cyclins were first discovered in the 1980s when scientists studying extracts of fertilized sea urchin eggs noted a decrease in a particular protein band that only occurred at the time of cell division. Although many other protein bands from the analyzed extracts increased gradually over time, this protein became more abundant as the cell cycle progressed, then decreased after mitosis (**Figure 1**).

Cyclin-dependent kinases (Cdks) are a family of serine/threonine protein kinases that must be activated by cyclins before they can phosphorylate downstream proteins to regulate progression through the cell cycle. Cdks are related to the *cdc* (cell division cycle) genes first identified in yeast. In contrast to the cyclical expression of the cyclins, the expression of the Cdks is stable throughout the cell cycle. It is the changes in

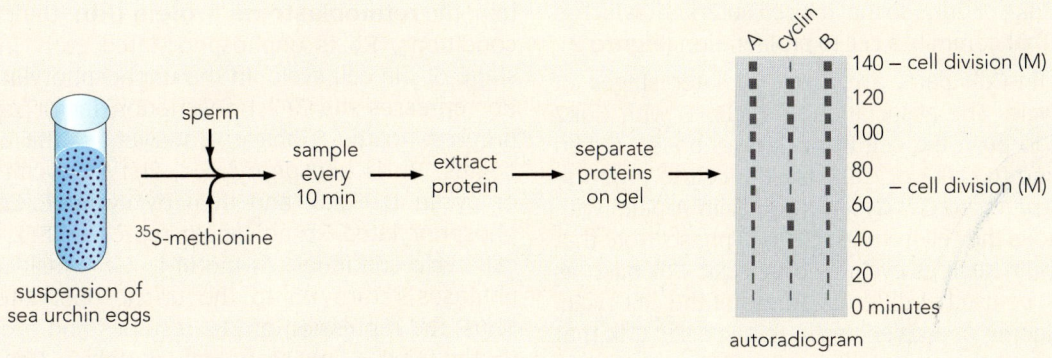

Figure 1 Cyclin proteins were identified in sea urchin eggs. Protein synthesis was measured from extracts of fertilized sea urchin eggs radiolabeled with ^{35}S-methionine. Samples were harvested at 10-minute intervals following fertilization and these extracts were then run on a gel. The density of the protein band represents the amount of protein synthesis. Most proteins gradually increased the amount of radiolabel, as shown by proteins A and B, indicating a linear increase in protein synthesis. However, one band showed a cyclical pattern of expression in which there was a gradual increase in radiolabel followed by a sudden decrease in signal intensity. This decrease always occurred at the mitotic phase (M) of the cell cycle. These proteins were named cyclins.

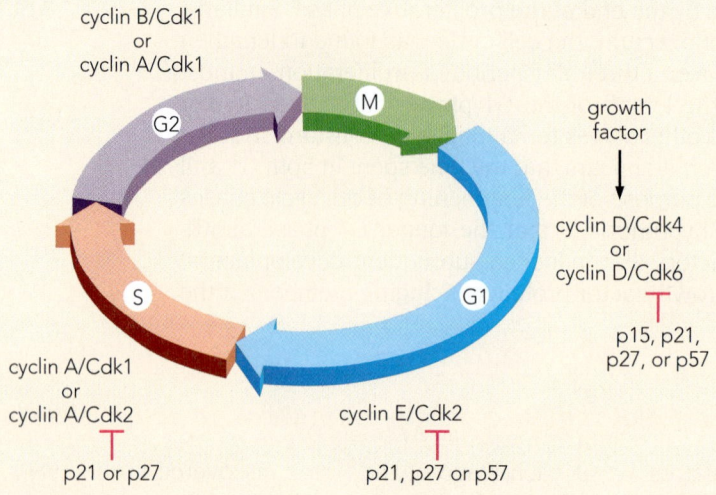

Figure 2 Cyclins and cyclin-dependent kinases regulate progression through the cell cycle. Cyclins are proteins that increase through the various stages of the cell cycle, then decrease after mitosis. Cyclin-dependent kinases (Cdks) are serine/threonine kinases that are activated by the cyclins to phosphorylate downstream proteins needed to drive the cell to the next phase of the cell cycle. At each cell cycle phase, specific cyclin/Cdk pairs are active. Growth factors activate cyclin D and Cdk4 or cyclin D and Cdk6 to initiate the G1 phase of the cell cycle. The cyclin E/Cdk2 pair drives the cell to the S phase. Cyclin A, coupled with Cdk1 or Cdk2 pushes the cell toward G2, from which cyclin B/Cdk1 or cyclin A/Cdk1 drives the cell to mitosis (M). The Cdk inhibitory proteins p15, p21, p27 and p57 can stop the progression of the cell cycle at different phases. All three of these Cdk inhibitory proteins can stop the activity of Cdk4 or Cdk6 associated with cyclin D in the G1 phase. In contrast, p21, p27 and p57 can halt the activity of Cdk2 associated with cyclin E or cyclin A at the G1 and S phases of the cell cycle.

cyclin levels that regulate the cyclical activity of these kinases.

Pairs of specific cyclins and Cdks are used at different cell cycle stages. A given cyclin binds to and activates an associated Cdk. Once activated, the Cdk influences subsequent substrate proteins, pushing the cell to the next phase of the cell cycle. For example, cyclin D activates Cdk4 or Cdk6 so that the cell enters G1, which is the phase that commits a cell to proliferation (**Figure 2**).

Other cyclin-Cdk pairs are active at later stages of the cell cycle. The association of cyclin E with Cdk2 is needed to push the cell from G1 to S phase. Cyclin A must bind to Cdk2 or Cdk1 for the cell to progress through S phase to G2. Cyclin B or cyclin A pairs with Cdk1 to move the cell from the G2 to M phase. Note that a given cyclin such as cyclin A or specific Cdk such as Cdk2, may be used at different stages of the cell cycle: It is the binding of a given cyclin to a specific Cdk that defines the pair's actions in the cell cycle.

The role of cyclin D in activating either Cdk4 or Cdk 6 at the G1 phase of the cell cycle is among the best-understood cyclin/Cdk interactions. As noted above, some Cdks can bind to more than one cyclin. However, Cdk4 and Cdk6 only interact with cyclin D, and the pairing of cyclin D with one of these Cdks is crucial for regulating cell proliferation. Cyclin D also appears to be unique in that it can be expressed continuously, as long as a growth factor is present to stimulate cyclin D synthesis. The specific growth factor required to stimulate cyclin D expression depends on the cell type and the stage of embryonic development.

The G1 stage is also impacted by another important protein, the **retinoblastoma protein (Rb)**. Under normal conditions, Rb is unphosphorylated early in the G1 stage of the cell cycle. In the unphosphorylated state, Rb represses the EF2 transcription factor needed to progress to the S phase. However, as the cell cycle begins, Rb is phosphorylated, first by cyclin D/Cdk4 or cyclin D/Cdk6, and then by cyclin E/Cdk2. The phosphorylated Rb no longer represses EF2, and the cell cycle continues. At the M-to-G1 transition, phosphatases return Rb to the unphosphorylated state. Thus, the repression of EF2 resumes and progression to the next S phase is halted unless Rb is again phosphorylated (**Figure 3**).

The importance of Rb regulation becomes apparent when the *Rb* gene is mutated. Mutations in *Rb* lead to continued cell proliferation, and thus tumor formation. *Rb* was first identified and named due to its role in a childhood retinal tumor. Children carrying two copies

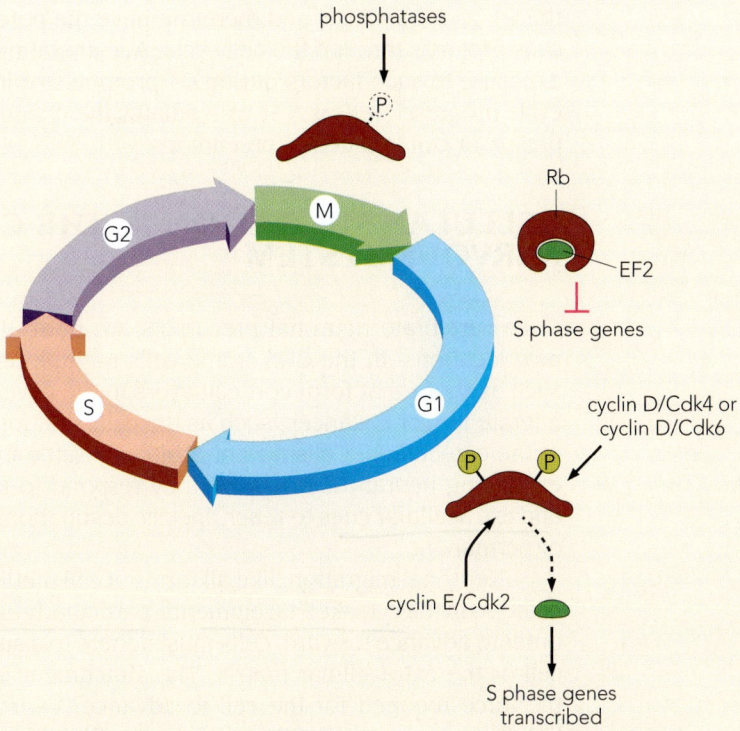

Figure 3 The retinoblastoma protein is used to modulate cell cycle activity. In the G1 phase of the cell cycle, retinoblastoma protein (Rb) is unphosphorylated and represses the activity of the EF2 transcription factor that is required for a cell to progress to the S phase of the cell cycle. As the cell cycle begins, Rb is phosphorylated first by cyclin D paired with either Cdk4 or Cdk6, then by cyclin E/Cdk2. Phosphorylated Rb no longer represses EF2, so the S-phase genes are transcribed and the cell progresses to the S phase of the cell cycle. In the M phase, phosphatases remove the phosphates and Rb returns to the unphosphorylated state, thus allowing it to again repress EF2 in the G1 phase. Mutations in *Rb* lead to excess cell production and tumor formation.

of the *Rb* mutation develop multiple retinal tumors in both eyes. Although Rb is often referred to as a single protein, there are actually several members of the Rb family, and each binds to specific transcription factors.

Additional proteins influence cell cycle progression and arrest

In all regions of the developing nervous system, cell proliferation must continue until the proper number of neural cells is generated. Cell cycle progression must also be inhibited at some point to prevent excess cell production, because too many neurons can be as disruptive to neural function as too few. Cell cycle progression can be halted by the action of **Cdk inhibitory proteins**, a group of small proteins that block Cdk activity. These Cdk inhibitors include proteins such as p15, p21, p27, and p57, with the number indicating the molecular weight of the protein. All of these Cdk inhibitors have the ability to inhibit the activity of Cdk4 or Cdk6 associated with cyclin D. In addition, p21, p27, and p57 inhibit Cdk2 associated with cyclin A and cyclin E (Figure 2). As discussed further in Chapter 6, these Cdk inhibitors not only negatively regulate cell proliferation, but also often link exit from the cell cycle with the onset of cellular differentiation.

The final number of neurons is achieved not only by regulating cell proliferation, but also through the process of naturally occurring, programmed cell death that takes place in the developing nervous system, a topic covered in Chapter 8.

Although this description of cyclin/Cdk pairs and substrate proteins provides only a simplified overview, it introduces the complex interactions needed to regulate the correct number of cells. The specific activities of cyclins, Cdks, Rb, Cdk inhibitors, and various interacting proteins continue to be investigated, both to shed light on the basic mechanisms of cell division during development and to uncover the cellular pathways that control tumor formation.

fibroblast growth factor (FGF), epidermal growth factor (EGF), and transforming growth factor α (TGFα) families, regulate cellular proliferation at the G1-to-S transition and therefore have the potential to impact whether cells progress through the cell cycle. Accumulating evidence suggests that exposure to such factors during G1 promotes or inhibits regulators of cell cycle progression (Box 5.1) or mediates the change from an uncommitted precursor cell to a neural precursor.

CELLULAR MIGRATION IN THE CENTRAL NERVOUS SYSTEM

Once vertebrate neuronal precursors are born, they begin to migrate to new locations. In the CNS, some cells migrate relatively short distances, such as those that form cortical layers in the developing forebrain. Others migrate longer distances, such as the cells that form the granule cell layer of the cerebellum. Cells do not arbitrarily settle in a nearby region of the expanding neural tube. Rather, they respond to a variety of intracellular and extracellular cues to reach specific destinations that may be temporary or permanent.

Neuronal migration, like all forms of cell motility, depends on dynamic changes in the cytoskeletal elements, particularly actin and microtubules. In order to advance forward, cells must adhere to a substrate, such as another cell or the extracellular matrix. The attachment is necessary to generate the force required for the cell to advance. Various proteins are used to link the cytoskeleton of the migrating cell to the substrate. The migration of neurons is similar to the advancement of the growth cone found at the tip of an extending axon or dendrite (see Chapter 7). Cell migration in the developing CNS and the extension of growth cones both rely on extracellular signals that allow for either passive growth or instructive guidance toward a target location. In many cases these extracellular signals initiate activity of intracellular Rho GTPase signaling pathways. The Rho family of small GTPases can activate effector proteins that then influence the actin and microtubule dynamics and cell adhesion required for cell advance. GTPases are activated by various guanine nucleotide exchange factors (GEFs) that convert GDP to GTP. GTPase activity is stopped by GTPase activating proteins (GAPS) that return Rho GTPase to the inactive GDP-bound form (**Figure 5.6**). Due to the ability of the activated Rho family of GTPases to influence cytoskeletal dynamics and cell adhesion, this intracellular pathway is often utilized in neuronal migration and growth cone advance, as detailed below and in Chapter 7, respectively.

The migration patterns of neurons in the mammalian neocortex and cerebellum are currently among the best characterized and most fully understood in the developing CNS. Therefore, these areas are highlighted to demonstrate similarities and differences in the cell migration

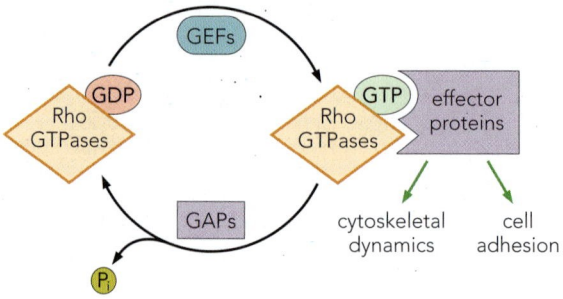

Figure 5.6 Rho family GTPases influence cell migration and adhesion. The Rho family of small GTPases are intracellular signals that are activated when guanine exchange factors (GEFs) exchange GTP for GDP. The activated Rho GTPases then influence various effector proteins, such as actin and microtubules, as well as cell adhesion molecules. The activity of Rho GTPases is stopped when GTPase activating proteins (GAPs) remove phosphate (Pi) and return the GTPases to the inactive GDP-bound state. Due to the ability of Rho GTPases to influence cytoskeletal dynamics and cell adhesion, these GTPases are often activated during neuronal migration.

mechanisms utilized. The cell migration mechanisms described below and the interkinetic nuclear migrations described earlier are entirely different. In this section of the chapter the migration of post-mitotic precursor neurons that have already undergone interkinetic nuclear migration and subsequent cell division are illustrated.

In the neocortex, newly generated neurons form transient layers

The adult mammalian neocortex is comprised of six cell layers (**Figure 5.7**), each of which is characterized by distinct cellular morphologies and projection patterns. In order to form these six layers, a number of transient layers are first formed to help guide migrating precursor cells out of the ventricular zone. These transient layers also guide the axons from other CNS regions that later synapse with neurons in the emerging cortical layers. In mice, these cortical layers form between embryonic days 11 and 18. In the developing human cerebral cortex, layering begins at about week six of gestation and continues through approximately week 20.

As early as the mid-1800s, scientists began detailing the layers of the cortex in both adult and embryonic tissues. As shown in **Figure 5.8**, a number of distinct embryonic layers form over time. Initially, the **ventricular zone (VZ)** consists of a simple pseudostratified neuroepithelial region with only a single layer of cells along the lumen of the neural tube (Figure 5.8A). However, as cells in the VZ undergo asymmetric cell division and some begin to migrate away, two layers arise. The VZ persists and the **preplate (PP)** forms adjacent to the pial surface (Figure 5.8B). The PP therefore contains the first neurons to migrate out of the VZ. As more cells leave the VZ, the cortex begins to thicken. Cells leaving the VZ migrate through an intermediate zone (IZ) that primarily contains axons arriving from neurons located in other regions of the developing nervous system, such as the thalamus. Eventually a new cell layer called the **cortical plate (CP)** is formed. With the formation of the CP, the cells of the former PP separate into two layers: the **marginal zone (MZ)** below the pial surface and the **subplate (SP)** adjacent to the IZ. Some subplate neurons are also found among the axons of the IZ (Figure 5.8C). Thus, the MZ and SP contain the first cells that exited the VZ, while the CP becomes the destination for most subsequent migrating precursor cells.

Each of these early formed layers of the neocortex—the MZ, CP, and SP—exhibit unique cellular characteristics. The MZ, which later forms layer I of the adult cortex, contains specialized, horizontally oriented cells named for Cajal and Gustaf Retzius, the two scientists who first described them in the 1890s. The **Cajal–Retzius (CR) cells** are thought to play a major role in helping to establish neuronal patterning in the CP. The neurons migrating into the CP are primarily the pyramidal excitatory projection neurons of the neocortex. The transient layer of subplate (SP) neurons (Figure 5.8C–E) associated with the IZ does not persist into adulthood, but SP neurons are thought to act as temporary guides and contacts for axons that arrive before the CP cells are present. This has been particularly well documented in the primary visual cortex. For example, in the cat, axons of cells arriving from the lateral geniculate nucleus (LGN) of the thalamus wait in the SP region of the IZ for about three weeks before their cortical target neurons reach the developing CP. Thus, the SP cells seem to attract and maintain the axons of LGN neurons until the correct target neurons are available.

In addition to the VZ, a second layer of proliferative cells forms between the VZ and the IZ. This **subventricular zone** (SVZ) lies adjacent to the VZ,

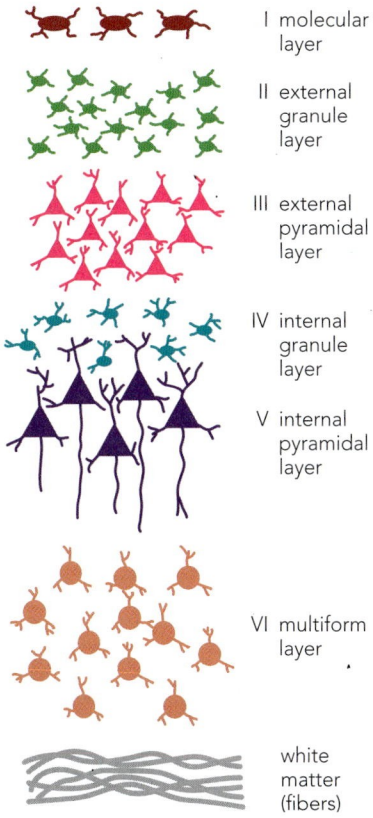

Figure 5.7 The adult cerebral cortex is organized into six layers. The cerebral cortex of mammals is characterized by six defined layers composed of different cell types, each with distinct projection patterns. Numbered consecutively from the pial (basal) surface to the ventricular (apical) surface are the molecular layer (layer I), the external granule layer (layer II) and external pyramidal layer (layer III), the internal granule layer (layer IV) and internal pyramidal layer (layer V), and the multiform layer (layer VI). The fibers of the white matter are located beneath layer VI.

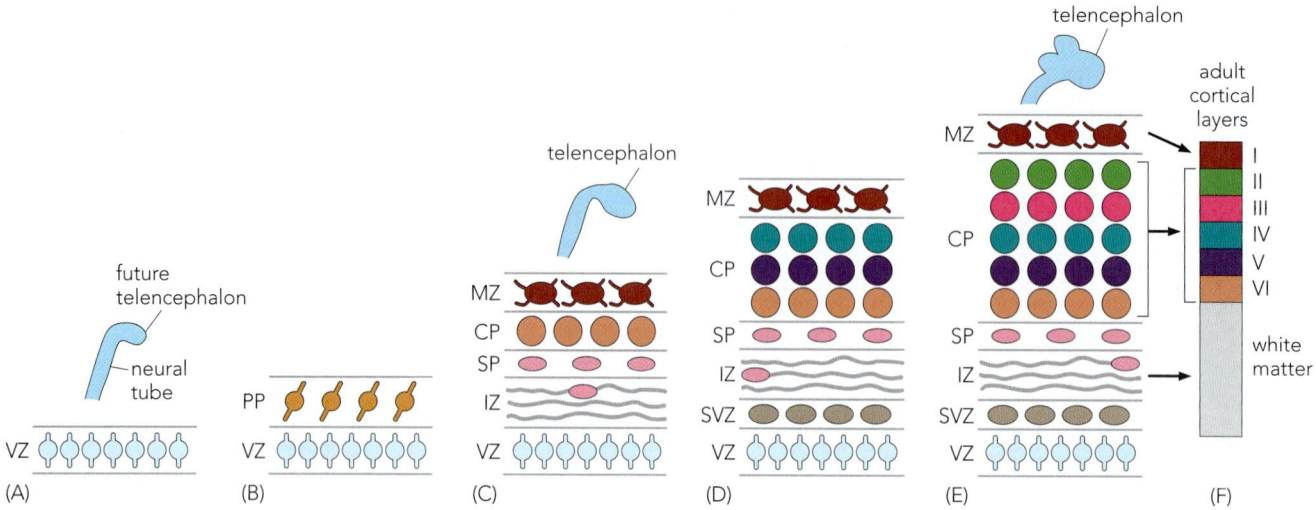

Figure 5.8 Transient embryonic layers are formed prior to the adult layers. Before the six layers of the adult cerebral cortex form, several transient layers emerge that provide guidance cues to migrating neurons and incoming axons. The formation of these transient layers is illustrated in panels A–E. Above panels A, C, and E, the corresponding stage of neural tube development is shown. (A) Initially the neural tube contains only the ventricular zone (VZ), consisting of a single layer of cells along the lumen of the ventricle. (B) As these proliferative cells divide, some daughter cells begin to migrate away and form a second layer called the preplate (PP). (C) With more and more cells migrating away from the VZ, the PP expands and generates a new layer called the cortical plate (CP). The formation of the CP splits the former PP into a marginal zone (MZ) closest to the pial surface and a subplate (SP) zone adjacent to the intermediate zone (IZ). The IZ is a layer that contains the axons of neurons originating in other areas of the developing nervous system as well as some subplate neurons. (D and E) As development continues, some cells from the VZ settle in a second zone of proliferation called the subventricular zone (SVZ), which lies adjacent to the VZ. Other cells from the ventricular zone migrate through the SVZ, IZ, and SP to reach the CP, where they form additional layers. Ultimately, five layers of cells are found in the CP region. (F) The transient layers of the developing cerebral cortex give rise to different regions of the adult cortex. The MZ will go on to form layer 1 of the adult cerebral cortex, while the CP will give rise to neurons of the other five layers (II–VI). The IZ will contribute to the white matter in the adult. With the exception of the SVZ that remains in some regions of the adult CNS, the other transient layers are lost as neuronal migration ends.

just one layer further from the lumen of the neural tube (Figure 5.8D, E). Therefore, some of the cells migrating out of the VZ settle in the SVZ, where they continue to divide. The VZ eventually disappears as more and more neurons exit that layer, and the SVZ then becomes a primary source of additional neurons and most of the glial cells of the cortex. Proliferation continues in the SVZ through the middle and late stages of cortical development. In the mouse, this proliferation continues into the second postnatal week, whereas in humans the proliferation continues at least through the second year of life. Most areas of the SVZ will also decrease in size as proliferation ceases. However, the SVZ persists into adulthood in the region of the lateral ventricles. In rodents, these cells have been shown to migrate to populate regions of the olfactory bulb and may contribute to olfactory function throughout life.

With the exception of the MZ that becomes layer I of the mature cerebral cortex, none of the transient layers described above has counterparts in the six neuronal layers of the adult cortex. However, as described below, the cells that migrate into the CP will go on to form layers II through VI of the adult cortex (Figure 5.8F). The axons of the IZ will contribute to the white matter that lies below the cortical layers.

Most neurons travel along radial glial cells to reach the cortical plate

Studies reveal two primary mechanisms by which neurons reach the PP and CP: somal translocation and migration along radial glia cells. Both forms are a type of radial migration in which neurons travel in a radial direction from the ventricular surface. Many neurons that are generated early in development reach the PP and MZ by **somal translocation**. In this mode of migration, a cell in the VZ extends a process to reach the pial surface. The cell body then migrates through its attached process toward the pial surface while also detaching from the ventricular surface. These early born neurons thus reach the PP as if being pulled to the pial surface (**Figure 5.9**)

As the cortex continues to thicken, somal translocation is no longer used. Instead, the primary mechanism by which later-generated neurons reach the CP is by traveling along **radial glial (RG) cells**. The RG cell body resides in the VZ but extends a long process to the pial surface of the neural tube. The RG are believed to be the first cells to differentiate in the cerebral region and act as scaffolds for the neurons that form subsequently. A newly generated neuron attaches to the vertically oriented process of an RG and crawls along this process toward the CP. As the neuron reaches the CP region, it enlarges, its soma becomes more globular in appearance, and the strength of its attachment to the RG decreases, allowing the neuron to settle into its layer and begin to make axonal connections (**Figure 5.10**).

The hypothesis that RG direct neural migration was first proposed by His in 1889. Cajal's work supported this hypothesis and suggested that the RG later differentiate into astrocytes. Only in recent decades, however, have scientists confirmed that astrocytes arise from RG. The RG are now recognized as a multipotent progenitor population that gives rise to neurons and most glia in the embryonic and perhaps the adult nervous system.

Migrating neurons adhere to RG through protein–protein interactions. These include the adherens junctions and glycoproteins such as **astrotactin**, which localizes to the neuronal cell surface and acts as a ligand for the corresponding receptors found on the RG of the cerebral cortex. Astrotactin was first characterized in the cerebellum by Mary E. Hatten and colleagues in 1988. In the cerebellum, astrotactin also facilitates neuronal migration on specialized glial cells, as described below.

Other molecules, particularly receptors of the integrin family, are expressed on neurons to mediate their attachment and migration along the glial fibers. Integrin receptors are heterodimers comprised of two subunits, designated α and β. There are several different α and β subunits and the particular combination of subunits determines binding specificity. For example, the α3β1 receptor combination binds to the extracellular matrix protein

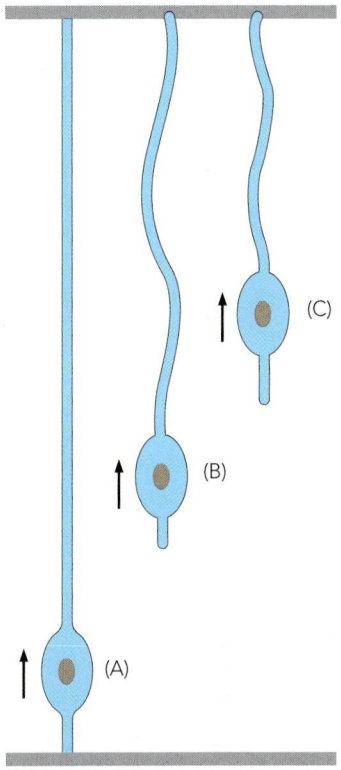

Figure 5.9 Somal translocation is one form of radial migration. Neurons generated during the earliest stages of cerebral development use somal translocation to migrate toward the pial surface. With somal translocation, the cell body is pulled toward the pial surface. First a cell in the ventricular zone extends a process to reach the pial surface (A). The attachment at the ventricular surface is lost (B) and the cell body is pulled toward the pial surface as it migrates through the remaining process (C), until it ultimately reaches the pial surface.

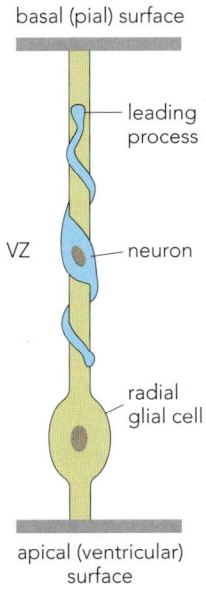

Figure 5.10 Neurons also migrate out of the ventricular zone by attaching to radial glial cells. A second form of radial migration requires radial glial cells (RG, yellow). The cell bodies of radial glial cells are found in the ventricular zone (VZ). Each RG cell extends a long process to reach the pial surface. Neurons (blue) migrating out of the VZ attach to the RG and crawl along the RG process. The neuron's leading process is used to help guide the cell toward the pial surface. Proteins present on the neurons and RG mediate the attachment, migration, and release of the neurons at the proper stage of development.

Table 5.1 Examples of selected integrin subunit pairs that bind to various extracellular matrix proteins encountered by neurons *in vivo*.

Integrin receptor subunits	Extracellular matrix molecule
α3β1	Laminin
α5β1	Fibronectin
α6β1	Laminin
α7β1	Laminin
α6β4	Laminin
α1β1	Laminin, Collagen
α2β1	Laminin, Collagen

Based on Reichert LF & Tomaselli KJ [1991] Extracelllular matrix molecules and their receptors: functions in neural development. *Annu Rev Neurosci* 14:531–570.

laminin. During cortical development the β1 subunit is highly expressed along with the α3, α5, or α6 subunits. These subunit combinations form receptors for the widely distributed extracellular matrix proteins laminin and fibronectin (**Table 5.1**). Differences in the integrin receptor subunits expressed on the migrating neurons provide a means for these cells to differentiate among the available extracellular proteins. In at least some regions, integrin receptors are needed for the neurons to attach to RG cells. To test the necessity of these receptors in cortical migration, subunits of the integrin receptors have been experimentally manipulated. In some studies, mutant mice lacking one or more subunits were generated. In others, antibodies that block the function of a particular integrin subunit were added to cultures of embryonic cortical neurons. These various studies confirmed that decreasing or removing the integrin subunits available to cortical neurons interfered with their ability to migrate normally. Together, the studies reveal the necessity of proper integrin receptor expression for developing cortical neurons to recognize the appropriate cues on RG and other cell surfaces.

Cells in the cortical plate are layered in an inside-out pattern

As cells migrate out of the VZ and SVZ to form layers in the CP, an interesting developmental pattern emerges known as "inside-out" or "inside first, outside last" patterning, because cells settle in the innermost layers of the CP prior to the outermost layers—a pattern that at first may seem counterintuitive. To form the layers of the CP, the first neurons to exit the cell cycle and leave the VZ settle in what will become the deepest layer of the CP, the future layer VI of the adult cerebral cortex (**Figure 5.11**A). As development continues, later-born cells divide and migrate past these earlier-born neurons to settle in a more superficial layer of the CP (Figure 5.11B–D). This pattern continues until the latest-born neurons originating in the SVZ reside in the outermost layer of the CP, the future layer II of the adult cerebral cortex, closest to the MZ (Figure 5.11E)

As neurons arrive in the deepest layers (that is, future layers VI and V), they begin to form connections with other neurons at the same time that cells are migrating to more superficial layers (for example, future layers III and II). Thus, multiple developmental events overlap each other, and the nervous system integrates these events simultaneously.

Inside-out patterning in the cerebral cortex was first reported in mice by Angelvine and Sidman. In their 1961 study they used tritiated thymidine to map cellular locations at different developmental stages in the developing rodent brain. Numerous studies throughout the 1960s

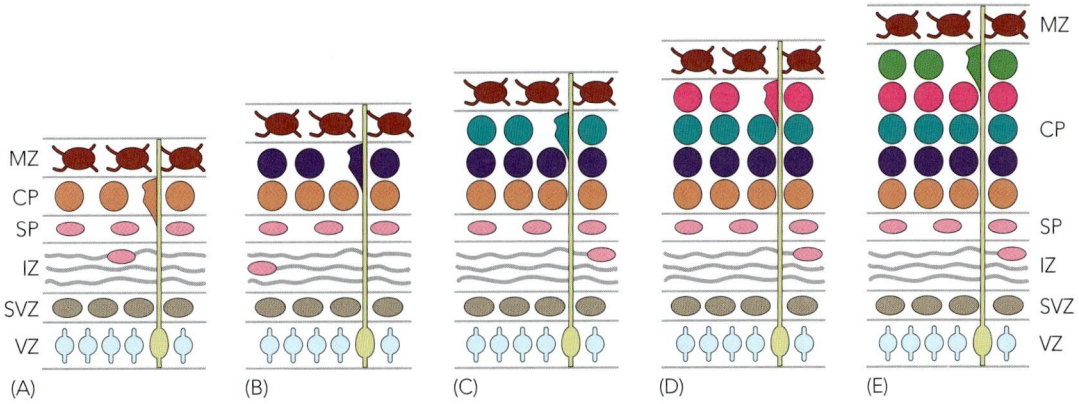

Figure 5.11 Cerebral cortical neurons rely on inside-out patterning to form layers. (A) As neurons leave the ventricular zone (VZ) by attaching to radial glial cells (RG, yellow), they begin to accumulate in the cortical plate (CP), detaching from the RG prior to reaching the marginal zone (MZ). The first cells to arrive at the CP will form the deepest, innermost layer (future layer VI). (B) The next layer of neurons (future layer V) will migrate past the first layer to form a new layer closer to the pial surface. (C) The third layer of cells to form (future layer IV) will migrate past the first and second layers. Similarly, the next layers (future layers III and II) will continue to migrate past existing layers until all six cortical layers have formed using this inside first, outside last pattern (D and E).

and 1970s helped further define the temporal sequence of events. For example, in 1965 Berry and Rogers confirmed this patterning in rats and in 1974 Rakic reported a similar migration pattern in primates. Thus, inside-out patterning is conserved across mammalian species. However, in other vertebrates, such as reptiles, which have fewer cortical layers, the pattern of neuronal migration is reversed and occurs in an "outside-in" manner. That is, the first cells to exit the VZ settle in the most superficial layers.

Changes in cortical migration patterns lead to clinical syndromes in humans

It is unclear why mammalian cerebral cortical neurons rely on an inside-out pattern of migration. Evolutionary change in the complexity of the cerebral cortex is one hypothesized reason. The early-born neurons that reach the PP and deep layers of the CP are thought to be phylogenetically older than the neurons that migrate to the more superficial layers of the CP. Whatever the underlying basis for this pattern of cell migration, inside-out patterning is necessary for normal development. Defects in cortical migration patterns cause congenital disorders in humans. Disruptions of the cortical migration patterns can result in mental retardation, seizures, and in some cases death. Some defects in neuronal migration result in **lissencephaly** ("smooth brain"), a condition in which the outer surface of the brain lacks the typical cortical gyri (ridges) and sulci (furrows). Rather than exhibiting an inside-out pattern of migration, the neurons in a lissencephalic brain stack one below another (**Figure 5.12**A). There are a number of genetic mutations that can cause lissencephaly, including mutations in the gene that codes for the protein reelin, described below. The clinical outcomes associated with lissencephaly are quite varied. Some children demonstrate near-typical developmental milestones and intelligence. Others have severe mental retardation, suffer frequent seizures, and die within the first decade of life.

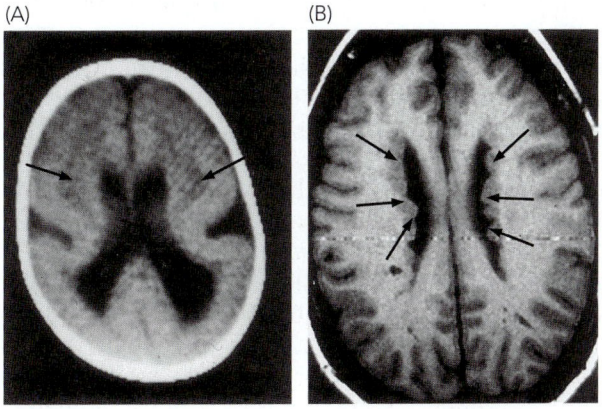

Figure 5.12 Defects in neuronal migration results in congenital disorders in humans. (A) A magnetic resonance imaging (MRI) scan illustrating the smooth brain appearance characteristic of lissencephaly. Compared to a typical brain or a brain with periventricular heterotopia (PH), the outer surface of the brain appears smooth and lacks the characteristic gyri and sulci. In addition, the ventricles are enlarged and some areas of misplaced neurons are observed in the white matter (arrows). (B) A brain scan illustrating PH. The arrows indicate nodules of neurons that cluster along the ventricles. (From Gressens P [2000] *Pediatric Res* 48:725–730.)

Another human neurological disorder resulting from altered neuronal migration is **periventricular heterotopia (PH)**. In this condition, nodules of neurons form throughout the periventricular region because these cells never leave the ventricular zone to migrate to the appropriate cortical layer. The nodules form in bilaterally symmetrical patterns and appear to be highly differentiated neurons located in the wrong regions of the brain (Figure 5.12B). The highly unusual pattern of cell formation results in a range of deficits. Many individuals experience epileptic seizures but exhibit no other neurological symptoms. In a less common form, PH is caused by an X-linked dominant mutation. The deficits in males that inherit this mutation are much more severe, because males have only one copy of the X chromosome and are therefore unable to compensate for the mutation. Many of these males die before birth due to the severity of the malformations that arise. Changes in proteins known to have roles in cell motility, migration, and adhesion are among those thought to underlie these clinical disorders.

The *Reeler* mutation displays an inverted cell migration pattern

In the 1950s, mice were described with a naturally occurring mutation that results in an easily observable gait disorder. The mutant mice are called reeler mice because they move in an unsteady, staggering fashion. The *Reeler* mutation is a recessive autosomal mutation that is only found in mice. Among the structural changes noted are defects in cell migration patterns in the cerebral cortex, hippocampus, and cerebellum. No known human homolog of this specific genetic mutation has been identified, although the clinical syndromes described above reveal some similarities in regard to cell migration patterns.

In the cerebral cortex, the *Reeler* mutation interferes with the normal association between migrating neurons and the RG. At early stages, cells in reeler mice form, undergo terminal mitosis, and begin to migrate out of the VZ—a sequence of events similar to what occurs in wild-type mice (**Figure 5.13**A, B). However, the cells do not migrate into the PP to form the CP and SP. Instead, cells that would normally form the subplate are found in the marginal zone with the CR cells. Together these cells form a unique layer called the **superplate** (Figure 5.13C, D). As additional neurons migrate out of the VZ, a disorganized CP forms with these later-migrating neurons unable to move past the existing layers. Thus, the newer layers continue to stack up, one below another (Figure 5.13C, D). The result is an "outside-in" pattern of neuronal migration, in which the first cells, those that become layer VI, migrate until they are stopped in the superplate. Subsequent cells stack up below this first layer so that layers V–II are arranged in an inverted pattern.

In reeler mice, even though the cortical layering is inverted, the overall gross organization of the brain appears relatively normal, and cell survival is not impacted. Further, cells of the peripheral nervous system and other organ systems appear unaffected, indicating that the *Reeler* mutation impacts specific cellular interactions in the developing CNS.

Cajal–Retzius cells release the protein Reelin, a stop signal for migrating neurons

Because RG function primarily as a passive system for neuronal attachment and migration, additional signals must be available to direct migrating

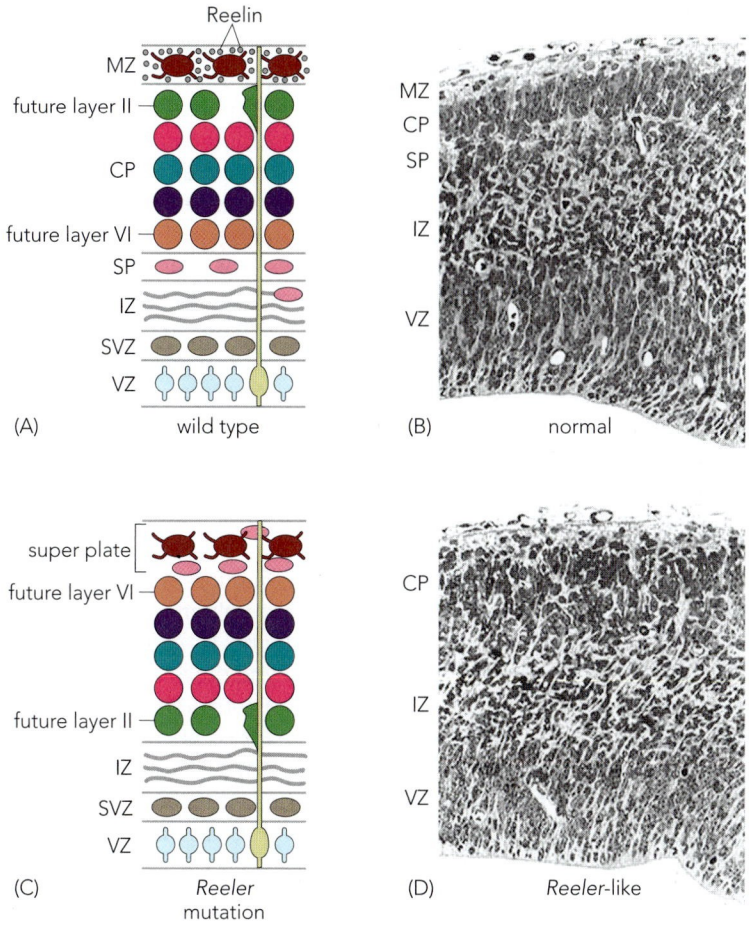

Figure 5.13 Loss of Reelin results an inverted pattern of neuronal migration. (A) A schematic diagram illustrates that in wild-type mice, the Cajal–Retzius (CR) cells of the marginal zone (MZ) produce the protein Reelin, which acts as a stop signal to prevent neurons of the cortical plate (CP) from entering the MZ. The Reelin signal instructs neurons to leave the RG (yellow) and settle in the correct layer of the expanding CP. Because the expanding cortical plate pushes the MZ, and thus the Reelin signal, outward, the newly arriving neurons are layered in the inside first, outside last pattern. (B) Photomicrograph of the cerebral cortex of a wild-type mouse. (C) The normal "inside-out" pattern of neuronal migration is reversed in mice with the *Reeler* mutation. In these mice the CR do not produce Reelin, so the cells from the ventricular zone (VZ) and subventricular zone (SVZ) continue to migrate as far as they can toward the pial surface. The cells of the subplate become mixed with the CR in the MZ, forming a layer called the super plate. Neurons that normally form the CP then migrate out of the VZ and travel as far as they can until existing layers of neurons prevent further migration. Therefore, the first cells to migrate out of the VZ (future layer VI) migrate until they are stopped at the super plate. Neurons of future layer V then migrate until they are stopped by this layer of neurons. This pattern continues, with each subsequent layer stacking up beneath the existing layers. Thus, the absence of Reelin results in CP layers that are organized in an "outside first, inside last" pattern, an orientation opposite of that found in the wild-type mice. (D) A photomicrograph of the cerebral cortex in a *Reeler* mutant mouse that demonstrates the disorganized MZ and CP that arises as a result of the inverse pattern of neural migration. (Adapted from Tissir F & Goffinet AM [2003] *Nat Rev Neurosci* 4:496–505.)

cells to detach from RG and settle in the proper cortical layer. Neurons may utilize several cues to initiate detachment and complete their migration. Experimental evidence accumulated over the years indicates that the RG surface changes as the cells reach the CP and that the axonal fibers from neurons already in the CP release extracellular signals that promote detachment from the RG.

Other evidence suggested cells adjacent to the CP provide the signal to end migration. For example, as scientists continued to investigate the *Reeler* mutation and cortical layering, a hypothesis began to emerge that the CR cells found just below the pial surface are necessary to halt migration. Unlike the rest of the cells in the emerging cortex, CR cells have a horizontal orientation with cell bodies and dendritic processes parallel to the pial surface. Under normal conditions, as new neurons migrate into the CP, the MZ and its CR cells are pushed further outward as the cerebral cortex thickens. Thus, the CR cells are always in the outermost layer, in a position to provide a stop signal to migrating cells.

In 1995 two teams of researchers succeeded in identifying a gene and protein expressed in CR cells. The gene was found to code for an extracellular glycoprotein that was termed **Reelin**. Mice lacking Reelin have the same phenotype as observed with the *Reeler* mutation. Without the Reelin signal, neurons migrate as far as they can, so that early-migrating neurons reach the outermost pial surface before they detach. Subsequent neurons migrate until they are stopped by an existing layer of

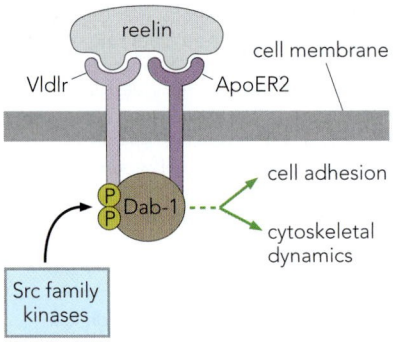

Figure 5.14 The Reelin signaling pathway influences cell adhesion and cytoskeletal dynamics. The Reelin protein binds to two low-density lipoprotein receptors, called the very-low-density lipoprotein receptor (Vldlr) and apolipoprotein E receptor 2 (ApoER2). To influence cell adhesion or cytoskeletal dynamics, Reelin is first internalized by the receptors. This leads to activation of the Src (sarcoma) family of kinases that in turn phosphorylate and activate the adaptor protein Disabled-1 (Dab-1). Dab-1 then activates signaling pathways that influence the expression of cell adhesion molecules that help the migrating neurons attach to the Cajal–Rezius (CR) cells. Dab-1 also influences cytoskeletal dynamics in migrating neurons to help attract the leading edge of migrating neurons to the marginal zone.

neurons, resulting in an "outside-in" pattern of neuronal layers similar to that observed in the *Reeler* mutation (Figure 5.13).

Subsequent studies suggested that the role of Reelin is more complex. The protein may actually provide both "stop" and "go" signals to the migrating neurons, depending on the timing and location of Reelin expression, and in some instances appears more effective in mediating somal translocation rather than radial glia-guided migration. Reelin has also been shown to have other functions, including orienting and stabilizing the cytoskeletal elements of neural processes. For example, once the cells reach their final position, Reelin influences the maturation of the dendritic processes and spines. Reelin may also attract the leading process of a migrating neuron and help attach, or anchor, that process to the MZ.

Reelin exerts its effects by binding to two low-density lipoprotein receptors: the very-low-density lipoprotein receptor (Vldlr) and apolipoprotein E receptor 2 (ApoER2). These receptors are enriched in the processes of cells nearing the MZ. Reelin is internalized by the receptors, leading to the activation of Src (sarcoma) family kinases that phosphorylate, in turn, the adaptor protein Disabled-1 (Dab-1) that is attached to the intracellular domains of Vldlr and ApoER2 (**Figure 5.14**). In migrating cortical neurons, Dab-1 then activates signaling pathways that influence the expression of cell adhesion molecules such as neural cadherin (N-cadherin) to promote interactions between the migrating neurons and the CR cell. Dab-1 activation has also been associated with mediating cytoskeletal dynamics, thus providing a means for Reelin released by the CR cells to attract the leading edge of a migrating neuron to MZ (Figure 5.14). In the absence of reelin, pyramidal cells in the neocortex and hippocampus lose the vertical orientation seen in wild-type mice and extend dendrites that are disoriented in several directions. Again, because reelin is thought to attract a migrating neuron's leading process, the dendrites are unable to orient properly in its absence.

Mice lacking both receptors have the same morphological and behavioral changes as those observed with *Reeler* and *Reelin* mutations. However, lack of only Vldlr or ApoER2 causes a milder phenotype. Studies suggest that the ApoER2 receptor is more important for the layering of neurons in the neocortex, whereas Vldlr is more important for guiding neuronal migration in the cerebellum. These differences may be mediated by differences in how the receptors process Reelin upon internalization.

Reelin expression in vertebrates has increased over the course of evolution. Comparisons of *Reelin* expression in different vertebrate species reveals that *Reelin* expression is increased in the mammalian CNS, both in the number of *Reelin*-expressing cells and the level of *Reelin* expressed in each cell. This suggests that reelin may have an expanded role in the development of the mammalian nervous system.

Cortical interneurons reach target areas by tangential migration

The inside-out patterning of cortical layers in the neocortex has been studied since the 1960s, and the importance of RG in guiding migrating neurons has been known since the 1970s. While other migration pathways and mechanisms had been proposed as early as the 1960s, it was not until the 1990s that other modes of migration were studied in detail. It is now known that the radial migration patterns described above are primarily utilized by the glutametergic, excitatory projection neurons (pyramidal neurons) of the cerebral cortex. These neurons arise from the VZ in the dorsal region of the telencephalon and comprise about 80% of the neurons in the neocortex. In contrast, the inhibitory, GABAergic interneurons that make up a much smaller percentage of neocortical neurons, arise from the

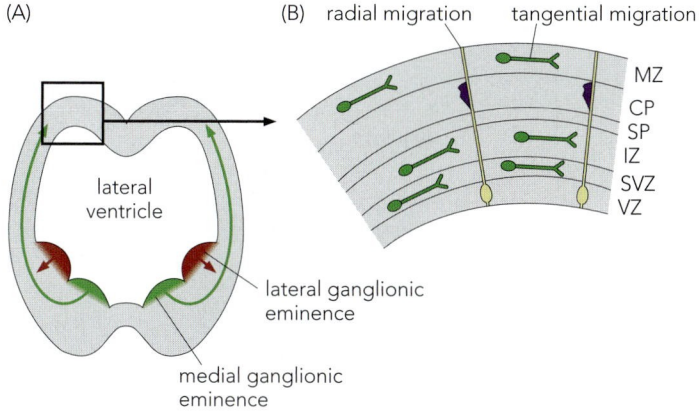

Figure 5.15 Cortical interneurons migrate tangentially in the neocortex. (A) Cortical interneurons originate from sites outside the ventricular zone and travel in a direction tangential (parallel) to the ventricular surface. A coronal section through the embryonic telencephalon reveals the location of the lateral and medial ganglionic eminences. Cells from the lateral ganglionic eminence (red) migrate to areas associated with the basal ganglia. Cells that migrate from the medial ganglionic eminence (green) travel along existing axons or cells to reach the neocortex. (B) The interneurons (green) that originate from the medial ganglionic eminence migrate tangentially in the neocortex, traveling parallel to the ventricular zone (VZ) through the subventricular zone (SVZ), intermediate zone (IZ), or marginal zone (MZ). Tangential migration often overlaps temporally with the radial migration of other neurons (purple) along radial glia (yellow).

ventrally located ganglionic eminences—structures that later give rise to components of the basal ganglia that contribute to the control of movement (**Figure 5.15**). Cells originating in the ganglionic eminences travel greater distances using a mode of migration called **tangential migration** in which neurons migrate along routes parallel to the VZ of the neocortex. It now appears that cells arising from the medial ganglionic eminence (MGE) give rise to interneurons in the neocortex, whereas those from the lateral ganglionic eminence (LGE) migrate to the caudate and putamen. Most of these interneurons travel in streams along existing cells or axons. Cells migrating out of the MGE reach the neocortex and travel through the IZ, SVZ, or MZ, where they respond to a variety of extracellular cues released by the cells they encounter. Such cues often direct cell migration by either attracting neurons to the correct location or repelling the neurons away from an incorrect pathway. Among the signals identified as attractive signals are Neuregulin-1 (Nrg1), which binds to Erb4 receptors on the interneurons, and the chemokine CXCL12 (C-X-C motif chemokine 12; also called stromal cell-derived factor 1), which binds to the CXCR4 (C-X-C chemokine receptor type 4) and CXCR7 receptors present on interneurons. In addition, the interneurons are directed away from non-target regions by the repellent proteins of the semaphorin family described in Chapter 7. Thus, multiple attractive and repellent cues are used to guide tangentially migrating interneurons so that they enter the correct cortical layer.

Tangential migration remains less well characterized than radial migration, but it has been observed in the forebrain, diencephalon, brainstem, and spinal cord. Originally considered a rare form of cell migration, it is now estimated that as many as 30% of the neurons in the developing nervous system migrate tangentially. Many of the extracellular cues utilized by tangentially migrating neurons appear to be similar to those used to guide newly formed axons to target cells (see Chapter 7). Considering the complexity of the cellular organization of the mammalian CNS, it is not surprising that multiple modes of migration may be used at different times or in different regions.

Cell migration patterns in the cerebellum reflect its distinctive organization

Like the neocortex, the cerebellum has been used to study cell proliferation and migration in the CNS for a number of years. Some of the mechanisms and signaling molecules used in the cerebral cortex are also used in the cerebellar cortex. Cerebellar development also features several interesting and noteworthy differences. Part of the reason that the cerebellum has unique patterns of proliferation and migration is because the organization of the cerebellar cortex is quite different from that seen in the cerebral cortex.

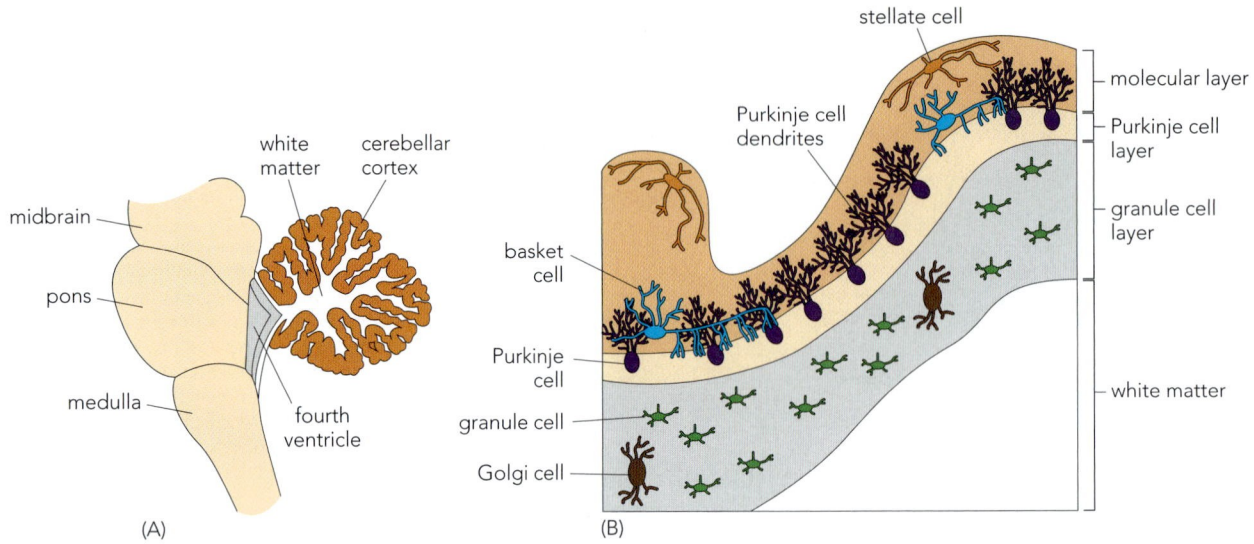

Figure 5.16 The adult cerebellar cortex is organized into the molecular, Purkinje cell, and granule cell layers. (A) The adult cerebellum is located dorsal to the pons and fourth ventricle. The cerebellar cortex is a highly convoluted structure consisting of three layers. (B) An enlarged region of the cerebellar cortex reveals the three adult layers. The molecular layer, which is the outermost of the cerebellum's three layers, primarily contains the elaborate dendritic trees of Purkinje cells. The middle, Purkinje cell layer, contains the Purkinje cell bodies. The granule cell layer is the innermost layer and contains the numerous, small granule neurons. Interneurons are also located in different layers of the adult cerebellum. These include the basket, stellate, and Golgi interneurons.

The three layers of the adult cerebellar cortex include the outer **molecular layer**, the middle **Purkinje cell layer**, and the inner **granule cell layer** (**Figure 5.16**). In the mature cerebellum, the cell bodies of the Purkinje cells are found in the Purkinje cell layer, while their elaborate dendritic trees extend to form the primary component of the molecular layer. The granule neurons found in the granule cell layer are notable for their small size and incredible number. Granule cells are the most numerous neurons in the mammalian brain; in fact, they may constitute half of the total number of neurons in the brain of some species. Also dispersed within the adult cerebellar layers are various interneurons such as Golgi, stellate, and basket cells.

Cerebellar neurons arise from two zones of proliferation

The cerebellum arises from the alar plate region in the caudal metencephalon. As described in Chapter 4, the alar plate primarily gives rise to sensory structures, whereas the basal plate gives rise to motor structures. Thus, the cerebellum is a unique derivative of the alar plate. In the caudal metencephalon, the alar plates expand in a dorsomedial direction, begin to cover the thin roof plate, and extend over the fourth ventricle. These expansions form the rhombic lips—transient structures at the edge of the fourth ventricle that extend along the posterior portion of the metencephalon (**Figure 5.17**A, B). The upper, or rostral, rhombic lip region produces specific cerebellar cell populations discussed below. In contrast, the lower, or caudal, rhombic lip region contributes cells of various pontine nuclei (Figure 5.17C). As the neural tube continues to develop and expand along the anterioposterior axis, the pontine flexure deepens and the rhombic lips gradually become compressed together to form the cerebellar primordium, the precursor to mature cerebellum (Figure 5.17D, E).

Similar to the cerebral cortex, at early developmental stages neuronal precursors are located in the VZ that surrounds the lumen of the neural tube—in this case the area that forms the fourth ventricle. The VZ is the site of production for Purkinje neurons, as well as the various interneurons. The Purkinje cells are the first cells produced in the cerebellum, and their precursors migrate radially out of the VZ to establish the Purkinje cell layer (PCL; Figure 5.17D). Initially, the newly formed PCL consists of several irregular rows of Purkinje cells, but it gradually thins to a characteristic single row as the cerebellar cortex expands. The Purkinje cells are the target cells for the

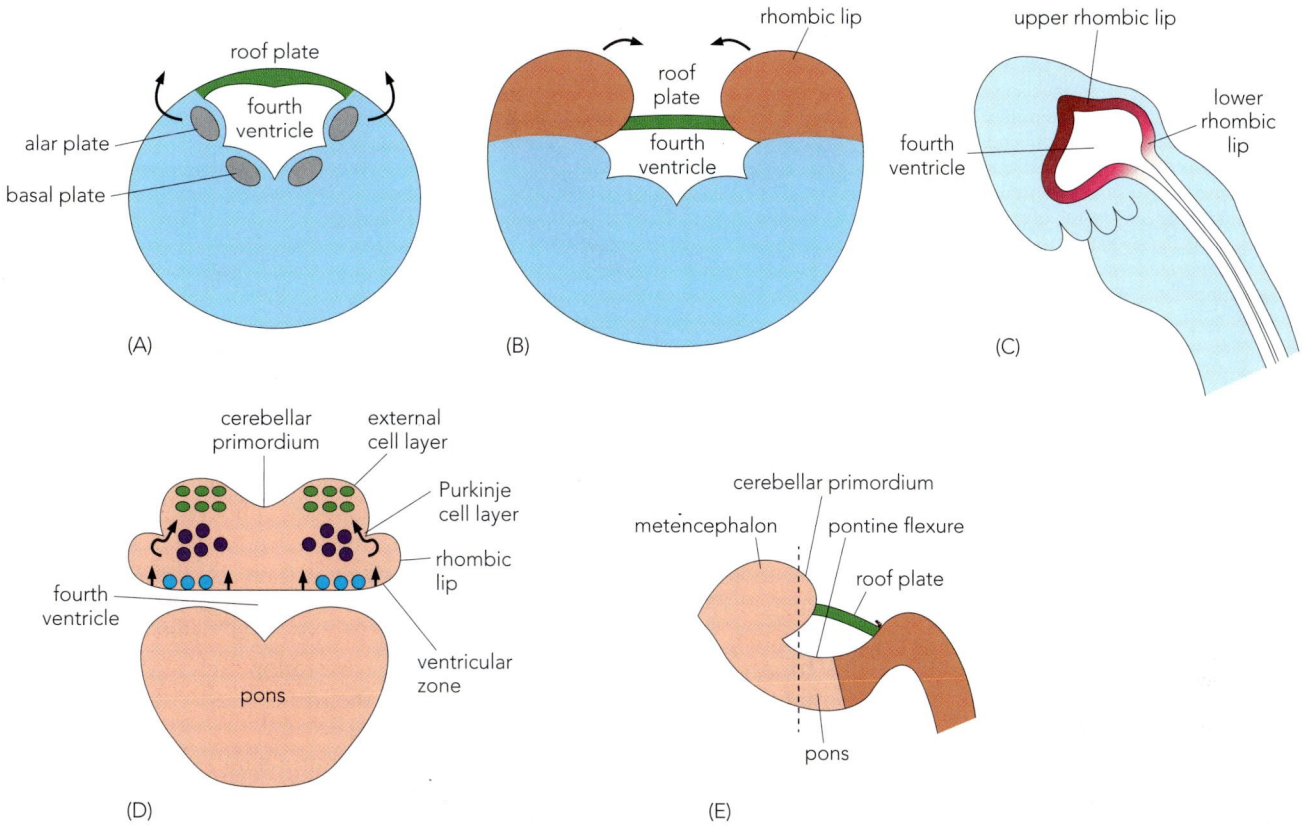

Figure 5.17 The cerebellum forms from the alar plate region of the metencephalon and generates two zones of proliferation. (A) The alar plate region of the caudal metencephalon gives rise to the cerebellum. (B) As the alar plate region extends in a dorsomedial direction, the expanded tissue region begins to extend over the thin roof plate that covers the fourth ventricle, thus forming the rhombic lips. (C) A dorsal view of an embryo at this stage shows the rhombic lips along the edges of the fourth ventricle. The upper (rostral) region of the rhombic lips is the source of granule cells of the cerebellum. The lower (caudal) region of the rhombic lips gives rise to various nuclei of the pons. (D and E) As the rhombic lips continue to extend medially, they eventually join together to form the cerebellar primordium, the precursor of the adult cerebellum. Panel D shows a cross section of the cerebellar primordium taken at the level indicated by the dashed line in panel E. Two areas of cell proliferation produce the neurons of the cerebellum. Cells of the ventricular zone (blue) are the first to migrate (arrowhead). These cells form the Purkinje cell layer (purple). Initially, the Purkinje cell layer is comprised of several irregular rows of cells, but it later thins to the single layer seen in the adult cerebellum. A second group of cells (curved arrows) migrates from the rostral region of the rhombic lips, traveling over the surface of the cerebellar primordium until they lie beneath the pial membrane. These cells form the transient external granule cell layers (green).

granule neurons that migrate to the internal granule cell layer by a different migratory pathway.

A second group of neuronal precursors originates from the rostral portion of the rhombic lips. These cells stream tangentially over the surface of the cerebellar plate to form the external granule cell layer (sometimes called the external germinal layer) that lies beneath the pial membrane. The external granule cell layer (EGL) becomes a second zone of proliferation in the cerebellum (Figure 5.17D).

The EGL is termed a "misplaced" or "displaced" germinal zone because the dividing cells originate at the cerebellar surface, rather than in the VZ. The cells of the EGL appear after mitotic activity has decreased in the VZ. In the chick embryo, for example, cells of the VZ begin to decrease mitotic activity around embryonic day eight and stop production of new cells by embryonic day 12. In contrast, the EGL is first detectable at embryonic day six and new cells are produced from embryonic days 8–15. In the mouse, proliferation and migration of Purkinje cells out of the VZ occurs between E11–14, while the granule cells migrate from the rhombic lip to proliferate in the EGL from approximately embryonic day 14 to postnatal day 15. Initially the EGL contains only one to two layers of proliferating cells, but over

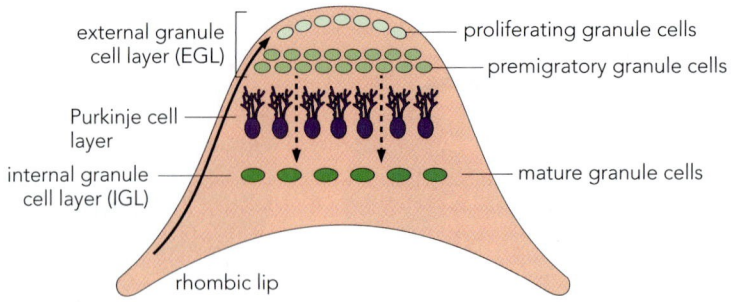

Figure 5.18 The external granule cell layer is a secondary source of proliferating cells. The cells that migrate from the rhombic lip (solid arrow) form several layers in the external granule cell layer (EGL). The outermost layer consists of proliferating granule cells. The proliferation of cells in the EGL continues into postnatal life in rodents and humans. The EGL also contains several layers of premigratory granule cells. These premigratory granule cells will later migrate inward (dashed arrows) past the existing Purkinje cells to form the internal granule cell layer (IGL).

time the cells expand in number and several rows of cells are observed in the EGL. The inner layers of cells in the EGL contain premigratory granule cell neurons (**Figure 5.18**). After sufficient granule cells are generated, these premigratory cells move inward past the Purkinje cells to reach the internal granule cell layer (IGL), where they undergo terminal differentiation.

Several proteins are known to support cell cycle activity within the EGL. For example, division of granule cell precursors in the EGL is promoted by the release of Sonic hedgehog (Shh) by the Purkinje cells. Jagged 1, a ligand for the Notch receptor, has also been shown to promote granule cell proliferation. As noted in the cerebral cortex, Notch receptor activity is necessary for continued proliferation. Shh and Jagged 1 may interact at downstream targets, such as the *Hes1* gene, to promote ongoing proliferation of precursors cells in the EGL.

The EGL continues to produce new neurons into early postnatal life. In mice and rats, the EGL produces new neurons through the first two to three postnatal weeks, whereas in humans the EGL produces new neurons through the first two postnatal years. The EGL is ultimately lost as the cells migrate to form the IGL.

Granule cell migration from external to internal layers of the cerebellar cortex is facilitated by astrotactin and neuregulin

A precise series of events takes place in order for the cells of the EGL to reach the IGL (**Figure 5.19**). As cells accumulate in the premigratory layers

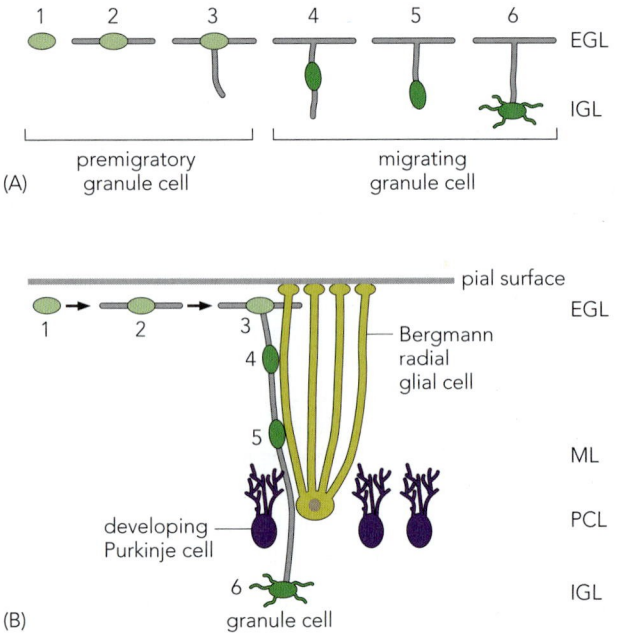

Figure 5.19 The migration of granule cell neurons from the external to the internal granule cell layer occurs over many steps. Panel A illustrates the extension of processes from the granule cells in the external granule cell layer (EGL), while panel B demonstrates how the granule cells use these processes to migrate inward along Bergmann glial cells. In panel A, a premigratory granule cell in the EGL (1) first extends bipolar processes parallel to the pial surface (2). The cell then extends a third process perpendicular to the pial surface (3). The cell body of a migrating granule cell then travels through this process (4, 5) to reach the internal granule cell layer (IGL), where it will differentiate (6). Panel B illustrates how Bergmann glia (specialized radial glial cells) interact with the perpendicular process of a granule cell to guide it from the EGL through the molecular layer (ML) and the Purkinje cell layer (PCL) to the IGL. The interaction between Bergmann glia and a granule cell are illustrated beginning with step 3 in panel A. (Adapted from Govek EE, Hatten ME, & Van Aelst L. [2011] *Dev Neurobiol.* 71:528–553.)

of the EGL, the cells in the deepest layer begin to extend bipolar processes parallel to the pial surface and perpendicular to the Purkinje dendrites. Each cell then extends a third process into the molecular layer—the layer that ultimately consists of Purkinje cell dendrites and interneurons. The soma of the granule cell moves along this third process through the molecular and Purkinje cell layers to reach the cell's final position in the IGL (Figure 5.19). Cells are guided through the layers along **Bergmann glia**, which are specialized RG with multiple parallel extensions. Bergmann glia originate in the VZ, then migrate to the Purkinje cell layer. From the Purkinje cell layer the Bergmann glia extend processes to the pial surface. Like the RG in the cerebral cortex, Bergmann glial cells later form astrocytes.

In the description of cell migration in the neocortex, it was noted that the protein astrotactin mediates adhesion between migrating neurons and RG. In the cerebellum, where astrotactin was first discovered, this interaction occurs between the astrotactin-expressing granule cells and the Bergmann glia. Hatten and colleagues first identified this integral membrane protein in the late 1980s by generating antibodies to granule cell proteins. To study the role of the newly identified protein in granule cell development, a cell culture method was developed. Using this preparation, it was found that although granule cells could normally adhere to and migrate along a Bergmann glial cell *in vitro*, granule cell adhesion and migration were blocked when anti-astrotactin antibodies were added to the cultures. Further evidence for the importance of astrotactin in cerebellar development was seen in mice that lack astrotactin. Granule cell migration in these mice is slow compared to wild-type mice due to the granule cells' decreased adhesion to Bergmann glia. The absence of astrotactin in mice also increases granule cell death and distorts the orientation of Purkinje cell dendrites (**Figure 5.20**). These changes result in behavioral deficits in balance and coordination. In recent years, scientists have identified a second member of the astrotactin family (ASTN2) that regulates the expression of astrotactin (now termed ASTN1) in the leading process of the migrating granule cell. Changes in ASTN2 expression have been linked to a number of developmental disorders, including autism. The mechanisms by which ASTN1 and ASTN2 mediate normal development are understandably important and expanding areas of research.

The ability of granule cells to continue migration, though at a slower rate, in the absence of astrotactin suggested that other molecules might also mediate adhesion and migration of granule cells in the cerebellum. The

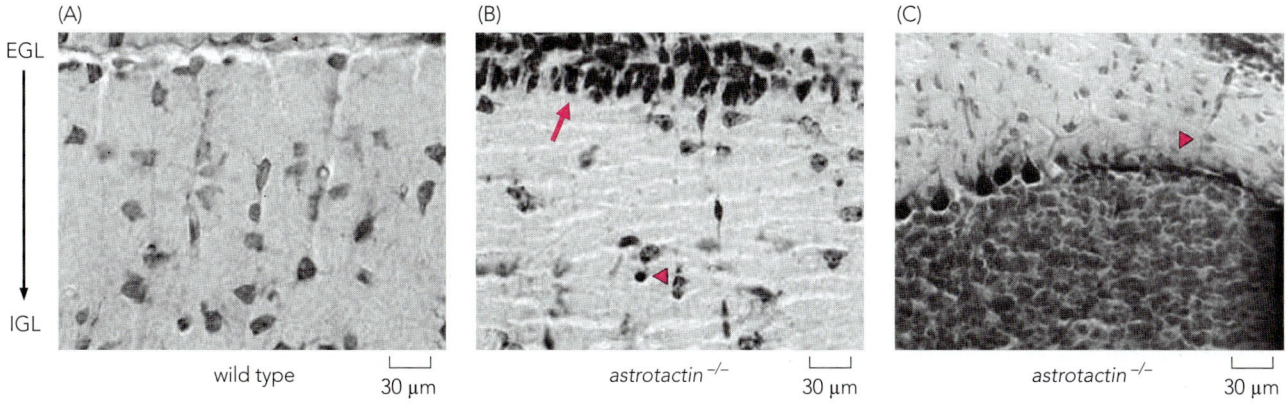

Figure 5.20 Astrotactin is necessary for normal granule cell migration in the cerebellum. (A) The cerebellum of a wild-type mouse at postnatal day 15 (P15) reveals granule cells migrating (arrow) out of the external granule cell layer (EGL) toward the internal granule cell layer (IGL). (B) A P15 mouse lacking astrotactin (*astrotactin*$^{-/-}$) shows cells remain in the EGL longer than wild-type mice (arrow) and some cells begin to die (arrowhead). (C) In the *astrotactin*$^{-/-}$ mice, the Purkinje cell layer is also distorted, with some Purkinje cells oriented properly while other adjacent Purkinje cells have cell bodies and dendrites disorganized and outside the plane of the tissue section (arrowhead). (Adapted from Adams NC, Tomoda T, Cooper M et al. [2002] *Development* 129:965–972.)

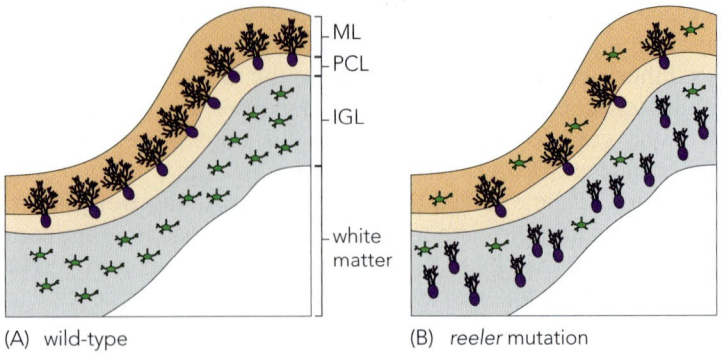

Figure 5.21 Cell layers are disorganized in reeler mice. In mice lacking the protein Reelin, the Purkinje cells fail to form the characteristic single Purkinje cell layer (PCL) seen in the cerebellum of wild-type mice (A). Instead, most Purkinje cells are clumped as aggregates in deeper layers of the cerebellum, closer to the ventricular surface. Granule cells, which normally secrete Reelin protein, are also fewer in number and many fail to migrate past the existing Purkinje cells (B). ML, molecular layer; IGL, internal granule cell layer.

cerebellar granule cells also interact with the Bergmann glia through ligand-receptor binding involving the ligand Nrg1 expressed on the granule cells and the receptor ErbB4 expressed on the Bergmann glia. Nrg1 signaling appears to regulate Bergmann glial cell numbers as well as the migration of the granule cells along the radial processes. Several experiments have demonstrated that disrupting ErbB4 receptor function or blocking Nrg1 expression leads to abnormal Bergmann glia development, including a disruption in the length of the glial processes. These changes in Bergmann glial cell development therefore lead to altered granule cell migration.

Mutant mice provide clues to the process of neuronal migration in the cerebellum

As noted earlier in the chapter, spontaneously occurring mutations in mice can provide insight to the mechanisms of cell migration and patterning. For example, reeler mice not only display altered layering in the cerebral cortex, as described above, but also show defects in Purkinje cell and granule cell migration in the cerebellar cortex. The majority of Purkinje cells do not migrate to form the characteristic single layer; instead, aggregations of Purkinje cells form deeper in the cerebellum (**Figure 5.21**). The granule cells of the EGL, which normally secrete Reelin protein prior to migrating to the IGL, are fewer in number, and most fail to migrate past the existing Purkinje cells. Thus, the overall cellular patterning and resulting function of the cerebellum are disrupted due to the absence of Reelin protein.

In the mouse mutant *Weaver*, named for its weaving gait with unsteady movements and tremor, the granule cell population is depleted in the cerebellum. This does not result from the decreased proliferation of granule cells, but rather from an inability of the cells, once generated, to migrate to the appropriate position. The mutation appears to arise either from a degeneration of the Bergmann glia fibers prior to granule cell migration or from aberrant orientation of the fibers across the cerebellum. The gait deficit shows the importance of properly located granule cells, as well as the necessity of Bergmann glia fibers for normal cerebellar development and function.

MIGRATION IN THE PERIPHERAL NERVOUS SYSTEM: EXAMPLES FROM NEURAL CREST CELLS

In 1868 Wilhelm His first described a unique population of cells in the chick embryo. These cells were named **neural crest cells** based on their site of origin—the crest of the neural folds. Neural crest cells are a transient cell population unique to vertebrates that are found along the length of the neural tube, extending from the posterior-most region to the area of the emerging diencephalon (see Figure 4.1). Neural crest cells give rise to a variety of cell types, including many neurons and supporting cells of the PNS. Specifically, neural crest cells give rise to PNS ganglia of the autonomic nervous system (that is, the sympathetic chain ganglia, parasympathetic ganglia, and enteric ganglia of the digestive tract), the dorsal root ganglia (the sensory neurons along the spinal cord), and some cranial nerve ganglia, as well as the Schwann cells and satellite cells of peripheral ganglia. Neural crest cells also give rise to melanocytes, smooth muscle cells of the aorta, chromaffin cells of the adrenal medulla, endocrine and paraendocrine cells, connective tissues, and components of the craniofacial skeleton. Thus, this single population of cells gives rise to numerous neural and nonneural cell types distributed throughout the body.

Much is now known about the formation, migration patterns, and fate determination of neural crest cells. As explained below, the pathways taken and the cell fates that arise from this population of migrating cells are remarkably diverse. The distances travelled by neural crest cells and the mechanisms that govern their migration in the embryo differ considerably from the radial migration described in the CNS. This section provides an overview of major migratory pathways used by neural crest cells and introduces some of the extracellular cues that influence their migration patterns.

Neural crest cells emerge from the neural plate border

Neural crest cells arise at the lateral edges of the neural plate—at the border that lies between the emerging neural and epidermal regions. As the lateral edges of the neural plate curl over to form the neural tube, the future neural crest cells are located at the dorsal surface of the neural tube, between the neural tube and overlying epidermis (**Figure 5.22**).

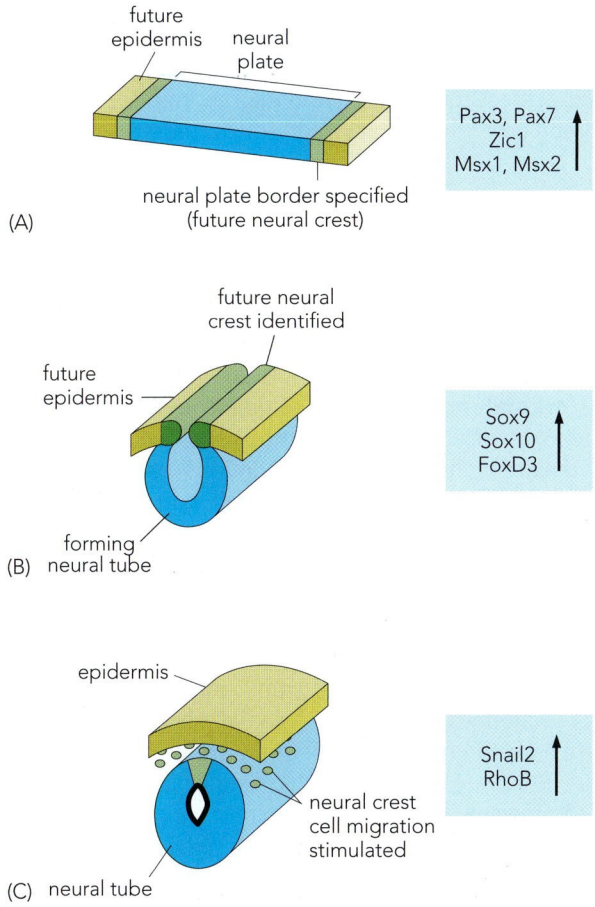

Figure 5.22 Neural crest cells arise from the dorsal surface of the neural tube. (A) Presumptive neural crest cells are specified at the neural plate border by transcription factors such as Pax3, Pax7, Zic1, Msx1, and Msx2. (B) Transcription factors such as Sox9, Sox10, and FoxD3 are up-regulated in the future neural crest cells located along the dorsal region of the forming neural tube. (C) Neural crest cells are considered a separate cell population once they start to migrate from the neural tube. The transcription factor Snail2 and GTPase RhoB stimulate the migration of neural crest cells.

The neural crest cells originate from neuroepithelial cells associated with the neural plate and are not considered a separate, unique population of cells until they emigrate, or **delaminate**, from the neural tube epithelium. Presumptive neural crest cells leave the dorsal edge of the neural tube just prior to, during, or after neural tube closure, depending on the animal species and the cells' location along the longitudinal axis. In most vertebrates, the neural crest cells undergo a cell type transition, detaching from neuroepithelium to become loosely packed, unconnected mesenchyme cells. This is the process known as epithelial-to-mesenchymal transition. The individual cells reaggregate at later time points as they coalesce into new structures, such as ganglia.

Several growth factors and transcription factors have been associated with inducing neural crest regions and specifying the neural crest cells as a separate cell population. The interactions of multiple molecules—particularly members of the BMP, Wnt, and FGF families—induce formation of the neural crest during gastrulation, thus establishing which regions have the ability to become neural crest cells as development proceeds. These molecules also regulate additional proteins that impact further specification of neural crest regions. Transcription factors such as Pax3, Pax7, Zic1, Msx1, and Msx2 are up-regulated as the neural border is specified, thus delineating a region that is distinct from adjacent neural tube and nonneural ectoderm. Another group of transcription factors that includes Sox9, Sox10, and FoxD3 is up-regulated at the time the future neural crest region is identifiable as distinct from the other regions of the neural tube. Later, Snail2 and RhoB are necessary for stimulating the neural crest cells to leave the neural tube (Figure 5.22). For example, Snail2, a transcriptional repressor, represses the expression of various cadherins. During normal neural crest development, the down-regulation of these adhesion molecules is necessary for the cells to migrate. The GTPase family member RhoB interacts with cytoskeletal elements to influence cell motility and migration. Thus, a number of transcription factors and signaling cascades must be integrated during the early stages of neural crest specification and migration to ensure the cells exit the neural crest and enter the correct migratory stream.

Neural crest cells from different axial levels contribute to specific cell populations

There are four primary anatomical regions from which neural crest cells arise, and cells from each of these regions migrate along limited pathways in the embryo. The regions are designated based on their axial level—that is, their location along the anteroposterior axis (**Figure 5.23**). For example, cells of the **cranial neural crest** originate between the midbrain and rhombomere 6 to give rise to structures associated with the head and neck. The remaining neural crest populations originate posterior to the hindbrain and are identified based on their corresponding somite level. Somites are paired blocks of mesoderm that are numbered sequentially from anterior to posterior. The somites are often used as anatomical landmarks to designate the axial level of other structures. For example, cells of the **vagal neural crest** arise from the region that extends from the posterior hindbrain to somite 7. The vagal neural crest produces cells that contribute to structures in both the head and trunk, including the sensory neurons of the glossopharyngeal and vagus cranial nerves and the sympathetic, parasympathetic, and enteric ganglia of the trunk. The **trunk neural crest** extends from the somites 8–28. Trunk neural crest cells give rise to the sympathetic, parasympathetic, and dorsal root ganglia, as well as the adrenal chromaffin cells. Neural crest cells that arise posterior to somite 28 give rise to parasympathetic and enteric ganglia. These most posterior crest cells are

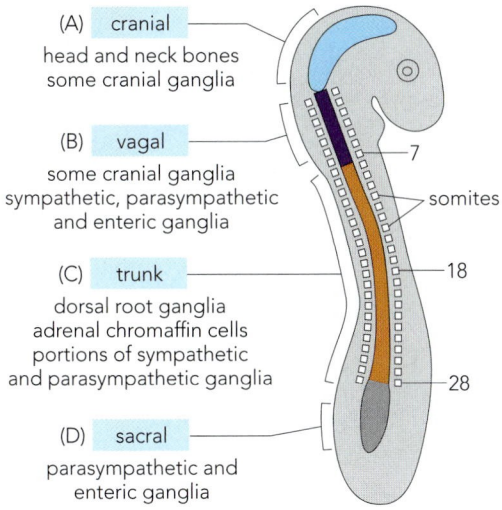

Figure 5.23 The location of neural crest cells along the anteroposterior axis determines the types of cells that develop. (A) Originating along the dorsal neural tube from the level of the midbrain to rhombomere 6, cranial neural crest (blue) gives rise to a variety of head and neck structures, including skeletal regions and neurons associated with some cranial ganglia. (B) The vagal neural crest (purple) arises from the level of the posterior hindbrain to somite 7. Vagal neural crest cells give rise to neurons associated with cranial and trunk ganglia, including sensory neurons of the glossopharyngeal and vagus cranial nerves and the sympathetic, parasympathetic, and enteric ganglia of the trunk. (C) The trunk neural crest (orange) arises at the level of somites 8 through 28. Trunk neural crest cells give rise to the dorsal root ganglia and posterior regions of the sympathetic and parasympathetic ganglia, as well as to adrenal chromaffin cells. (D) Sacral neural crest cells (gray) arise from the regions posterior to somite 28 and give rise to parasympathetic and enteric ganglia.

the **sacral neural crest**, though it is now common to include the sacral neural crest as part of the trunk region.

In addition to specific cell types associated with each region, at all levels neural crest cells contribute melanocytes, glial Schwann cells, ganglion satellite cells, and endocrine cells. Many neural crest cells have the potential to become any one of a diverse set of neural crest derivatives; as a result of this multipotency, some defects in neural crest development can promote cancers, such as melanoma.

Cranial neural crest forms structures in the head

The neural crest cells at the most anterior regions give rise to head structures, including cranial ganglia, endocrine cells, pigment cells, and cranioskeletal structures. The cranial neural crest cells are unique, in that they are the only crest cells to give rise to skeletal components under normal conditions. Skeletal derivatives of the cranial neural crest include the lower jaw (mandible), bones of the face, the hyoid bone of the neck, and the three bones of the middle ear: the malleus, incus, and stapes. Because so many facial structures arise from the cranial neural crest, altered neural crest development often leads to birth defects such as cleft palate.

Although neural crest cells are multipotent, some populations appear to be more restricted in their fate options. For example, cranial neural crest cells transplanted to trunk regions can form sympathetic and dorsal root ganglia as well as cells of the adrenal medulla and Schwann cells. However, neural crest cells transplanted from the trunk region to the cranial region do not readily form cartilage. As described in the following sections and in Chapter 6, neural crest cells encounter a number of extracellular cues that not only direct the migration of the cells, but also regulate these cell fate options.

The cranial neural crest cells are also the source of many of the neurons that comprise the cranial ganglia. Cranial neural crest derivatives include the sensory neurons of the trigeminal, facial, glossopharyngeal, and vagus nerves (**Figure 5.24**). The only populations of cranial ganglia that are not formed from neural crest cells are those that arise from the embryonic **placodes**, thickened patches of ectoderm located in the head region (**Figure 5.25**). Lens epithelia cells, olfactory, auditory, and vestibular sensory cells, as well as various supporting cell types associated with these structures, arise from their corresponding placodes. Other placodes contribute all of the neurons to a single type of ganglion. For example, the statoacoustic (also called the vestibulocochlear) ganglia of the eighth cranial nerve is derived from the otic placode. In addition, Le Douarin

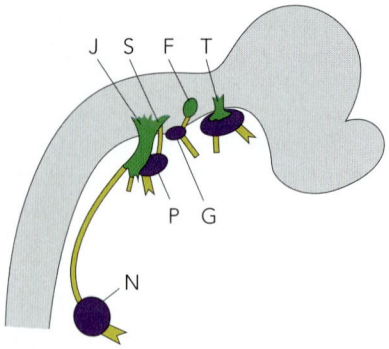

cranial nerve		proximal (neural crest)	distal (placode)
trigeminal	V	trigeminal (T)	trigeminal (T)
facial	VII	facial (F)	geniculate (G)
glossopharyngeal	IX	superior (S)	petrosal (P)
vagus	X	jugular (J)	nodose (N)

Figure 5.24 Some cranial nerve ganglia contain neurons of neural crest and placodal origin. Most peripheral ganglia originate from the neural crest. However, some cranial nerve ganglia contain a mixture of neurons derived from both the neural crest and embryonic placodes (see also Figure 5.25). Among the cranial nerve ganglia with a mix of neural crest-derived and placode-derived neurons are those associated with the trigeminal, facial, glossopharyngeal, and vagus nerves. In these cranial ganglia of mixed origin, the proximal regions of these ganglia are of neural crest origin (green), while the distal portions are of placode origin (purple). (Adapted from Ayer-Le Lievre CS & Le Douarin NM [1982] *Dev Biol* 94(2):291–310.)

and colleagues determined that some placodes generate the neurons located in the distal portions of the cranial ganglia of the trigeminal, facial, glossopharyngeal, and vagus nerves (Figure 5.24). Thus, some cranial ganglia consist of both neural crest and placode-derived neurons.

Multiple mechanisms are used to direct neural crest migration

The neural crest cells form as a large group along the length of the neural tube, but individual cells or small groups of cells soon begin to emigrate

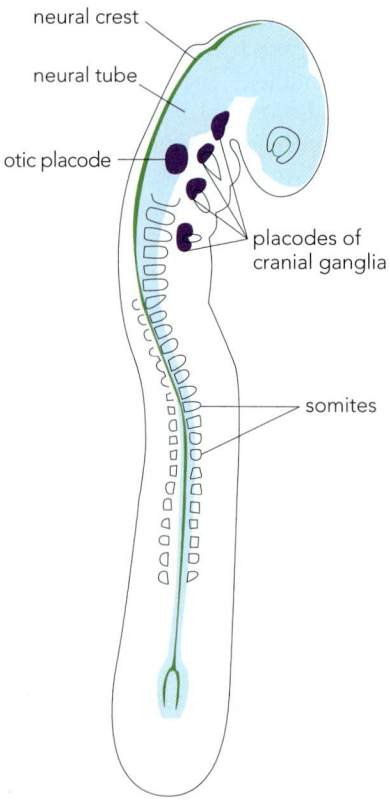

Figure 5.25 Embryonic placodes give rise to some cranial nerve ganglia. Placodes (purple) are patches of ectoderm located in the head region of the embryo. The otic placode gives rise to the embryonic inner ear (otocyst) as well as the associated statoacoustic (vestibulocochlear) ganglia of the eighth cranial nerve. Other placodes gives rise to neurons found in the distal portion of cranial nerve ganglia, as shown in Figure 5.24.

and migrate in restricted directions. Cranial crest cells migrate in lateral and ventral directions to form structures in the head and neck, while trunk crest cells migrate through or around somites to form melanocytes, peripheral ganglia, and related cells. In order to form the many different cell types and tissues, neural crest cells often travel considerable distances throughout the embryo to reach the proper destination and aggregate into tissues.

Unlike the cells of the CNS, neural crest cells continue to divide as they migrate throughout the periphery. At any level, the first cells to emigrate are generally those that travel the farthest. Thus, the neural crest cells that give rise to more ventral structures, such as sympathetic ganglia in the trunk, emigrate prior to progressively more dorsal structures, such as the dorsal root ganglia.

Among the signals for inducing emigration along a particular pathway are neural cell adhesion molecule (NCAM) and N-cadherin. Both of these adhesion molecules are expressed in neural crest cells prior to migration, but their expression decreases at the time migration is initiated. These molecules are expressed again once the neural crest cells aggregate into ganglia. Changes in the expression of other molecules are also implicated in regulating the initial migratory events of neural crest cells. One recent example is the calcium-binding protein annexin 6, which has been associated with initiating neural crest cell emigration from cranial neural crest regions in chick.

Once migration begins, the extracellular environment plays a large part in determining where the neural crest cells will travel. The neural crest cells produce hyaluronic acid, which is believed to alter and expand the extracellular spaces through which neural crest cells must migrate. Thus, the neural crest cells themselves alter their local environment to permit migration.

Additionally, permissive substrates are located on the extracellular matrix and on cell surfaces throughout the embryo to allow for attachment and migration of neural crest cells. Such permissive cues include the extracellular matrix (ECM) molecules fibronectin, laminin, and collagen. Neural crest cells express the corresponding integrin receptors that are needed to interact with these molecules (Table 5.1). The composition of the ECM differs based on the location in the embryo and often changes as development progresses. Thus, even though permissive substrates do not attract or specifically direct the migration of neural crest cells, the relative levels of certain ECM molecules may preferentially support the migration of different neural crest cell populations at different times to help guide them toward their ultimate target location.

However, the neural crest cells need more than widely distributed permissive substrates to reach a target. For example, if a cell expressed the corresponding receptors for an encountered substrate, it could easily be misguided to a different region. Therefore, other cues must be present to direct the migration of the neural crest cells as they migrate along permissive pathways. Several molecules have now been identified at the different axial levels to provide such instructive, directional cues. Some of the best characterized to date come from studies of trunk neural crest.

Trunk neural crest cells are directed by permissive and inhibitory cues

The neural crest cells that arise in the trunk region of the embryo give rise to numerous cell types, including the dorsal root ganglia and sympathetic chain ganglia. These ganglia are easily identified in the adult by their ladder-like segregation along the body axis. This patterning arises from the

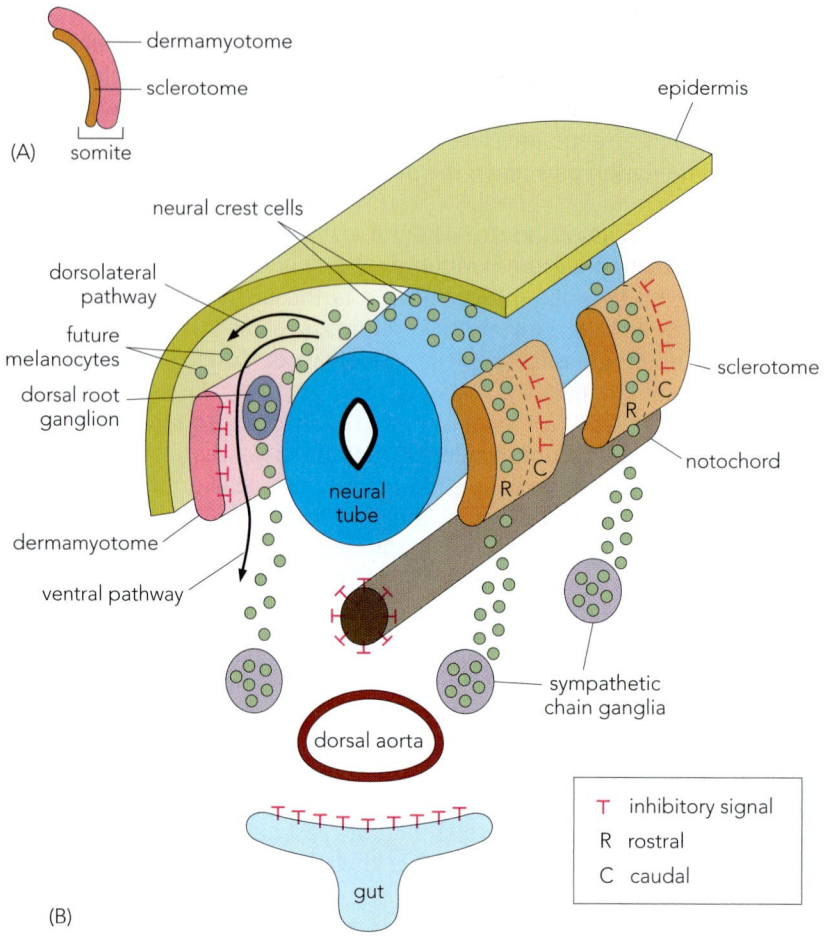

Figure 5.26 Multiple guidance cues direct neural crest cells to target regions in the trunk. (A) Several passive and inhibitory cues used to direct migration of neural crest cells originate in either the dermamyotome or the sclerotome of the somites, the paired blocks of mesoderm tissue along the length of the neural tube. (B) Trunk neural crest cells migrate along two main pathways: the ventral pathway and the dorsolateral pathway. Most cells enter the ventral pathway, where they coalesce into various ganglia, such as the dorsal root ganglia (DRG) and sympathetic chain ganglia. The neural crest cells are directed along this pathway by a combination of permissive and inhibitory cues. In the caudal half of each sclerotome (orange), inhibitory cues direct the neural crest cells through the rostral half of the sclerotome to produce the ladder-like patterns of dorsal root and sympathetic ganglia. Inhibitory cues also direct neural crest cells away from the dermamyotome (pink), notochord (brown), and developing gut (aqua) to ensure that the cells arrive at the correct target region. Only the cells that will form melanocytes enter the dorsolateral pathway. Melanocytes migrate later than the other neural crest populations and inhibitory cues are believed to be expressed in this region during the early stages of neural crest migration to prevent neural crest cells from accidentally entering the melanocyte pathway prematurely.

migratory pathways taken by the neural crest cells early in development as they migrate in streams through limited portions of the somites. Somites are divided into two segments, the dermamyotome and the sclerotome (**Figure 5.26**A). The **dermamyotome** (also spelled dermomyotome) will later give rise to dermis, skeletal muscle, and vascular tissues, while the **sclerotome** later generates the cartilage and bone of the axial skeleton and rib cage.

In studies of mouse and chick embryos, the neural crest cells that migrate ventrally through the sclerotome go on to form the dorsal root ganglia, the sympathetic ganglia, the Schwann cells and satellite cells of these ganglia, neurons around the aorta, and chromaffin cells of the adrenal medulla. Neural crest cells that migrate through the dorsolateral pathway, located between the epidermis and dermamyotome, will form melanocytes. These general patterns of migration are largely conserved across species. However, differences have been noted in some species. For example, in *Xenopus laevis*, but not other amphibians, melanocytes migrate mainly through the ventral pathway.

In experiments in mouse and chick, it has been noted that neural crest cells entering the ventral pathway preferentially migrate through the rostral (anterior) portion while avoiding the caudal (posterior) region of the sclerotome (Figure 5.26B). By limiting the pathway options for migration, the segmented, ladder-like patterns of the ganglia emerge.

There appear to be multiple mechanisms that both permit and direct neural crest cell migration in the rostral portion of the sclerotome. The extracellular surface of the rostral half of the sclerotome is favorable for neural crest cell migration, because substrate molecules such as the protein tenascin are present. In addition, the caudal segment produces inhibitory proteins that direct neural crest cells away from that region of the sclerotome. Several experiments have suggested that inhibitory cues may be the primary mechanism used to direct this trunk neural crest migration. The inhibitory cues in the caudal segment are provided by the cell surface ligands of the ephrin family of molecules, particularly ephrin-B1 and ephrin-B2. The neural crest cells express the associated EphB1 and EphB2 tyrosine kinase receptors and therefore avoid regions expressing the inhibitory signals generated by the ephrin ligands.

In addition to the Eph family of molecules, other inhibitory guidance cues direct migrating neural crest cells. For example, in order for sympathetic ganglia to migrate to their final location, the notochord, dermamyotome, and dorsal portion of the developing gut all produce inhibitory cues that direct sympathetic chain ganglia to coalesce near the dorsal aorta (Figure 5.26B). Inhibitory proteins known to influence neural crest cell migration include aggrecan and other chondroitin-sulfate proteoglycans, semaphorin 3A, slit, and peanut agglutinin-binding glycoproteins. Many of these same inhibitory signals also direct outgrowth of axonal processes, which are described in Chapter 7.

Melanocytes take a different migratory route than other neural crest cells

Melanocytes, the pigment cells of the embryo, are directed along a dorsolateral pathway between the somites and overlying ectoderm (see Figure 5.26). In most vertebrates, the neural crest cells that become melanocytes enter the dorsolateral pathway after other neural crest cells have entered the ventral pathway. There is accumulating evidence to suggest that subsets of neural crest cells are prespecified to become melanocytes and enter the dorsolateral pathway. Endothelin 3, and its associated EDNRB receptor, are thought to be involved in the proliferation, migration, and delayed differentiation of melanocytes. Evidence also suggests that inhibitory cues, such as peanut agglutinin-binding glycoprotein and chondroitin-sulfate proteoglycan molecules, are expressed at the early stages of neural crest migration to prevent the other, earlier-migrating neural crest cells from entering the dorsolateral pathway. When it is time for the presumptive melanocytes to enter this pathway, the expression of the inhibitory molecules decreases to permit cell migration. It also has been proposed that presumptive melanocytes express molecules that permit their migration on otherwise inhibitory substrates.

Determining the mechanisms that specify and guide melanocyte migration remains an active area of research, both to address issues of basic science and to understand clinical conditions that arise from altered melanocyte production and migration. Alterations in melanocyte migration can contribute to tumor formation in adulthood as well as cause birth defects. Waardenburg syndrome is a well-characterized congenital disorder that results from changes in the migration patterns of the pigment-producing melanocytes. Waardenburg syndrome is a rare condition, occuring in 1:10,000–1:40,000 births. Individuals have a range of symptoms that often include changes in skin pigmentation, a white forelock of hair, and eyes that may be very pale or of two different colors. In addition, many individuals with Waardenburg syndrome have hearing loss that results because inner ear melanocytes are needed to regulate the ionic balance of inner ear fluids. Disrupted migration of these melanocytes alters inner ear function and impairs hearing. Changes in the expression of genes identified in animal models of neural crest migration, such as *Pax3* and *Sox10*, contribute to Waardenburg syndrome.

SUMMARY

This chapter highlights some of the common mechanisms used during the proliferation and migration of neurons, glia, and neural crest cells in the vertebrate nervous system. Some mechanisms are specific to a given region of the developing nervous system, such as the radial migration of neurons into the cerebral and cerebellar cortices. Others are more commonly used, such as permissive and inhibitory cues that guide the migration of numerous neuronal precursor populations in the CNS and PNS. As seen in earlier chapters, the same signaling molecules are often used throughout development, including during proliferation and migration. Extracellular matrix proteins, Rho GTPases, and ephrin-Eph interactions are just a few examples of signals used throughout different stages of neural development.

As noted at the beginning of the chapter, neuronal fate in vertebrates can be determined at the time a cell is born or by the extracellular environment that the cell encounters during migration. Some cells have the ability to take on a number of different fates given the right conditions, others are more restricted in cell fate options, and others are specified at the time of terminal mitosis. Chapter 6 provides several examples of how cell fate is determined in neuronal cell populations in vertebrates and invertebrates.

FURTHER READING

Adams NC, Tomoda T, Cooper M et al. (2002) Mice that lack astrotactin have slowed neuronal migration. *Development* 129(4):965–972.

Angevine JB Jr. and Sidman RL (1961) Autoradiographic study of cell migration during histogenesis of cerebral cortex in the mouse. *Nature* 192:766–768.

Anthony TE, Klein C, Fishell G, & Heintz N (2004) Radial glia serve as neuronal progenitors in all regions of the central nervous system. *Neuron* 41(6):881–890.

Ayer-Le Lievre CS & Le Douarin NM (1982) The early development of cranial sensory ganglia and the potentialities of their component cells studied in quail-chick chimeras. *Dev Biol* 94(2):291–310.

Berry M & Rogers AW (1965) The migration of neuroblasts in the developing cerebral cortex. *J Anat* 99(4):691–709.

Butts T, Green MJ & Wingate RJ (2014) Development of the cerebellum: simple steps to make a "little brain." *Development* 141(21):4031–4041.

Caviness VS Jr, Takahashi T & Nowakowski RS (1995) Numbers, time and neocortical neuronogenesis: a general developmental and evolutionary model. *Trends Neurosci* 18(9):379–383.

D'Arcangelo G, Miao GG, Chen SC et al. (1995) A protein related to extracellular matrix proteins deleted in the mouse mutant reeler. *Nature* 374(6524):719–723.

Fishell G, Mason CA & Hatten ME (1993) Dispersion of neural progenitors within the germinal zones of the forebrain. *Nature* 362(6421):636–638.

Duit S, Mayer H, Blake SM et al. (2010) Differential functions of ApoER2 and very low density lipoprotein receptor in Reelin signaling depend on differential sorting of the receptors. *J Biol Chem* 285(7):4896–4908.

Frotscher M (1997) Dual role of Cajal-Retzius cells and reelin in cortical development. *Cell Tissue Res* 290(2):315–322.

Ghashghaei HT, Lai C & Anton ES (2007) Neuronal migration in the adult brain: are we there yet? *Nat Rev Neurosci* 8(2):141–151.

Hack I, Hellwig S, Junghans D et al. (2007) Divergent roles of ApoER2 and Vldlr in the migration of cortical neurons. *Development* 134(21):3883–3891.

Hatakeyama J, Wakamatsu Y, Nagafuchi A et al. (2014) Cadherin-based adhesions in the apical endfoot are required for active Notch signaling to control neurogenesis in vertebrates. *Development* 141(8):1671–1682.

Kaltezioti V, Kouroupi G, Oikonomaki M et al. (2010) Prox1 regulates the notch1-mediated inhibition of neurogenesis. *PLoS Biol* 8(12):e1000565.

Katsuyama Y & Terashima T (2009) Developmental anatomy of reeler mutant mouse. *Dev Growth Differ* 51(3):271–286.

Lee GH & D'Arcangelo G (2016) New insights into reelin-mediated signaling pathways. *Front Cell Neurosci* 10:122.

López–Bendito G, Cautinat A, Sánchez JA et al. (2006) Tangential neuronal migration controls axon guidance: a role for neuregulin–1 in thalamocortical axon navigation. *Cell* 125(1):127–142.

Martynoga B, Drechsel D & Guillemot F (2012) Molecular control of neurogenesis: a view from the mammalian cerebral cortex. *Cold Spring Harb Perspect Biol* 4(10).

Morin X & Bellaiche Y (2011) Mitotic spindle orientation in asymmetric and symmetric cell divisions during animal development. *Dev Cell* 21(1):102–119.

Noden DM (1975) An analysis of migratory behavior of avian cephalic neural crest cells. *Dev Biol* 42(1):106–130.

Ogawa M, Miyata T, Nakajimat K et al. (1995) The reeler gene-associated antigen on Cajal–Retzius neurons is a crucial molecule for laminar organization of cortical neurons. *Neuron* 14(5):899–912.

Paridaen JT & Huttner WB (2014) Neurogenesis during development of the vertebrate central nervous system. *EMBO Rep* 15(4):351–364.

Rakic P (1974) Neurons in rhesus monkey visual cortex: systematic relation between time of origin and eventual disposition. *Science* 183(4123):425–427.

Rakic P (2002) Neurogenesis in adult primates. *Prog Brain Res* 138:3–14.

Robinson V, Smith A, Flenniken AM & Wilkinson DG (1997) Roles of Eph receptors and ephrins in neural crest pathfinding. *Cell Tissue Res* 290(2):265–274.

Sauer FC (1935) Mitosis in the neural tube. *J Comp Neurol* 62:377–405.

Sauer ME & Chittenden AC (1959) Deoxyribonucleic acid content of cell nuclei in the neural tube of the chick embryo: evidence for intermitotic migration of nuclei. *Exp Cell Res* 16(1):1–6.

Sidman RL, Miale IL, & Feder N (1959) Cell proliferation and migration in the primitive ependymal zone: an autoradiographic study of histogenesis in the nervous system. *Exp Neurol* 1:322–333.

Shimojo H, Ohtsuka T & Kageyama R (2011) Dynamic expression of notch signaling genes in neural stem/progenitor cells. *Front Neurosci* 5:78.

Spear PC & Erickson CA (2012) Interkinetic nuclear migration: a mysterious process in search of a function. *Dev Growth Differ* 54(3):306–316.

Tissir F & Goffinet AM (2003) Reelin and brain development. *Nat Rev Neurosci* 4(6):496–505.

Wang Y, Li G, Stanco A et al. (2011) CXCR4 and CXCR7 have distinct functions in regulating interneuron migration. *Neuron* 69(1):61–76.

Wu CY & Taneyhill LA (2012) Annexin a6 modulates chick cranial neural crest cell emigration. *PLoS One* 7(9):e44903.

6 Cell Determination and Early Differentiation

A wide range of cell types is needed to perform the many diverse functions of the adult nervous system. Each neuron, glial cell, sensory cell, and support cell must acquire highly specialized characteristics in order to contribute to the functions of the adult nervous system. The previous chapter discussed how vertebrate neuroepithelial cells divide, establish neural precursors, and migrate to new locations where they will ultimately differentiate into fully mature neurons. This chapter focuses on some of the common mechanisms by which cells of the invertebrate and vertebrate nervous systems transition from a precursor stage to acquire a particular cell fate. Processes regulating cell fate determination of subtypes of neurons, glial, and specialized sensory cells are considered.

Cell fate is established over the course of development. During early embryogenesis, neuroepithelial cells have the potential to form numerous cell subtypes. As development progresses, however, cells are exposed to various signals that restrict their cell fate options. Depending on the specific precursor and the signals available, a given cell may remain **multipotent**—that is, retain the ability to develop into more than one cell type—for an extended period. However, this ability only persists up until the time of **cellular determination**, the stage at which further embryonic development or experimental manipulation can no longer alter the type of cell that forms. Thus, the determined cell has acquired its fate. A determined cell will then begin to differentiate and ultimately acquire the unique cellular characteristics associated with a particular cellular subtype.

For some cell types, cell fate options become restricted early in the cell cycle in response to intrinsic cues, such as those that arise from nuclear or cytoplasmic signals inherited from a precursor cell. For other cells, fate is largely regulated by extrinsic cues encountered during migration or at the final destination. These extrinsic cues are often the same types of signals discussed in earlier chapters, such as extracellular matrix molecules and diffusible factors. A previously held view was that the fate of invertebrate precursors relied on intrinsic cues, whereas vertebrate precursors relied primarily on extrinsic cues. Although these generalizations apply to some cells in these model systems, it is now recognized that such distinctions do not apply to all cells. Further, many intrinsic and extrinsic cues overlap temporally and spatially to influence cell fate, making it difficult to establish what cues predominate for any given cell population. Despite the inherent challenges of sorting out the types of cues that direct cell fate decisions, several animal model systems have provided considerable insight into the signaling pathways that establish cell fate.

Here in Chapter 6, examples from selected regions of invertebrate and vertebrate nervous systems illustrate how undifferentiated precursor cells develop as specialized neuronal, glial, or sensory cells. While the examples provided are by no means all-inclusive, they represent some of the most common and best-understood mechanisms underlying cellular determination. Many of these basic mechanisms are conserved across species, as well as across different regions of the nervous system in a given animal model. Common mechanisms include lateral inhibition, Notch signaling, and temporally regulated transcription factor cascades. In recent years the importance of epigenetic modifications in regulating cell fate options has also been highlighted. Epigenetic modifications that lead to changes in the accessibility of DNA binding sites provide an additional means for the nervous system to utilize the limited number of signaling pathways available to achieve a wide range of developmental outcomes.

LATERAL INHIBITION AND NOTCH RECEPTOR SIGNALING

A cell passes through several stages prior to adopting a particular cell fate. As introduced in Chapter 5, during early neurogenesis selected cells within the neuroepithelium begin to express **proneural genes**—the genes that provide a cell with the potential to become a neural precursor. The expression of proneural genes leads, in turn, to the activation of transcription factors and neuron-specific genes that influence the particular characteristics of a neuron. Cells that do not express proneural genes later become one of the surrounding glial or other nonneuronal cell types of the nervous system. One common mechanism for specifying neuronal versus nonneuronal cells is **lateral inhibition**, a process that relies on the level of Notch receptor activity in a given cell. This process has been observed in invertebrate and vertebrate animal models, indicating it is an evolutionarily conserved mechanism for neural specification.

Lateral inhibition designates future neurons in *Drosophila* neurogenic regions

In the developing *Drosophila* nervous system, the areas of ectoderm that ultimately give rise to the neurons are called the neurogenic regions. Cells within the neurogenic region begin to express low levels of proneural genes, such as *atonal* and members of the *achaete-scute* complex (*achaete*, *scute*, *lethal of scute*, and *asense*). The cells that express these genes make up a **proneural cluster** (**PNC**), and at this stage of development each cell in the cluster has the potential to become a neuron. Thus, at the earliest stages the cells are equivalent, with each cell expressing low levels of proneural genes.

Through cell–cell interactions, one cell in the PNC becomes specified as a neural precursor, while the surrounding cells in the cluster become nonneuronal cells. An example of how this occurs involves the expression of the ligand Delta and the receptor Notch in cells of the PNC. In this example, the proneural genes of the *achaete-scute* complex (*AS-C*) initiate the expression of the ligand Delta in all the cells of the proneural cluster (**Figure 6.1**A). The same cells also express the receptor Notch. Thus, all

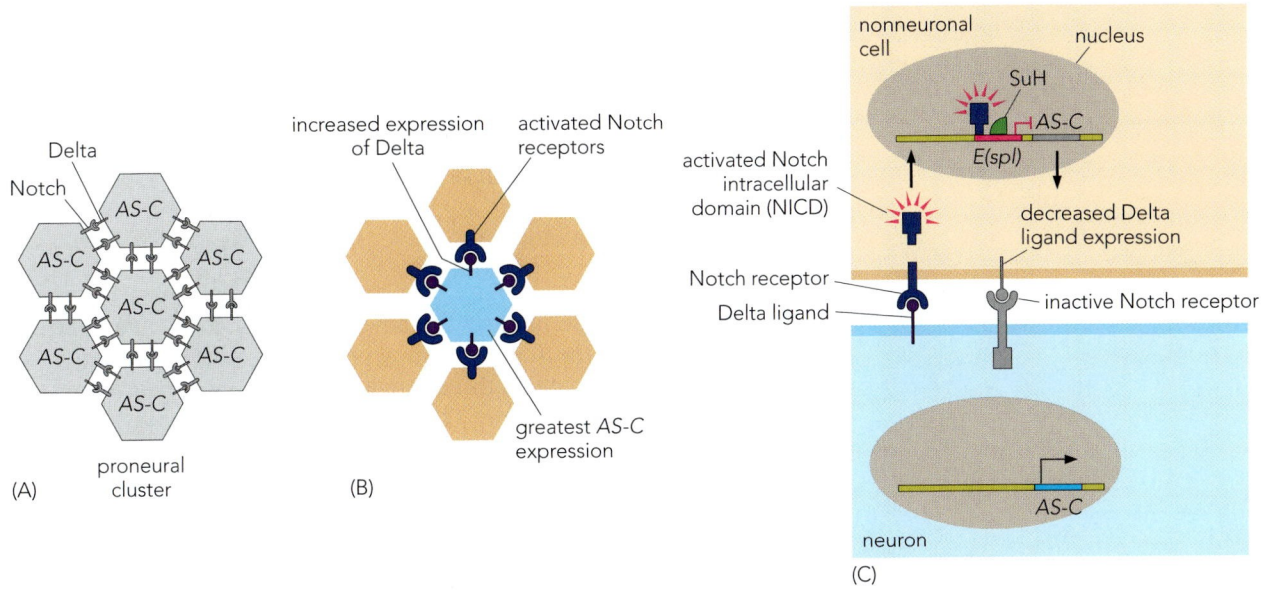

Figure 6.1 Specification of neural precursors in *Drosophila* neuroectoderm. (A) Low levels of proneural genes, such as those of the *achaete-scute* complex (*AS-C*), begin to be expressed in a subset of neuroectoderm cells called the proneural cluster (PNC). All cells in the PNC express *AS-C* genes that promote expression of Delta ligands. Notch receptors are also expressed in all cells of the PNC, so at this stage all have the potential to become neurons. (B) Some cells within the PNC begin to express higher levels of the Delta ligand. In this example, the center cell (blue) expresses a sufficient level of Delta to activate the Notch receptors in surrounding cells. (C) An enlargement of one ligand-receptor pair. When Notch is activated in an adjacent cell, the intracellular portion of the Notch receptor is cleaved and the now-activated Notch intracellular domain (NICD) travels to the nucleus, where it interacts with the DNA binding protein Suppressor of Hairless (SuH). SuH then turns on expression of *Enhancer of split (E[spl])*, which inhibits the expression of the proneural *AS-C* genes in the Notch-activated cell, thus leading to a nonneuronal cell fate. The inhibition of *AS-C* genes also causes a decrease in the expression of Delta ligand in that cell, thus preventing Notch activation in the adjacent (blue) cell. Because Notch is not activated in this cell, *AS-C* genes continue to be expressed at higher levels, so the cell is directed to a neuronal fate.

cells initially express both the ligand and receptor. However, an imbalance in Delta expression begins as proneural genes lead one cell to start expressing a slightly higher level of Delta ligand (Figure 6.1B).

How the initial increase in Delta expression occurs is still under investigation. What is clear is that once sufficient Delta expression is attained, the ligand can bind to Notch on an adjacent cell, initiating a signal transduction cascade that ultimately leads one cell in the pair to a neuronal fate and the other cell to a nonneuronal fate. The signaling pathway is initiated when the bound Notch receptor undergoes proteolysis. The resulting Notch intracellular domain (NICD) is then transported to the nucleus, where it forms a complex with other proteins and interacts with Suppressor of Hairless (SuH; Figure 6.1C, top). In the nucleus, SuH acts as a DNA-binding protein that increases the expression of *Enhancer of split* [*E(spl)*], which functions, in turn, as a suppressor of neural fate by inhibiting the expression of proneural *AS-C* genes. Thus, Delta binding to the Notch receptor initiates the pathway for inhibiting neural fate in the Notch-activated cell. In addition, the Notch-activated cell decreases its own expression of Delta ligand, so it is unable to activate the Notch receptor on a neighboring cell. Therefore, because the Notch signaling pathway is not initiated in that cell, the proneural *AS-C* genes continue to be expressed and direct that cell to continue to differentiate as a neuron (Figure 6.1C, bottom).

Through this balance of Delta expression and Notch activation, the cells of the PNC become designated to adopt nonneuronal or neuronal cell fates. Cells that have the Notch signaling pathway activated become nonneuronal cells, whereas those cells that do not have the Notch signaling pathway activated become neurons. This balance must be properly maintained so that the correct number of neurons and nonneuronal cells are generated. Experimental manipulations highlight the importance of this balance. In *Drosophila* mutants that lack *AS-C* genes, the majority of neurons are absent in both the central nervous system (CNS) and peripheral nervous system (PNS). Conversely, extra copies of these genes result in extra neurons in the *Drosophila* nervous system.

Lateral inhibition designates stripes of neural precursors in the vertebrate spinal cord

Lateral inhibition also impacts the development of cells within vertebrate neuroectoderm. In the *Xenopus* neural plate, for example, the region of the future spinal cord contains three longitudinal stripes of neural precursors on each side of the midline. The stripes will ultimately give rise to the motor neurons (medial rows), intermediate zone neurons (center rows), or dorsal sensory interneurons (lateral rows) in the adult spinal cord (see Chapter 4). The adjacent interstripes do not produce neurons (**Figure 6.2**A).

However, before these stripe regions are established in the neural tube, proneural genes are expressed to establish which cells become neural precursors. During the late stages of gastrulation, the proneural

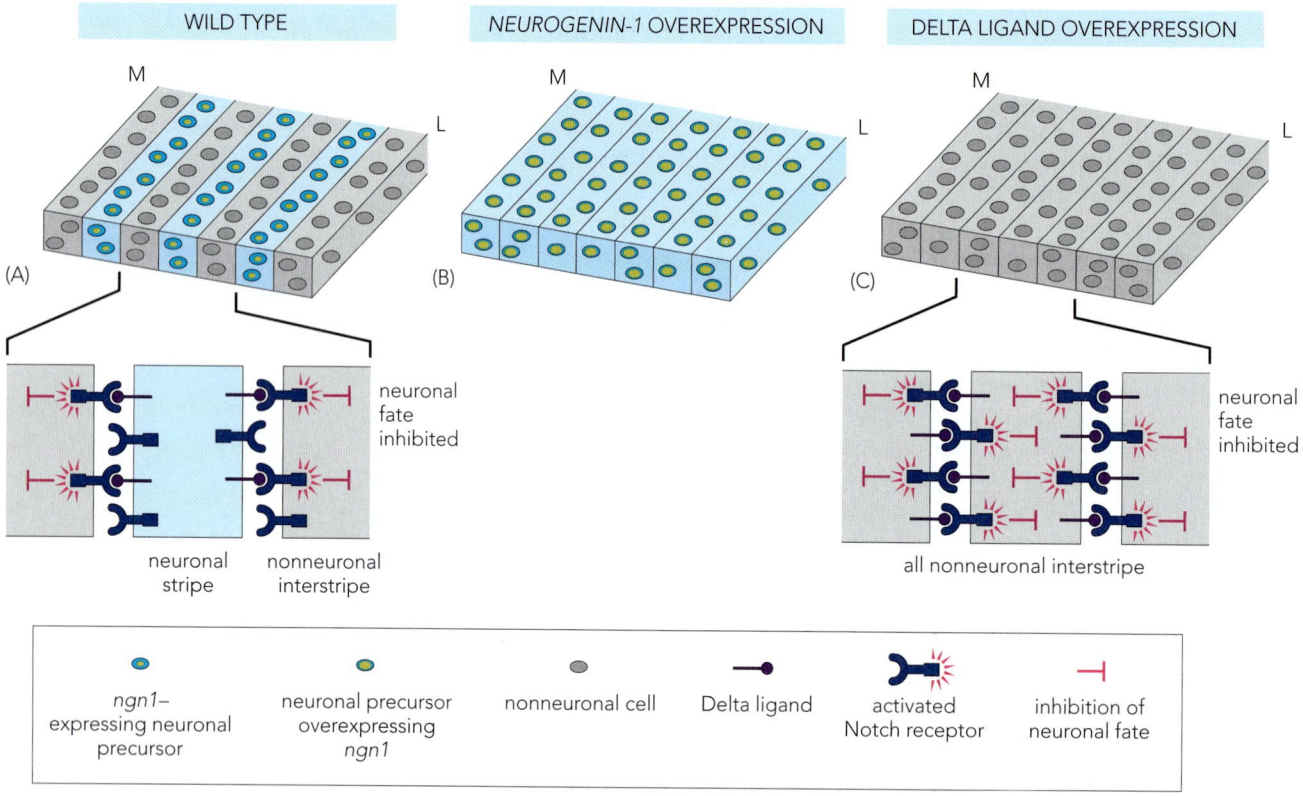

Figure 6.2 Neurons are restricted to stripes along the *Xenopus* neural plate. Half segments of the neural plate illustrate how Notch signaling regulates the formation of neuronal stripes in *Xenopus*. Each segment represents the stripes found on one side of the neural plate where three stripes of neural precursor cells (blue) emerge during late gastrulation. The blue row on the medial (M) side will later form motor neurons, the row in the center will form intermediate zone neurons, and the row on the lateral side (L) will form dorsal interneurons (see also Chapter 4). (A) The three stripes of the neural precursor cells express the proneural gene *neurogenin-1* (*ngn1*) (yellow), a member of the *atonal* gene family. Between the stripes of neural precursors are interstripes containing cells that do not express *ngn1* and ultimately develop nonneuronal fates (gray). Normally, Delta is restricted to the stripe regions so that only Notch receptors expressed on the cells of interstripes are activated. Activation of Notch inhibits proneural genes and represses neuronal fate. (B) Overexpression of the proneural gene *ngn1* leads to an increase in neurons in both the stripe and interstripe regions of the neural plate. (C) Overexpression of the ligand Delta leads to production of nonneuronal cells in all stripe regions. When Delta is overexpressed, Notch is activated on cells of both stripe and interstripe regions, thus directing more cells to a nonneuronal fate.

bHLH gene *neurogenin-1* (*ngn1*) is expressed in cells that will form the stripe regions. Thus, *ngn1*, a member of the *atonal* gene family, is necessary for establishing which cells have the potential to develop into neurons. Experimental overexpression of *ngn1* led to an increase in the number of neurons in the *Xenopus* neural plate so that neurons were found in both stripe and interstripe regions (Figure 6.2B). The *ngn1* gene induces downstream expression of *NeuroD*, another homolog in the *atonal* gene family, which is needed to regulate further development of the neurons. A direct link between *ngn1* and *NeuroD* expression was seen in studies in which overexpression of *ngn1* led to overexpression of *NeuroD* as well.

Once *ngn1* designates cells in the stripe regions as the neural precursors, lateral inhibition ensures that the further development of neurons is restricted to the stripe regions. In the *Xenopus* spinal cord, it appears that the Delta ligand is expressed only in cells within the stripes, and this expression may be regulated by *Xenopus achaete-scute* homolog (*Xash*) genes, such as the *Xash1* or *Xash3*. In contrast to the Delta ligand, the Notch receptor is expressed in cells of both the stripe and interstripe regions, though only the Notch receptors in the interstripe region will receive Delta signals. Thus, as Notch-bearing cells in the interstripe regions are activated by Delta-expressing cells in the stripe regions (Figure 6.2A), neuronal fate remains suppressed. This process ensures that interstripe cells go on to develop a nonneuronal fate. The importance of restricted Delta expression was seen in experiments in which the overexpression of Delta led to activation of Notch receptors in both stripe and interstripe regions, leading to an increase in the number of nonneuronal cells and the production of fewer neurons in the neural plate (Figure 6.2C).

Thus, similar to what was observed in the PNC of *Drosophila*, the *Xenopus* neural plate uses Delta and Notch signaling to pattern regions of neuronal and nonneuronal cells. The neural precursors within the stripes subsequently receive additional signals to become specific neural types, such as motor neurons and sensory interneurons. As described in Chapter 4, these signals include the ventrally derived protein Sonic hedgehog (Shh) and the dorsally derived proteins of the transforming growth factor β (TGFβ) and Wnt families that lead to the activation of the various transcription factors (the transcription factor code) that induce the unique characteristics of the various neuronal cell types of the mature spinal cord.

CELLULAR DETERMINATION IN THE INVERTEBRATE NERVOUS SYSTEM

Notch signaling activity remains important after lateral inhibition. In a number of regions of the *Drosophila* nervous system, the uneven distribution of Notch and Numb proteins further restricts cell fate options in precursor cells. Subsequently, the temporal expression of specific transcription factors often provides additional cues to influence the fate options available to the neuronal precursors.

Cells of the *Drosophila* PNS arise along epidermal regions and develop in response to differing levels of Notch signaling activity

The *Drosophila* peripheral nervous system consists of **sensory organ progenitors** (**SOPs**) that arise at various locations across the epidermis (**Figure 6.3**A). The SOPs give rise to various sensory organs, including the mechanosensory and chemosensory organs, as well as the chordotonal organs that contain stretch receptors. The fate of the different cell types

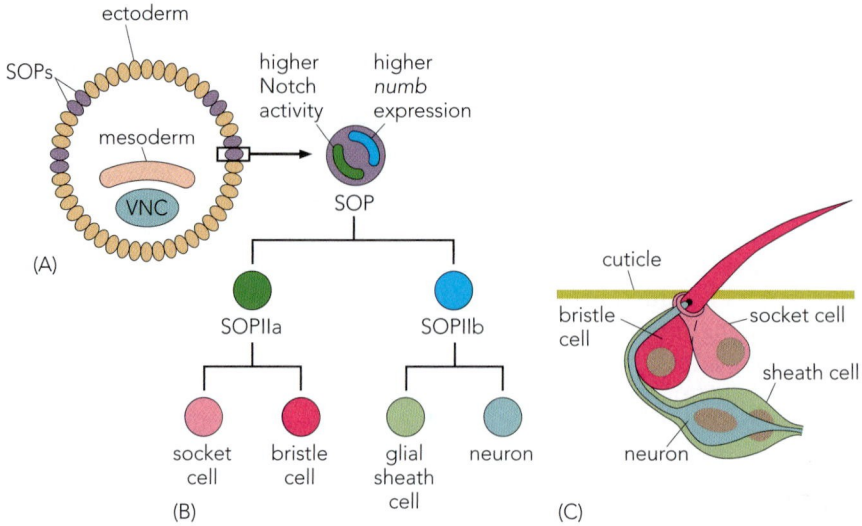

Figure 6.3 Cells of the *Drosophila* PNS arise from sensory organ progenitors (SOPs). (A) A cross section through the ventral stripe of the *Drosophila* ectoderm shows that SOPs originate at various locations along the epidermal ectoderm. The SOPs produce cells associated with PNS structures. SOPs typically arise after the neuroblasts of the CNS have formed, such as those that contribute to the ventral nerve cord (VNC) that lies between the mesoderm and ectoderm (see Figure 6.5). (B) Each SOP has an unequal distribution of Numb (green), a protein that inhibits the Notch receptor and ultimately promotes a neuronal cell fate. When the SOP divides, the SOPIIb cell inherits higher levels of Numb. Because SOPIIa does not inherit sufficiently high levels of Numb, its Notch receptor can be activated by local ligands to initiate downstream signaling pathways that result in nonneuronal cell fates. Thus, the division of SOPIIa produces the nonneuronal socket and bristle cells. In contrast, when the SOPIIb divides one daughter cell expresses *numb* at a higher level and forms a neuron. The other daughter cell does not inherit sufficient *numb* and becomes a glial sheath cell. (C) The mature sensory bristle complex is made up of a bristle cell with a hair that extends above the cuticle, an associated socket cell, a sensory neuron, and a glial sheath cell that surrounds the neuron.

depends, in part, on the distribution of Notch and Numb proteins in the precursor cells.

In *Drosophila*, the SOPs typically arise after the CNS cells are established. The process of lateral inhibition determines which cells from a PNC will become SOP cells. Each SOP then divides asymmetrically to produce two intrinsically different daughter cells, SOPIIa and SOPIIb (Figure 6.3B), which give rise to the four distinct cell types of the touch-sensitive sensory bristle complex. As introduced in Chapter 5, the differential distribution of Numb and Notch can influence cell development, with Numb inhibiting Notch receptor activation. The precursor cell has an asymmetrical distribution of Numb protein so that only one daughter cell, the SOPIIb, inherits high levels of the protein. In the SOPIIa cell, which does not inherit high levels of Numb, activation of Notch signaling remains. Notch signaling initiates downstream pathways that suppress neural fate, so the SOPIIa cell divides, instead, to produce two nonneuronal cells: a bristle and socket cell (Figure 6.3B). The Numb and Notch proteins also become distributed asymmetrically in the SOPIIb cell. When this cell divides, the daughter cell with greater Notch signaling becomes a type of glial cell called a sheath cell, but the daughter cell containing high levels of Numb goes on to form a neuron (Figure 6.3B). Thus, by segregating Numb protein, the four different types of cells that make up the mature sensory bristle complex can form (Figure 6.3C).

The importance of Notch and Numb expression levels in SOPs was seen when levels of either Notch or Numb were experimentally altered (**Figure 6.4**). When *Notch* was repressed, SOPIIa cells were unable to activate the signaling pathways that promote formation of nonneuronal socket and bristle cells. Instead, in the absence of high levels of Notch signaling activity, the SOPIIa produced only neurons (Figure 6.4B). Conversely, when *numb* was absent, no sensory neurons formed because there was insufficient Numb present to block Notch activity in the SOPIIb cell. The *numb* mutants were unresponsive to touch, thus behaving as if they were numb. This behavioral phenotype occurred because instead of sensory neurons, the *numb* mutants now produced either socket and bristle cells or only socket cells (Figure 6.4C, D).

Notch signaling is also critical in sensory organs of the vertebrate nervous system. Examples are given later in this Chapter that describe the differentiation of sensory cells in the organ of Corti of the inner ear and in the retina of the eye. Thus, the same general signaling pathways are used to establish structurally diverse sensory regions in multiple species.

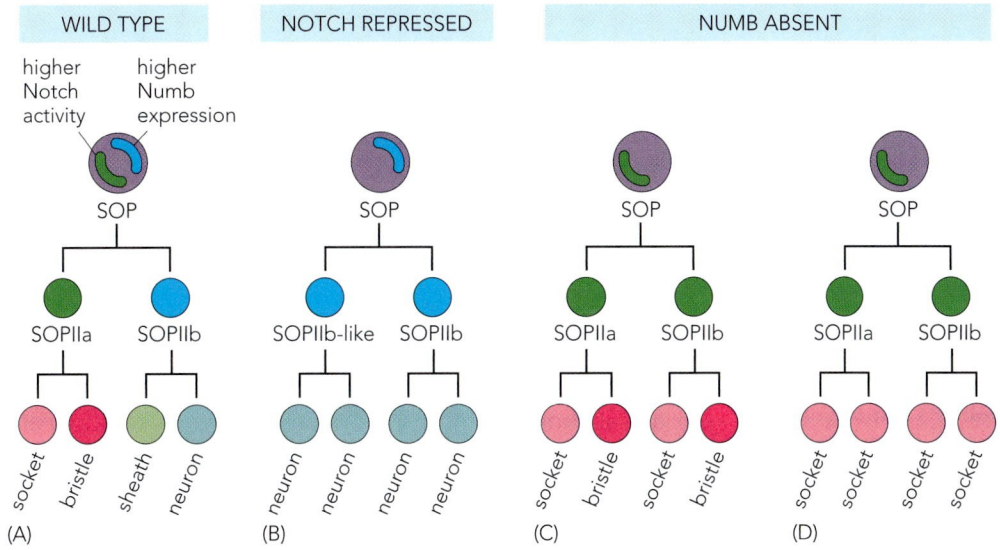

Figure 6.4 Altering Notch or Numb expression changed cell fate options of SOPIIa and SOPIIb descendants. As in other neuronal populations, the level of Notch receptor activation influences cell fate. (A) SOP cell fates in wild-type *Drosophila*. Under normal conditions SOPIIa cells have the Notch receptor activated at high levels and go on to produce the nonneuronal socket and bristle cells. In contrast, SOPIIb cells have higher Numb expression at levels sufficient to inhibit Notch signaling and therefore produce a neuron and glial sheath cell. (B) When the Notch receptor is experimentally repressed, the signaling pathways that lead to the formation of socket and bristle cells cannot be activated. Thus, in the absence of Notch signaling, the SOPIIa cells function like SOPIIb cells and only produce neurons. (C, D) In the absence of *numb*, Notch receptors are activated in both SOPIIa and SOPIIb cells. Therefore, only nonneuronal cells are produced. The absence of Numb leads to the production of socket and bristle cells (C) or only socket cells (D).

Ganglion mother cells give rise to *Drosophila* CNS neurons

Cellular determination of many neurons in the *Drosophila* brain and **ventral nerve cord** (**VNC**), a structure that is functionally analogous to the vertebrate spinal cord, also relies on the unequal distribution of Notch and Numb proteins in the precursor cells.

VNC cells originate from a ventral stripe of ectoderm in a defined manner. The cells of the proneural cluster that were previously designated to become neurons through lateral inhibition enlarge and delaminate by moving inward to form neuroblasts (**Figure 6.5**A). Each neuroblast then divides unequally, forming one large and one small daughter cell (Figure 6.5B). The smaller cell is a ganglion mother cell (GMC). The larger cell is a neuroblast that continues to proliferate, producing another GMC and neuroblast with each cell division (Figure 6.5C). As successive divisions occur, the new GMCs are situated between the first GMC and the current neuroblast. The GMCs ultimately divide equally to produce cells with either neural or glial fates. The number of GMCs generated from a neuroblast varies from a few to over 20, depending on the neuroblast lineage. Neurons in the *Drosophila* brain divide in a similar manner to those of the VNC.

Apical and basal polarity proteins are differentially segregated in GMCs

As described in Chapter 5, neuronal precursors in the vertebrate CNS distribute specific proteins to the apical and basal poles of the daughter cells. When the cells divide asymmetrically, differences in the distribution of these proteins establishes whether the daughter cell continues to proliferate or becomes a basal progenitor cell that migrates away from the ventricular surface (see Figure 5.5). Many of the proteins segregated to the apical or basal poles of vertebrate CNS precursors were first discovered in *Drosophila*. The homologous *Drosophila* proteins function to designate which cell will continue as a proliferating neuroblast and which cell will form a GMC. As in the vertebrate CNS, the cell in which Notch activity remains high will continue to proliferate.

The apical pole proteins include the Par (Partitioning defective) complex, which consists of Par3 (Bazooka) and Par6, atypical protein kinase C (aPKC), Inscuteable (Insc), and partner of inscuteable (Pins). The basal proteins include Numb, Brat (brain tumor), Prospero, Partner of

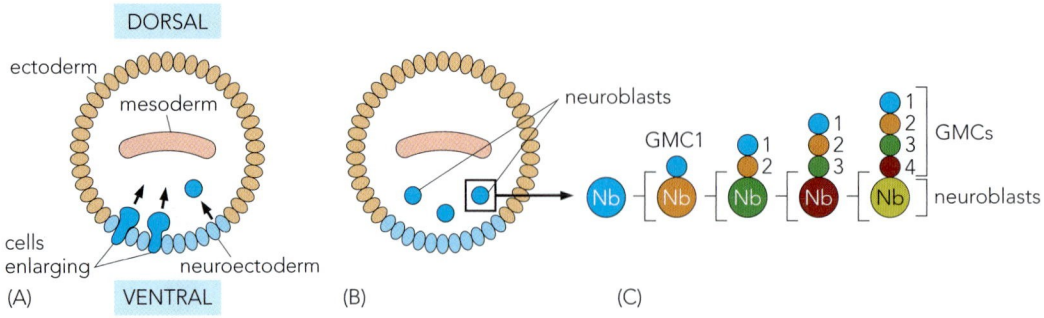

Figure 6.5 The *Drosophila* ventral nerve cord arises from neuroblasts that originate in the ventral ectoderm. (A) A cross section through the ventral stripe of *Drosophila* ectoderm shows that cells of the ventral neuroectoderm enlarge and migrate inward (arrows) after invagination of the mesoderm has been completed. The areas of epidermal ectoderm located more dorsally do not give rise to CNS neurons. (B) The cells that migrate inward form neuroblasts that will later coalesce to form the ventral nerve cord. (C) Each neuroblast (Nb) divides unequally to produce two daughter cells, a new neuroblast and the first ganglion mother cell (GMC1). The resulting neuroblast (orange) again divides unevenly, forming another neuroblast (green) and a second GMC (2, orange). These asymmetric divisions continue for various lengths of time, depending on the Nb lineage. Ultimately each GMC divides equally and produces a neuron and glial cell or two neurons (not shown).

Numb (Pon), and Miranda (**Figure 6.6**). Similar to the vertebrate neurons, the apical proteins are needed to direct the orientation of the mitotic spindles that determine the plane of cell division as well as direct basal proteins to the opposite pole. As a neuroblast divides, the new neuroblast inherits the apical proteins and the GMC inherits the basal proteins. The new neuroblast is able to divide again due to the availability of sufficient Notch signaling activity. In contrast, the GMC stops proliferating because the concentration of Numb in that cell prevents high levels of Notch signaling. Furthermore, the basal protein Prospero, now concentrated in the GMC, represses proliferation genes while activating determination genes. Thus, as in the *Drosophila* PNS, the level of Notch signaling activity first regulates the specification of neural and nonneural regions during the process of lateral inhibition and then governs whether a cell proliferates or becomes committed to a neural fate. In the CNS, only the cells that inherit proteins that interfere with Notch signaling are able to commit to the GMC fate.

Cell location and the temporal expression of transcription factors influence cellular determination

Intrinsic cues also help direct the fate of *Drosophila* CNS neurons. In *Drosophila*, neurons that arise from the original neuroblast do not migrate away. As a result, like most invertebrate neurons, a cell's origin is closely linked to its final position in the embryo. Thus, the first GMCs produced are found in the deeper layers of the CNS and have longer axons. In contrast, the GMCs from later cell divisions are located more superficially and have shorter axons.

A temporal sequence of transcription factor expression has been observed in neuroblasts and GMCs. These transcription factors are called **temporal identity factors** (**TIFs**). The TIFs that a cell expresses do not

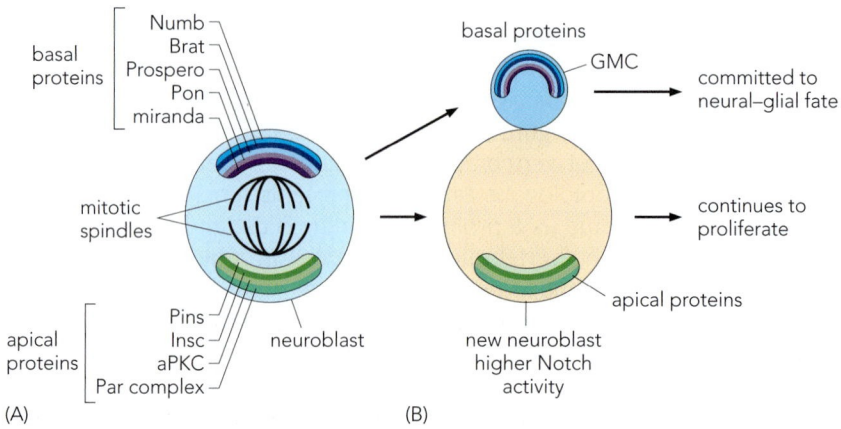

Figure 6.6 Ganglion mother cells inherit the basal protein complex to commit to neural fate. (A) During asymmetric division of the *Drosophila* neuroblast, proteins are segregated to the apical and basal poles. Apical proteins include those of the Par complex (Par3 and Par6), atypical protein kinase C (aPKC), inscuteable (Insc), and partner of inscuteable (Pins). The apical proteins help orient the mitotic spindles to determine the plane of cell division. The proteins also help direct basal proteins to the opposite pole of the cell. Basal proteins include Numb, brain tumor (BRAT), Prospero, Partner of Number (Pon), and miranda. (B) The concentration of Numb in the GMC prevents high levels of Notch activity and therefore prevents continued proliferation. Prospero represses proliferation genes and activates determination genes so that the GMC is able to commit to the neural–glial fate. In the apical cell, Numb levels are not high enough to inhibit Notch receptor activity, so the new neuroblast continues to proliferate.

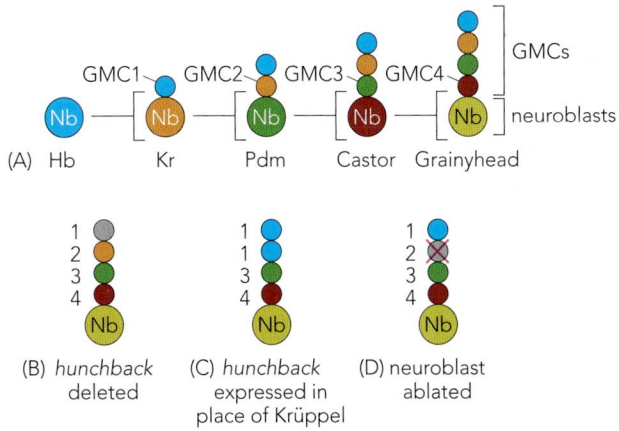

Figure 6.7 Transcription factors are expressed in a temporal sequence in neuroblasts of the *Drosophila* ventral nerve cord. (A) The first neuroblast (Nb, blue) that arises in the ventral nerve cord expresses the transcription factor Hunchback (Hb). When this neuroblast divides, the resulting GMC (GMC1, blue) inherits Hb. The new Nb (orange) now expresses a second transcription factor called Krüppel (Kr) that is inherited by next GMC produced (GMC2). The third neuroblast (green) expresses Pdm, while the next (red) expresses Castor, and the final Nb in this lineage (yellow) expresses Grainyhead. Each of these transcription factors is also expressed in the corresponding GMC. (B–D) Experimental manipulations reveal how the timing of transcription factor expression impacts the cells that arise. In the first example (B), *hunchback* was deleted from the first Nb. Only that cell failed to form (1, gray). Because the other transcription factors were not altered in dividing Nbs, the remaining GMCs (2–4) and final Nb (yellow) formed at the correct time. (C) When *hunchback* expression was sustained and took the place of Krüppel, the resulting GMC now had the characteristics of GMC1 (blue cells). The other GMCs (3 and 4) and final Nb (yellow) expressed the correct transcription factor and developed as expected. (D) When a neuroblast was experimentally ablated (cell 2), only that cell failed to form. The other GMCs (1, 3, and 4) and Nb (yellow) expressed the correct transcription factors and developed at the correct time. (Adapted from Isshiki T, Pearson B, Holbrook S & Doe CQ [2001] *Cell* 106:511–521.)

appear sufficient to designate its fate. Rather, cell fate is determined by a combination of transcription factor expression and cell location. In the VNC, for example, five transcription factors are expressed in sequence—namely, Hunchback, Krüppel, Pdm, Castor, and Grainyhead. This same sequence is used by other neuroblast lineages in the *Drosophila* CNS, although Grainyhead may not act as a TIF in all regions.

A neuroblast first expresses Hunchback; this expression is inherited by GMC1 when the neuroblast divides (**Figure 6.7**A). The daughter neuroblast now expresses Krüppel and divides to generate the Krüppel-expressing GMC2 and a daughter neuroblast that expresses Pdm. Subsequent neuroblasts express the remaining TIFs in sequence during each subsequent division.

The timing of TIF expression is critical, as can be shown under experimental conditions. If one transcription factor is absent, only the cell type arising at that stage will be eliminated. For example, a series of experiments by Chris Doe and colleagues found that when *hunchback* is absent, only the GMC generated during the first cell division is missing (Figure 6.7B). If a transcription factor is experimentally maintained, then those cell types will persist longer. Continued expression of *hunchback* during the period that Krüppel-expressing cells would normally be produced, for example, led to the formation of GMCs with characteristics of the earliest cells (Figure 6.7C). In another study, one neuroblast was experimentally ablated. Although that cell never formed, the subsequent neuroblasts continued to arise in order and express the transcription factors normally present during those cell divisions (Figure 6.7D). Again, the transcription factors alone are not believed to regulate cell fate, but their presence appears to establish which cell types can form during different stages of development in the *Drosophila* CNS. As described later in this chapter, homologs of some of these TIFs have been identified in the cerebral cortex and mammalian retina, where they appear to serve similar functions.

MECHANISMS UNDERLYING FATE DETERMINATION IN VERTEBRATE CNS NEURONS

In the vertebrate nervous system, reduced Notch receptor activity, environmental cues, and the temporal expression of specific transcription factors also coordinate to influence neuronal fate options and initiate cellular differentiation. Examples of how such cues contribute to the development of cerebellar granule cells, cerebral cortical neurons, neural-crest-derived neurons, and PNS and CNS glial cells are described in the following sections.

Changes in transcription factor expression mediate the progressive development of cerebellar granule cells

Because the developmental events that lead to the formation and migration of cerebellar granule cells are so well documented (see Chapter 5), a number of studies have focused on the signals that regulate development of this highly specialized group of cells. Similar to other regions of the nervous system, the level of Notch receptor activity regulates whether precursors in the external granule cell layer (EGL) continue to proliferate or commit to the granule cell fate. Notch receptors are expressed on granule cell precursors in the EGL (**Figure 6.8**). Manipulations of Notch activity *in vivo* revealed that if Notch activity is experimentally increased, granule cells proliferate longer. Conversely, if Notch receptor activity is inhibited, cells stop proliferating early and begin to express Math1 (mouse Atonal homolog 1), a transcription factor characteristic of committed granule cells.

The importance of Math1 in granule cell fate was seen in cell cutures of **embryonic stem (ES) cells**—cells harvested from the blastocyst stage embryo that have the potential to develop into any cell type. *In vitro*, experimentally induced, transient expression of *Math1* was sufficient to specify ES cells as committed granule precursor cells. For example, transient *Math1* expression led to the increased expression of the transcription factors Zic1 and Zic2, as well as other markers of early differentiating, premigratory granule cells (Figure 6.8). However, expression of *Math1* alone could not induce markers of mature granule cells.

Several *in vitro* studies have demonstrated that ES cells can develop as granule cells when treated sequentially with many of the same molecules that induce their formation *in vivo* (see Chapter 3). Among the signals for specifying granule cell characteristics *in vitro* are FGF8 (fibroblast growth factor 8), Wnts, BMPs (bone morphogenetic proteins), GDF7 (growth differentiation factor 7), Shh (sonic hedgehog), and the Notch ligand Jagged 1. When embryonic stem cells were grown in a culture media containing signals that support granule cell development, experimentally induced expression of *Math1* was then able to increase the number of cells expressing markers of mature granule cells, such as GABAα6r (gamma-amino butyric acid type A receptor α6 subunit; Figure 6.8). Thus, a combination of extrinsic signals appears to regulate the expression of transcription factors and proteins that mediate the progressive development of cerebellar granule cells.

In vivo studies have also shown that extrinsic signals such as brain-derived neurotrophic factor (BDNF) are required to support later developmental events, including granule cell survival, the differentiation of granule cell processes, and the migration of the cells to the internal

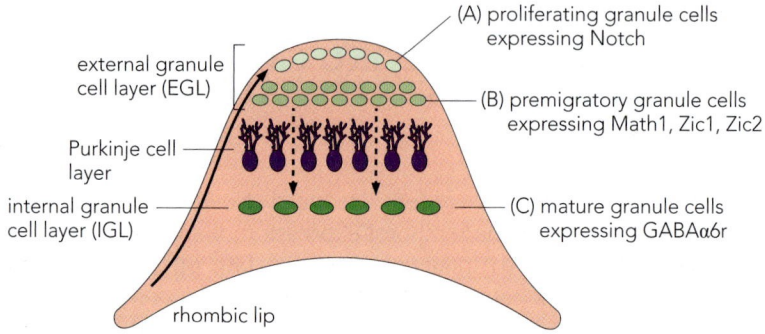

Figure 6.8 Multiple signals influence determination and differentiation of cerebellar granule cells. (A) In the external granule cell layer (EGL), granule cells express high levels of the Notch receptor, thus permitting ongoing proliferation of these cells. (B) The transcription factor Math1 is expressed in premigratory granule cells. Math1 expression is indicative of a committed granule cell fate and induces the expression of other transcription factors, including Zic1 and Zic2. (C) As granule cells migrate from the EGL to the internal granule cell layer (IGL), a sequence of extrinsic signals is needed to induce expression of proteins characteristic of mature granule cells such as the receptor GABAα6r. Known extrinsic signals include fibroblast growth factors (FGFs), Wnts, bone morphogenetic proteins (BMPs), brain-derived neurotrophic factor (BDNF), and sonic hedgehog (Shh).

granule cell layer (IGL). Cerebellar granule cells must therefore integrate multiple signals to progress from the granule cell precursor stage to a fully differentiated granule cell neuron.

Temporal cues help mediate the fate of cerebral cortical neurons

In the vertebrate cerebral cortex, the time of neurogenesis is linked to the migratory route and fate of newly generated neuronal precursors (see Chapter 5). Those cells born early migrate to the deepest layers of the emerging cortical plate, while later-born neurons migrate to more superficial layers, thereby creating the "inside first, outside last" pattern of cortical development. The cells in each layer are fated to become a particular type of cortical neuron. The link between the time migration is initiated and the cortical layer destination suggested that there are temporal and environmental cues to direct the neurons to the correct cortical layer. *In vivo* and *in vitro* studies have confirmed that both types of cues are important for cell fate determination of cortical neurons.

A series of transplantation studies done in the cerebral cortex of ferrets by Susan McConnell and colleagues demonstrated that cortical neural progenitors become progressively restricted in their cell fate options. The ferret is a popular animal model for studies of CNS development and the time of cortical layer formation has been well documented. To monitor the fate of cortical neurons, the researchers harvested neurons from the ventricular zone of a donor cortex at one stage of development. The cells were then dissociated, labeled with tritiated thymidine, and injected into a host cortex at a different stage of development. The host animals were then allowed to develop for several days. The studies found that early-generated progenitors—those that would normally migrate to the deepest layer (layer VI)—were now able to migrate to more superficial layers (layers II/III) when transplanted to an older host cortex (**Figure 6.9**A). Thus, the early-born neurons could respond to environmental cues in the host environment and migrate to a new destination. This effect was only seen, however, when the cells were harvested early in the cell cycle, prior to the final mitotic division. Cells harvested in the late stage of the cell cycle still migrated to the deeper layer (layer VI). This finding suggested there were also intrinsic temporal cues present to influence fate options of cortical neurons. Thus, by the time a cell is post-mitotic, cell fate is established and cannot be altered, even when placed in a new host environment. In contrast to early stage neurons injected into older hosts, late-stage progenitors that normally migrate to layer II/III did not to migrate to the deeper layer VI when transplanted to younger hosts, even when the cells were harvested early in the cell cycle (Figure 6.9B).

A third set of experiments confirmed that the fate potential of the cortical neurons becomes gradually restricted during normal development. Mid-stage progenitors, those that would migrate to layer IV, were only able to migrate to a new location (layer II/III) if transplanted to an older host. However, these progenitors were unable to migrate to the deeper layer VI when transplanted to a younger host. Together, the transplantation experiments revealed early-mid-stage progenitors are multipotent early in the cell cycle and can adopt new cortical fates when placed in an older host environment. Yet, the progenitors gradually lose this ability to change fate. As cells reach mid-late stages of development they become restricted in cell fate options. Thus, the progenitors arising from mid-late stages of cortical development are unable to adopt the fates of younger progenitors and remain committed to the fate corresponding to their time of migration.

A number of subsequent studies have identified transcription factors that are specific to neurons located in different cortical layers. For example,

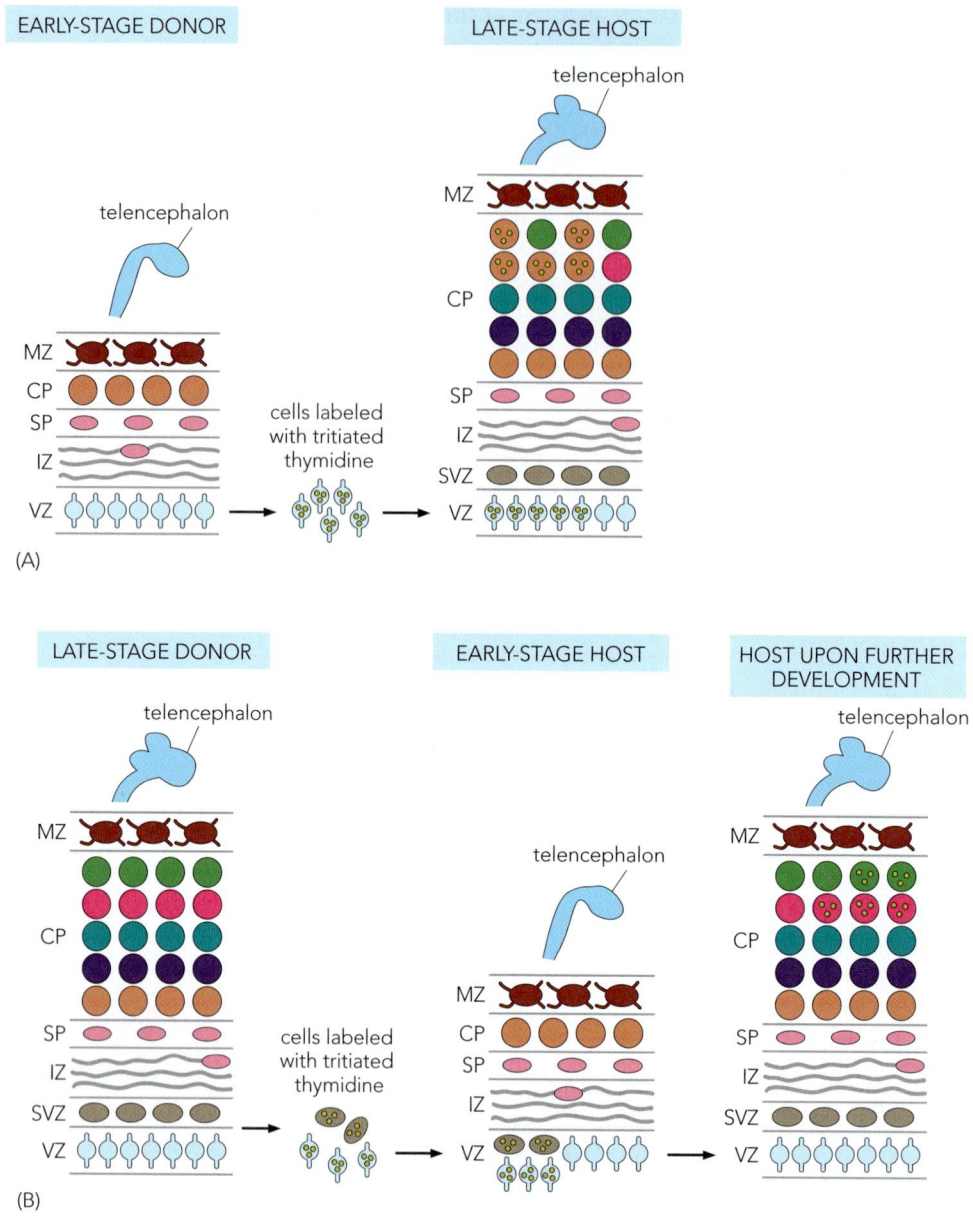

Figure 6.9 Cell fate options of cortical neurons become restricted as development progresses. (A) Neuronal progenitors were harvested from the ventricular zone (VZ) at an early stage of cortical development. The cells were dissociated, labeled with tritiated thymidine, and injected into the VZ of an older host embryo. In the older host embryo, the donor cells migrated past the deep layer (layer VI), where they would normally settle, and instead migrated to superficial layers (layers II/III) consistent with the age of the host embryo. Thus, early-stage cortical progenitors are able to alter their fate and migrate to a new layer. However, this effect was only seen if the cells were harvested early in the cell cycle, prior to the final mitotic division. Thus, cell fate could not be altered in post-mitotic cortical neurons. (B) Cortical progenitors were harvested from the VZ and subventricular zone (SVZ) of a late-stage donor embryo at a time the progenitors would migrate to superficial layers II/III. The labeled cells from the older donor were injected into an early-stage host embryo at a stage when the progenitors would migrate to deep layer VI. The injected donor neurons continued to migrate to the superficial layers (layers II/III), thus indicating that older cortical neurons could not switch fate even in a new environment. MZ, marginal zone; CP, cortical plate; SP, subplate; IZ, intermediate zone.

Tbr1 (T-box brain 1) and FoxP2 (Forkhead box protein P2) are layer-VI-specific transcription factors, whereas Cux1 (cut like homeobox 1), Satb2 (SATB homeobox 2), and Brn2 (brain-2; also called POU class 3 homeobox 2, POU3f2) are specific for more superficial layers (layers II–IV). It has been suggested that some combinations of transcription factors are mutually repressive to prevent cells from adopting the fate of cells in adjacent layers. Ongoing studies seek to identify how transcription factors determine cortical layer cell fate. In many cases it has been difficult to clearly separate the impact of transcription factor expression from environmental cues. In at least some cases, the identified transcription factors regulate differentiation of neurons once a layer-specific fate has been established.

Recent evidence has also indicated that temporal identity factors homologous to those found in *Drosophila* play a role in cerebral cortical fate potential. For example, *Ikaros*, the mammalian ortholog of *hunchback*, is expressed in early-stage cortical progenitors. In mice, *Ikaros* is detected in early-stage progenitors of the ventricular zone, but is decreased at later stages. When *Ikaros* was overexpressed in mice, the number of progenitor neurons was increased. If *Ikaros* expression was sustained, more early-born, Tbr1-positive neurons were generated and fewer late-born neurons were present (**Figure 6.10**). Thus, cells expressing markers for layer VI were increased, while those for layers III and IV were decreased. If *Ikaros* was misexpressed in later-born progenitors, early-born fates could not be generated, consistent with the idea that *Ikaros* encodes a temporal factor utilized only by early-generated progenitor cells, similar to the function of *hunchback* in *Drosophila* neuroblasts. *Ikaros* appears to provide the early-generated neurons with the ability to adopt deep-layer cortical fates. The expression of other transcription factors is then needed for the cells to differentiate into mature cortical neurons with the characteristics typical of cells in that layer.

Epigenetic factors influence determination and differentiation in vertebrate neurons

In recent decades studies have also begun to focus on how **epigenetic** factors influence cell fate options in the developing nervous system. Epigenetic mechanisms play a very important role in regulating gene activation and repression by controlling the accessibility of DNA binding sites to transcription factors. Common epigenetic modifications include DNA methylation—the process by which methyl groups are added to DNA at the promoter region, often to repress gene transcription; noncoding RNAs—RNAs that are not translated into protein but instead influence gene expression at transcriptional or post-transcriptional stages; and histone modifications—post-translational modifications to the histone proteins that wrap around the DNA strand. Histone modifications include the recruitment of histone

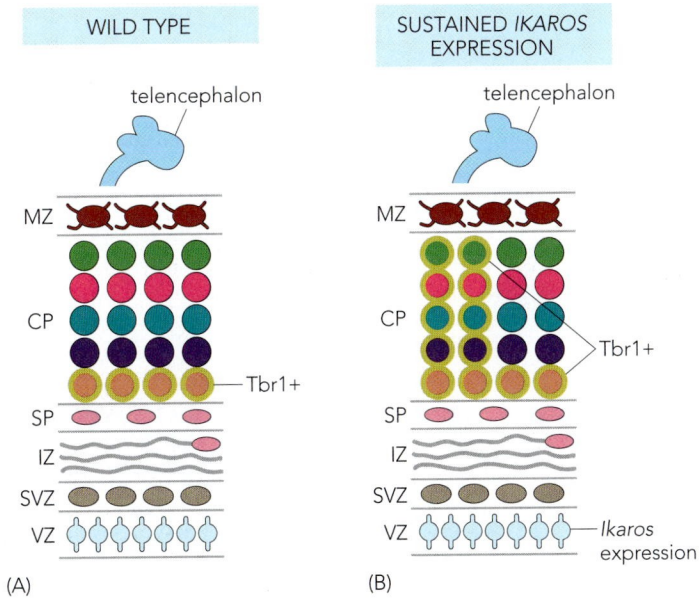

Figure 6.10 The temporal identity factor Ikaros influences the fate of early generated cortical neurons. (A) *Ikaros*, the mammalian ortholog of the *Drosophila hunchback*, is expressed in early-stage, but not later-stage progenitor neurons. Early-born neurons that settle in future layer VI are positive for the transcription factor Tbr1 (Tbr1+). (B) When *Ikaros* was overexpressed in mice so that *Ikaros* expression continued through the stages that later-born neurons are generated, the number of early-born Tbr1+ cells increased and migrated throughout more superficial layers. However, if *Ikaros* was expressed in later-born progenitors, cell fate was not altered (not shown). These results support the hypothesis that the temporal identify factor Ikaros, like Hunchback, only influences the fate of early-generated progenitor cells.

Figure 6.11 Developmental changes in the subunit composition of BAF complex influence proliferation and fate options of vertebrate neurons. Subunits of the BAF (brahma-related gene 1 and brahma-associated factors) complex change as neurons progress from proliferative to post-mitotic stages. The subunit composition influences the accessibility of DNA binding sites for transcription factors and therefore whether target genes are expressed. The BAF complex is comprised of multiple subunits. (A) In embryonic stem (ES) cells, the BAF complex (esBAF) includes two 155 subunits, as well as a 45a and a 53a subunit that are important during neural development. (B) In neural progenitor (np) cells, the BAF complex (npBAF) continues to express the 45a and 53a subunits, but exchanges one of the 155 subunits with 170. (C) In post-mitotic neurons (n), the BAF complex (nBAF) no longer expresses the 43a and 53a subunits, but instead expresses the 45b and 53b subunits. nBAF continues to express the 170 subunit, which alters the chromatin state so that transcription factors such as Pax6 can access their target genes and induce cellular characteristics of nonproliferating and post-mitotic neurons. In esBAF, the lack of the BAF 170 subunit means the chromatin is tightly packed, in which case Pax6 and other binding sites are less accessible in ES cells, thus preventing them from adopting a neural fate prematurely. (Adapted from Yoo AS & Crabtree GR [2009] *Curr Opin Neurobiol* 19:120–126.)

modifiers or alterations in chromatin structure. Chromatin is comprised of the DNA strand and the histone proteins. Changes in chromatin structure influence the accessibility of target genes. For example, when chromatin is in a lightly packed (euchromatin) state, the corresponding promoter region of a target gene becomes accessible to transcription factors.

Epigenetic mechanisms are widely used throughout the developing embryo. Research in the developing vertebrate CNS has revealed several important epigenetic modifications that determine whether a cell remains in the proliferative state or begins the determination process. One example comes from genes related to the *Drosophila* gene *brahma* (*brm*). The group of related factors in vertebrates includes ATP-dependent chromatin-remodeling enzymes of the BAF complex. BAF stands for Brg1 (brahma-related gene 1) and Brm- (brahma-)associated factors. This group of proteins determines the accessibility of DNA binding sites.

Brg1 appears to be particularly important during neural proliferation, whereas Brm is required for cell fate determination of progenitors and differentiation of post-mitotic neurons. Brg1 and other subunits are needed to maintain Notch signaling in order to repress proneural genes and keep cells in a proliferative state. In contrast, Brm and other subunits activate the transcription of neuron-specific genes such as *Neurogenin* and *NeuroD*.

The BAF complex is comprised of at least 15 subunits whose composition changes as cells progress from a proliferative to post-mitotic state. In the vertebrate CNS, the neural embryonic stem (ES) cells express subunits that comprise the esBAF complex (**Figure 6.11**), whereas neural progenitor (np) cells express slightly different subunits in the npBAF complex. Post-mitotic neurons (n) express a third group of subunits to make up the nBAF complex. The changes in subunit composition correlate with the transition to each stage of neuronal development. Thus, the

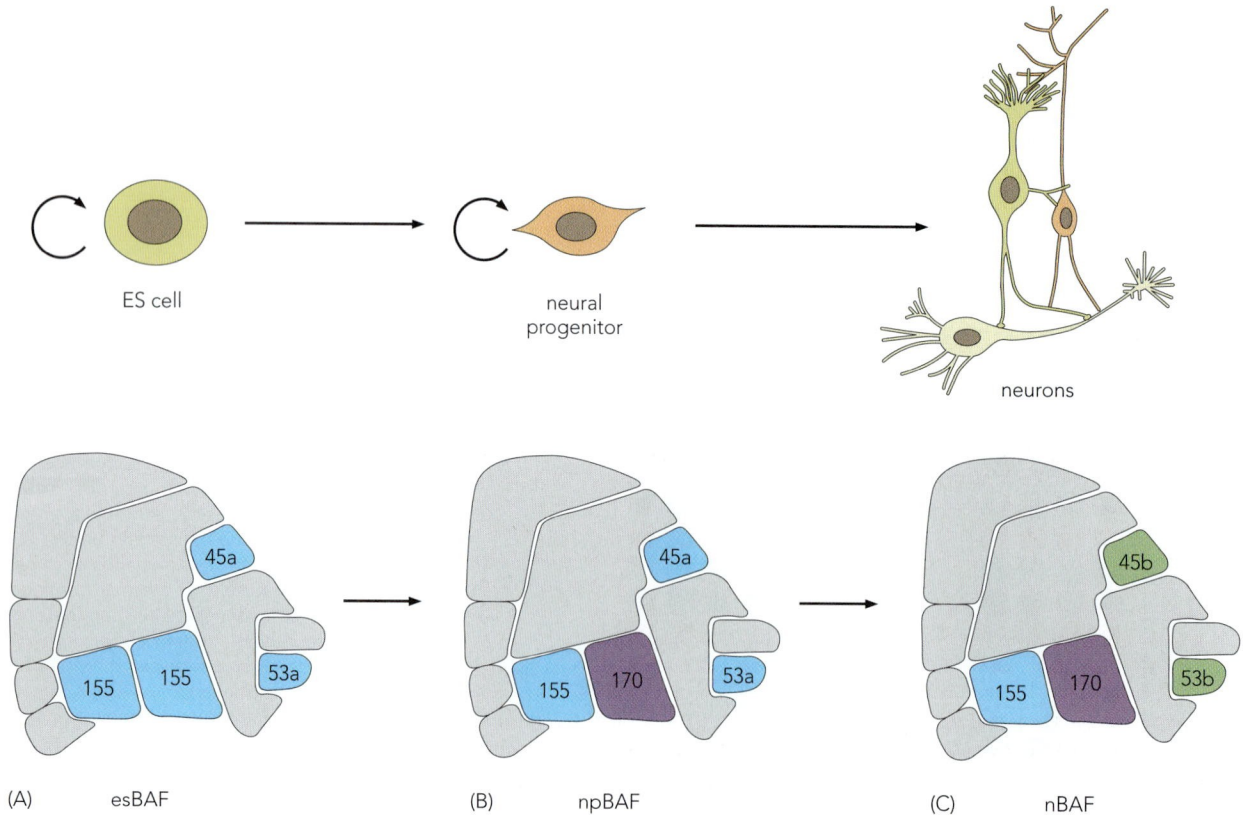

subunit composition of the BAF complex, as well as the dosage of individual subunits, influences whether the cells continue to proliferate or progress to the post-mitotic state. For example, in the developing cerebral cortex and cerebellum, the BAF subunits BAF45a and BAF53a are required for the continued proliferation of neural progenitor cells. In post-mitotic neurons, these subunits are exchanged with BAF45b and BAF53b (Figure 6.11). The changes in the expression of these subunits is correlated with the transition from proliferating progenitors to committed cortical and granule cell fates.

Further, changes in BAF subunit composition have been linked to changes in the activity of the transcription factor Pax6. In the developing CNS, Pax6 plays a wide range of roles throughout the developing forebrain (see Chapter 3). In the cerebral cortex, Pax6 activates target genes such as *Tbr2* (T-box brain 2), which is detected in nonproliferating progenitors (for example, the basal progenitor cells). Pax 6 then activates *Cux1*, which is detected in post-mitotic neurons, and *Tle1* (transducin-like enhancer protein-1), which is needed for the survival of post-mitotic neurons. During early stages of neurogenesis (E12.5–E14.5 in mouse), two BAF155 subunits are highly expressed (the esBAF complex). BAF155 inhibits the euchromatin state of Pax 6 target genes. This means the DNA promoter regions for *Tbr2*, *Cux1*, and *Tle1* are not easily accessible. Thus, Pax6 cannot readily bind to the target genes and initiate their expression in proliferating cells. As subunits in the npBAF complex begin to be expressed, one of the BAF155 subunits is replaced with BAF170 (Figure 6.11). The decreased expression of BAF155 and concomitant increase in BAF170 expression leads to greater accessibility to DNA promoter regions so that Pax6 can initiate the expression of the target genes in the neural progenitor and post-mitotic neurons at the times they are needed.

These examples indicate one of the ways epigenetic regulation of transcription factor binding sites can influence whether genes necessary for determination and subsequent differentiation are expressed. In addition to the role of the BAF complex in CNS neurons, other BAF subunit complexes are associated with the differentiation of Schwann cells and oligodendrocytes. Thus, epigenetic modifications provide another means by which the limited number of available transcription factors can exert a wide range of effects in the developing nervous system.

DETERMINATION AND DIFFERENTIATION OF NEURAL-CREST-DERIVED NEURONS

The experimental accessibility of the neural crest has allowed investigators to study the fate of a number of neural-crest-derived cell populations. As discussed in Chapter 5, the fate options available to neural crest cells are probably the most varied in the nervous system, and each neural crest cell population relies on specific signals for determination and differentiation. Most neural crest cells appear to be particularly influenced by extrinsic signals encountered as they migrate from the neural folds toward their final destinations.

Environmental cues influence the fate of parasympathetic and sympathetic neurons

Neural crest cells from the caudal hindbrain through the sacral region are divided into vagal and trunk populations. Among the derivatives of the vagal and trunk neural crest cells are neurons in the parasympathetic and sympathetic divisions of the autonomic nervous system. The vagal neural crest gives rise mainly to parasympathetic neurons that innervate the gut and utilize the neurotransmitter acetylcholine. In contrast, sympathetic neurons that innervate smooth muscle cells and utilize the neurotransmitter norepinephrine (also called noradrenaline) are derived from the trunk neural crest. The vagal and trunk populations of neural crest cells have been utilized extensively to evaluate whether neural crest cell fate is predetermined or regulated by cues from the extracellular environment.

The influence of the environment on the fate options of parasympathetic and sympathetic neurons from the vagal and trunk regions of the neural tube was first described in the now-classic studies of Nicole LeDouarin and colleagues in the 1970s. These studies relied on transplantation techniques pioneered by LeDourain in which the neural crest cells of quail were transplanted to a chick embryo at a similar stage of development. In such cases, the quail cells integrate into the chick host and differentiate as if they were chick cells. Because the quail cells can be identified histologically by the increased heterochromatin in the nucleus (see Figure 3.10), investigators are able to determine the fate of the transplanted quail cells.

Using this chick–quail chimera method, LeDourain and colleagues transplanted neural crest cells from one region of the neural tube of a quail embryo to a different region of a chick neural tube. Because neural crest development occurs in a rostral-to-caudal progression, cells from the vagal region develop prior to those of the trunk region. Vagal-to-trunk and trunk-to-vagal transplantations were performed with these progressive developmental differences taken into account so that neighboring donor and host cells were at similar stages of development (**Figure 6.12**).

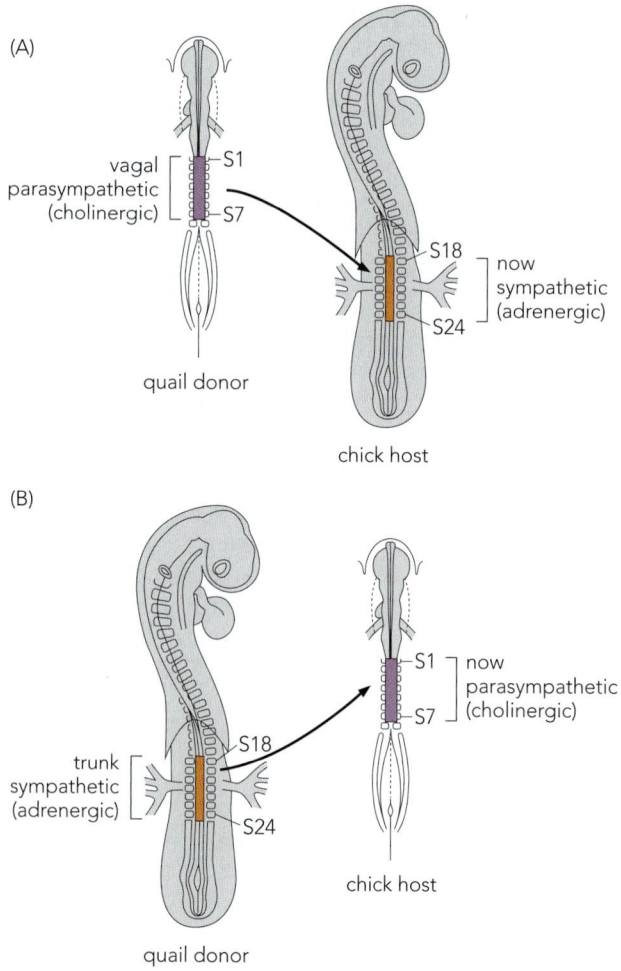

Figure 6.12 Transplantation experiments in which quail cells were grafted into a chick embryo demonstrated that quail cells adopted the fate of the host tissue. Histological differences between quail and chick cells allowed researchers to trace the fate of donor cells in the resulting chimeras. (A) When neural crest cells from the vagal region of a quail donor—cells that normally differentiate as parasympathetic, cholinergic neurons—were grafted to the trunk region of the chick host, the transplanted cells migrated along the typical route of trunk cells and adopted a sympathetic, adrenergic fate. (B) Conversely, when quail trunk neural crest cells—cells that normally become sympathetic, adrenergic neurons—were grafted to the vagal region of the chick embryo, the cells migrated along routes typical of vagal neural crest cells and adopted a parasympathetic, cholinergic fate. The size difference between the quail donors and chick hosts reflects differences in the developmental stage of the embryos. Development in the vagal region precedes trunk development, so to ensure that the transplanted donor cells are at the same developmental stage as their neighboring host cells, cells from the vagal region of a younger donor were transplanted into the trunk region of an older host (A), whereas cells from the trunk region of an older donor were transplanted into the vagal region of a younger host (B). (Adapted from Le Douarin NM [1980] *Nature* 286:663–669.)

When cells from the vagal neural crest were transplanted to trunk regions, the majority of the vagal cells now migrated along the route of trunk-derived neural crest cells and developed into sympathetic neurons. Similarly, when trunk neural crest cells were transplanted to vagal regions, the trunk cells took the expected migratory route for vagal crest cells and became parasympathetic neurons. Thus, the fate of parasympathetic and sympathetic neurons was not predetermined, but appeared to depend on cues encountered along the migratory route of the neural crest cells.

Experiments have also indicated that the differences in fate options do not result from the selective survival of neural crest cells in different regions, but are largely caused by environmental cues produced in different tissues at each axial level. Thus, it appears that neural crest cells arise from a multipotent precursor population that can give rise to a number of different cell types. More recent studies have supported the finding that environmental cues can alter neural crest fate by regulating the expression of specific transcription factors, at least during defined stages of development. Among the identified extrinsic signals are Wnt and BMP family members. Wnt signaling activates the expression of Neurogenins (Ngns) that are important for development of dorsal root ganglion (DRG) neurons, while BMPs activate transcription factors to specify subtypes of sympathetic neurons.

Wnt influences the expression of Ngn2 and Ngn1 in early- and late-migrating DRG neurons, respectively. These Ngns then activate the expression of the neuronal differentiation marker, NeuroD. Cells expressing Ngns also increase their expression of a Notch ligand called Delta-like ligand 1 (Dll1). Dll1 binds to Notch receptors expressed on surrounding cells, thus preventing those cells from developing as neurons. As a result, only those cells expressing Ngns will be able to adopt a neural fate.

Sympathetic neurons can change neurotransmitter production later in development

During normal development, all sympathetic neurons originate as adrenergic neurons that produce norepinephrine (noradrenaline) (**Figure 6.13**A). The majority of sympathetic neurons go on to innervate tissues such as skin

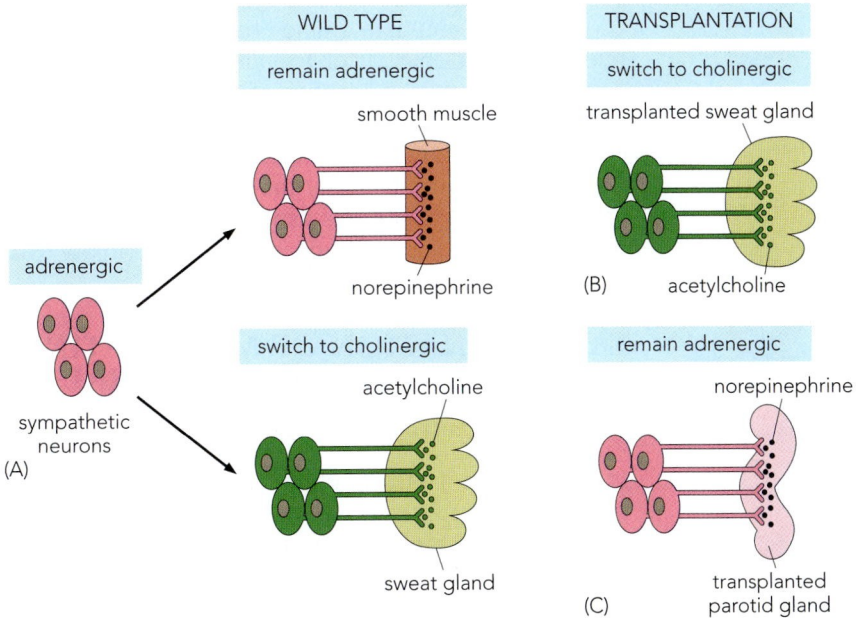

Figure 6.13 Neurotransmitter production can change during postnatal development. (A) All sympathetic neurons are initially adrenergic, producing the neurotransmitter norepinephrine (also called noradrenaline). At the time of innervation, most sympathetic neurons, such as those that innervate smooth muscle, continue to produce norepinephrine. These sympathetic neurons remain adrenergic into adulthood. However, some sympathetic neurons switch neurotransmitter fate to become cholinergic. For example, at the time that sympathetic nerve fibers innervate sweat gland tissues, their neurons begin to produce acetylcholine. (B) Transplantation studies revealed that changing the target tissue altered the neurotransmitter production of sympathetic neurons. When sweat gland tissue (a cholinergic target) replaced smooth muscle tissue (an adrenergic target), the sympathetic neurons switched from their normal adrenergic fate and became cholinergic. (C) Conversely, when the sweat gland was replaced with the parotid gland (another adrenergic target), the sympathetic neurons failed to become cholinergic, as they normally would, and instead remained adrenergic.

and smooth muscle cells, and the neurons that make these connections remain adrenergic. BMPs from the dorsal aorta appear to provide the local environmental signal that induces the production of the enzymes needed to synthesize norepinephrine. BMPs also activate various combinations of transcription factors such as members of the Phox (paired-like homeobox) family and GATA3 (GATA binding protein 3). The particular combinations of transcription factors expressed in a given cell lead to further development and differentiation of the various subtypes of sympathetic neurons.

A smaller population of sympathetic neurons innervates other target cells such as sweat glands. Although these sympathetic neurons initially produce norepinephrine, once they begin to innervate their target tissues, they stop expressing the enzymes needed for norepinephrine synthesis and begin to express the enzymes needed to make acetylcholine. Thus, these sympathetic neurons switch from adrenergic to cholinergic (Figure 6.13A). In rats, this takes place during the second postnatal week at the time of sweat gland innervation—a relatively late stage to switch an aspect of cell fate. A number of transplantation, lesion, and cell culture experiments have demonstrated that the cells that innervate sweat glands arise from the same population of sympathetic neurons, so the change in neurotransmitter is not a result of differential survival of a subset of neurons.

The signal to switch neurotransmitter production appears to come from the target tissue itself. The target tissue provides a signal that instructs the neurons to stop producing norepinephrine and start producing acetylcholine. Evidence for the role of the target tissue again comes from multiple experiments. For example, if a tissue that contains sweat glands, such as the footpad of rat, is transplanted to a region that does not have many sweat glands, such as the skin of the thoracic region, the arriving sympathetic neurons innervate the transplanted footpad and over a period of three to six weeks switch neurotransmitters, becoming cholinergic sympathetic neurons (Figure 6.13B). Conversely, when an adrenergic target, the parotid gland, is transplanted to the footpad region, the innervating sympathetic neurons remain adrenergic and do not switch to cholinergic neurons (Figure 6.13C). Similar results were found in tissue culture studies. Co-culturing adrenergic sympathetic neurons with sweat gland tissue caused the neurons to switch to cholinergic sympathetic neurons. In contrast, sympathetic neurons remained adrenergic when cultured with an adrenergic target.

Although multiple studies have confirmed that sweat gland tissue releases a diffusible factor to induce the change in sympathetic neuron neurotransmitter, the identity of the factor remains uncertain. Several candidate molecules have been identified, including cytokines such as ciliary neurotrophic factor (CNTF) and leukemia inhibitory factor (LIF). Scientists continue to study the growth factors and signal transduction cascades that regulate this aspect of sympathetic neuron differentiation, both during development and in response to injury (**Box 6.1**).

DETERMINATION OF MYELINATING GLIA IN THE PERIPHERAL AND CENTRAL NERVOUS SYSTEM

In addition to generating multiple subtypes of neurons, the nervous system also produces numerous subtypes of glial cells. Many of the glial cells adopt their cell fates after neuronal fates are specified. Among the glial types produced in the vertebrate peripheral and central nervous systems are the myelinating glia that wrap around axons to speed the conduction of action potentials.

Box 6.1 Developing Neuroscientists: Gp130 Cytokines Play Key Roles in Regulating Transmitter Phenotype During Development and in Response to Injury

Richard Zigmond received his undergraduate degree from Harvard University in 1966 and his Ph.D. from Rockefeller University in 1971. Since 1985 he has been a faculty member in the Neurosciences Department at Case Western Reserve University, where his research focuses on the changes in gene expression in sympathetic and sensory neurons after axotomy. He is currently investigating the roles of gp130 cytokines, other growth factors, and macrophages that regulate the growth of axotomized neurons.

The vast majority of sympathetic neurons use norepinephrine as their primary neurotransmitter. In the 1970s and 1980s, a group of researchers that included Edwin Furshpan, Story Landis, Paul Patterson, David Potter, and colleagues discovered that when neonatal rat sympathetic neurons from the superior cervical ganglion (SCG) were cultured with certain nonneuronal cells—for example, heart cells—they underwent a switch in the transmitter they synthesized and released. The neurons switched from producing norepinephrine to producing acetylcholine. Patterson's laboratory went on to determine that the factor released by cultured heart cells responsible for this cholinergic differentiation is leukemia inhibitory factor (LIF), a protein previously known primarily in the immune system. Landis examined whether a similar "cholinergic switch" ever occurs in sympathetic neurons *in vivo*. Basing her studies on previous observations that sympathetic innervation of adult sweat glands uses acetylcholine as its transmitter, her lab discovered that, *prior* to innervating their targets, these neurons synthesize norepinephine, but that *after* contact with sweat glands, the neurons synthesize acetylcholine—remarkably analogous to the situation that had been found in cell culture. Elegant tissue transplant studies followed and firmly established that it was the target sweat glands that acted on the sympathetic neurons to trigger the cholinergic switch.

Whether the sweat glands acted on the neurons through LIF was not determined immediately. A major advance came from the availability of animals in which the gene for LIF had been knocked out ($LIF^{-/-}$). Studies on these animals produced some rather surprising results. Landis's laboratory found that, in $LIF^{-/-}$ mice, the cholinergic switch occurred just as in wild-type animals. How do we account for the fact that while LIF can trigger the cholinergic switch in culture, it is not required *in vivo*?

It is now recognized that LIF belongs to a family of peptides that does not have a lot of amino acid sequence homology, but does have a common three-dimensional structure and acts through a common receptor system that includes the signaling subunit gp130 (see Chapter 8). These LIF-related cytokines are often referred to as **gp130 cytokines**. Further studies in neonatal sweat glands and in adult SCGs established that in fact other members of the gp130 family, in addition to LIF, were present and almost certainly are involved in switches in transmitter expression.

In adult animals, changes in the neurotransmitters that sympathetic neurons synthesize and release can also be dramatically altered in response to severing the cells' axons (axotomy). For example, SCG neurons begin to express several additional neuropeptides after axotomy, including vasoactive intestinal peptide (VIP) and galanin. Our lab found that the increases in VIP and galanin were significantly reduced, though not totally abolished in $LIF^{-/-}$ mice. This led us to question why this may be so.

Here again the development of mutant mice allowed the research to move forward. The laboratory of Hermann Rohrer made a conditional knockout of the gp130 receptor subunit in neurons synthesizing norepinephrine. The researchers found that these mice did not undergo a cholinergic switch. Instead, the neurons innervating sweat glands remained adrenergic. These studies demonstrated that these neurons required the binding of LIF-related cytokines to gp130 in order to induce changes in neurotransmitter synthesis.

Using gp130-knockout mice, our laboratory then found that the changes in SCG neurons that occur in response to neuronal injury also depend on gp130 signaling. For example, the increases in expression of VIP and galanin that are normally observed after axotomy were completely abolished in the absence of gp130. Further, increases in nerve fiber outgrowth that are typically seen *in vitro* following nerve injury were absent in neurons harvested from gp130-knockout mice. Together, these studies support the hypothesis that gp130 cytokines are necessary for the inducing characteristics of sympathetic neurons during development and in response to injury.

Neuregulin influences determination of myelinating Schwann cells in the PNS

The myelinating glia of the peripheral nervous system are the Schwann cells. Schwann cell determination and differentiation typically occur after neural fate specification has begun. Signals to induce the formation of Schwann cells and initiate the myelination of peripheral axons arise from the associated neural-crest-derived neurons. These signals include members of the Neuregulin (Nrg) family of proteins. There are four members of the Nrg family (Nrg1–Nrg4) as well as several isoforms of Nrg1. In the PNS, Nrg1 stimulates Schwann cell proliferation, survival, migration, and myelination. As discussed in Chapter 5, Nrg1 also plays a role in the development of Bergmann glia in the cerebellum.

Nrg1 was initially called glial growth factor (GGF) due to its ability to stimulate the production of glial Schwann cells *in vitro*. Since the 1990s, numerous studies have demonstrated the importance of Nrgs in Schwann cell determination and myelin formation. For example, in cell cultures of neural crest precursor cells from the dorsal root ganglia (DRG), the addition of Nrg1 led to a decrease in the number of neurons and an increase in the number of Schwann cells. Because the total number of cells remained the same, the results suggested that Nrg1 signaling suppresses neural fate while promoting Schwann cell fate (**Figure 6.14**).

The Nrgs signal through ErbB receptors that are found on neural and Schwann cell precursors. Expression studies demonstrated that the Nrg1 ligands and ErbB receptors are distributed on neuronal and Schwann cell precursors at the correct developmental stages to not only initiate Schwann cell fate but also stimulate myelination. One current model proposes that once developing neurons begin to express sufficient amounts of Nrg1, they are able to activate ErbB receptors on adjacent neural crest cells, signaling them to become Schwann cells. Axon-derived Nrg1 then stimulates the myelination process, inducing the Schwann cells to extend cytoplasmic processes to wrap around the axon (Figure 6.14). Studies continue to explore the mechanisms that govern the determination and differentiation of peripheral glial cells. Efforts are also focused on understanding the similarities and differences that underlie myelination in the PNS and CNS.

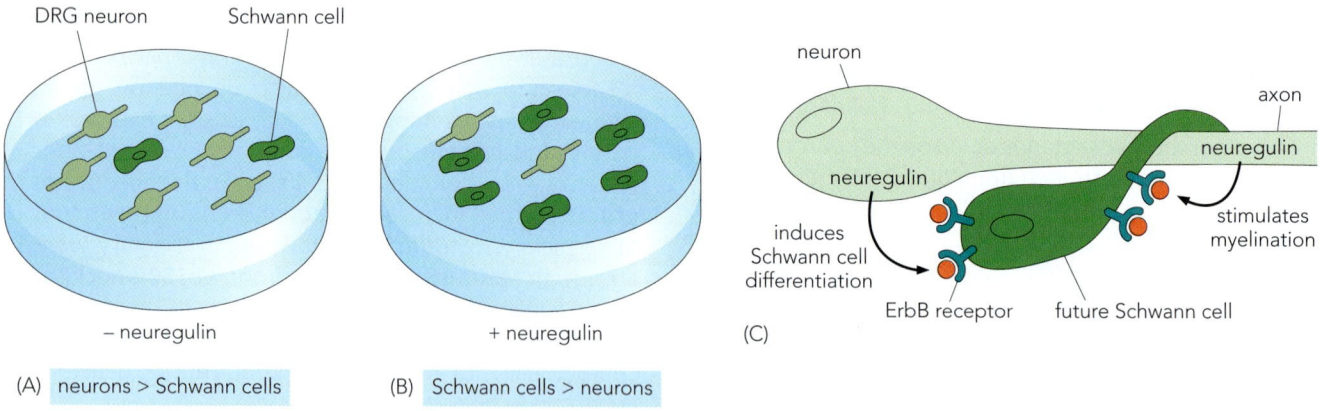

Figure 6.14 Neuregulin suppresses neuronal fate and stimulates Schwann cell determination and myelination. (A) When precursor cells from dorsal root ganglia (DRG) were placed in a cell culture dish lacking neuregulin, the majority of the precursor cells differentiated into DRG neurons. (B) When neuregulin was added to these cell cultures, the number of Schwann cells increased, although the total number of cells remained the same. These results suggested that neuregulins stimulate differentiation of Schwann cells while suppressing differentiation of neurons. (C) Peripheral neurons produce neuregulin, which binds to ErbB receptors on adjacent neural crest precursor cells, signaling those precursors to differentiate into Schwann cells. Additionally, neuregulin released by the axon initiates the myelination process so that extensions from the Schwann cell begin to wrap around the peripheral axon.

Precursor cells in the optic nerve are used to study oligodendrocyte development

The myelinating glia of the CNS are the oligodendrocytes. To investigate the mechanisms regulating determination and differentiation of oligodendrocytes, many studies have utilized preparations of the optic nerve. The optic nerve became a useful experimental system due to its relatively simple composition. The optic nerve contains the axons of the retinal ganglion cells (RGCs) and two types of glial cells: type 1 astrocytes, which contact blood vessels that run through the optic nerve, and oligodendrocytes. The type 1 astrocytes arise from the epithelial cells in the optic stalk, whereas the oligodendrocytes arise from **oligodendrocyte precursor cells** (**OPCs**) that originate in the ventricular zone near the third ventricle and migrate into the optic nerve.

Because the optic nerve contains no neuronal cell bodies, scientists have been able to design experiments that focus specifically on glial cell differentiation. Since the 1970s, scientists have made numerous discoveries showing that the glial cells in the optic nerve proliferate then form in response to a combination of extrinsic cues and intrinsic timing mechanisms. *In vitro* analysis of optic nerve cells has proved particularly useful for studying the development of oligodendrocytes. In contrast to the cellular composition of the optic nerve *in vivo*, early studies of rat optic nerve cultures revealed that three different glial types were present (**Figure 6.15**). The cultures not only included type 1 astrocytes and oligodendrocytes, but also type 2 astrocytes. A single precursor population of cells gave rise to both the oligodendrocytes and type 2 astrocytes, whereas the type 1 astrocytes were derived from a separate progenitor pool. Because of these initial observations, OPCs were originally called O2A cells—that is, these precursors gave rise to either oligodendrocytes (O) or type 2 astrocytes (2A). However, later studies revealed that type 2 astrocytes were generated only under certain cell culture conditions and did not normally contribute to the optic nerve *in vivo*. The observation that OPCs can give rise to type 2 astrocytes in cell culture demonstrated that OPCs have a degree of plasticity that allows them to differentiate into type 2 astrocytes when provided with the proper signals. *In vivo*, OPCs produce signals during development to suppress the production of type 2 astrocytes.

The process of deciphering the signaling mechanisms that regulate glial fate in the optic nerve was advanced by the identification of proteins and other antigens that are selectively present on the surfaces of specific cell types. For example, antibodies against rat neural antigen-2 (RAN-2) selectively bind to type 1, but not type 2, astrocytes. Anti-galactocerebroside (GC) binds only to differentiated oligodendrocytes, while antibodies against A2B5 bind OPCs and type 2 astrocytes. The restricted production of these antigens led to the development of a cell culture method to selectively harvest a given cell population (**Figure 6.16**). Cell culture dishes are coated with an antibody to one of the cell-type-specific antigens. When dissociated optic nerve cells are added to the dish, the cells that produce that antigen bind to the antibody and adhere to the culture surface. The adherent cells can then be studied and the loose cells can be transferred to another dish coated with a different antibody. When the final dish is coated with A2B5, a purified population of OPCs results.

By successfully isolating OPCs *in vitro*, researchers were able to determine that OPCs rely on the growth factors platelet-derived growth factor (PDGF) or neurotrophin-3 (NT-3) during the proliferative phase of development. Once the proliferation phase ends, a second signal is

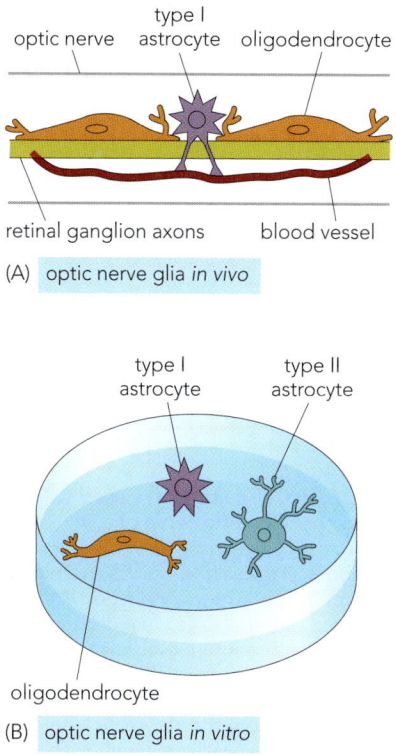

Figure 6.15 Identification of optic nerve glial types *in vivo* and *in vitro*. (A) *In vivo*, the adult optic nerve contains two glial cell types: oligodendrocytes, which myelinate axons of the retinal ganglion cells, and type 1 astrocytes, which contact blood vessels. (B) *In vitro*, optic nerve progenitor cells generate three types of glial cells: type 1astrocytes, oligodendrocytes, and type 2 astrocytes. The type 1 astrocytes are derived from one progenitor pool, while oligodendrocyte precursor cells (OPCs) give rise to the oligodendrocytes and type 2 astrocytes found in cell culture. Type 2 astrocytes are not found *in vivo* due to the presence of molecules that suppress the formation of type 2 astrocytes in the optic nerve.

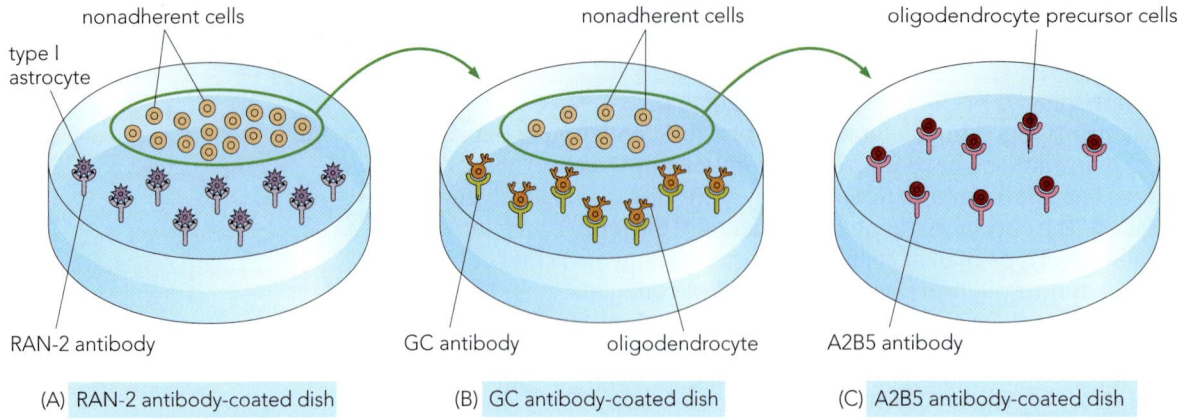

Figure 6.16 Cell cultures are used to harvest specific optic nerve glia. In this method, cell culture dishes are coated with an antibody that selectively binds an antigen on one of the optic nerve glial types. When dissociated optic nerve cells are added to the dish, cells that produce the corresponding antigen bind to the culture dish. Scientists then study the cells that adhere to the dish, or remove the loose cells and transfer them to a culture dish coated with a different antibody. In this example, the first dish (A) is coated with RAN-2 antibody, which selectively binds type 1 astrocytes. The loose cells are transferred to a second dish (B) coated with GC antibody to bind any differentiated oligodendrocytes. The final dish is coated with A2B5 antibody to bind OPCs (C). Thus, at the end of the three culture preparations, only OPCs are present in the culture dish. These cells can then be used to identify molecules that regulate differentiation of oligodendrocytes or type 2 astrocytes.

required to promote either the oligodendrocytes or type 2 astrocyte fate. Oligodendrocytes require signals such as thyroid hormone (TH) or retinoic acid (RA), whereas the type 2 astrocytes require signals such as BMPs.

In vivo studies have confirmed the necessity of the above factors for oligodendrocyte development. For example, PDGF and NT-3 are produced by type 1 astrocytes in the optic nerve to promote OPC proliferation. TH is present throughout the developing nervous system and the receptors for TH are localized to OPCs and oligodendrocytes. The necessity of TH in oligodendrocyte development was seen in rats and mice that are hypothyroid. The optic nerves of these animals revealed a decrease in the number of oligodendrocytes. Further, hypothyroid mice revealed delayed myelination throughout the CNS. Conversely, myelination is accelerated in hyperthyroid mice, further demonstrating the importance of this hormone in regulating oligodendrocyte development in the CNS.

Internal clocks establish when oligodendrocytes will start to form

Several studies have demonstrated that OPCs proliferate for a specific number of cycles before they are directed to an oligodendrocyte fate. Under normal conditions, the OPCs typically divide a maximum of eight times before they stop proliferating. This observation suggested that the cells use an internal clock to monitor the length of the proliferation phase. In recent years, scientists have identified intracellular mechanisms that help regulate this clock. For example, cyclin-dependent kinase (Cdk) inhibitors signal when cells should exit the cell cycle and initiate differentiation. The expression of two Cdk inhibitors, p27 and p57, increases in OPCs over the course of the proliferation phase and reaches a plateau at the time cells commit to the oligodendrocyte fate. These Cdk inhibitors normally inhibit the cyclinE/Cdk2 pair that drives the G1–S transition of the cell cycle (see Box 5.1). Thus, OPCs continue to proliferate until their intracellular levels of Cdk inhibitor are sufficient to end the cell cycle. When the expression of these Cdk inhibitors is kept below a certain threshold, proliferation continues.

Signals that regulate levels of p27 and p57 in OPCs have also been identified. For example, a protein called inhibitor of differentiation 4 (Id4) is expressed at high levels in proliferating OPCs, but diminishes as the cells switch from the proliferation to determination phase (**Figure 6.17**). If Id4 is overexpressed in OPCs, proliferation is extended and determination does

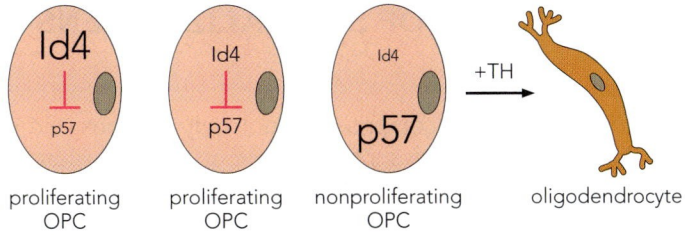

Figure 6.17 Levels of Id4 and p57 interact to regulate differentiation of oligodendrocytes. In OPCs, expression of the inhibitor of differentiation 4 (Id4) protein is highest during the proliferation phase, decreases gradually, and reaches its lowest level when the cells stop proliferating. In contrast, the expression of the Cdk inhibitor p57 is initially low, increases gradually during the proliferation phase, and peaks just prior to differentiation. Id4 normally suppresses p57 expression. Thus, when Id4 expression falls below a critical level, p57 is no longer suppressed, and cells switch from the proliferative phase to the nonproliferative phase. The nonproliferating OPC is then able to respond to differentiation factors, such as thyroid hormone (TH), and become a mature oligodendrocyte.

not occur. Id4 also interacts with p57. When Id4 is expressed at a sufficient level, it is able to suppress p57. As Id4 expression decreases over time, p57 levels are able to rise. Thus, proliferation ceases, and cells begin to respond to fate determination signals such as TH.

Identifying the signals that regulate the development of these myelinating glia is important not only in terms of understanding normal development, but also for potential therapeutic treatments in demyelinating diseases such as multiple sclerosis (**Box 6.2**). Therefore, research continues to explore the various signals and intracellular pathways that regulate the survival, proliferation, and differentiation of oligodendrocytes in the optic nerve and other regions of the CNS.

Box 6.2 The clinical significance of oligodendrocytes

Oligodendrocytes wrap around axons in the central nervous system (CNS), thus providing the myelin needed for normal transmission of signals throughout the body. There are a number of genetic diseases (such as leukodystrophies) and acquired diseases (such as multiple sclerosis) that lead to a progressive loss of myelin. When myelin is damaged or missing, nerve conduction is impaired, resulting in a variety of functional deficits. In severe cases, particularly in some of the genetic forms, death may result. The degree of functional deficit in any patient is quite variable, depending on which areas of the CNS have lost myelin. An active area of research involves studying ways to produce new oligodendrocytes or activate oligodendrocyte precursor cells that remain in the adult central nervous system. By studying the signals that regulate formation of oligodendrocyte precursor cells in the embryo and the adult, scientists hope to one day develop targeted treatments to repair areas of damage and halt the progression of these currently incurable demyelinating diseases.

In the case of demyelinating diseases, having more healthy oligodendrocytes could lead to improved function of the nervous system. However, in other cases, oligodendrocytes impede repair of the CNS damage. Unlike axons of the peripheral nervous system (PNS), the axons of CNS neurons cannot regenerate after damage. Thus, any injury to CNS axons, such as occurs in spinal cord and traumatic brain injuries, is permanent. Scientists have noted that one reason CNS axons do not regrow after damage is because of the presence of inhibitory molecules at the site of injury. Many of the inhibitory signals are found on the oligodendrocytes that axons encounter. Such inhibitory signals include myelin-associated glycoprotein (MAG), a member of the immunoglobulin superfamily, and isoforms of neurite outgrowth inhibitor (NOGO), a member of the reticulon family of membrane proteins. Both MAG and NOGO bind the same axonal receptor: NOGO-66 receptor (NgR), also called the Reticulon 4 receptor (RTN4R). Such inhibitory proteins are not found on PNS Schwann cells. In fact, studies have shown that CNS axons can regenerate and grow across Schwann cells but not oligodendrocytes.

It is not clear why oligodendrocytes in the CNS would produce inhibitory proteins. One idea is that due to the complexity of synaptic connections in the CNS, any attempt to regenerate axonal connections could result in faulty innervation patterns that might lead to undesirable behavioral consequences. Scientists are working to develop ways to overcome innate inhibitory signals so that damaged areas of CNS can be selectively treated to regrow new, functional axonal connections. These examples show how understanding the biology of just one cell type in the CNS, the oligodendrocyte, has the potential to impact a number of disease processes.

DEVELOPMENT OF SPECIALIZED SENSORY CELLS

The nervous system is comprised of not only neurons and glia but also specialized sensory cells such as those used for vision and hearing. Many of the same signaling pathways that regulate determination and initial differentiation of neurons and glia in other regions of the nervous system also influence development of these sensory cell populations. This section provides examples of how the fates of cells are determined in the *Drosophila* eye, vertebrate ear, and vertebrate retina. Each example begins with an overview of the anatomical organization of the sensory system, then explains how processes such as lateral inhibition, transcription factor cascades, or internal timing mechanisms are used to direct development of these specialized sensory cell types.

Cell–cell contact regulates cell fate in the compound eye of *Drosophila*

The *Drosophila* compound eye is made up of about 750–800 hexagonal units called **ommatidia** organized into vertical columns. Each ommatidium contains several cell types organized in a precise pattern (**Figure 6.18**). Among the cells in each ommatidium are eight photoreceptor cells located at the center. These special sensory neurons are **retinula cells**, commonly called **R cells**. Each of the eight R cells is characterized by its spectral sensitivities and connections within the brain. Additional cells in each ommatidium include the cone cells that cover the R cells and two primary pigment cells located just outside the photoreceptor cluster. There are also secondary and tertiary pigment cells and bristle cells located at the edges of the ommatidium. These cells are shared with adjacent ommatidia. The cells of the ommatidia arise from the imaginal disc, a sheet of about 20,000 cells. These cells initially are equivalent and have the potential to become any of the cell types of the ommatidium (**Box 6.3**).

In order for cells of the imaginal disc to become photoreceptor cells, the future retinula cells require signaling through receptors of the epidermal growth factor (EGF) family. If EGF signaling is blocked, the cells become one of the nonphotoreceptor cell types. However, the EGF pathway does not determine which type of photoreceptor cell (R1–R8) a cell will become. Additional local signals establish final photoreceptor cell fates. The precise location and order of photoreceptor cell determination has been recognized for over 35 years. However, because the retinula cells

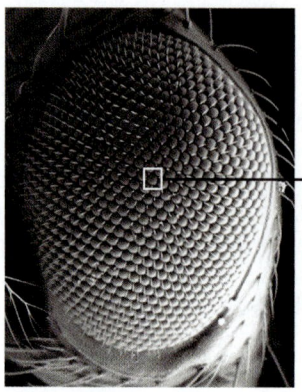

(A)

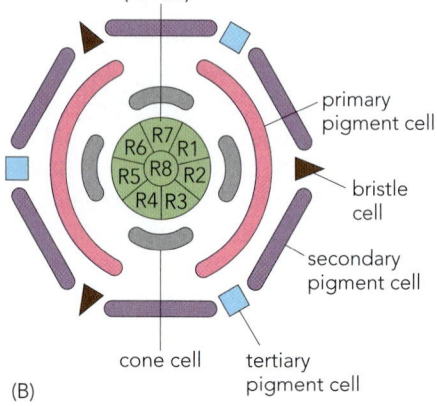

(B)

Figure 6.18 Cells types of the *Drosophila* **ommatidium.** (A) The compound eye of the adult *Drosophila* is seen in this scanning electron micrograph. Each eye consists of several hundred hexagonally shaped ommatidia. A single ommatidium is highlighted. (B) The cells of each ommatidium are arranged in a precise order. Each ommatidium is comprised of eight photoreceptor cells called retinula cells (R1–R8). The photoreceptors are surrounded by four cone cells. Two primary pigment cells lie adjacent to the cone cells. These cells are surrounded by secondary pigment cells, tertiary pigment cells, and bristle cells. (A, from Jackson GR (2008) *PLoS Biol 6*:e53.)

Box 6.3 Developing Neuroscientists: Mosaic Analysis of Cell Fate Specification

Adam Haberman is an assistant professor of Biology at the University of San Diego. He received his bachelors degree in biochemistry from the University of Texas at Austin and his Ph.D. in Cell Biology from the Johns Hopkins University School of Medicine. Using Drosophila, *Dr. Haberman's research focuses on the cellular mechanisms that differentially promote neuronal survival or neuronal degeneration.*

Concepts we now think of as biological principles were once hotly debated questions. In the 1970s, many scientists were trying to determine if cell fates were determined by a cell's lineage or its environment. If lineage determined cell fate, then the two daughter cells produced by a particular mitosis would always have the same two cell fates, no matter what cells were neighboring them. If environment determined cell fate, then a cell could only adopt a certain fate if it received specific signals from neighboring cells. The development of the eyes of the fruit fly *Drosophila melanogaster* turned out to be a useful system for addressing this question.

Fly eyes contain about 800 identical units, called ommatidia, that each have more than 20 specific cells. All of these cells come from the eye imaginal disc, a small patch of cells that are specified early in development. By just watching the patterning of the imaginal disc, it was not possible to determine if cell fates were determined by lineage or by environment. However, a genetic technique called mitotic recombination made it possible to map cell fate. Cells were irradiated to cause rearrangement of chromosome arms during mitosis. While the parental cell was heterozygous for a mutation, its two daughter cells were each different. One daughter carried two wild-type copies of the gene, and the other was homozygous for the mutation. As those two daughter cells underwent mitosis, they created copies of themselves with their unique genetics. Since little cell migration occurs during eye development, these groups of cells stayed near each other in patches called clones. The resulting eyes were a mosaic of clones, and every cell in a clone shared a lineage, since they all derived from a single cell.

These clones were only useful if they could be identified. Therefore, rearranged chromosomes carried mutations called markers, which resulted in cellular changes that were easy to see under a microscope. The most common marker was a mutation in the *white* (*w*) gene. Eye cells with a wild-type *w* gene (w^+ cells) created pigment granules filled with easily seen pigments that give the eye its red color. Cells with two mutant copies of the *w* gene (w^- cells) made no pigment granules, so the eyes appeared white. Mosaic eyes were mostly red but contained white patches created by w^- clones.

When X-rays induced mitotic recombination in random parts of the eye, each resulting mosaic eye was unique. Researchers looked a hundreds of mosaic eyes and mapped which cell fates could come from the same clone. Under the microscope, it was easy to see which cells in each ommatidium were w^-. The researchers made diagrams showing the location of w^- cells in each ommatidium and looked for patterns.

What they discovered was that there was no relationship between cell fate and lineage in the fly eye. There were no rules stating, for instance, that if an R5 cell came from a clone, the neighboring R4 cell had to come from the same clone. The fate of each cell was independent of whether or not it shared a lineage with another cell. Therefore, cell fate could not be determined by lineage, but had to be determined by environment. Today we know that cell fate decisions are determined by signaling between cells, but proof that environmental signaling occurred was a significant result at the time.

Fly biologists also used mosaic analysis to understand how these environmental signals worked. Researchers had identified mutations in two genes, named *sevenless* (*sev*) and *bride of sevenless* (*boss*), in which eyes had no R7 cells. However, it was unclear how these genes worked. To determine the signals encoded by these genes, scientists placed a marker mutation on the same chromosome arm as the mutation they wanted to follow. Then they could make clones that were all mutant for *sev* or *boss* and which could be easily distinguished from the rest of the eye. After analyzing hundreds of mutant ommatidia, some patterns became clear. There were no ommatidia containing R7 cells that lacked *sev* (**Figure 1**A). This meant that *sev* was only required in the R7 cell and must be a signal-receiving gene. Additional studies found that

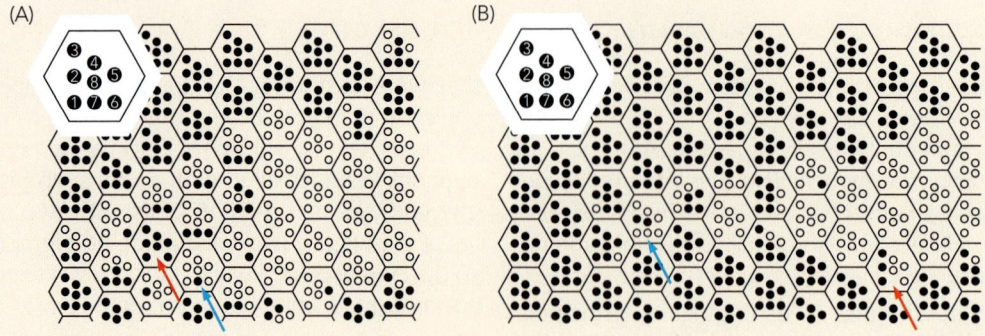

Figure 1 Mosaic analysis comparing the effects of mutations in *sevenless* and *boss* led to an understanding of how these genes functioned in R7 formation. (A) The diagram represents a cross section through many ommatidia, with each hexagon representing an ommatidium. Within each hexagon, the circles represent the R1–R8 photoreceptors. Open circles represent mutant cells that lack the *sevenless (sev)* gene. Mutant cells are detected by their lack of the cell marker, the white gene (*w–*). Filled circles represent wild-type cells (*w+*) that carry both the white marker and the *sev* gene. By analyzing hundreds of ommatidia, scientists established that the R7 cell must express the *sev* gene for an ommatidium to have an R7 cell (the middle cell in the bottom row within each ommatidium in this example). This gene was named *sevenless* because cells lacking the gene did not form an R7 photoreceptor. The arrows indicate the different cell patterns that were observed in these experiments. The red arrow points at an ommatidium in which R1–R5 and R8 are all wild type, yet R7 is missing. The blue arrow points to an ommatidium in which all cells except R7 are mutant, yet R7 is present. Collectively, these data indicated that *sev* is needed only in the presumptive R7 for R7 cell fate determination. (B) Similar methods were used to determine the role of *bride of sevenless* (*boss*) in R7 formation. The open dots indicate cells that lack the wild-type gene *boss*, while the filled circles represent those that carry the *boss* gene. In this panel, the blue arrow points to an ommatidium in which only R2 and R8 are wild type, yet R7 is present. The red arrow points to an ommatidium in which R2 is wild type, but R8 is mutant for *boss*. In this case, R7 is absent. Together, these data indicated that *boss* is specifically required in R8 for the R7 cell to form. (A, adapted from Tomlinson A & Ready DF [1987] *Dev Biol* 123:264–275. With permission from Elsevier, Inc. B, adapted from Reinke R & Zipursky SL [1988] *Cell* 55:321–330. With permission from Elsevier, Inc.)

no ommatidia that had an R8 cell lacking *boss* had an R7 cell. Therefore *boss* had to be a signal-sending gene, and it could only send the signal from the R8 cell (Figure 1B).

We now know that *boss* encodes a membrane-bound signaling protein and that *sev* encodes its receptor on the surface of the presumptive R7 cell. However, the basic principles of how these genes worked were determined without knowing what kinds of proteins they encoded. The ability to decipher so much information through manipulation of fly genetics demonstrates the power of mosaic analyses.

were identified and numbered prior to the discovery of the order in which they are determined, the numbering does not reflect the order of retinula production. In fact, the R8 cell forms first and is required in order for the other cells to develop. Following R8, the R2 and R5 cells are the next to be determined and are formed at the same time. The next pair to form at the same time is R3 and R4, followed by the R1 and R6 pair, and finally R7 (**Figure 6.19**).

The R8 cell fate arises largely as the result of lateral inhibition. The expression levels of proneural genes such as *atonal1* and the activity of transcription factors Senseless and Rough contribute to the determination of R8. The *atonal1* gene and Senseless positively regulate one another. The transcription factors Senseless and Rough can repress one another, with the transcription factor expressed at the higher level prevailing (**Figure 6.20**A). Initially, the *atonal1* gene is widely expressed in stripes of 12–15 cells in the epithelial sheet of the imaginal disc (Figure 6.20B). The *atonal1* gene then activates the expression of the *senseless* gene. The expression of *atonal1* and *senseless* gradually become restricted to about 10 cells arranged in a rosette pattern (Figure 6.20C). Lateral inhibition through the Notch–Delta signaling pathway leads to activation of *E(spl)*

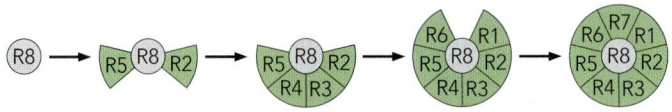

Figure 6.19 The eight photoreceptor cells are determined in a precise order. R8 is the first photoreceptor cell fate to be determined and is required for the formation of the other retinula cells. Following determination of R8, the R2 and R5 cells develop at the same time. The next pair to be determined is R3 and R4, followed by R1 and R6. R7 is the last photoreceptor cell fate to be determined.

in a subset of cells (see Figure 6.1). *E(spl)* inhibits *atonal1* expression in those cells, thus preventing them from becoming R8 cells. Three cells then remain in what is called the R8 equivalence group (Figure 6.20D). Two of the three cells will produce the transcription factors Rough and Senseless. The level of Rough in these cells is sufficient to inhibit Senseless. Thus, the cell in the equivalence group that only expresses Senseless is able to increase the expression of *atonal1* and become the R8 photoreceptor.

Cell–cell contacts and gene expression patterns establish R1–R7 photoreceptor cell types

The signaling pathways that regulate the determination of the R7 photoreceptor were among the first to be discovered. In the 1970s, Seymour Benzer and colleagues identified mutants that lacked the R7 photoreceptor cell. These mutants developed a cone cell in the usual location of the R7 photoreceptor. Benzer's lab, as well as Gerald Rubin's lab, then discovered that the mutants lack a tyrosine kinase receptor found in wild-type R7 cells. This receptor, named Sevenless (Sev) after the mutant phenotype, activates a Ras signal transduction cascade that leads to the activation of transcription factors that induce the expression of genes necessary for the R7 fate (**Figure 6.21**A). If the Ras pathway is

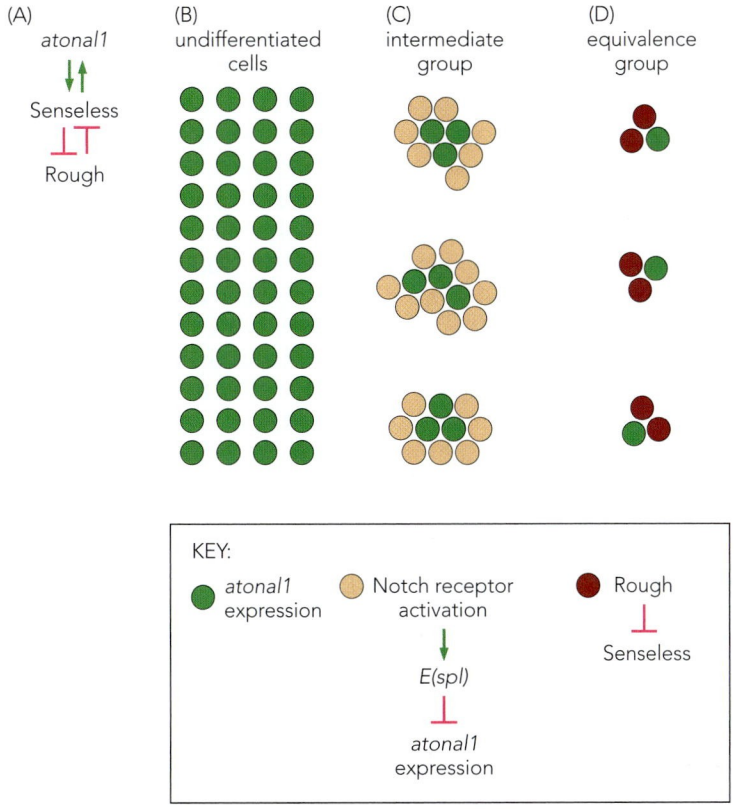

Figure 6.20 Transcription factor levels and *atonal1* expression identify cells that have the potential to form the R8 photoreceptor cell. (A) *Atonal1* and the transcription factor Senseless positively regulate the expression of one another while the transcription factors Senseless and Rough repress one another. When levels of Senseless are higher than those of Rough, Senseless inhibits Rough. When Rough levels are higher, Senseless becomes repressed. (B) *Atonal1* and *senseless* are initially expressed in stripes of 12–15 cells in the epithelial sheet that gives rise to the *Drosophila* eye (green cells). However, as development continues, *atonal1* and *senseless* expression become progressively more restricted. (C) Expression of *atonal1* and *senseless* are first limited to a rosette containing about 10 cells. In a subset of cells, Notch receptor activation leads to increased expression of *Enhancer of split* (*E*[*spl*]), which inhibits, in turn, the expression of *atonal1* (tan cells), thus leading to a decrease in *senseless* expression as well. These cells are then prevented from becoming an R8 cell. (D) Three cells remain in the R8 equivalence group. Two cells express the transcription factor Rough (red cells) that is present at levels sufficient to inhibit Senseless and prevent formation of an R8 cell. In cells lacking high levels of Rough, Senseless levels are sufficient to promote expression of *atonal1* and those cells adopt the R8 fate (green cell in each equivalence group). (B–D, adapted from Tsachaki M & Sprecher SG [2012] *Dev Dyn* 241:40–56.)

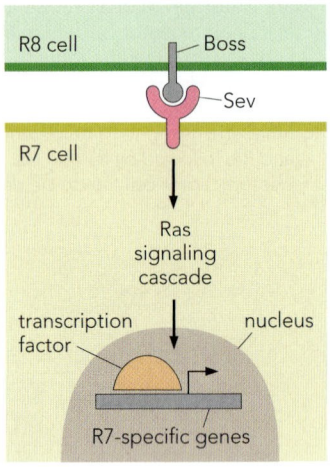

(A)

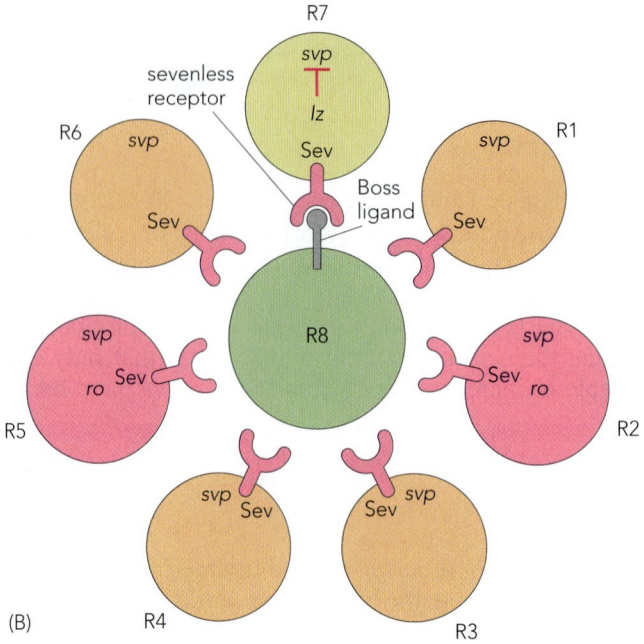

(B)

Figure 6.21 Signaling pathways that lead to the formation of R7 and the remaining photoreceptors. (A) The discovery of a *Drosophila* mutant that lacked the R7 cell led to the identification of a tyrosine kinase receptor called Sevenless (Sev). The Sev receptor is expressed by the R7 photoreceptor cell and activated by the Bride of Sevenless (Boss) ligand expressed on R8. When Boss binds to Sev, the Ras signal transduction cascade is activated, which leads to the formation of R7. In the absence of the Sev receptor or if the Ras pathway is experimentally inactivated, a cone cell is produced instead. Thus, activation of the Ras pathway is necessary for R7 formation. (B) R1–R6 cells also contact R8 and transiently express the Sev receptor. However, the expression of other genes directs these cells to different photoreceptor fates. For example, the R1, R3, R4, and R6 cells (orange) express *seven up* (*svp*), which prevents continued expression of the Sev receptor and therefore those cells do not adopt an R7 fate. In R7 (yellow), *lz* inhibits *svp* so that Sev can remain expressed. R2 and R5 (pink) require the expression of other transcription factor encoding genes, such as *rough* (*ro*), to differentiate as non-R7 cells.

experimentally inhibited, then the prospective R7 cell becomes a cone cell instead, just as when the Sev receptor itself is absent. The Sev receptor is activated by a membrane-bound ligand found on the surface of the centrally located R8 cell. This ligand, identified by Lawrence Zipursky and colleagues, was named Boss (Bride of Sevenless). Boss activates the Sev receptor on R7 to initiate the Ras signaling pathway. In the absence of Boss, the R7 cell fails to form.

Perhaps surprisingly, even though only one cell adopts the R7 fate, the Sev receptor is expressed transiently in other cells, including the R1–R6 photoreceptors and the cone cells. Because the cone cells are not normally in contact with R8 and Boss is not diffusible, it makes sense that these cells do not normally adopt an R7 fate despite expressing the Sev receptor. However, the R1–R6 photoreceptors that also express the Sev receptor are in contact with R8. Yet, these photoreceptors still do not adopt an R7 fate. One reason R1–R6 cells do not acquire the R7 fate is because these photoreceptors also express additional genes for transcription factors, such as *seven up* (*svp*), which blocks cells from adopting the R7 fate. For example, R1, R3, R4, and R6 express the Sev receptor but also express *svp* and therefore do not form R7 cells even though they are in contact with R8 (the orange cells in Figure 6.21B). The importance of *svp* in normal photoreceptor formation is seen in mutants lacking the gene. When *svp* is absent, additional photoreceptors adopt the R7 fate.

The importance of *svp* in preventing R7 formation is further noted by the expression of another transcription factor encoding the gene *lozenge* (*lz*). During normal R7 development, *lz* represses the expression of *svp* in R7 cells, thus preventing them from adopting any other photoreceptor fate (the yellow cell in Figure 6.21B). In R2 and R5, different transcription factors encoding genes, such as *rough* (*ro*), are present that appear necessary to prevent the formation of R7 cells (the pink cells in Figure 6.21B). Thus, specific patterns of transcription factor expression are needed to ensure that only one R7 forms and the other photoreceptors adopt the correct R1–R6 fates.

These are just a few examples of the transcriptional networks that are utilized during differentiation of the fly ommatidium. Additional networks have been identified, and more continue to be discovered, highlighting the complexity of the signaling cascades used to determine even a single cell type in the developing nervous system.

Cells of the vertebrate inner ear arise from the otic vesicle

The vertebrate inner ear processes information related to hearing and balance. All of the sensory cells, supporting cells, and neurons of the inner ear derive from the **otic vesicles**, or **otocysts.** As described in Chapter 5, each otocyst arises from the otic placode, a thickened patch of ectoderm adjacent to rhombomere 5 on each side of the hindbrain (see Figure 5.25). The otic placode invaginates to produce the otic pit, which then closes and sinks below the surface ectoderm to form the otocyst. The otocyst ultimately develops into the mature vertebrate inner ear that consists of an auditory region that senses sound and a vestibular region that processes information related to balance (**Figure 6.22**A). Both regions convey

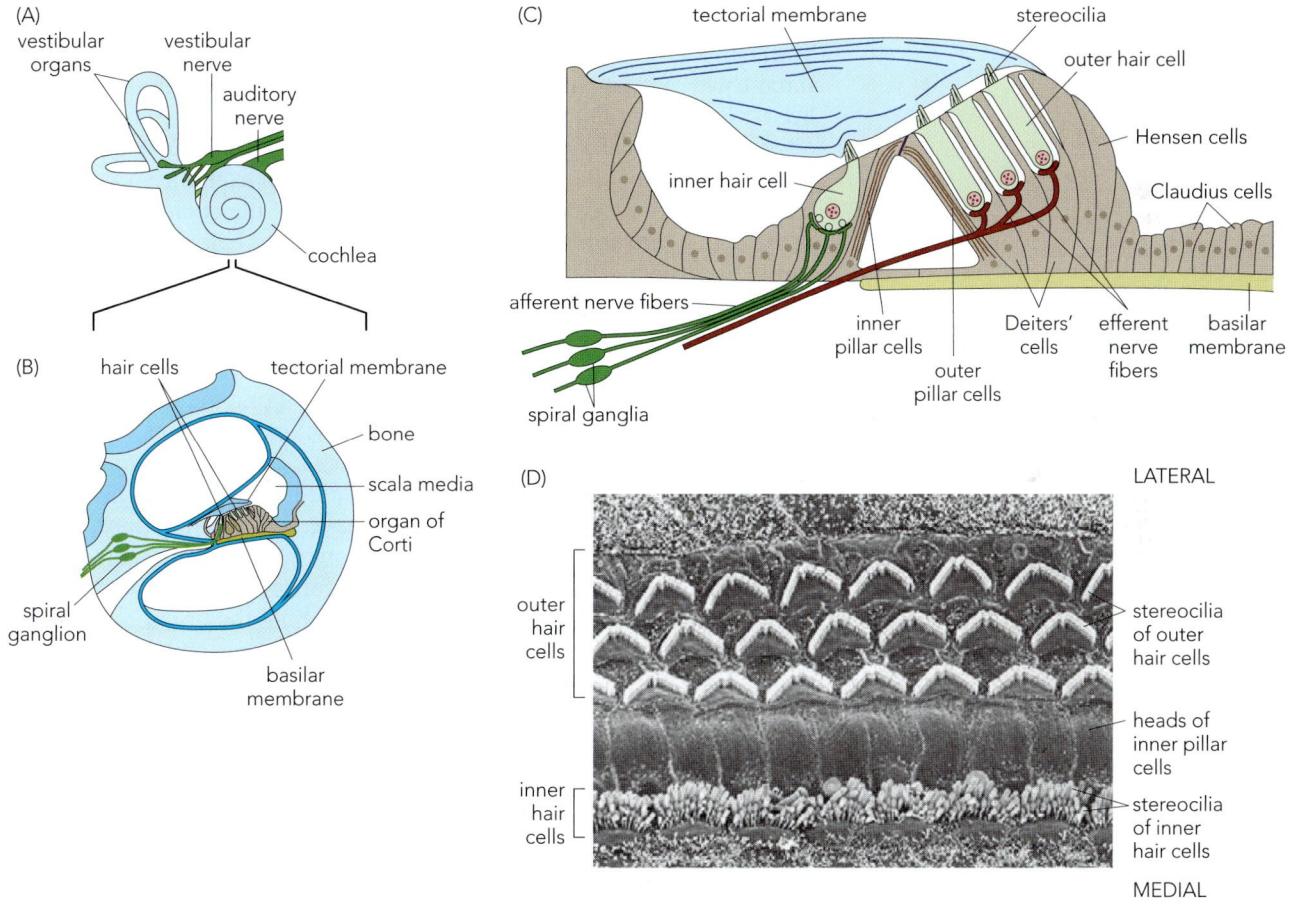

Figure 6.22 The mature inner ear is comprised of auditory and vestibular regions that have precisely organized cellular arrangements. (A) The inner ear is comprised of an auditory portion that processes sounds and a vestibular portion that processes information related to balance. Each region sends information from sensory hair cells through the corresponding segment of the eighth cranial nerve to the brain. The entire inner ear is encased in a bony capsule (blue). (B) A cross section through the cochlea illustrates the location of the auditory sensory epithelium, the organ of Corti. The organ of Corti runs the length of the cochlear spiral and lies on the basilar membrane. Much of the organ of Corti is covered by the tectorial membrane. The organ of Corti is located in the fluid-filled scala media. The spiral ganglion neurons that relay afferent information are located in the central bony core of the cochlea. (C) The cells of the organ of Corti are precisely organized. The inner hair cells are arranged in a single row located more medially, closest to the spiral ganglion neurons. The outer hair cells are organized in three rows closer to the lateral edge of the organ of Corti. The tectorial membrane lies above the stereocilia of both types of hair cells. The inner and outer hair cells are surrounded by various supporting cells, including the inner and outer pillar cells, respectively. Outer hair cells are also surrounded by the cells of Deiters. Additional supporting cells are found lateral to the outer hair cells, including Hensen and Claudius cells. (D) A scanning electron micrograph of the surface of the organ of Corti shows the organized patterning of the inner hair cells, outer hair cells, and inner pillar cells. This surface view also illustrates the organization of the stereocilia. The outer hair cell stereocilia are arranged in a "W"-like pattern, whereas the inner hair cell stereocilia have a more shallow "U"-like pattern. (D, adapted from Hudspeth AJ [2013] *Neuron* 80:536–537.)

sensory information to the brain via the corresponding branch of the eighth cranial nerve.

The sensory epithelium in both the auditory and vestibular regions of the inner ear consists of sensory hair cells surrounded by various supporting cells. The hair cells have rows of stereocilia, or "hairs," that project from the apical surface. Each stereocilium has a dense core of actin and contains ion channels that are active in signal transduction. The inner ears of species as diverse as zebrafish, chicks, and mice all feature precisely organized patterns of hair cells and supporting cells. Although the anatomical organization and patterning of these cells differs somewhat across these animal models, the signaling pathways that determine hair cell or supporting cell fate are largely conserved.

In the mature mammalian auditory system, the sensory and supporting cells lie within the organ of Corti that extends the length of the cochlear spiral (Figure 6.22B). The organ of Corti lies on the basilar membrane and is surrounded by the scala media, a fluid-filled chamber that lies between two other scalae: the scala vestibuli and scala tympani. The hair cells are aligned in distinct rows along this epithelium (Figure 6.22C). The cochleae of mammals typically contain one row of inner hair cells and three rows of outer hair cells, although there can be as many as four or five rows of outer hair cells in some regions. The tectorial membrane lies above the stereocilia of the hair cells. The inner hair cells are innervated by the majority of afferent nerve fibers that extend from the spiral ganglion neurons that are located in the central region of the cochlea. One branch of the spiral ganglion neurons extends to the inner hair cells and the other to the brainstem. The outer hair cells, in contrast, receive fewer afferent fibers, but are innervated by most of the efferent auditory nerve fibers (Figure 6.22C). The cell bodies of the efferent nerve fibers originate in brainstem nuclei. This efferent innervation influences motility of the outer hair cells and helps modify acoustic vibrations along the organ of Corti.

The various supporting cells, which include the pillar cells and the cells of Deiters, Henson, and Claudius, are located below and adjacent to hair cells and may extend cellular processes to surround the hair cells (Figure 6.22C). These supporting cells have various functions, such as providing structural support to the organ of Corti, maintaining ionic homeostasis, and clearing excess neurotransmitters from around hair cells. One of the most striking observations about the organ of Corti is the very precise and consistent organization of the hair and supporting cell populations (Figure 6.22D).

Notch signaling specifies hair cells in the organ of Corti

The patterning of cell types within the mammalian organ of Corti is established using many of the same signaling mechanisms that specify cells in the proneural regions of *Drosophila*. The cells that develop along the cochlear duct are initially homogeneous in appearance. However, the transcription factor Sox2 (Sry-related HMG box 2) soon begins to be expressed in a restricted area, giving rise to the prosensory region. Similar to the role of the PNC in *Drosophila*, the prosensory region designates the cells that have the ability to become hair cells or supporting cells of the organ of Corti (**Figure 6.23**A). Cells within this prosensory region first proliferate to ensure that a sufficient pool of precursor cells is obtained. Once sufficient precursors are available, the cells require additional signals to differentiate into sensory hair cells or supporting cells.

Hair cells are the first cell type to differentiate in the developing organ of Corti and rely on the proneural bHLH transcription factor Atonal homolog 1 (Atoh1; also called Math1 in mice). *Atoh1/Math1* is first expressed at the

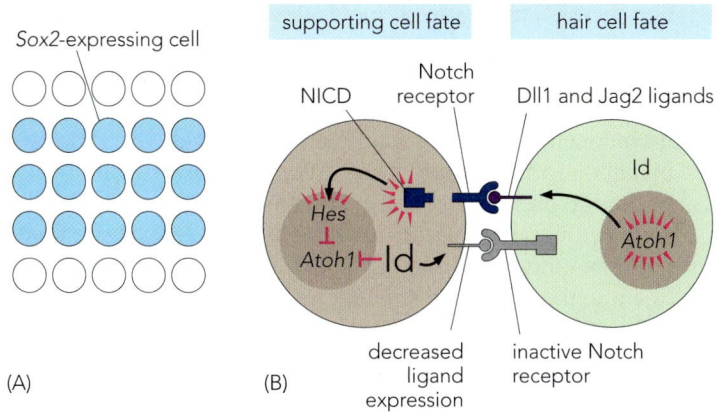

Figure 6.23 Lateral inhibition establishes hair cell and supporting cell fates. (A) *Sox2* is expressed in a subset of cells in the developing cochlear epithelium. The *Sox2*-expressing cells (blue) form a prosensory region that is capable of differentiating into hair cells and supporting cells. (B) Lateral inhibition determines which cells in the prosensory region will become supporting cells (tan) and which will become hair cells (green). Hair cells require the activation of the *atonal homolog-1* (*Atoh1*) gene that promotes expression of the ligands Delta-like1 (Dll1) and Jagged2 (Jag2), which bind and activate the Notch receptor on an adjacent cell. The activated Notch intracellular domain (NICD) travels to the nucleus, where it activates *Hairy/Enhancer of Split (Hes)* genes that inhibit *Atoh1*, thus suppressing hair cell fate and leading to a supporting cell fate. In addition, elevated levels of Inhibitors of differentiation (Id) proteins inhibit *Atoh1*, leading to decreased Dll1 and Jag2 expression and thus preventing supporting cells from adopting the hair cell fate. In contrast, cells with low levels of Id proteins are able to maintain expression of *Atoh1* and differentiate as hair cells. (Adapted from Puligilla C & Kelley MW [2009] in Encyclopedia of Neuroscience [LR Squire ed], pp 999-1004.

time of terminal mitosis in the subset of cells that will become hair cells. Studies evaluating mice that lack the gene have shown that *Atoh1/Math1* is required for hair cell differentiation. In these mice, there was a loss of hair cells, as well as a secondary, indirect loss of supporting cells. Further evidence for the importance of *Atoh1/Math1* in hair cell development was seen when scientists misexpressed the gene in regions of the organ of Corti that do not normally produce hair cells. The misexpression led to hair cell formation in these regions. Thus, *Atoh1/Math1* activation is needed to specify the hair cell fate.

Similar to the early events described above that initiate neuronal cell fate in *Drosophila* and *Xenopus, atonal*-related genes initiate the sensory hair cell fate by regulating the expression of mammalian cell surface ligands that activate the Notch receptor. In the inner ear *Atoh1/Math1* induces expression of the ligands Jagged2 (Jag2) and Delta-like1 (Dll1), both of which bind and activate the Notch receptor pathway. Notch receptors are initially expressed throughout the developing sensory epithelium. However, Jag2 and Dll1 ligands are expressed only in those cells that become hair cells. Ligand expression in the future hair cells appears to result from a decrease in the expression of a group of helix-loop-helix proteins called the Inhibitors of differentiation (Id). Id proteins block *Atoh1/Math1* expression so that cells with elevated Id levels cannot produce Jag2 or Dll1 ligands. However, in the future hair cells, reduced Id expression allows for sufficient *Atoh1/Math1* that is needed to induce ligand expression. The Jag2 and Dll1 ligands then bind to the Notch receptors on adjacent cells and activate the pathways that lead to formation of one of the supporting cell types. Therefore, in the inner ear, lateral inhibition initiated by the ligands Jag2 and Dll1 expressed in future hair cells prevents adjacent cells from adopting a hair cell fate (Figure 6.23B).

Under normal conditions, Notch signaling promotes supporting cell differentiation by activating *Hairy/Enhancer of split-1 (Hes1)* and *Hes5*, two downstream targets of mammalian Notch that function similar to the *En(spl)* gene in *Drosophila*. *Hes1* and *Hes5* also inhibit *Atoh1/Math1* to prevent hair cell fate in Notch-activated cells. Evidence for the importance of this pathway was seen when the *Notch* gene was disrupted in mice. Decreased Notch signaling led to increased production of hair cells—particularly inner hair cells. Conversely, overexpression of *Hes1* led to a decrease in hair cells and an increase in supporting cells. Thus, much like the patterning of neuronal and nonneuronal cells in *Drosophila* neuroepithelium (see Figure 6.1), the sensory hair cells and nonsensory supporting cells of the organ of Corti rely on the Notch signaling pathway and associated *Hes* genes to establish cell types in a precise and orderly array.

While the necessity of the Notch signaling pathway in hair cell determination is well established in the organ of Corti, additional transcription factors and signaling molecules are produced in the developing sensory epithelium and play various roles in regulating the survival and differentiation of hair cells. Thus, lateral inhibition establishes which cells are fated to become hair cells, but other signals are needed for differentiation into mature hair cells. Scientists continue to investigate the various signals involved in hair cell and supporting cell differentiation in a variety of animal models. In mammals, hair cells are unable to regenerate after terminal mitosis. However, in other species, such as birds, hair cells can regenerate following trauma induced by excessive noise or exposure to certain drugs. By investigating the mechanisms that regulate hair cell differentiation during development, scientists hope to one day stimulate similar pathways in the adult to produce new hair cells and alleviate hearing loss in humans. This is one more example of how discoveries stemming from basic experimental research can shape potential therapeutic treatments.

Cells of the vertebrate retina are derived from the optic cup

The sensory epithelium of the vertebrate eye processes visual stimuli. The components of the mature eye derive largely from extensions of the embryonic forebrain and surrounding tissues. The sensory epithelium of the eye first begins to form in the anterior region of the neural plate. The neural plate forms bilateral optic grooves as the neural folds begin to curve upward. Once the neural tube begins to close, these grooves evaginate, extending outward. With the closure of the neural tube, these extensions form the optic vesicles (**Figure 6.24**). As the diencephalon continues to expand, the optic vesicles also extend outward, eventually contacting the surface ectoderm. This contact induces the surface ectoderm to form the lens placode. The lens placode invaginates to form the lens pit then ultimately pinches off to form the lens vesicle that is the precursor to the adult lens. At the same time, the optic vesicles invaginate to form the optic cups. The anterior portion of the optic cup gives rise to the iris, the colored portion of the adult eye. In the posterior region of the optic cup, two distinct layers are formed—namely, the neural retina at the inner layer and the future pigment epithelium at the outer layer. The optic stalk represents the remaining connection between the forebrain and eye and later forms the optic nerve. The optic nerve contains the axons of the retinal ganglion cells (RGCs) that transmit visual information to the CNS. Structures such as the cornea and ciliary muscles are not derived from the optic cups, but instead originate from multiple tissues, including the surrounding surface ectoderm and mesoderm, as well as neural crest and mesenchymal cells.

Visual stimuli in the form of light enters the front of the eye and passes through the cornea and lens to reach the retina at the back of the eye (**Figure 6.25**A). The retinal sensory epithelium is organized into a laminar structure consisting of six neural cell types and a specialized population

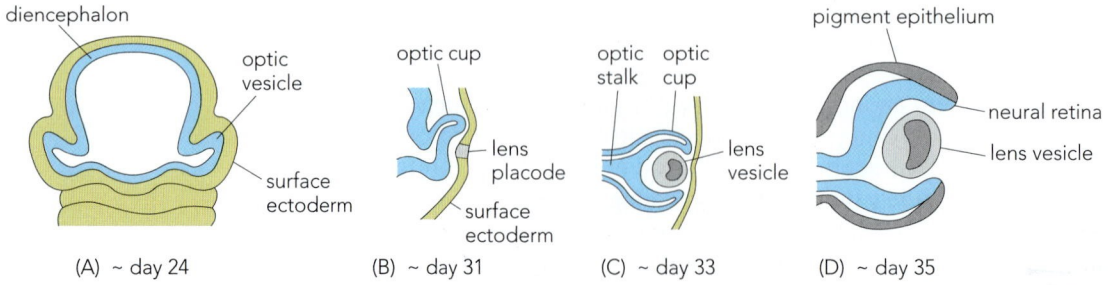

Figure 6.24 The optic vesicles give rise to the structures of the vertebrate eye. (A) The structures of the vertebrate eye arise from optic vesicles that form as extensions of each side of the diencephalon. (B) Each optic vesicle continues to expand outward until it reaches the surface ectoderm where it induces the ectoderm to form the lens placode. The optic vesicle also begins to invaginate, thus forming the optic cup. (C) The lens placode subsequently invaginates to form a lens pit (not shown), which then pinches off to form the lens vesicle. The optic cups continue to expand and the optic stalk, the remaining connection between the forebrain and the developing eye, is established. The optic stalk later forms the optic nerve. (D) The inside layer of the optic cup gives rise to the neural retina, while the outer layer forms the pigment epithelium. The lens vesicle gives rise to the lens of the adult eye. In the human embryo, these events take place at approximately gestation days 24–35. (Adapted from Larsen, WJ [1993] Human Embryology. With permission from Churchill Livingstone.)

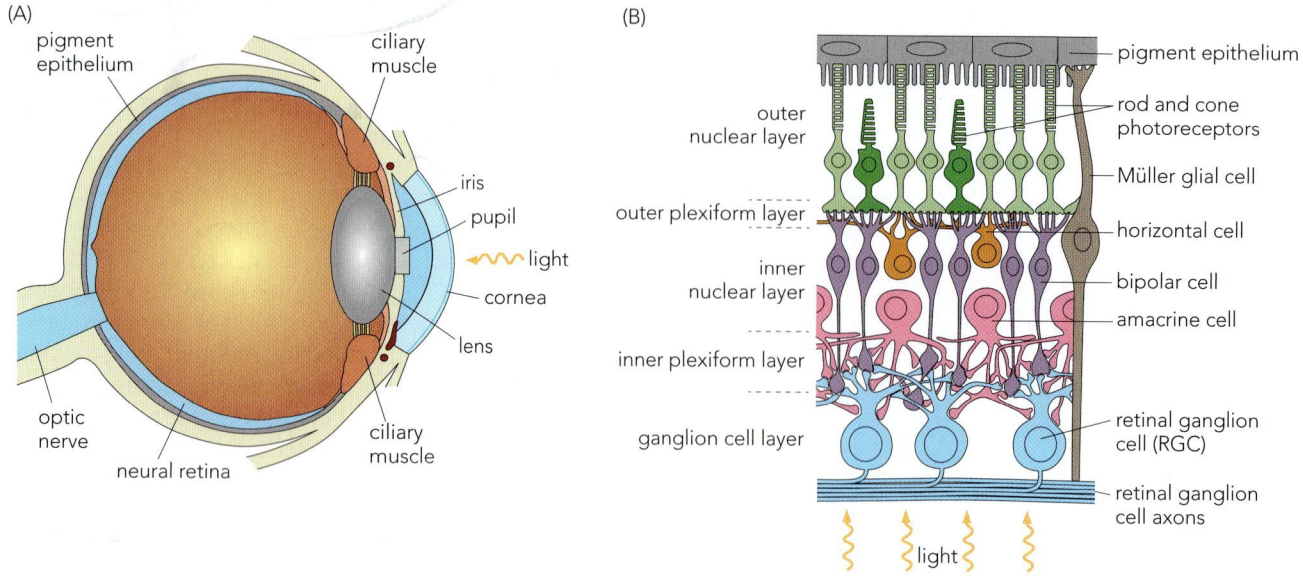

Figure 6.25 The mature eye is comprised of precisely organized structures designed to process light stimuli. (A) The adult eyes respond to light stimuli that are then processed as visual information. Light enters the eye through the cornea and lens. The iris regulates the diameter of the pupil in response to the intensity of light, and the ciliary muscles modify the shape of the lens for near and far vision. The light then reaches the neural retina, the visual sensory epithelium that lies just in front of the pigment epithelium at the back of the eye. Axons of the retinal ganglion cells form the optic nerve. (B) The sensory epithelium of the neural retina consists of several precisely organized cellular layers: the outer nuclear layer, the inner nuclear layer, and the retinal ganglion cell layer. Between these cellular layers are areas of synaptic contacts called plexiform layers. The outermost cellular layer, closest to the pigment epithelium, is the outer nuclear layer. The outer nuclear layer contains the cell bodies of the rod and cone photoreceptor cells that detect light stimuli. The pigment epithelium is closest to these cells because they help remove and recycle molecules necessary for photoreceptor function. The inner nuclear layer contains cell bodies of the bipolar, horizontal, and amacrine interneurons and the Müller glia. The processes of the Müller glia span the retinal layers. The retinal ganglion cells are found in the innermost layer, closest to the lens. This means light must pass through the ganglion cell layer to reach the rods and cones. The rods and cones then relay information back through the interneurons in the inner nuclear layer to stimulate the retinal ganglion cells. The axons of the retinal ganglion cells travel in the optic nerve to the brain.

of glial cells called the Müller glia. In the mature retina, RGCs are located in the layer closest to the lens and furthest from the retinal pigment epithelium (Figure 6.25B). The rod and cone photoreceptors are found in the outermost retinal layer, adjacent to the pigment epithelium. Situated between these layers are interneurons. Light enters the eye and passes through the layers of ganglion cells and interneurons before reaching the rod and cone photoreceptors. The photoreceptors then relay signals through the interneurons to reach the ganglion cell layer. The RGCs in turn project nerve fibers to the brain via the optic nerve. The organization of the retinal layers can seem counterintuitive, since the light must travel through the retina to reach the photoreceptors that then signal back to ganglion cells at the innermost layer. This seemingly flipped arrangement appears to reflect the importance of having the photoreceptors adjacent to the retinal pigment epithelium that helps remove and recycle molecules needed for photoreceptor function.

The names of the retinal layers are based on their histological appearance. Cell bodies are found in nuclear layers, and synaptic contacts are found in plexiform layers (Figure 6.25B). The outer nuclear layer contains the cell bodies of the rods and cones, whereas the inner nuclear layer contains cell bodies of the various interneurons and the Müller glia. The interneurons of the retina include horizontal, bipolar, and amacrine cells. These cells relay or modify information transmitted from photoreceptors to RGCs. Spanning the layers of the retina are the processes of the Müller glia cells. Like other glia, Müller glia have diverse roles, such as serving as a source of stem cells, providing nutritional support to retinal cells, and contributing to signal transduction cascades.

The vertebrate retina cells are generated in a specific order and are organized in a precise pattern

All of the retinal cell types arise from a single population of epithelial cells, the **retinal progenitor cells** (**RPCs**). In most vertebrates, the order in which the retinal cells differentiate is similar. Typically, the first cells generated are the retinal ganglion cells (RGCs) followed in sequence by the horizontal cells, cone cells, and amacrine cells. These four cell types are the early-generated cell types. The late-generated cell types include the rod cells, bipolar cells, and lastly the Müller glia (**Figure 6.26**). Though the RPCs give rise to cells of different fates in a specific order that are categorized as "early" or "late," the timing of production for the different cell types often overlaps. Thus, early-generated cells, such as ganglion cells and amacrine cells, are still being produced as the late-generated cells, such as the rod photoreceptors and bipolar cells, begin to be generated. Numerous studies in various species have demonstrated that the neuroepithelial cells of the developing retina have the capacity to produce any of the seven cell types, at least during a limited period of development. The sequential production of the different types of retinal cells suggests that the cellular environment into which a cell is born will change over time. Therefore, retinal cells produced at different developmental stages are exposed to different extrinsic factors that activate, in turn, various combinations of transcription factors to determine retinal cell fate.

As often occurs in neural development, many of the mechanisms that regulate retinal development are also used in other regions of the nervous system. For example, Notch and the downstream signaling molecules, Hes1 and Hes5, play a role in retinal cell development. As in other neural regions, one function of these signals is to keep the RPCs in a proliferative state until a sufficient progenitor pool is available. Once Notch signaling decreases in a group of RPCs, those cells begin to express various transcription factors that direct the cells to a particular retinal cell fate.

Within the group of cells that gives rise to the early-generated retinal cells, *Math5*, another member of the *atonal* gene family, has been shown to be important for the differentiation of RGCs, whereas *NeuroD* and *Math3* have been shown to be important for the differentiation of amacrine cells. Mice lacking the *Math5* gene revealed a decrease in the number of RGCs and an increase in amacrine cells. Conversely, when both *NeuroD* and *Math3* genes were mutated, the number of amacrine cells decreased and the number of RGCs increased (**Figure 6.27**).

The formation of later-generated bipolar cells also requires expression of bHLH transcription factors, but in this case the cells require *Mash1* and *Math3*. Deletion of *Mash1* and *Math3* genes in mice results in the loss of bipolar cells and an increase in the latest born cells, the Müller glial cells. Together, these observations demonstrate that while RPCs have the capacity to become multiple cell types, the expression of specific transcription factors at each stage of embryonic development ensures that

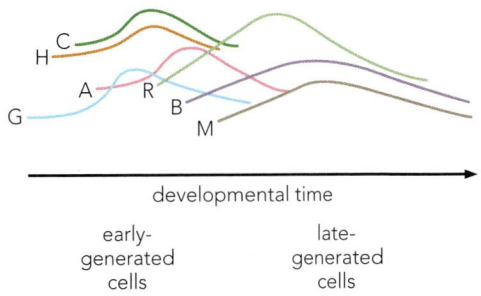

Figure 6.26 Temporal sequence of retinal cell type formation. All retinal cells arise from a single population of retinal progenitor cells. Determination of retinal cell subtypes depends largely on the temporal order in which they arise. Early-generated cells include the retinal ganglion cells (G), horizontal (H), cone (C), and amacrine (A) cells. The peak production of these cell types occurs during embryonic development. Late-generated fates include the rod (R), bipolar (B), and Müller (M) cells. The production of these cell types peaks in late embryogenesis and in some species, such as rat, continues into early postnatal life. (Adapted from Napier HRL & Link BA [2009] in Developmental Neurobiology [G Lemke ed], pp 251–258. With permission from Elsevier.)

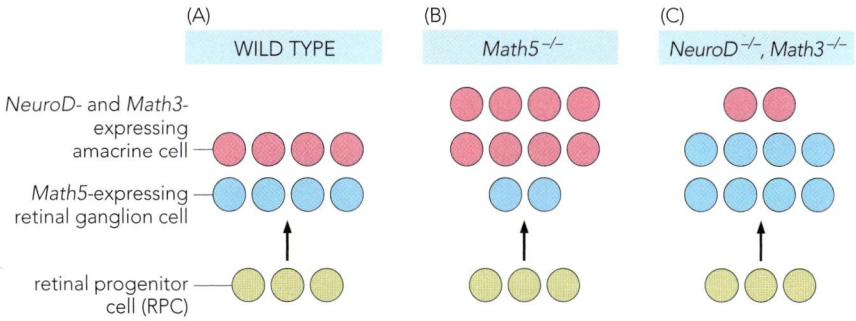

Figure 6.27 Transcription factors help regulate the fate of early-generated retinal cell types. The retinal progenitor cells (RPC) give rise to all cell types of the retina, including the early-born retinal ganglion cells and amacrine cells. Each cell type expresses specific transcription factors that lead to a particular retinal cell fate. (A) In wild-type mice, retinal progenitor cells give rise to retinal ganglion cells (RGCs) that express *Math5* and amacrine cells that express *NeuroD* and Math3. (B) In mice lacking the gene for *Math5* (*Math5*$^{-/-}$), there is a decrease in the number of RGCs produced, but an increase in the number of amacrine cells. (C) Conversely, in mice lacking genes for both *NeuroD* and *Math3* (*NeuroD*$^{-/-}$, *Math3*$^{-/-}$), the number of amacrine cells decreases but the number of RGCs increases. Together these studies show the importance of proper transcription factor expression in generating early-born cell types of the retina.

the correct fate is generated at the proper time. Any alteration in transcription factor expression can lead to a dramatic shift in the types of retinal cells produced.

Temporal identity factors play a role in vertebrate retinal development

The precise timing of expression for a number of transcription factors, in addition to those illustrated above, is likely required to ensure that the proper cell types are generated at the correct times in retinal development. Studies of the mouse retina have recently detected genes related to the *Drosophila hunchback* and *castor* (see Figure 6.7). As noted earlier, the ortholog of *hunchback* is *Ikaros*, also called *Ikaros/Znfn1a1* (*Ikzf1*). The ortholog of *castor* is *Casz1*. In mice, *Casz1* is downstream of *Ikaros*, just as *castor* is downstream of *hunchback*. *Casz1* expression is not detected during the early stages of retinogenesis, but it begins to be expressed after E14.5. *In vitro* and *in vivo* studies found that Casz1 is necessary for the formation of later-generated, but not earlier-generated, RPCs, particularly for rod cells. Casz1 appears to regulate the progression to later cell fates while also preventing continued generation of early cell fates. In the absence of Casz1, early-born cells, such as horizontal, cone, and amacrine cells, increased in number, whereas the number of rod cells decreased. Ikaros represses the expression of *Casz1* in RGCs so these early cell fates are established. However, in later born cells, *Ikaros* is no longer expressed, so *Casz1* expression directs cells to later retinal cell fates. These results suggest that these transcription factors may provide temporal identity cues to vertebrate RGCs similar to the cues utilized by *Drosophila* neuroblasts.

SUMMARY

This chapter introduces some of the many ways in which cell fates are determined in the nervous system. Many of these cellular mechanisms are conserved across invertebrate and vertebrate species. The determination of cell fate typically begins with the differential expression of proneural genes, directing some cells toward neural or sensory cell fates and others toward nonneuronal, supporting fates. Subsequent development often results from a combination of environmental signals and temporally regulated transcription factor cascades that change over time. In both invertebrates and vertebrates, the timing of expression for specific transcription factors is often critical in directing a cell toward a particular fate. While illustrating the determination of many diverse cell types in various regions of the invertebrate and vertebrate nervous systems, this discussion is incomplete both in scope and detail, representing only a small selection of the ways in which nervous system cell fates are determined. Nevertheless, these selected examples provide insight into the numerous steps required for a generic neural precursor to establish a particular cell fate.

FURTHER READING

Alsiö JM, Tarchini B, Cayouette M & Livesey FJ (2013) Ikaros promotes early-born neuronal fates in the cerebral cortex. *Proc Natl Acad Sci USA* 110(8):E716–725.

Banerjee U, Renfranz PJ, Hinton DR et al. (1987) The Sevenless protein is expressed apically in cell membranes of developing *Drosophila* retina; it is not restricted to cell R7. *Cell* 51(1): 151–158.

Banerjee U, Renfranz PJ, Pollock JA & Benzer S (1987) Molecular characterization and expression of sevenless, a gene involved in neuronal pattern formation in the *Drosophila* eye. *Cell* 49(2):281–291.

Bhatt S, Diaz R, & Trainor PA (2013) Signals and switches in Mammalian neural crest cell differentiation. *Cold Spring Harb Perspect Biol* 5(2).

Billon N, Jolicoeur C, Tokumoto Y et al. (2002) Normal timing of oligodendrocyte development depends on thyroid hormone receptor alpha 1 (TRα1). *EMBO J* 21(23):6452–6460.

Brown NL, Patel S, Brzezinski J & Glaser T (2001) Math5 is required for retinal ganglion cell and optic nerve formation. *Development* 128(13):2497–2508.

Chitnis AB (1995) The role of Notch in lateral inhibition and cell fate specification. *Mol Cell Neurosci* 6(4):311–321.

Desai AR & McConnell SK (2000) Progressive restriction in fate potential by neural progenitors during cerebral cortical development. *Development* 127(13):2863–2872.

Dugas JC, Ibrahim A & Barres BA (2007) A crucial role for p57(Kip2) in the intracellular timer that controls oligodendrocyte differentiation. *J Neurosci* 27(23):6185–6196.

Francis NJ & Landis SC (1999) Cellular and molecular determinants of sympathetic neuron development. *Annu Rev Neurosci* 22:541–566.

Frantz GD & McConnell SK (1996) Restriction of late cerebral cortical progenitors to an upper-layer fate. *Neuron* 17(1):55–61.

Hafen E, Basler K, Edstroem JE & Rubin GM (1987) Sevenless, a cell-specific homeotic gene of Drosophila, encodes a putative transmembrane receptor with a tyrosine kinase domain. *Science* 236(4797):55–63.

Hirono K, Kohwi M, Clark MQ et al. (2017) The Hunchback temporal transcription factor establishes, but is not required to maintain, early-born neuronal identity. *Neural Dev* 12(1):1.

Isshiki T, Pearson B, Holbrook B & Doe CQ (2001) *Drosophila* neuroblasts sequentially express transcription factors which specify the temporal identity of their neuronal progeny. *Cell* 106(4):511–521.

Kelley MW (2006) Hair cell development: commitment through differentiation. *Brain Res* 1091(1):172–185.

Le Douarin NM (1980) The ontogeny of the neural crest in avian embryo chimaeras. *Nature* 286(5774):663–669.

Lemke GE & Brockes JP (1984) Identification and purification of glial growth factor. *J Neurosci* 4(1):75–83.

Leone DP, Srinivasan K, Chen B et al. (2008) The determination of projection neuron identity in the developing cerebral cortex. *Curr Opin Neurobiol* 18(1):28–35.

Lessard J, Wu JI, Ranish JA et al. (2007) An essential switch in subunit composition of a chromatin remodeling complex during neural development. *Neuron* 55(2):201–215.

Livesey FJ & Cepko CL (2001) Vertebrate neural cell-fate determination: lessons from the retina. *Nat Rev Neurosci* 2(2):109–118.

Lütolf S, Radtke F, Aguet M et al. (2002) Notch1 is required for neuronal and glial differentiation in the cerebellum. *Development* 129(2):373–385.

Manuel MN, Mi D, Mason JO & Price DJ (2015) Regulation of cerebral cortical neurogenesis by the Pax6 transcription factor. *Front Cell Neurosci* 9:70.

Mattar P, Ericson J, Blackshaw S & Cayouette M (2015) A conserved regulatory logic controls temporal identity in mouse neural progenitors. *Neuron* 85(3):497–504.

Narayanan R & Tuoc TC (2014) Roles of chromatin remodeling BAF complex in neural differentiation and reprogramming. *Cell Tissue Res* 356(3):575–584.

Raff M (2011) Looking back. *Annu Rev Cell Dev Biol* 27:1–23.

Reinke R & Zipursky SL (1988) Cell–cell interaction in the *Drosophila* retina: the bride of sevenless gene is required in photoreceptor cell R8 for R7 cell development. *Cell* 55(2):321–330.

Solecki DJ, Liu XL, Tomoda T et al. (2001) Activated Notch2 signaling inhibits differentiation of cerebellar granule neuron precursors by maintaining proliferation. *Neuron* 31(4):557–568.

Stanke M, Duong CV, Pape M et al. (2006) Target-dependent specification of the neurotransmitter phenotype: cholinergic differentiation of sympathetic neurons is mediated in vivo by gp130 signaling. *Development* 133(1):141–150.

Srivastava R, Kumar M, Peineau S et al. (2013) Conditional induction of Math1 specifies embryonic stem cells to cerebellar granule neuron lineage and promotes differentiation into mature granule neurons. *Stem Cells* 31(4):652–665.

Tuoc TC, Narayanan R & Stoykova A (2013) BAF chromatin remodeling complex: cortical size regulation and beyond. *Cell Cycle* 12(18):2953–2959.

Tsachaki M & Sprecher SG (2012) Genetic and developmental mechanisms underlying the formation of the *Drosophila* compound eye. *Dev Dyn* 241(1):40–56.

Willardsen MI & Link BA (2011) Cell biological regulation of division fate in vertebrate neuroepithelial cells. *Dev Dyn* 240(8):1865–1879.

Yang Y, Lewis R & Miller RH (2011) Interactions between oligodendrocyte precursors control the onset of CNS myelination. *Dev Biol* 350(1):127–138.

Yiu G & He Z (2006) Glial inhibition of CNS axon regeneration. *Nat Rev Neurosci* 7(8):617–627.

Yoo AS & Crabtree GR (2009) ATP-dependent chromatin remodeling in neural development. *Curr Opin Neurobiol* 19(2):120–126.

Zigmond RE (2011) gp130 cytokines are positive signals triggering changes in gene expression and axon outgrowth in peripheral neurons following injury. *Front Mol Neurosci* 4:62.

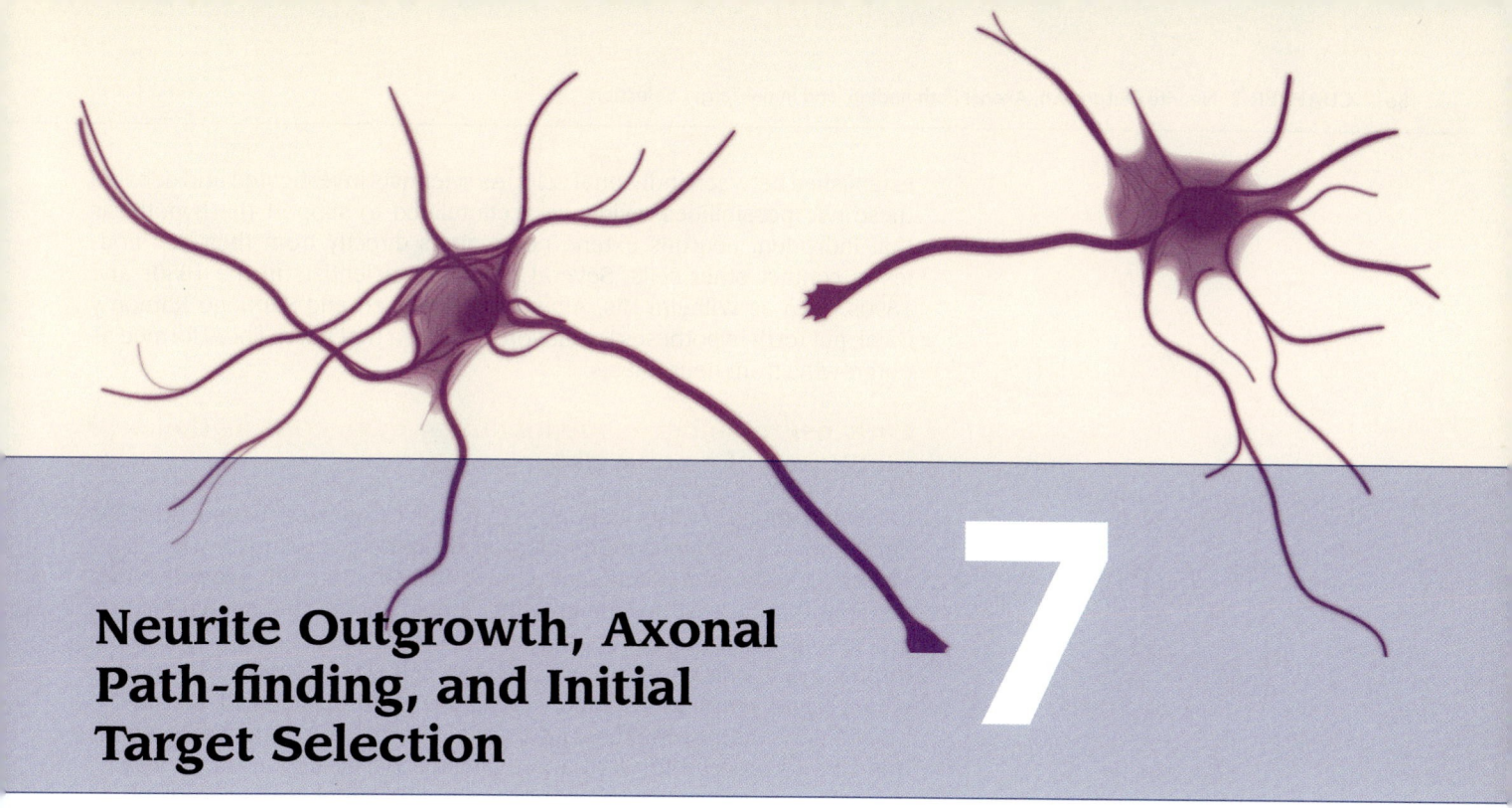

7 Neurite Outgrowth, Axonal Path-finding, and Initial Target Selection

Chapter 6 described how various cells within the nervous system first acquire a cellular fate and begin to differentiate. There is a further aspect of cellular differentiation that is unique to neurons. Unlike other embryonic cells, neurons extend long processes—axons and dendrites—from their cell bodies. These processes, or nerve fibers, are also referred to as **neurites**, particularly in cell culture preparations where it is not always obvious whether the process is an axon or a dendrite.

Because all aspects of nervous system function rely on synaptic communication, it is crucial that axons and dendrites make connections with the appropriate target cells. For this neural connectivity, or wiring, to be established, the extending nerve fibers must first be guided to the correct target tissue, where they will begin to seek an appropriate partner cell. Once the nerve fibers contact the proper target cell, they then begin to differentiate the cellular machinery needed to form a presynaptic nerve terminal and develop a mature synaptic connection.

Chapter 7 describes examples of commonly used signals that regulate the initial outgrowth and guidance of nerve fibers in different invertebrate and vertebrate animal models. Examples of well-characterized guidance cues that influence the growth of axons from spinal motor neurons to limb skeletal muscle, commissural interneurons to the midline of the developing spinal cord, and the retinal ganglion axons to the optic tectum are provided to illustrate common ways nerve fibers navigate within the embryo using the local environment, intermediate targets, and final target cells, respectively. Chapters 9 and 10 describe the mechanisms used to form mature synaptic connections between partner cells.

GROWTH CONE MOTILITY AND PATHFINDING

While it is now established that axons and dendrites extend from the neuronal cell body, this concept was not readily accepted by scientists of the nineteenth century. As discussed in Chapter 5, scientists in the late nineteenth century debated whether neurons were connected through a syncytial network of cytoplasmically interconnected cells or by contacts established between individual cells. As scientists investigated and debated these two possibilities, evidence accumulated to support the hypothesis that individual neurons extend nerve fibers directly from their cell bodies to contact other cells. Several prominent scientists in the 1880s and 1890s, such as Wilhelm His, Albert von Koelliker, and Santiago Ramón y Cajal, put forth hypotheses that favored the idea that nerve fibers formed as outgrowths from neuroblasts.

Early neurobiologists identify the growth cone as the motile end of a nerve fiber

Cajal and other scientists working at that time recognized that the tip of the nerve fiber was a unique morphological structure necessary for nerve fiber extension. Cajal is credited with naming this structure the **growth cone**, the term that he used for the growing, motile end of an extending neural process. In 1890, Cajal proposed how a neuron might extend a process through embryonic tissues to reach a target cell. Cajal's detailed description of a growth cone as "...endowed with exquisite chemical sensitivity, rapid amoeboid movements, and a certain motive force, thanks to which it is able to proceed forward and overcome obstacles met in its way..." turned out to be remarkably accurate, despite being based entirely on his observations of sections of fixed tissue (**Figure 7.1**). Today, time-lapse video recordings can be used to directly monitor growth cone motility and extension under different conditions. Such recordings reveal that a growth cone is in constant motion as it extends and samples the environment to identify growth-promoting and growth-inhibiting regions. The movements observed in these videos are largely consistent with Cajal's first description of growth cone behavior. Growth cone motility is similar to that of other highly motile cells such as fibroblasts. The similarities in motility have led some to refer to the growth cone as a "fibroblast on a leash."

In vitro and *in vivo* experiments confirm neurite outgrowth from neuronal cell bodies

Cajal's identification of growth cones and his description of how nerve fiber outgrowth might occur still did not settle the debate about whether nerve fibers extended directly from the cell body. However, 20 years later, in 1910, Ross G. Harrison published a paper describing the tissue culture method he developed to grow embryonic frog spinal cords in a hanging droplet of clotted frog lymph. The lymph provided nutrients and structural support for the neurons. Using this method, Harrison was able to view the extension of an axon directly from the cell body as well as the formation and branching of a growth cone (**Figure 7.2**). Although some scientists at the time criticized the work and felt tissue culture did not replicate what was happening *in vivo*, this finding is now seen as a pivotal moment in developmental neurobiology, demonstrating axonal outgrowth from individual cells while pioneering the use of tissue culture for experimental analysis—a method that continues to be of great value today.

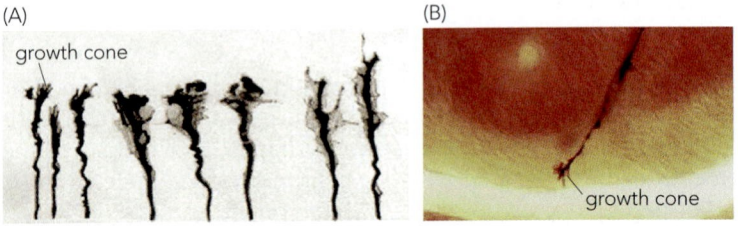

Figure 7.1 Cajal observed growth cones in fixed tissue and predicted their movements. (A) Cajal viewed growth cones in embryonic tissue sections and made detailed drawings of his many observations. In this panel of drawings, he drew the growth cones observed in an embryonic day 4 chick spinal cord. Based on his observations, in 1890 Cajal proposed that growth cones were highly motile and actively sampled the local environment to reach a target cell. (B) A photomicrograph of a single axon and growth cone extending in the embryonic chick spinal cord. This image was taken from one of Cajal's original slides. (From De Carlos, JA & Borrell, J [2007] *Brain Res Rev* 55:8–16.)

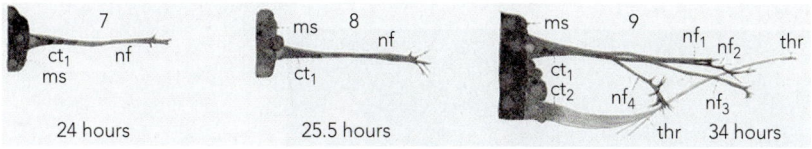

Figure 7.2 Harrison developed a tissue culture method that demonstrated the outgrowth of nerve fibers from neuronal cell bodies. In 1910 Ross Harrison reported a tissue culture method to study frog spinal cord neurons. With this method he was able to view growth cones and the extension of nerve fibers directly from neuronal cell bodies as shown in his original drawings from experiments conducted in 1908. These original sketches document the increasing length and branching of the nerve fibers at 24 hours (his panel 7), 25.5 hours (panel 8), and 34 hours (panel 9). During this period, the length of the nerve fiber (nf) increased (panel 7 and 8), then formed four distinct branches (nf1–nf4) by 34 hours. The labels he used in his drawings refer to cells (ct1, ct2), masses of cells (ms), nerve fibers (nf), and fimbrin filaments (thr). (From Harrison RG [1959] *J Exp Zool* 142:5–73.)

In vivo evidence of nerve fiber outgrowth was later reported in studies by Carl C. Speidel. In the 1930s Speidel published his first descriptions of axonal outgrowth in the tail of the tadpole. Because the tails are transparent, he was able to view neurite outgrowth in the intact animal. Thus, results from both *in vitro* and *in vivo* preparations revealed that neural processes extend from the neuronal cell body and proved that neural connections did not arise as part of a syncytial network of preformed connections.

Substrate binding influences cytoskeletal structures to promote growth cone motility

As first noted in the hanging droplet cultures conducted by Harrison, growth cones must adhere to a substrate in order to advance. In these preparations, Harrison observed that the nerve fibers attached to the fibrillar components of the lymph. Other scientists soon confirmed that growth cones could not extend in a fluid medium or on substrates that do not support attachment. It is only when the neurons are attached to a substrate that tension is established, thus allowing the growth cone to advance and the neurite to extend.

The structural components of the growth cone play a pivotal role in creating the initial attachment and extension of nerve fibers. Like many other highly motile cells, growth cones contain abundant actin filaments and microtubules organized in a manner that permits cellular movements. The actin and microtubules overlap in the cell body and proximal portion of the growth cone, but are differentially concentrated in the axon and distal portion of the growth cone. Microtubules are concentrated in the axon, whereas actin is concentrated in the distal portion of the growth cone (**Figure 7.3**A). The growth cone contains all of the cellular machinery required for motility, as demonstrated in cell culture experiments in which the fibroblast-like movements of growth cones were observed to continue for a limited time after the growth cone was severed from the axon. Thus, growth cone motility does not depend entirely on the cellular components localized to the cell body.

Within the growth cone itself, distinct regions are identified based on morphologic appearance and the cytoskeletal elements present. In the region of the growth cone that lies closest to the neurite shaft, the **central mound** or **central domain** (**C-domain**) is identified. This central region contains many organelles as well as bundles of stable microtubules that originate in the neurite shaft (Figure 7.3B). Some microtubules, called **dynamic microtubules**, also extend into the distal region of the growth cone, where the individual polymers are motile and appear to help establish the direction of growth cone advancement (Figure 7.3B, C).

The leading edge of the growth cone, a region also called the **peripheral domain** (**P-domain**), consists of sheetlike projections. These flattened projections are the **lamellipodia**—motile "veils" that contain a fanlike meshwork of filamentous (fibrillar) actin (F-actin) and subunits of globular actin (G-actin), but few organelles. The lamellipodia are found between thin projections containing rodlike bundles of F-actin. These projections, called **filopodia** or microspikes, have receptor proteins that sample the surrounding environment. If even a single filopodium touches an attractive cue, the entire growth cone will turn toward that cue. In contrast, filopodial

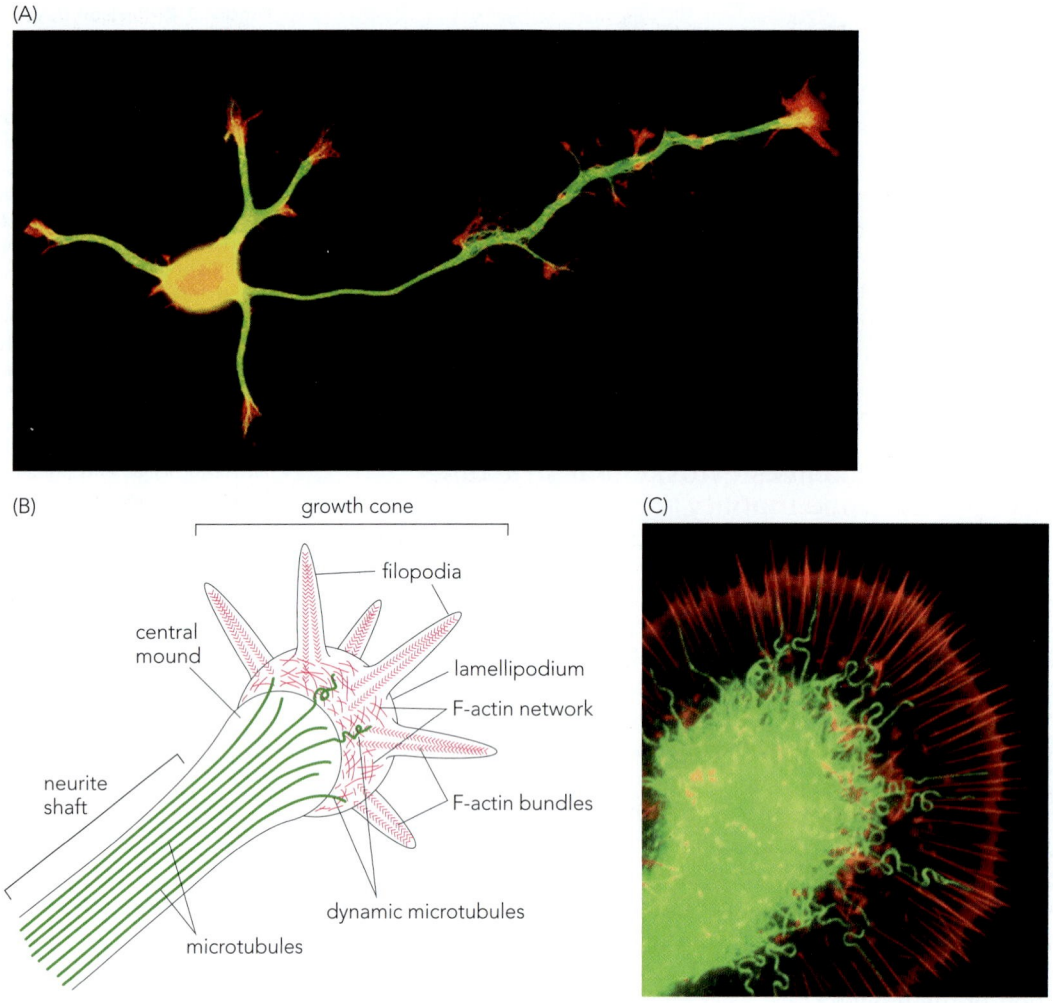

Figure 7.3 Cytoskeletal elements regulate growth cone motility and axon extension. Distinct cytoskeletal elements are concentrated in the neuronal cell body, axon, and growth cones. (A) The axon shaft (green) is characterized by microtubules. The distal region of the growth cone and emerging axonal branches are characterized by actin (red). Microtubules and actin overlap the cell body and proximal growth cone (yellow). (B) A schematic representation of the distribution of cytoskeletal elements reveals that the neurite shaft contains microtubules that extend into the central mound region at the proximal end of the growth cone. A smaller number of dynamic microtubules extend past the central mound and help direct growth cone advancement. In the distal region of the growth cone, mesh-like F-actin networks fill the flattened lamellipodium. Thin projections called filopodia extend outward from the lamellipodium. Filopodia, which contain bundled F-actin, sample substrates and determine which pathway a growth cone will follow. (C) Photomicrograph of a growth cone showing microtubules labeled green and actin labeled red. Note the dynamic microtubules extending into the distal region of the growth cone and the actin bundles concentrated in the filopodia. (A, from Kalil K, Szebenyi G & Dent EW [2000] *J Neurobiol* 44:145–158; B, adapted from Lowery LA & Van Vactor D [2009] *Nat Rev Mol Cell Biol* 10:332–343; C, courtesy of Elizabeth Hogan.)

contact with an inhibitory cue causes the growth cone to turn away or temporarily collapse back toward the axon shaft.

In order for a growth cone to advance, the neuronal cell body and emerging nerve fibers must be attached to a substrate. Attachment allows the microtubules in the central mound region to stabilize. This stability, coupled with the temporary movement of the dynamic microtubules into the peripheral region of the growth cone, helps advance the growth cone forward. However, the movement and turning of the growth cone requires polymerization of actin subunits located at the ends of the actin bundles in the lamellipodia and filopodia. As with other aspects of growth cone motility, the mechanisms underlying actin subunit polymerization are similar to that described in other motile cells such as fibroblasts.

Actin subunits are characterized by a barbed (plus end) and a pointed (minus end). During polymerization, actin subunits are added

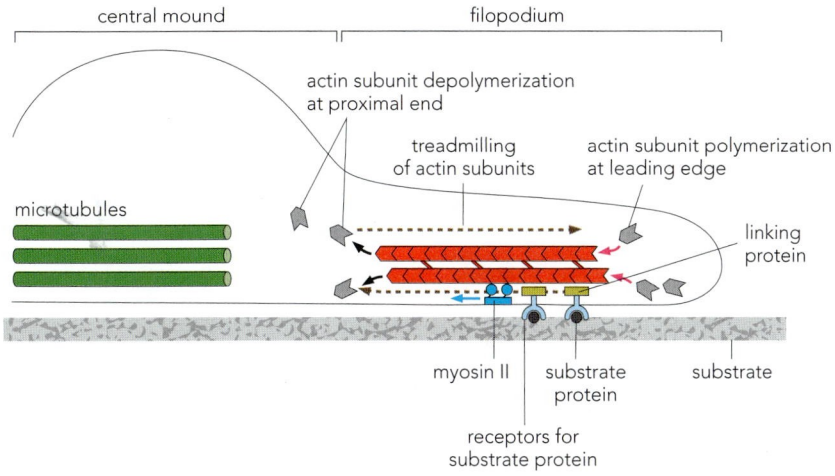

Figure 7.4 Recycling of actin subunits and substrate attachment lead to forward movement of growth cones. In a filopodium, actin subunits polymerize and depolymerize in a cycle that results in net forward movement of the growth cone. Actin subunits are added to the barbed, distal end of the actin filaments (red) at the leading edge of the growth cone (polymerization of subunits, pink arrows), while other subunits are released at the pointed, proximal ends of the filaments (depolymerization of subunits, black arrows). The released actin subunits are recycled and used again (dashed arrows). Myosin II acts as a protein motor to help move actin in a retrograde direction (blue arrow), away from the leading edge of the growth cone. Receptors for substrate proteins attach the growth cone to the substrate. These receptors interact with linking proteins to stabilize the actin filament bundles and temporarily slow the retrograde flow of actin. This provides the tension that allows the growth cone to move forward. In such cases, depolymerization proceeds at a slower rate than polymerization, so the recycling, or treadmilling, of subunits contributes to the net forward movement of the growth cone. (Adapted from Letourneau PC [2009] In Encyclopedia of Neuroscience [LR Squire ed], pp 1139–1145.)

to the barbed ends of actin filaments. Actin polymerization begins with **nucleation**—the formation of a small complex of approximately three G-actin monomers. Continued addition of these actin monomers leads to the elongation of the actin filament. In the growth cone, this occurs at the leading (distal) edge of the lamellipodia and filopodia (**Figure 7.4**). The addition of the actin subunits pushes the cell membranes at the leading edge forward. Depolymerization occurs when actin subunits are removed from the pointed ends of filaments that are located closest to the central mound, in the proximal region of the growth cone.

The movement of actin subunits is further facilitated by myosin II, which acts as a motor protein to move actin rearward, toward the central mound region where the subunits are removed. The released subunits are recycled for later addition to the leading edge, creating the cyclical or treadmilling action of actin subunits. During neurite outgrowth, the addition of actin subunits at the leading edge occurs at a rate faster than the depolymerization near the central mound, causing forward extension of the growth cone.

Actin-binding proteins regulate actin polymerization and depolymerization

A number of actin-binding proteins interact to influence the polymerization and depolymerization of the actin filaments. These proteins may promote the addition of new actin subunits at a barbed end, sever existing actin filaments, or cap the barbed end of a filament. Understanding how actin-binding proteins interact to influence growth cone motility is currently an active area of research. In many cases, actin-binding proteins appear to induce the formation of new barbed ends that lead to branching of existing actin filaments. Among the actin-binding proteins shown to influence the creation of new barbed ends are the Arp2/3 complex (actin-related proteins 2 and 3) and formins. Arp2/3 binds to the side of an existing actin filament. Such binding initiates nucleation of G-actin subunits to promote branching of the filament. This is often seen in the lamellipodia (**Figure 7.5**A). In contrast, formins function primarily in filopodia, where they bind and add new actin subunits to the barbed ends of the existing actin filaments (Figure 7.5B). New barbed ends can also be exposed by severing an existing actin filament. The actin-binding proteins ADF (actin depolymerizing factor) and cofilin are two examples of such severing proteins that create new sites for the addition of G-actin subunits (Figure 7.5C). Another group of actin-binding proteins are the capping proteins such as CapZ that bind to the barbed end of the actin filaments to prevent the loss

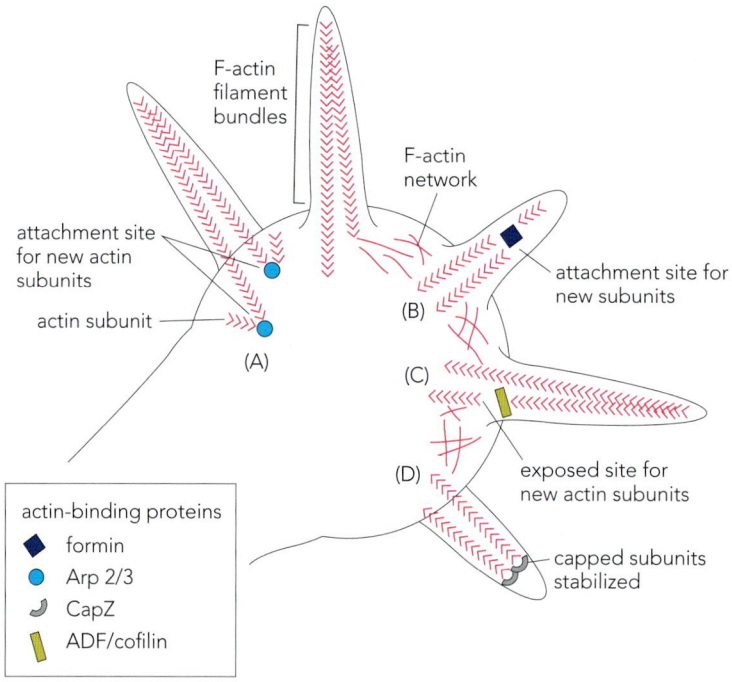

Figure 7.5 Actin-binding proteins help regulate actin polymerization and depolymerization. Different actin-binding proteins attach to or sever existing actin filaments to create new sites for actin polymerization. (A) New barbed ends are formed along existing actin filaments when the actin-binding protein complex Arp2/3 attaches to an existing actin filament and initiates nucleation of G-actin subunits at that site. This branching is often seen in the lamellipodia of the growth cone. (B) The actin-binding protein formin typically attaches to the barbed ends of actin filaments in the filopodia to help recruit additional actin subunits to the elongating filopodia. (C) New barbed ends can be created by the actin-severing proteins ADF or cofilin. The exposed ends create new sites for the addition of G-actin subunits to promote growth cone extension. In some cases, the levels of ADF and cofilin exceed the levels of available G-actin, so the severing causes a breakdown of the actin filaments and growth cone collapse. (D) Capping proteins such as CapZ attach to the barbed end of actin filaments to prevent the loss of the existing subunits and stabilize the growth cone.

of existing actin subunits and stabilize the growth cone (Figure 7.5D). Such capping proteins can also be inhibited by additional actin-binding proteins to allow for continued polymerization of the actin filaments during periods of growth cone extension.

In many cases, it appears that the balance of actin-binding protein activity and the concentration of available G-actin subunits determines whether the growth cone will stabilize, branch, or extend. Thus, a single actin-binding protein may have multiple functions to mediate actin dynamics and growth cone movements. For example, depending on the cellular context, the severing of actin filaments may result in decreased or increased growth cone extension. When the levels of ADF/cofilin are higher, increased severing occurs and the actin filaments break down. Thus, growth cone extension is prevented. However, if the available levels of G-actin are high enough, the additional barbed ends created by ADF/cofilin provide, instead, multiple new sites for polymerization. Thus, growth cone movement and extension are increased.

Rho family GTPases influence cytoskeletal dynamics

Additional intracellular signaling events that involve members of the Rho family of small GTPases are also thought to impact growth cone motility and neurite extension by influencing actin polymerization. As noted in Chapter 5, Rho family GTPases activate effector proteins that influence cytoskeletal dynamics (see Figure 5.6). In some cases, specific Rho family GTPases enhance actin polymerization, whereas other Rho family GTPases increase depolymerization. Such differences are associated with promoting growth cone advancement or retraction, respectively (**Table 7.1**). For example, the Rho family GTPase Cdc42 (cell division control protein 42) activates N-WASP (neural-Wiskott-Aldrich syndrome), whereas Rac1 (Ras-related C3 botulinum toxin substrate 1) activates WAVE (WASP-family verprolin-homologous protein). Either of these proteins can then activate the Arp2/3-dependent polymerization of actin filaments to promote growth cone advancement. In contrast, RhoA (Ras homolog family member A) activity is most often associated with the retraction of growth cones by activating ROCK (Rho-associated coiled-coil-containing protein kinase).

Table 7.1 Rho family small GTPases influence cytoskeletal proteins and growth cone behavior

Rho family small GTPase	Protein activated	Effect on actin	Growth cone response
Cdc42	N-WASP	↑ Arp2/3-mediated actin polymerization	Extension
Rac1	WAVE	↑ Arp2/3-mediated actin polymerization	Extension
RhoA	ROCK	↑ myosin II-mediated retrograde flow of actin	Retraction

Arp2/3, actin-related proteins 2 and 3; cdc42, cell division control protein 42; N-WASP, Neural-Wiskott-Aldrich syndrome; Rac1, Ras-related C3 botulinum toxin substrate 1; RhoA, Ras homolog family member A; ROCK, Rho-associated coiled-coil-containing protein kinase; WAVE, WASP-family verprolin-homologous protein.

ROCK then increases the activity of the myosin II motor protein that regulates retrograde flow of actin. With an increase in retrograde flow, growth cone advancement is prevented, so the growth cone retracts.

A number of intracellular signals have also been implicated in regulating the cytoskeletal organization and motility of growth cones. For example, the level of intracellular calcium can determine whether a growth cone is active or inactive. Studies have found, for many growth cones, that motility requires an optimal window of calcium concentration. If the cytosolic calcium level is too low or too high, the growth cone is inactive. The mechanism by which this occurs is unclear, but it appears that calcium helps regulate processes such as actin flow and the addition of plasma membrane to the extending growth cone. In at least some contexts, specific levels of calcium may be associated with activating different actin-binding proteins or Rho GTPases.

GROWTH CONE SUBSTRATE PREFERENCES *IN VITRO* AND *IN VIVO*

In the developing embryo, the extending growth cone will encounter a number of extracellular cues that direct the growth cone along a particular pathway. The directed growth occurs as the growth cone cytoskeleton is reorganized in response to such cues. Over the years, scientists have observed that the morphology of a growth cone changes under different growth conditions. A complex morphology with a broad lamellipodium and multiple filopodia is characteristic of an extending, motile growth cone exploring the surface on which it is growing, actively seeking the most favorable substrate on which to advance. A smaller, less elaborate growth cone is typical of a neurite extending on a favorable substrate. Some actively growing growth cones are described as torpedo-shaped, such as those growing along the surface of other neurites. The lamellipodia of these growth cones wrap around the neurites, giving the torpedo-like appearance. In contrast to the morphologies associated with motile growth cones, a simple growth cone with few or no filopodia is observed when a growth cone has stopped extending and is about to collapse or change direction. Finally, when a neurite reaches its target cell, the previously motile growth cone adopts a simpler morphology as it prepares to transform into a presynaptic connection with the target cell. How the extracellular environment influences these changes in growth cone morphology and behavior has been the focus of research for several decades.

In vitro studies confirm that growth cones actively select a favorable substrate for extension

Conclusive experimental evidence that the growth cone itself was actively involved in selecting a pathway for extension was not obtained until the 1970s. *In vitro* studies conducted at that time established that growth

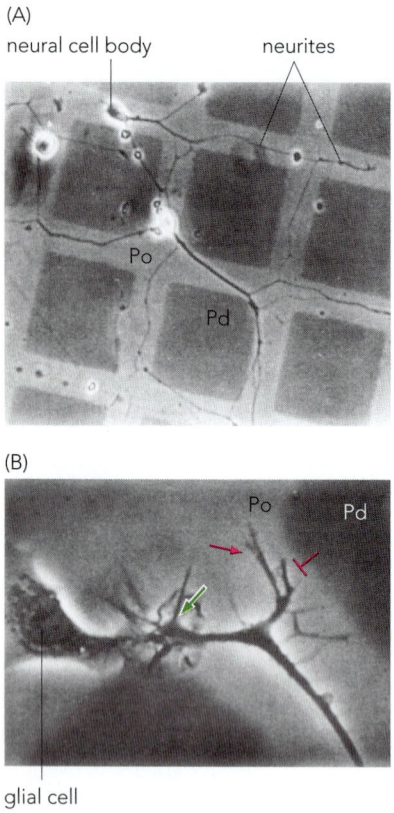

Figure 7.6 Dorsal root ganglion (DRG) neurons detect differences in substrates and preferentially grow along substrates that are more adherent. When DRG neurons were placed in cell culture with squares and lanes of different substrates, the DRG axons preferred to grow along favorable, more adherent substrates. (A) For example, when presented with a choice of palladium (squares, Pd) or polyornithine (lanes, PO), the neurites extended along the polyornithine-coated lanes. (B) Growth cones were observed to extend filopodia to each surface and sample the choices available before choosing the more adherent substrate. Here, one branch of the growth cone extends across the surface of a glial cell (green arrow), while the other continues along the lane of polyornithine (PO, pink arrow). The growth cone avoids the square covered with Pd. (From Letourneau PC [1975] *Dev Biol* 44(1):92–101.)

cones are able to distinguish among different substrates on a petri dish and actively select to grow along particular pathways. In one classic *in vitro* study from 1975, Paul Letourneau demonstrated that embryonic chick dorsal root ganglion (DRG) neurons were able to distinguish and preferentially grow along lanes of specific substrates. In these experiments, a pattern was made on cell culture dishes by placing a small grid in a dish before coating the dish with the metal palladium (Pd). When the grid was removed, the dish contained squares of Pd and lanes without Pd (the areas once covered by the grid). Depending on the type of petri dish used, the Pd-free lanes consisted of either untreated petri dish plastic or tissue culture plastic that is positively charged to attract the net negative charge of neuronal membranes. In other cases, the lanes were coated with polyornithine (a poly-cationic amino acid) or collagen, two substrates shown to be favorable for the adhesion of other cell types.

When DRG neurons were cultured in the various dishes, the neurons did not adhere to the untreated petri dish plastic, but were able to adhere to and extended neurites along areas of tissue culture plastic or Pd. However, when the tissue culture lanes were coated with the other substrates, the neurites avoided the areas of Pd and instead grew along the polyornithine or collagen pathways (**Figure 7.6**A). The neurites also grew on top of any glia encountered in the dish. Thus, the neurites extended along whichever substrate was most favorable for neuronal attachment and neurite extension.

These studies also demonstrated the importance of the growth cone filopodia in selecting a substrate pathway for extension. As a growth cone was advancing along one substrate, the filopodia were observed to extend in multiple directions. As the filopodia sampled the environment, a single filopodium would periodically contact an adjacent substrate (Figure 7.6B). If a filopodium contacted a surface that was less adherent, it rapidly withdrew. However, if even one filopodium contacted the substrate that was more adherent, the entire growth cone would turn and extend in that direction. Thus, the growth cone filopodia sampled available substrates and guided the growth cone along the most favorable of the choices.

Subsequent investigations determined that the growth cones from different neuronal populations often have specific substrate preferences. It has also been noted that the most adherent substrate is not always best for optimal growth; if a substrate is too sticky, neurite extension is not possible because the growth cone cytoskeletal elements are unable to move freely.

Extracellular matrix molecules and growth cone receptors interact to direct neurite extension

Many of the substrates favorable for neurite attachment and outgrowth *in vitro*, such as polyornithine, are not found in the embryo. However, growth cones also show pathway preferences *in vivo*. Several proteins expressed in the cellular environment through which a growth cone extends have been shown to provide guidance cues that lead a developing nerve fiber to its correct target cell. Within the embryo, such pathway selection is mediated by specific proteins present on the embryonic cell surfaces or in the surrounding extracellular matrix. These proteins bind to the corresponding receptors expressed on the growth cones. Among the pathway cues encountered *in vivo* are the extracellular matrix (ECM) molecules, such as the laminins, collagens, and fibronectins. These large extracellular molecules are not diffusible and remain near the cell that secretes them, thereby providing local cues to extending neurites.

In order to bind to some substrates, growth cones express integrin receptor proteins. Each integrin receptor is a heterodimer of two different integrin subunits. As discussed in Chapter 5, the pairing of specific subunits produces the binding specificity for a particular ECM molecule. For example, in vertebrates, the α5β1 subunits form a receptor to bind fibronectin, the α6β1 subunits bind laminins, and the α1β1 subunits bind both collagens and laminins (see Table 5.1). Growth cones extending from different neuronal populations express distinct integrin receptor combinations at specific developmental stages to respond to spatially and temporally regulated guidance cues.

Net forward movement of the growth cone can only occur if the integrin receptors that bind to a substrate also interact with the actin filaments. Such an interaction is possible because the integrin receptors are also associated with various linkage, or clutch, proteins. Linkage proteins include vinculin and talin, which cluster near the site where the growth cone contacts the substrate. Because these proteins link the integrin receptor to the actin filaments, the actin filaments are stabilized at the site of substrate contact and actin polymerization at the leading edge of the growth cone occurs (**Figure 7.7**). In this way a favorable permissive or instructive substrate promotes actin polymerization and growth cone advancement.

In some regions of the embryo, a growth cone may encounter an unfavorable or inhibitory substrate. Such substrate cues can block integrin-mediated adhesion, causing the growth cone to turn away from the inhibitory cue or to collapse, rather than extend forward. Growth cone turning and collapse result from a combination of changes in the cytoskeletal dynamics, which include cessation of actin polymerization at the barbed ends of actin filaments; an increase in the retrograde flow of actin; greater severing of actin filaments, leading to actin breakdown; and the removal of capping proteins that would otherwise stabilize the barbed ends of the actin filaments.

Roles of pioneer axons and axonal fasciculation in target selection

The spatially and temporally restricted expression of specific integrin receptors on growth cones and the differential expression of ECM proteins throughout the embryo provides one mechanism by which growth cones preferentially select *in vivo* pathways for extension. A limitation of ECM molecules as guidance cues is that they are broadly expressed in embryonic tissues, and growth cones from neurites arising from different neural populations often encounter the same ECM molecule at the same location. Thus, additional cues are required to ensure neurites from a particular

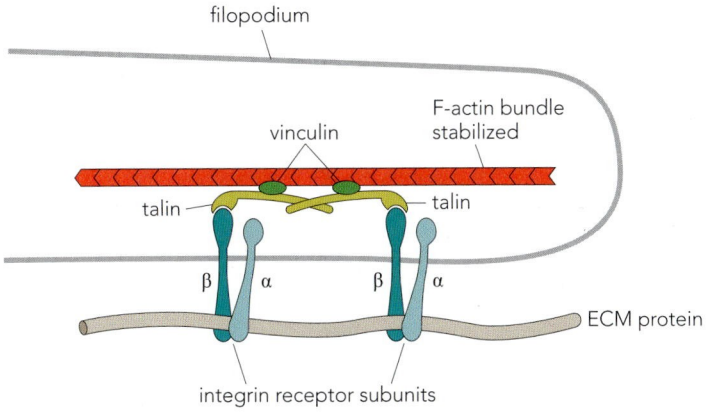

Figure 7.7 Integrin receptors interact with clutch proteins to link actin filaments to substrate. Growth cones must adhere to a substrate in order to advance forward. Integrin receptors in the cell membrane of the growth cone bind to extracellular matrix (ECM) proteins on the substrate. The receptors also interact with various clutch (linking) proteins, such as talin and vinculin, which contact actin filaments to stabilize F-actin bundles so that additional actin subunits can be added at the growth cone's leading edge. (Adapted from Alberts B, Johnson A, Lewis J et al. [2014] Molecular Biology of the Cell, 6th ed. Garland Science.)

neuronal population travel together toward the correct target tissue. For many extending axons, guidance cues are found on adjacent nerve fibers. Initially, growth cones extend toward a target by following cues located on these nerve fibers; then, as outgrowth continues, the growth cones respond to specific cues located on intermediate and final target cells.

In any given neuronal population, a single axon, or a small set of axons, must be the first to extend in the direction of a target tissue. These first nerve fibers, typically called **pioneer axons**, often respond to local ECM guidance cues. Pioneer axons then express other guidance cues along their axon shafts for subsequent nerve fibers to follow, thereby directing axons from the same neuronal population toward the same target tissue. Additionally, during the initial stages of outgrowth, axons from the same neuronal population tend to preferentially bundle together (fasciculate), further facilitating the extension of a group of nerve fibers to the same target.

Fasciculation is promoted by adhesion molecules, such as neural cell adhesion molecule (NCAM) and neural cadherin (N-cadherin), which are expressed on adjacent axons (**Figure 7.8**). In 1987, NCAM was the first neural adhesion molecule identified. Additional adhesion molecules have since been discovered, including L1, neuron-glia cell adhesion molecule (NgCAM), and transient axonal glycoprotein-1 (TAG-1). Members of the IgG superfamily, these molecules contain IgG domains and fibronectin type III repeats. The adhesion molecules often use homophilic binding in which they recognize a binding partner of the same structure on an adjacent cell. There are also some instances of heterophilic binding with other molecules in the IgG superfamily, other ECM molecules, or members of the semaphorin family (**Box 7.1**). Unlike the IgG superfamily, members of the cadherin family, including N-cadherin, are calcium-dependent. Calcium helps stabilize the ectodomains of these adhesion molecules. Thus, adhesion between axons expressing N-cadherin takes place only in the presence of calcium (Figure 7.8).

In addition to promoting fasciculation of axons bundles, in some instances neural adhesion molecules decrease axonal bundling. For example, a post-translational modification of NCAM occurs in which polysialic acid (PSA) is added to the NCAM molecule. As described later in this chapter, PSA NCAM is expressed on motor neuron axons as they reach their target muscle. This highly sialylated form of NCAM interferes with the bundling of axons, so the axons are able to branch away from each other and spread across the muscle surface to contact individual target cells.

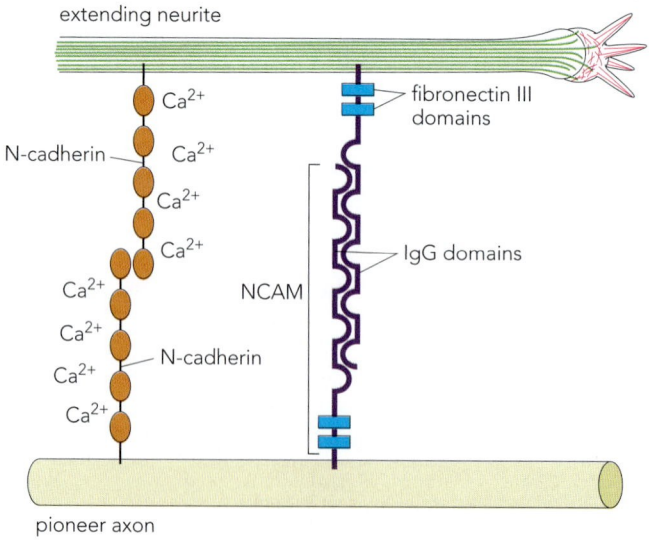

Figure 7.8 Pioneer axons use cell adhesion molecules to guide other axons. Cell adhesion molecules, such as neural cell adhesion molecule (NCAM) and neural cadherin (N-cadherin), are located along the axon shafts of pioneer axons to promote fasciculation and direct related axons to the same target region. Homophilic binding to NCAM or N-cadherin molecules on adjacent, extending neurites allows the axons to recognize each other. NCAM is a member of the IgG superfamily and consists of IgG domains and fibronectin III domains. N-cadherin is a member of the cadherin family and requires calcium (Ca^{2+}) for binding to take place.

Box 7.1 Semaphorins provide repulsive cues to direct dorsal root ganglion axons entering the spinal cord

Throughout the developing nervous system, axons rely on repulsive guidance cues to avoid inappropriate axonal pathways and target regions. Among the most widely expressed and best-characterized repulsive guidance cues are those of the semaphorin family. The first semaphorin (Sema) protein was identified as an inhibitory cue for grasshopper sensory axons in 1992. The name derives from its function in directing axons to a particular pathway, similar to the role of a semaphore. In the 1980s, a similar molecule was purified from chick brain. This protein, called collapsin, caused the stalling and retraction of dorsal root ganglion (DRG) axons *in vitro*. As additional members of the semaphorin family were identified, it became apparent that collapsin was homolgous to Sema3A (originally called SemaIII). *In vivo*, Sema3A was found to be expressed in the ventral spinal cord at the time DRG axons innervate target laminae in the dorsal horn of the spinal cord, suggesting a role in repelling axons away from the ventral region (**Figure 1**A). *In vitro*, COS cells expressing Sema3A also repelled DRG axons (Figure 1B). Thus, ventrally expressed Sema3A provides a repulsive guidance cue for sensory axons to guide them to correct targets in the dorsal region.

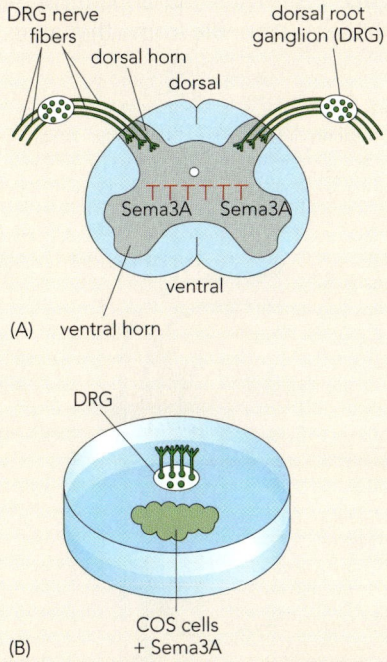

Figure 1 Semaphorin 3A provides a repulsive cue to direct growth of dorsal root ganglion (DRG) axons. (A) The axons of DRG neurons that enter the central nervous system project to cells within the dorsal horn of the spinal cord. The ventral spinal cord expresses Semaphorin 3A (Sema3A) to prevent DRG axons from entering the ventral spinal cord. (B) *In vitro*, DRG neurites are repelled by Sema3A-expressing COS cells.

The Sema family consists of a large number of proteins, some of which, such as Sema3A, are secreted proteins. Others are membrane-associated through either a transmembrane domain or GPI linkage. There are currently 21 known semaphorins in vertebrates, five in invertebrates, and two in viruses. Semaphorins are found in numerous tissues, but some of the best-characterized functions have been identified in the nervous system, where they are found in both neuronal and nonneuronal cells. Semaphorins have multiple roles in neural development and may also be active during neural injuries and pathologies.

To identify receptors for Sema3A, scientists made fusion proteins that contained a segment of Sema3A fused to alkaline phosphatase (AP), an enzyme that can be visualized using a colorimetric reaction. With this method scientists could identify DRG neurons that bound the Sema3A-AP fusion protein (**Figure 2**). This method ultimately led to the identification of the receptor neuropilin-1. Neuropilins are now known to be a receptor component for many semaphorins.

Around the same time, other scientists studying viruses identified a different receptor called plexin. Plexins were also subsequently detected in neurons. It now appears that both neuropilin and plexin components are required to transduce the Sema3A signal that directs axons away from inappropriate target regions. The activation of the receptors appears to alter cytoskeletal organization, thus leading to growth cone retraction and collapse. The large size of the semaphorin family and the associated receptor families, together with the distribution of these molecules in numerous tissues, suggests there are multiple, diverse functions for these molecules. Scientists continue to sort out intracellular mechanisms that lead these molecules to regulate a variety of cellular events, including axon retraction, synapse formation, cell migration, and vascular development.

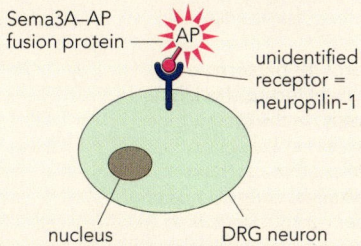

Figure 2 Neuropilin is identified as a receptor for semaphorin 3A. To identify receptors for semaphorins, scientists made a fusion protein consisting of Sema3A and alkaline phosphatase (AP). These fusion proteins were incubated with DRG neurons. Using a colorimetric assay to detect the AP, scientists identified neurons that bound the Sema3A and ultimately isolated the previously unidentified receptor as neuropilin-1, a receptor component for members of the semaphorin family.

Research in invertebrate models leads to the labeled pathway hypothesis

Much of what is known about how pioneer axons function has come from studies in three invertebrate nervous systems: grasshopper, *Drosophila,* and *C. elegans*. Following individual neurites in an invertebrate system is relatively simple compared to tracking nerve fibers in vertebrate systems. Invertebrate neurons are larger, fewer in number, and easily identified by their location in the nervous system. Together these properties have allowed scientists to observe the initial outgrowth and turning of subsets of neurons as they travel through different embryonic regions.

Studies of pioneer axons by Corey Goodman and colleagues tracked the normal pathways taken by specific neurons in developing grasshoppers. The grasshopper embryo consists of 17 segments divided into half segments on each side of the body. Each half segment contains ganglia and within these ganglia are a number of identified neurons. For example, in each half of a thoracic segment identified neurons include Q1, Q2, Q5, Q6, G, and C. The scientists determined that the axons of the G and C neurons connect to the opposite side of the body by traveling along a transverse-oriented bundle of nerve fibers called a **commissure**. Having reached the opposite side, the G and C axons join the A/P (anterior/posterior) fascicle, but none of the other longitudinally oriented axon bundles in the area.

To determine how these axons reach the opposite side and choose the correct fascicle, the researchers completed a number of studies using both electron microscopic reconstructions of developing growth cones and *in vivo* manipulations. Among their findings, it was noted that the growth cones of G neurons are able to detect differences between the axons of the A and P neurons traveling within the A/P fascicle. The growth cones of G neurons were found to prefer to extend along the P axons, to which they adhered more tightly (**Figure 7.9**). Thus, the filopodia of the G neuron growth cones could distinguish the A/P bundle from other axon bundles in

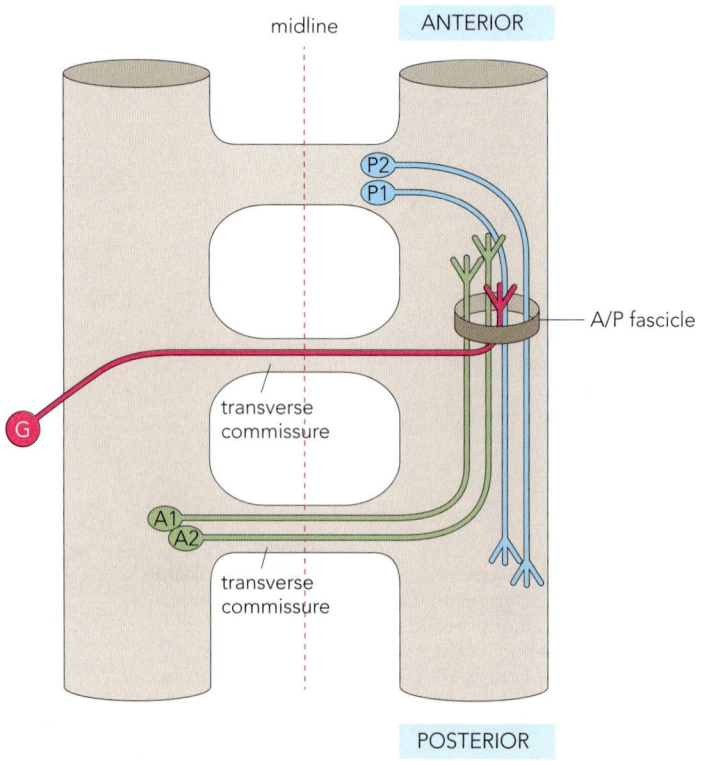

Figure 7.9 The growth patterns of grasshopper neurons led to the labeled pathway hypothesis. In the grasshopper embryo, G neurons extend axons across a transverse commissure to reach the contralateral longitudinal anterior–posterior (A/P) fascicle. Other neurons, including A1, A2 and P1, P2, extend axons within the A/P fascicle. Studies revealed that the axons of G neurons preferentially grow along A and P axons, showing the greatest adherence to P axons. The axons of G neurons ignored axons in other fascicles located in the area (not shown), even when A and P axons were missing. (Adapted from Bastiani MJ, Raper JA & Goodman CS [1984] *J Neurosci* 4(9): 2311–2328.)

the region, then further differentiate P axons from A axons. Together, the observations suggested that axons express molecules that guide growth cones to join and travel in the correct fascicle. These observations led to the **labeled pathway hypothesis** described by Goodman and colleagues in 1982. This hypothesis states that the axons of pioneer neurons differentially express molecules, or labels, on their surfaces. Later-extending growth cones then choose which axon pathway to follow based on the expression patterns of these labels.

To test the labeled pathway hypothesis directly, finely sharpened microelectrodes were used to selectively ablate specific A and P axons within the A/P fascicle, then the growth of individual G axons was documented. The axons of G neurons crossed the commissure as expected, but upon reaching the opposite side, the axons grew in a random, searching manner as if seeking the axonal guidance cues normally present in the intact A/P fascicle. The ablation of either A or P axons disrupted G axon outgrowth, with the loss of P axons being particularly disruptive. Even though there were over 20 other axon bundles in the area, the G axons failed to show a preference for any of these longitudinal fascicles, further demonstrating that the G axons recognized a cue specific to axons located in the A/P fascicle.

Fasciclins are expressed on axonal surfaces

Subsequent studies in invertebrates identified a family of molecules that helps early axons associate with the correct fascicles as they travel throughout the embryo. These molecules, called **fasciclins**, are expressed on the surfaces of early axons in animal models such as the grasshopper and *Drosophila*. In *Drosophila,* fasciclin II (Fas II) was later determined to be an ortholog of NCAM. Like NCAM, the fasciclins are members of the IgG superfamily and use homophilic binding between adjacent cells. Fasciclins are often expressed on discrete regions of axons to direct growth cones to join selected pathways.

Axons travel with a particular bundle until they reach a **choice point**, where they select the next axon bundle to join. For example, the axons of G neurons in the grasshopper first choose to grow along axons in the transverse commissure, then they choose to grow along axons in the A/P fascicle. In order for the neurites to switch the axonal bundle in which they travel, there must be temporally regulated changes in the expression of axonal cues encountered at the choice point. Fasciclins are among the cues used at such choice points.

In grasshopper and *Drosophila*, axons traveling along a longitudinal pathway express Fas II. However, axons preparing to cross a transverse commissure down-regulate Fas II expression and begin to express fasciclin I (Fas I), allowing the extending axons to interact with axons in the transverse commissure that also express Fas I. Once the axons exit the transverse commissure on the opposite side, they again express Fas II in order to adhere to and grow along the axons present in the longitudinal pathway (**Figure 7.10**).

Guidance cues on axonal surfaces are used throughout the developing nervous system and appear particularly important for early-extending axons to help direct axons along a path leading toward a target tissue. Additional guidance cues specific to a given neuronal population are then used to direct axons to specific target tissues.

Vertebrate motor neurons rely on local guidance cues

Guidance cues that direct the growth of axons from spinal motor neurons to limb skeletal muscles have been the focus of investigations for

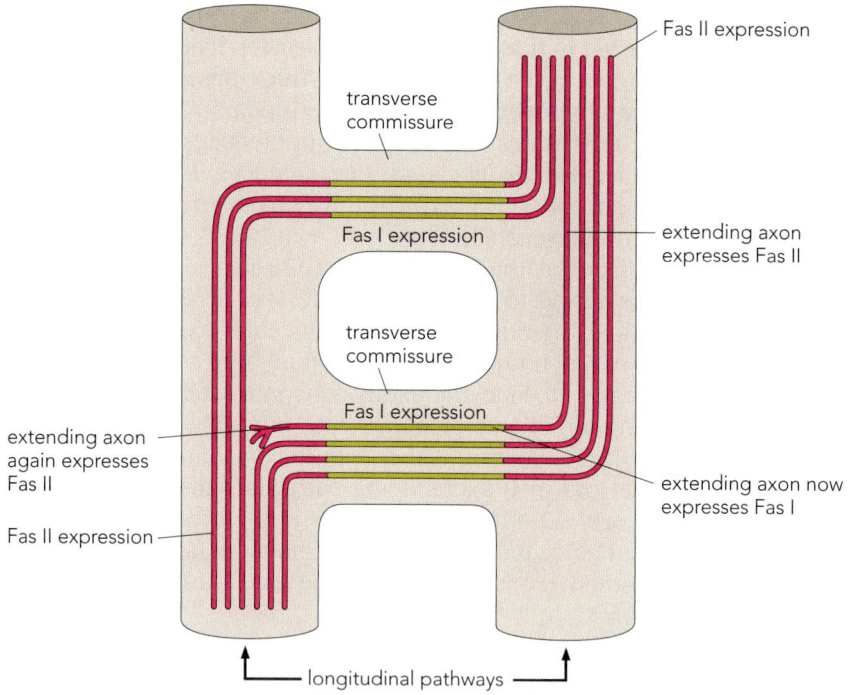

Figure 7.10 Axons adjust fasciclin expression patterns to cross the midline. Identical fasciclins bind to each other via homophilic binding mechanisms. In this example from invertebrates, fasciclin I (Fas I) is expressed in the transverse commissures, whereas fasciclin II (Fas II) is expressed in longitudinal pathways. In order to cross the midline, axons decrease expression of Fas II and increase expression of Fas I, allowing the axons to interact with other Fas-I-expressing axons in the commissures. Once the axons have reached the contralateral side, they again express Fas II in order to interact with other longitudinally projecting axons. (Adapted from Bastiani MJ, Harrelson AL, Snow PM & Goodman CS [1987] *Cell* 48(5):745–755).

over 100 years. As described in Chapter 4, motor neurons are located in the ventral half of the spinal cord. These motor neurons extend axons to innervate the skeletal muscles that control movement. As the spinal motor nerve axons exit the spinal cord, they form discrete axon bundles at intervals along the length of the spinal cord (**Figure 7.11**). The motor axons in the ventral root join the sensory axons of the dorsal root ganglia to form the spinal nerves. Several studies over the past century have shed light on how the stereotyped, segmental patterns of axons emerge as the motor neurons project fibers toward different limb muscles.

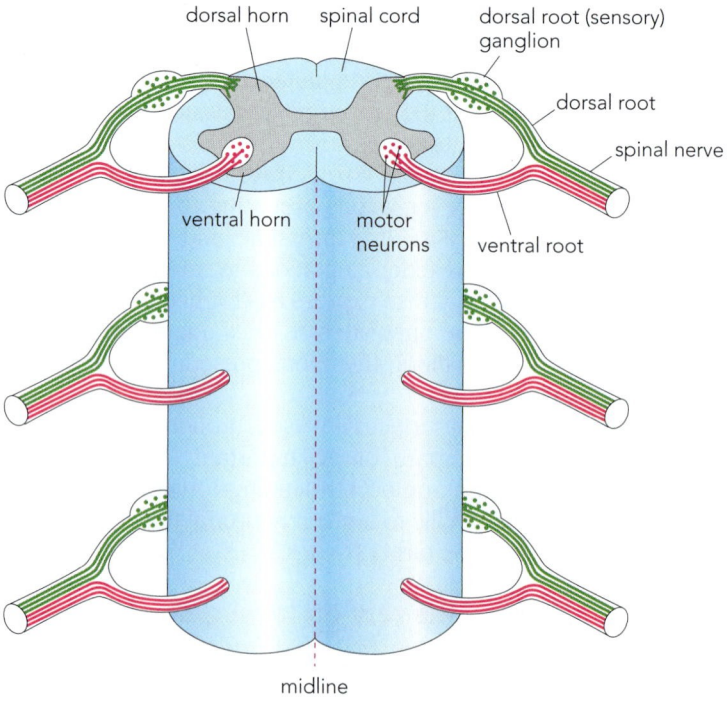

Figure 7.11 Spinal nerves form discrete axonal bundles along the length of the spinal cord. Motor neurons located in the ventral horn of the spinal cord extend bundles of axons at intervals along the length of the spinal cord, creating a ladder-like pattern. The motor axons contribute to the spinal nerves that are comprised of the motor axons in the ventral root and the sensory (dorsal root) axons in the dorsal root. A number of guidance cues establish the patterning of sensory and motor axons along the spinal cord.

During the first half of the twentieth century, scientists examining the role of limb musculature in guiding motor axons to individual muscles formulated seemingly conflicting hypotheses. Some scientists concluded that muscle-derived cues were necessary to guide motor axons to the correct target muscle. Work by other scientists suggested that motor axons were guided to the appropriate target region by local environmental cues and then innervated muscles in a random pattern, making contact with the closest available muscle. Some of these conflicting findings were likely due to differences in the experimental approaches used. In some cases, scientists were studying re-innervation to regenerated amphibian limb muscles, while in others, scientists were evaluating innervation to embryonic chick limb tissue grafted to new sites.

In the late 1970s and early 1980s, several key experiments by Lynn Landmesser and colleagues helped resolve the differing hypotheses for how motor neuron axons are guided to a target muscle region. Their findings indicated that under normal developmental conditions, local guidance cues, rather than muscle-derived guidance cues, directed motor axons to the correct target region. Their findings also demonstrated that if an experimental manipulation caused the motor axons to be too far away from the local guidance cues, the axons could not respond appropriately and failed to reach their proper targets.

Landmesser's group utilized the chick embryo for studies of motor neuron innervation of muscles. This was a particularly useful model system, because by the second half of the twentieth century scientists had shown through anatomical and electrophysiological studies that specific groups of motor neurons within the spinal cord innervated particular limb muscles (**Figure 7.12**). For example, motor neurons located in the ventromedial motor column innervate axial (trunk) muscles, whereas those in the lateral motor column innervate distal (limb) muscles. Thus, it was possible to map axonal growth from specific spinal cord regions to specific muscles.

The scientists also took advantage of useful anatomical features of the lumbosacral region of the chick spinal cord. Normally, lumbosacral segment 1 (LS1) is located rostral (anterior) to LS2 and LS3, and the size of the motor neuron pools in the lateral motor column (LMC) increases from rostral to caudal. Thus, LS3 has a larger pool of motor neurons in the LMC than does LS1 (Figure 7.12).

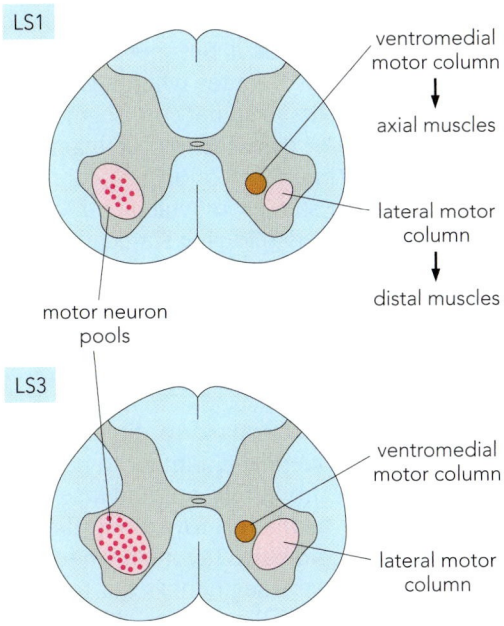

Figure 7.12 Specific pools of motor neurons vary in size and innervate different muscles. The motor neurons in the ventral horn are found in groups called motor neuron pools. Within the motor neuron pools, the region comprising the ventromedial motor column contains motor neurons that innervate axial muscles. In contrast, the lateral motor column contains motor neurons that innervate distal muscles. The size of the lateral motor column increases from lumbosacral segment 1 (LS1) to lumbosacral segment 3 (LS3), providing an anatomical marker for scientists investigating axonal guidance from different spinal cord levels.

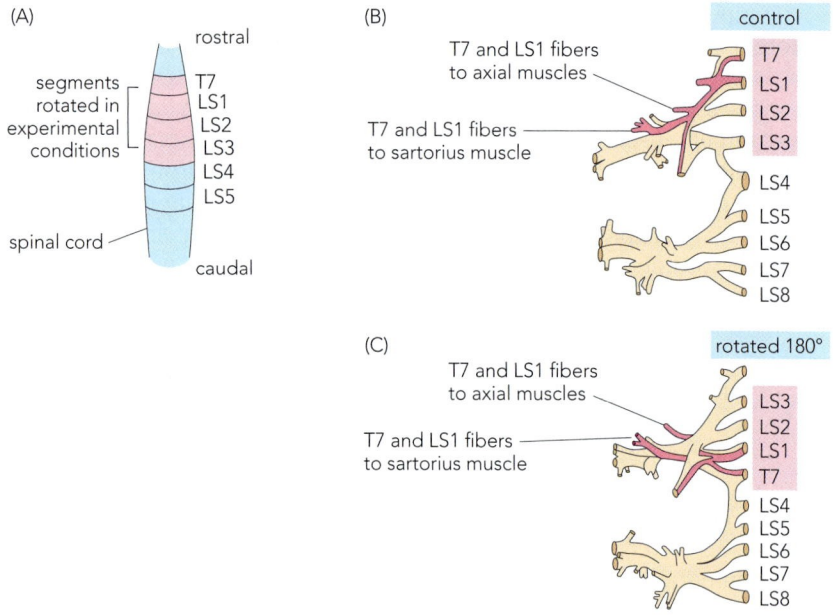

Figure 7.13 Motor axons grow in specific patterns to reach target muscles. (A) A segment of the chick spinal cord from thoracic level 7 (T7) to lumbosacral level 3 (LS3) was rotated 180 degrees along the rostral–caudal (anterior-posterior) axis. The growth of motor axons to axial and sartorius muscles was then compared in control and experimental preparations. (B) A schematic outline of projections (dark pink) in control conditions. Axons from T7 and lumbosacral segment 1 (LS1) normally project to axial muscles and the sartorius muscle located in the hindlimb of the chick embryo. (C) The correct motor axon projections persisted following spinal cord rotation. When the motor axon pathways were visualized approximately 6 days after rotation, the motor axons (dark pink) were found to innervate the correct muscles. That is, in both control and reversed spinal cord segments, motor axons from T7 and LS1 grew to the axial and sartorius muscles. Motor axons from the reversed spinal cord had to cross over inappropriate muscles to reach the correct target. This suggested that nerve fibers did not simply grow to the closest muscle, but used guidance cues to locate the correct muscle to innervate. (B and C, adapted from Lance-Jones C & Landmesser L [1980] *J Physiol* 302:581–602.)

These features were used to test what happened to motor neurons and their axons when a portion of chick spinal cord was rotated 180 degrees, inverting the rostral–caudal orientation of that spinal cord segment (**Figure 7.13**A). When a spinal cord segment from thoracic segment 7 (T7) through LS3 was rotated, the size of the motor neuron pools did not change. The LS3 motor neuron pool remained larger than the LS1 pool, even when rotated to a more rostral position. The scientists then used anatomical tracing methods to see if the axons from the inverted motor neuron pools were still guided to the correct target or simply innervated the closest muscle.

The researchers found that axons extending from the rotated spinal segments still projected to the correct muscle with very few axons projecting to a non-target muscle. For example, axons from T7 and LS1 grew along new paths in order to reach the correct axial and sartorius (distal) muscles (Figure 7.13B,C). Electrophysiological recordings confirmed that the muscles were innervated by axons originating in the correct spinal cord segment. Thus, the axons did not grow passively to the closest muscle group as some earlier studies suggested, but specifically sought the correct target muscle and grew along novel routes when necessary.

Subsequent studies noted that if spinal cord segments were displaced too far from their original site, then the motor axons were unable to contact the correct muscle. This observation helped clarify some of the contradictory findings of earlier studies, in which both correct and random innervation patterns were reported following surgical manipulations. Thus, if spinal cord segments are placed too far away from local guidance cues, motor axons cannot find their correct target.

Several molecules help direct motor axons to muscles

Many of the molecular mechanisms used to direct motor axons to the correct muscle have now been identified. Similar to neural crest cells (see Chapter 5), axons of motor neurons rely on guidance cues found in the somites. These cues lead the axons to grow along the rostral (anterior) segment of the somites—the region also favored by neural crest cells. This preferential growth along the rostral segments leads to the ladder-like segmentation of motor axons seen in Figure 7.11. Studies have found that

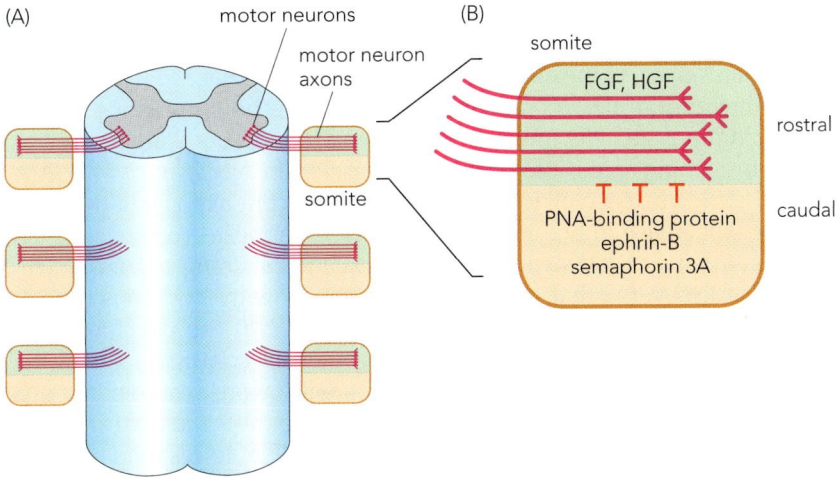

Figure 7.14 Multiple cues direct the motor axons through the rostral segments of somites. (A) A ladder-like pattern of motor axons arises as motor axons project through the rostral (green), but not the caudal (tan), segment of the somite. (B) Several studies have detected inhibitory cues in the caudal (posterior) segments of somites, including peanut-agglutinin-binding protein (PNA-binding protein) and members of the ephrin and semaphorin families. These proteins repel motor axons from the caudal regions of somites. Fibroblast growth factor (FGF) and hepatocyte growth factor (HGF) are candidate molecules that are thought to attract motor axons through the rostral (anterior) portions of the somites.

repulsive cues produced by members of the ephrin and semaphorin families, as well as peanut-agglutinin-binding proteins, direct axons away from the caudal segments of the somites (**Figure 7.14**). Other cues promote the preferential growth of axons along the rostral portion of the somite. Current evidence suggests that fibroblast growth factor (FGF) and hepatocyte growth factor (HGF) may direct the axons to the rostral segment.

As in other systems, adhesion molecules are also present at key locations to direct growth of motor neuron axons. For example, the axons of the LMC that innervate limb muscles are divided into different branches, or plexuses. The cervical and brachial plexuses innervate forelimb muscles, whereas the lumbar and sacral plexuses innervate hindlimb muscles. The fibers of each plexus first travel together until they reach the proximal portion of the limb bud. At this choice point, the fibers from each plexus separate to travel along either a dorsal or ventral path to reach specific extensor or flexor muscles, respectively (**Figure 7.15**A). Various cues present at the proximal limb choice point regulate the binding and branching of motor axons into the dorsal or ventral pathway. For example, L1 and

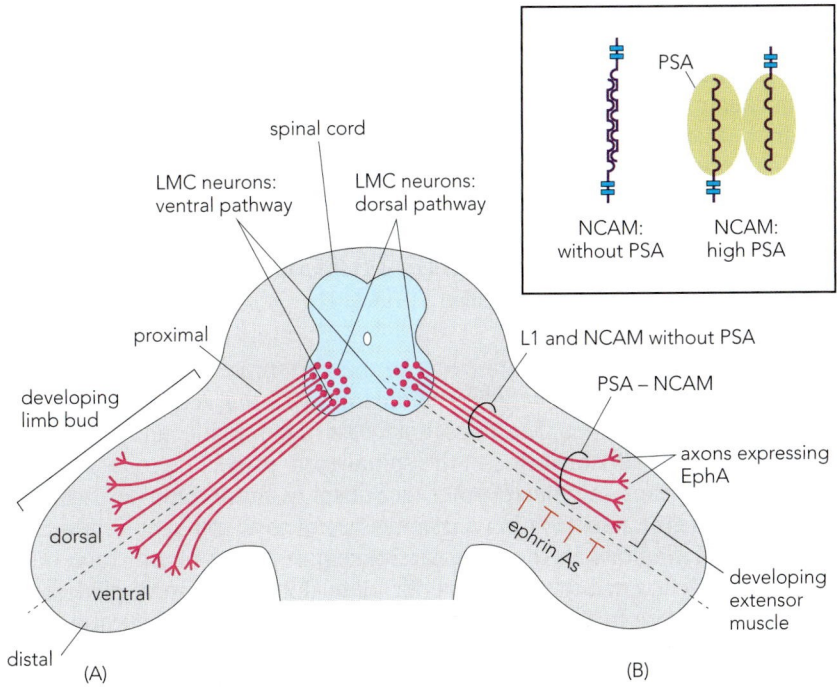

Figure 7.15 Axons of motor neurons in the lateral motor column are directed along the dorsal pathway by multiple cues. (A) Axons from motor neurons in the lateral motor column (LMC) initially extend together as a single bundle until they reach the proximal region of the developing limb bud. At this choice point the axons separate and travel along two distinct pathways: the dorsal pathway, which directs motor axons from the lateral region of the LMC, and the ventral pathway, which directs motor axons from the medial region of the LMC. (B) Several guidance cues interact to direct motor axons along the dorsal pathway. Adhesion molecules such as L1 and NCAM keep these axons grouped together as they approach the limb bud. Ephrin A molecules in the ventral region of the limb bud then provide inhibitory cues to prevent EphA-expressing motor axons from entering the ventral pathway. NCAM on axons and developing muscle cells (myotubes) further helps direct these axons appropriately as they approach the developing extensor muscle tissue. In contrast to the NCAM expressed in more proximal regions, which kept the axons bundled, NCAM expresses high levels of polysialic acid (PSA) as the axons approach the muscle. The presence of PSA decreases fasciculation and promotes branching of axons across the muscle surface by interfering with homophilic binding between adjacent axons (inset). (Adapted from Bonanomi D & Pfaff SL [2010] *Cold Spring Harb Perspect Biol* 2(3): a001735.)

NCAM adhesion molecules are important in the bundling and guidance of motor axons in the dorsal pathway (Figure 7.15B). L1 is associated with the axons themselves; in the absence of L1, there is a decrease in fasciculation and an increase in nerve branching. NCAM, found both on axons and on developing muscle cells (myotubes), influences axon bundling as well as axonal branching across the surfaces of muscles. In the absence of NCAM, there is some decrease in adhesion between adjacent axons, but the main effect is a decrease in branching along muscle surfaces. The level of PSA associated with NCAM determines how much branching occurs. Scientists have discovered that NCAM without PSA, or with low levels of PSA, promotes fasciculation and keeps axons bundled at the developmental stages when the axons need to travel together. In contrast, NCAM with higher levels of PSA (PSA-NCAM) interferes with axon–axon binding and thus increases the branching of nerve fibers. It is hypothesized that this decrease in fasciculation helps individual nerve fibers to respond better to local guidance cues in the area of the target muscle (Figure 7.15). In recent years, scientists have also noted that the level of spontaneous electrical activity present in developing muscles regulates levels of PSA-NCAM. If electrical activity is blocked *in ovo* at the time when motor axons choose the dorsal or ventral pathway, PSA levels decrease, causing the axons to remain bundled together and dorsal–ventral pathfinding errors to occur.

A decrease in electrical activity also decreased the expression of EphA4 receptors normally found on axons projecting along the dorsal pathway. Under normal conditions, ephrin A ligands are expressed in the mesenchyme of the ventral portion of the limb, where they provide inhibitory cues to direct EphA4-expressing dorsal axons away from the ventral pathway (Figure 7.15B). If spontaneous electrical activity is blocked at the time the axons reach the choice point, however, then the levels of EphA4 decrease, so the axons are no longer inhibited by ephrin A ligands in the ventral pathway, thereby contributing to the pathfinding errors observed in these experiments. Thus, spontaneous electrical activity influences the expression of a number of proteins necessary for normal pathfinding.

Other molecules have been identified as important for motor axon growth to target muscles. These include members of the glial-cell-derived neurotrophic factor (GDNF) and semaphorin families (see Box 7.1). Thus, many factors interact to contribute to the normal pathfinding choices of motor neuron axons as they travel to the correct limb muscle. Once the motor axons reach the correct target muscle, a number of other cellular events occur to transform the growth cone into a presynaptic nerve terminal. The molecular mechanisms regulating synaptic connections between motor neurons and muscles are discussed in Chapter 9.

INTERMEDIATE, MIDLINE TARGETS FOR SPINAL COMMISSURAL AXONS

As noted in Chapter 4, the dorsal half of the vertebrate neural tube gives rise to multiple cell types, including spinal interneurons. A subset of axons from the spinal interneurons does not cross the midline and therefore remains on the **ipsilateral** side of the spinal cord, projecting to targets on the side where they originated. Another subset of these axons crosses to the **contralateral** side, forming a commissure. These spinal commissural interneurons originate in the dorsolateral region of the neural tube and grow ventrally toward the midline floor plate (FP)—a group of glial cells transiently present in the embryonic spinal cord. The contralateral axons never cross back to the side of the spinal cord from which they originate; instead, they project anteriorly or posteriorly until they innervate their

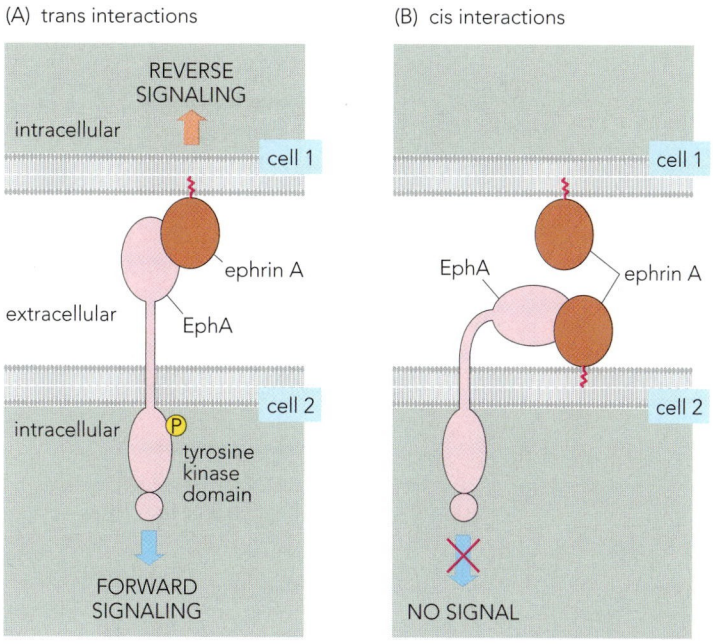

Figure 7.31 Multiple mechanisms for Eph/ephrin signaling. (A) Eph/ephrin signaling occurs by trans interactions when a receptor and ligand located on adjacent cells interact. Trans interactions can result in forward signaling when the ligand initiates signaling in the receptor-bearing cell (cell 2) or reverse signaling when the receptor initiates signaling in the ligand-bearing cell (cell 1). (B) When an EphA receptor and ephrin A ligand are located in the same cell (cell 2), the ligand can bind to the receptor. These cis interactions do not initiate signaling in the receptor-bearing cell (cell 2), but instead prevent the EphA receptor from interacting with ephrin A ligands on an adjacent cell (cell 1) so that no signal is transduced. (Adapted from Egea J & Klein R [2007] *Trends Cell Biol* 17:230–238.)

Another surprising finding is that ephrin ligands and Eph receptors participate in cis interactions. When ephrin ligands and Eph receptors are expressed in the same cell, the ligands can attenuate the response of the receptors to extrinsic sources of ephrins (Figure 7.31B). The ligands acting in cis do not activate the receptors, but rather interfere with receptor activation by any ephrins expressed on adjacent cells (trans interactions). For example, whereas the EphA3 receptor gradient is expressed from a low nasal to high temporal gradient on retinal axons, the ephrin A5 ligand is expressed in the opposite gradient on the retinal axons. Thus, ephrin A5 is found in a high nasal to low temporal gradient across the retina. The higher ephrin A5 on the nasal axons can bind the existing EphA3 receptors on those axons and thus attenuate their response to tectal-derived ephrin A ligands. The gradient of the cis-acting ligand would therefore further sharpen the targeting of the retinal axons to the correct location along the mid-region and posterior areas of the tectum.

Axonal pathfinding in other regions of the nervous system also appears to be influenced by ephrin A and EphA family members signaling in cis. Recent studies of motor neurons, for example, found that when levels of ephrin A ligands acting in cis were experimentally increased, the cis interactions were able to attenuate receptor signaling. However, when ephrin A ligands acting in cis were lower, trans interactions dominated to provide the inhibitory cue to the motor neuron axons approaching an incorrect target region. Thus, subtle changes in the expression levels of ephrin A ligands acting in cis may be regulated at various stages of development to carefully control axonal pathfinding and mapping to target cells. This is another example of how the developing nervous system uses the limited number of signaling molecules available to full advantage to ensure precise wiring of the billions of neural connections required in the vertebrate nervous system.

Recent studies also suggest a mechanism by which at least some ephrins and Ephs may be able to influence cells at a distance, rather than solely through direct cell–cell contact. Both ephrin ligands and Eph receptors are found in **exosomes**. Exosomes are small membrane-bound vesicles that can incorporate full-length proteins, including ephrins and Ephs. The exosomes are packaged into multivesicular bodies within the

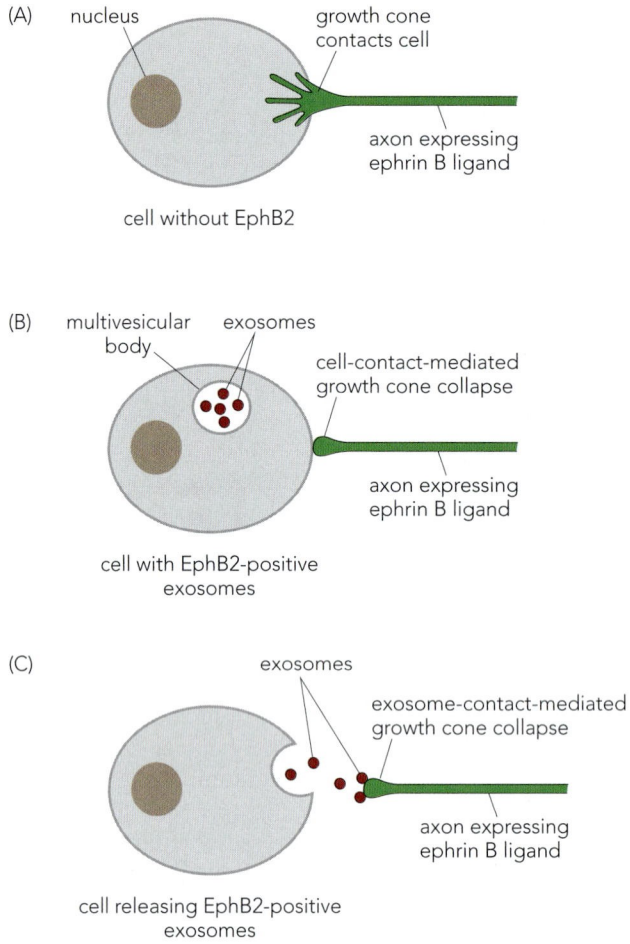

Figure 7.32 Exosomes may be used to mediate Eph/ephrin signaling at a distance. Exosomes are small membrane-bound vesicles packaged into multivesicular bodies that incorporate full-length proteins such as Eph receptors and ephrin ligands. Cells that contain or release such exosomes provide another mechanism by which Eph/ephrin signaling can influence neurite outgrowth. *In vitro* studies reveal the effects of EphB2-positive exosomes on the outgrowth of embryonic mouse cortical neurons expressing ephrin B ligands. (A) In control conditions, when a neurite from an embryonic mouse cortical neuron encounters a cell that does not contain EphB2, the growth cone readily contacts the cell surface. (B) When a cortical neurite encounters a cell with exosomes containing EphB2, the neurite collapses upon cell contact. (C) If a cell with exosomes containing EphB2 is stimulated to release the exosomes, the approaching cortical neurite collapses upon contact with the EphB2-positive exosomes. This suggests that exosomes released *in vivo* may provide a means for Eph-reverse signaling to occur at a distance. (B and C, adapted from Pasquale EB [2016] *J Cell Biol* 214:5–7.)

cell. The multivesicular bodies can fuse with the cell membrane to release their contents into the extracellular space. Studies by Rüdiger Klein and colleagues found that when ephrin-B-expressing cortical neurites contacted a cell with exosomes containing EphB2, the growth cone collapsed. In addition, if the cells were stimulated to release the exosomes containing EphB2, the approaching growth cones collapsed when they encountered the exosomes (**Figure 7.32**). These findings suggest that exosomes could transport EphB signals to distant targets and mediate growth cone behavior through reverse signaling, even in the absence of direct cell–cell contact. How exosome-associated Eph/ephrin family members may contribute to axon guidance *in vivo* is just beginning to be explored, but these recent findings suggest yet another mechanism whereby Eph/ephrin signaling could influence axonal pathfinding and target cell recognition.

Axonal self-avoidance as a mechanism for chemoaffinity

Other forms of chemoaffinity have been proposed to contribute to the highly complex yet highly ordered arrangements of connections in the nervous system. In the 1980s, for example, scientists studying leech neurons described the process of self-avoidance in which nerve fibers from the same neuron were repelled from one another, whereas nerve fibers from different neurons were able to intermingle. Self-avoidance ensured that axons and dendrites from the same neuron were able to uniformly spread across a target field while also allowing axons and dendrites from different neurons to intermingle and overlap within the target field. Subsequent studies in *Drosophila* found that Dscams (Down syndrome cell

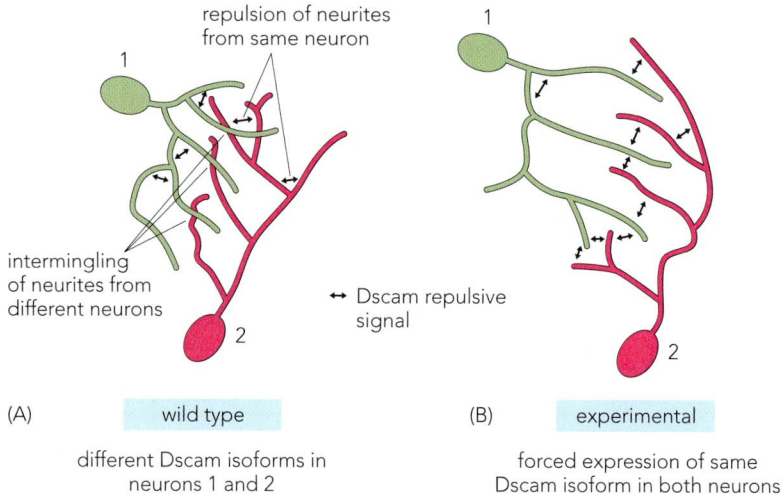

Figure 7.33 Self-avoidance of axons results in patterning of sensory axons in the body wall of *Drosophila*. (A) Different Dscam isoforms are expressed in different sensory neurons. The Dscams normally provide repulsive cues (double-headed arrows) that prevent neurites from the same neuron (1 or 2) from intermingling. However, the neurites from different neurons (1 and 2) overlap and cross one another because each expresses a different isoform of Dscam. This allows two neurons to innervate a shared target region. (B) In experimental conditions in which two neurons were forced to express the same Dscam isoform, the neurites from both cells now avoided one another, as well as their own neurites (double-headed arrows). The large number of Dscam isoforms present in the nervous system suggests these molecules may provide a type of chemoaffinity signal to help individual neurites branch and map to precise target locations. (Adapted from Zipursky SL & Sanes JR [2010] *Cell* 143(3):343–353.)

adhesion molecules), members of the IgG superfamily, were important for regulating this self-avoidance (**Figure 7.33**A). The most remarkable aspect of Dscams is the vast number of isoforms that exist. To date, in *Drosophila*, over 19,000 isoforms of Dscams that differ in the structure of their ectodomains have been identified. Axons and dendrites from the same neuron express the same isoform and are therefore repelled from one another. In contrast, the axons and dendrites from other neurons express different isoforms, allowing for the overlapping distribution of the nerve fibers from the different cells. Although fewer Dscam isoforms have been identified in vertebrates so far, it is thought that the Dscams and related proteins function in a similar manner to those found in *Drosophila*.

The numerous Dscam isoforms available suggest a means for a large number of molecular cues to be generated from a single gene. The importance of Dscams in mediating axonal patterning across a target tissue was seen when Dscam levels were altered in *Drosophila*. When two different sensory neurons were forced to express the same Dscam isoform, the axons from the different neurons would no longer intermingle and cross, but instead avoided one another (Figure 7.33B). The repulsive interactions between neuronal processes expressing different Dscam isoforms led to the suggestion that these molecules act as a form of chemoaffinity, directing individual axons and dendrites to specific target cells.

SUMMARY

Chapter 7 describes just some of the many signals that a nerve fiber must integrate as it extends from the neuronal cell body to reach a target cell. Growth cones actively grow along different embryonic pathways and select regions based on available permissive, attractive, and repulsive cues. Based on the current understanding of axonal guidance cues, it may no longer seem surprising that a single protein can have different functions in different regions of the developing embryo, such as being attractive to one set of axons at one stage of development and repulsive to another group of neurons at a different age. It may not even seem unusual that a single protein family can have both secreted and cell surface-bound forms. Yet, many of these findings that were first described in the past few decades have significantly changed how neurobiologists view guidance cues. Such findings, for example, address the long perplexing question of how the nervous system can generate all of the proteins necessary to establish the billions of neural connections required for proper function. As some scientists have long suspected, there does not need to be a single protein for each axon, or even for a group of axons. Instead, evidence now supports the hypothesis that the nervous system effectively uses the limited number of proteins available to yield subtle, cell-specific responses in different contexts.

FURTHER READING

Axel R (2005) Scents and sensibility: a molecular logic of olfactory perception (Nobel lecture). *Angew Chem Int Ed* 44:6110–6127.

Bastiani MJ, Raper JA & Goodman CS (1984) Pathfinding by neuronal growth cones in grasshopper embryos. III. Selective affinity of the G growth cone for the P cells within the A/P fascicle. *J Neurosci* 4(9):2311–2328.

Bastiani MJ, Harrelson AL, Snow PM & Goodman CS (1987) Expression of fasciclin I and II glycoproteins on subsets of axon pathways during neuronal development in the grasshopper. *Cell* 48(5):745–755.

Buck L & Axel R (1991) A novel multigene family may encode odorant receptors: a molecular basis for odor recognition. *Cell* 65:175–187.

Bushdid C, Magnasco MO, Vosshall LB & Keller A (2014) Humans can discriminate more than 1 trillion olfactory stimuli. *Science* 343:1370–1372.

Cypher C & Letourneau PC (1992) Growth cone motility. *Curr Opin Cell Biol* 4(1):4–7.

de Castro F, Lopez-Mascaraque L & De Carlos JA (2007) Cajal: lessons on brain development. *Brain Res Rev* 55(2):481–489.

Drescher U, Kremoser C, Handwerker C et al. (1995) In vitro guidance of retinal ganglion cell axons by RAGS, a 25 kDa tectal protein related to ligands for Eph receptor tyrosine kinases. *Cell* 82(3):359–370.

Gomez TM & Letourneau PC (2014) Actin dynamics in growth cone motility and navigation. *J Neurochem* 129(2):221–234.

Gong J, Körner R, Gaitanos L & Rüdiger K (2016) Exosomes mediate cell contact-independent ephrin-Eph signaling during axon guidance. *J Cell Biol* 214(1):35–44.

Goodman CS (1996) Mechanisms and molecules that control growth cone guidance. *Annu Rev Neurosci* 19:341–377.

Hanson MG & Landmesser LT (2004) Normal patterns of spontaneous activity are required for correct motor axon guidance and the expression of specific guidance molecules. *Neuron* 43(5):687–701.

Imai F and Yoshida & (2015) "Axon guidance in the spinal cord". In: A. Kumanogoh (Ed.) Semaphorins pp. 39-63. Springer: Japan.

Kennedy TE, Serafini T, de la Torre JR & Tessier-Lavigne M (1994) Netrins are diffusible chemotropic factors for commissural axons in the embryonic spinal cord. *Cell* 78(3):425–435.

Kao TJ & Kania A (2011). Ephrin-mediated cis-attenuation of Eph receptor signaling is essential for spinal motor axon guidance. *Neuron* 71(1):76–91.

Klein R (2012) Eph/ephrin signalling during development. *Development* 139(22):4105–4109.

Kolodkin AL, Levengood, DV. Rowe EG, Tai YT, Giger RJ & Ginty DD (1997) Neuropilin is a semaphorin III receptor. *Cell* 22 (90): 753–762.

Komiyama T & Luo L (2006) Development of wiring specificity in the olfactory system. *Curr Opin Neurobiol* 16:67–73.

Lance-Jones C & Landmesser L (1980) Motoneurone projection patterns in the chick hind limb following early partial reversals of the spinal cord. *J Physiol* 302:581–602.

Lepelletier L, Langlois SD, Kent CB et al. (2017) Sonic Hedgehog guides axons via zipcode binding protein 1-mediated local translation. *J Neurosci* 37(7):1685–1695.

Letourneau PC (1975) Cell-to-substratum adhesion and guidance of axonal elongation. *Dev Biol* 44(1):92–101.

Letourneau PC (2009) Axonal pathfinding: extracellular matrix role. In Encyclopedia of Neuroscience (Squire LR ed), pp 1139–1145. Academic Press.

Lin CH, Thompson CA & Forscher P (1994) Cytoskeletal reorganization underlying growth cone motility. *Curr Opin Neurobiol* 4(5):640–647.

Lowery LA & Van Vactor D (2009) The trip of the tip: understanding the growth cone machinery. *Nat Rev Mol Cell Biol* 10(5):332–343.

Tessier-Lavigne M & Kolodkin AL (eds) (2011) Neuronal Guidance: The Biology of Brain Wiring. Cold Spring Harbor Laboratory Press.

Messersmith EK, Leonardo ED, Shatz CJ et al. (1995) Semaphorin III can function as a selective chemorepellent to pattern sensory projections in the spinal cord. *Neuron* 14(5):949–959.

Mombaerts P (2006) Axonal wiring in the mouse olfactory system. *Annu Rev Cell Dev Biol* 22:713–737.

Mueller BK (1999) Growth cone guidance: first steps towards a deeper understanding. *Annu Rev Neurosci* 22:351–388.

Nakamoto M, Cheng HJ, Friedman GC et al. (1996) Topographically specific effects of ELF-1 on retinal axon guidance in vitro and retinal axon mapping in vivo. *Cell* 86(5):755–766.

Pasquale EB (2016) Exosomes expand the sphere of influence of Eph receptors and ephrins. *J Cell Biol* 214(1):5–7.

Placzek M, Tessier-Lavigne M, Yamada T et al. (1990) Guidance of developing axons by diffusible chemoattractants. *Cold Spring Harb Symp Quant Biol* 55:279–289.

Serafini T, Kennedy TE, Galko MJ et al. (1994). The netrins define a family of axon outgrowth-promoting proteins homologous to C. elegans UNC-6. *Cell* 78(3):409–424.

Sperry RW (1947) Nature of functional recovery following regeneration of the oculomotor nerve in amphibians. *Anat Rec* 97(3):293–316.

Roche FK, Marsick BM & Letourneau PC (2009) Protein synthesis in distal axons is not required for growth cone responses to guidance cues. *J Neurosci* 29(3):638–652.

Rutishauser U & Landmesser L (1991) Polysialic acid on the surface of axons regulates patterns of normal and activity-dependent innervation. *Trends Neurosci* 14(12):528–532.

Udin SB & Fawcett JW (1988) Formation of topographic maps. *Annu Rev Neurosci* 11:289–327.

Vitriol EA & Zheng JQ (2012) Growth cone travel in space and time: the cellular ensemble of cytoskeleton, adhesion, and membrane. *Neuron* 73(6):1068–1081.

Walter J, Henke-Fahle S & Bonhoeffer F (1987) Avoidance of posterior tectal membranes by temporal retinal axons. *Development* 101(4):909–913.

Walter J, Kern-Veits B, Huf J et al. (1987) Recognition of position-specific properties of tectal cell membranes by retinal axons in vitro. *Development* 101(4):685–696.

Yazdani U & Terman JR (2006) The semaphorins. *Genome Biol* 7(3):211.

Zipursky SL & Sanes JR (2010) Chemoaffinity revisited: dscams, protocadherins, and neural circuit assembly. *Cell* 143(3): 343–353.

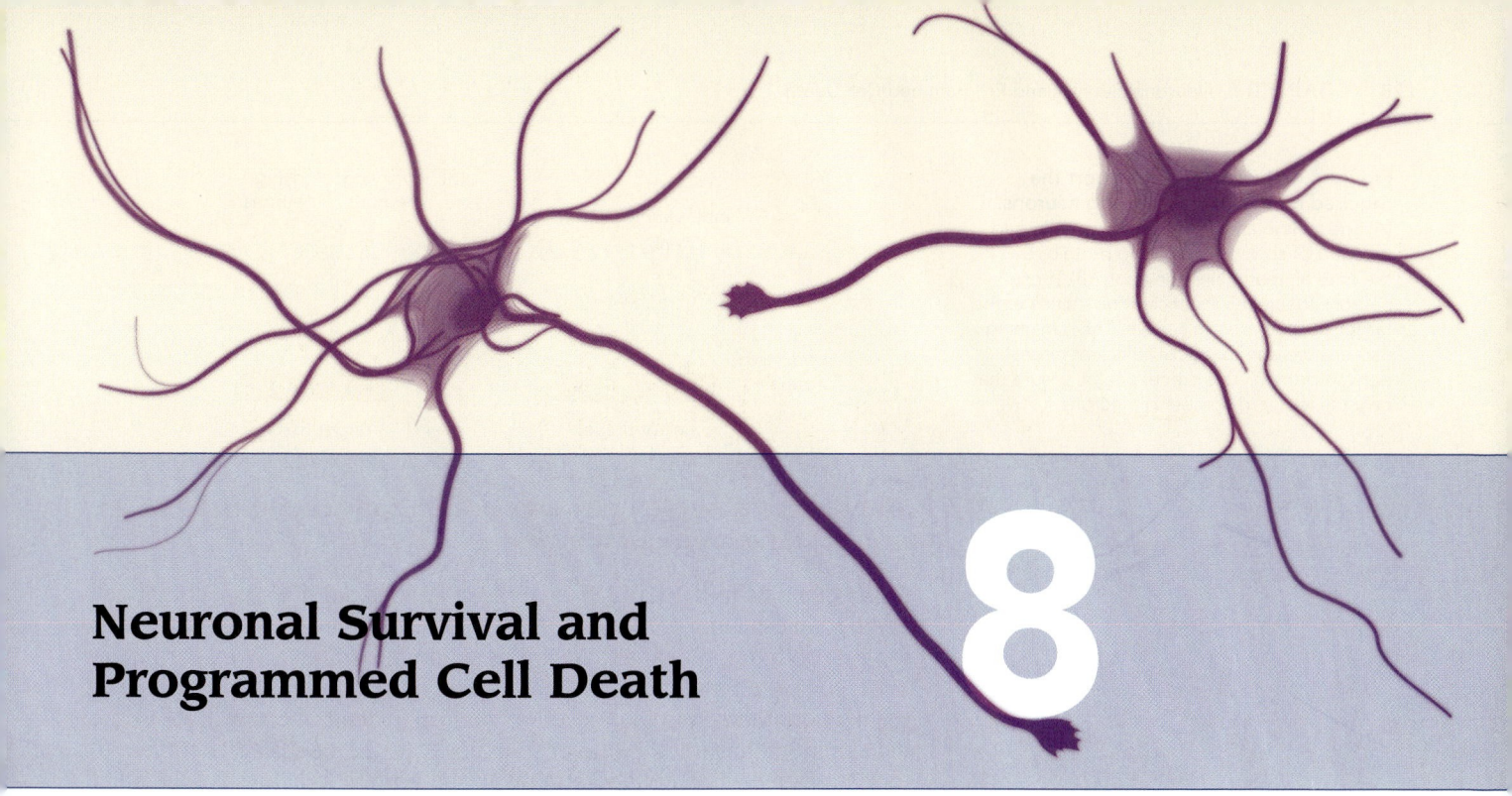

8 Neuronal Survival and Programmed Cell Death

Chapter 7 described mechanisms by which neurons extend nerve fibers along distinct pathways to reach target tissues. Once a nerve fiber reaches its final target region, the target tissue plays an important role in determining whether that neuron will survive. Scientists now know that many target tissues, particularly those in the peripheral nervous system (PNS), secrete survival-promoting (trophic) proteins that help regulate the final number of innervating neurons. As proposed by the **neurotrophic hypothesis**, the release of limited amounts of these proteins at the time of neural innervation helps ensure that only the correct number of neurons survive in a particular neuronal population. The neurons that do *not* obtain the target-derived proteins die off, thus preventing survival of any excess neurons (**Figure 8.1**).

Another means by which neuronal numbers are regulated in the developing nervous system is through **programmed cell death** (**PCD**), a process by which subsets of neurons produced during neurogenesis are later eliminated. Programmed cell death is now understood to be a developmentally necessary event that further helps establish the final, correct number of neurons in the nervous system. This chapter focuses on the discovery and role of target-derived proteins, their receptors, and the intracellular signaling pathways that intersect to regulate neuronal survival and programmed cell death during the course of neural development.

GROWTH FACTORS REGULATE NEURONAL SURVIVAL

From the late nineteenth through to the mid-twentieth century, neuroembryologists debated what role, if any, target tissues served in regulating aspects of neural development. The importance of target tissues was gradually accepted after a number of detailed studies documented that neuronal survival was altered by the presence or absence of target tissues. By the 1950s, the view that target-derived growth factor proteins were important in promoting neuronal survival and preventing neuronal death during embryonic development was firmly established.

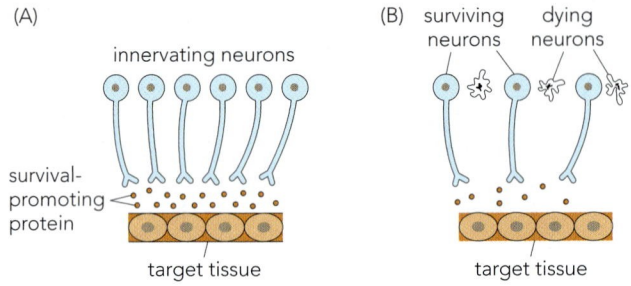

Figure 8.1 Target tissues support the required number of innervating neurons. (A) The cells in a target tissue secrete limited amounts of survival-promoting proteins at the time of neuronal innervation. (B) Those neurons that receive sufficient quantities of the target-derived protein survive. The remaining neurons die due to lack of sufficient target-derived protein. This process helps ensure that only the required number of neurons survive.

The death of nerve cells was not initially recognized as a normal developmental event

Most scientists in the late 1800s and early 1900s did not suspect that cell death is an essential part of neural development. There were many reasons for this, both theoretical and technical. Although Wilhelm Roux, writing in 1881, suggested that neuronal death was a normal part of embryogenesis, Cajal and others felt that, even if neuronal death occurred for some developing neurons, "the immense majority of the neuroblasts survive to term and succeed in collaborating with the normal structures of the adult nervous system." This view was widely held into the mid-twentieth century, as scientists seemed reluctant to consider the possibility that healthy neurons were initially overproduced, with a percentage being eliminated later in development. It seemed much more plausible that the nervous system would produce only the number of neurons needed.

The failure to consider cell death as a normal part of neural development was further influenced by technical considerations. On the slides of fixed tissue sections examined by early neuroembryologists, it was difficult to detect cell death at different developmental stages because any dying cells were quickly removed by phagocytosis. Therefore, cell death might not be evident in any given tissue section viewed. Cell death was later documented in fixed tissues by examining serial tissue sections and comparing detailed cell counts at different developmental stages. However, because the early researchers did not expect cell death, they were not comparing neuronal numbers in the tissue sections they collected. By the 1940s, scientists began to compare neuronal numbers at different developmental stages, leading more and more investigators to report a decrease in neuronal number over the course of embryogenesis. From these observations it was gradually accepted that cell death was a normal aspect of neural development.

Studies reveal that target tissue size affects the number of neurons that survive

Studies in the early twentieth century first described changes in cell number following various surgical manipulations of target tissues. In 1909, for example, Márian Shorey reported that removing the limb buds from chick or amphibian embryos resulted in **hypoplasia** of the motor and dorsal root ganglion (DRG) neurons innervating the limb bud. This hypoplasia resulted from a decrease in the size of the motor neuron pools and dorsal root ganglia, as well as the individual neurons within these populations. Others noted that grafting extra limb tissue resulted in **hyperplasia**, an increase in the size of the innervating neuronal populations and individual neurons. At the time, it was thought that target tissues, such as the limb buds, somehow regulated neuronal proliferation, migration, or differentiation to establish the size of the innervating neuronal populations. However, neuronal survival was not yet suspected to be influenced by target tissues.

In the late 1930s and early 1940s, Viktor Hamburger, a pioneer in the study of neuron–target interactions, found a direct correlation between the size of the target tissue and the extent of hypo- or hyperplasia observed in the innervating neural populations. Determining whether the cue originating in the target tissue controlled neuronal proliferation, migration, or differentiation became the focus of subsequent studies conducted by Hamburger and colleagues. However, the scientists found that none of the proposed mechanisms accounted for the changes in cell number. While they too observed hypoplasia of the innervating motor and DRG neurons in the absence of target limb bud tissue, they also noted a new key detail: When target tissues were removed, the nuclear material in many of the innervating neurons became **pyknotic** (irreversibly condensed), indicating that the cells were dying in the absence of target tissue. In contrast, when extra limb bud tissue was grafted into an embryo, hyperplasia of these innervating neurons was observed, suggesting an increase in the number of surviving neurons (**Figure 8.2**). Through careful analysis of neuronal cell numbers, the idea emerged that target tissues regulated neuronal survival, rather than influencing the proliferation, migration, or differentiation of innervating neurons, as originally hypothesized.

Another major finding from these studies was the observation that a certain percentage of cells died even under normal conditions. The number of neurons exhibiting this type of cell death varied with the cell population examined, but several neuronal types—including spinal motor neurons and neurons from dorsal root and sympathetic ganglia—revealed a decrease in neuronal number at the time of innervation, even in the presence of target tissues. As described later in the chapter, the significance of this observation became clear over subsequent decades as studies of neuron–target interactions continued.

Some tumor tissues mimic the effect of extra limb buds on nerve fiber growth

Studies from Hamburger's lab were among the most influential in explaining neuron–target interactions, and the 1940s and 1950s proved to be pivotal decades for revealing how target tissues regulate the

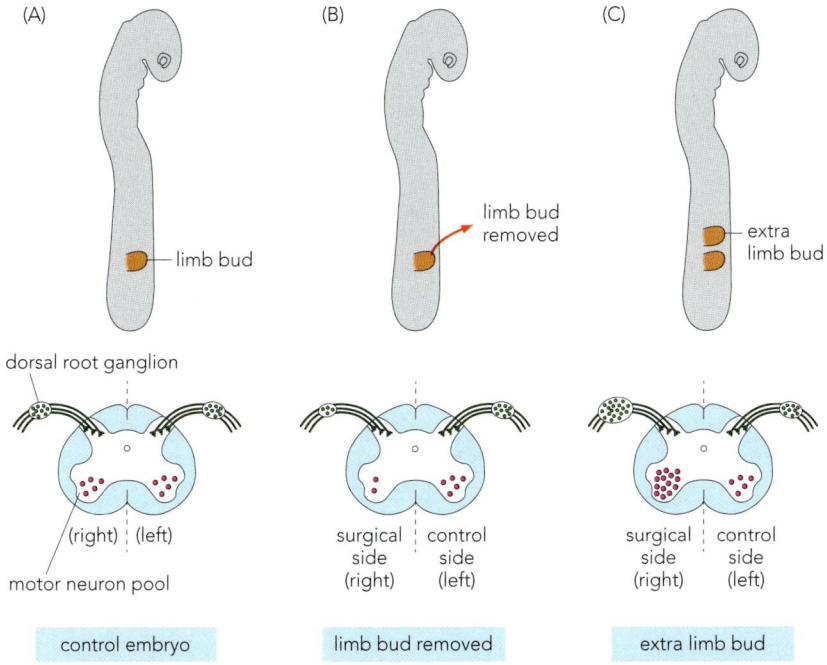

Figure 8.2 Target tissue size regulates the number of surviving neurons. In chick embryos, the amount of limb bud tissue ultimately determines how many motor and dorsal root ganglion neurons survive. (A) In the control embryo, the neuronal populations of the motor neuron pools and the dorsal root ganglia (DRG) are of equal size on either side of the embryo. (B) Removing the limb bud on one side, in this example the right (surgical) side of the embryo, results in a decrease in the number of motor and dorsal root ganglion neurons only on that side. The left (control) side continues to support the same number of neurons as found in the control embryos. (C) Grafting an extra limb bud to the right (surgical) side of the embryo increases the number of motor and dorsal root ganglion neurons on that side compared to the contralateral, control side. Together, these experiments demonstrate that the size of the target tissue directly influences the number of neurons that survive.

innervation and survival of associated neurons. It was during this period that the neurotrophic hypothesis emerged (see Figure 8.1) and the first target-derived, survival-promoting nerve growth factor was identified. The discovery of this first nerve growth factor is one of the best documented in the history of developmental neurobiology. There are many reasons for this, including the importance of the finding in understanding how neuronal numbers are established, the fortunate events that led to its purification, and the impact this discovery had on subsequent studies, including those that identified other nerve growth factors. The discovery of the first nerve growth factor also illustrates how the techniques available at any given period shape the types of discoveries that can be made at that time, and demonstrates how detailed and careful studies from one decade can be of ongoing benefit to scientists in future decades.

The first major finding that altered how scientists approached studies of target-derived factors was reported by Elmer Bueker, a student in Hamburger's lab during the 1940s. Because embryonic limb buds contain dividing cells and gradually increase in size during development, it was thought that other tissues containing rapidly dividing cells, such as tumors, might prove to be a better experimental tool than grafted limb buds. Therefore, Bueker grafted different tumors onto chick embryos to see if neuronal populations were influenced by the size of the tumor graft.

Bueker tested three tumors: an adenocarcinoma, the Rous sarcoma, and sarcoma 180. He implanted each into the body wall of early-stage chick embryos, prior to the period of neural innervation. Only the sarcoma 180 tumor survived and integrated into the body wall of the embryo. After five days *in ovo*, nerve fibers from DRG neurons extended into the growing tumor. In addition, the ganglia appeared larger on the side of the tumor graft compared to ganglia on the opposite side. Bueker proposed that chemical properties of the tumor promoted the growth of nerve fibers and increased neuronal numbers.

These findings, published in 1948, influenced subsequent studies by Hamburger and another member of his lab, Rita Levi-Montalcini, who followed up on this effect. She first repeated Bueker's studies, extending the survival times of the embryos. She also tested additional mouse sarcoma tumors (sarcoma 37 and sarcoma 1). She confirmed that the tumors did in fact lead to extensive nerve fiber growth into the tumor and an increase in the size of dorsal root ganglia on the side of the sarcoma graft. She also noted that sympathetic ganglia on that side showed an even greater increase in volume than the dorsal root ganglia (**Figure 8.3**). In contrast to what was observed with limb bud grafts, however, motor neuron number was *not* affected by the presence of the tumor.

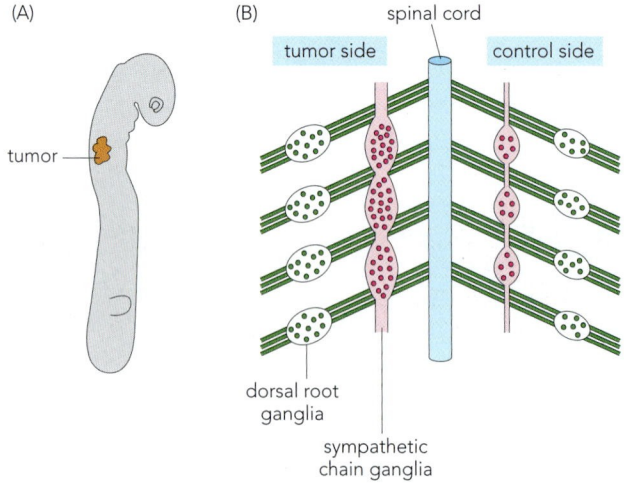

Figure 8.3 Tumor tissue promotes survival of dorsal root and sympathetic neurons. (A) To test whether a rapidly dividing tissue produced a factor to promote neuronal survival or outgrowth, a sarcoma 180 tumor was implanted onto the body wall of a chick embryo. (B) Studies by Bueker and Levi-Montalcini revealed that the DRG, as well as the sympathetic chain ganglia, increased in size on the side of the tumor graft. The size of the ganglia on the contralateral side did not change. These studies suggested that more neurons survived in the presence of the tumor. Unlike grafts of limb bud tissue, tumor tissue did not influence the size of the motor neuron pools. (Adapted from Levi-Montalcini R & Hamburger V [1951] *J Experimtl Zool* 116:321–361.)

In vitro studies led to a bioassay method to study nerve growth factors

As Levi-Montalcini and Hamburger continued to investigate the developmental effects of the sarcoma tumors, they discovered that sympathetic nerve fibers not only invaded the tumor, but also grew into veins and other tissues located away from the tumor. This led to the idea that the tumors released a factor into the embryonic circulatory system that could influence nerve fiber growth. To test this hypothesis, fragments of tumor were implanted on the chorioallantoic membrane surrounding early-stage embryos. This method prevented direct contact between the tumor and embryo, but allowed any factors released by the tumor to enter the circulatory system. In these preparations, sympathetic ganglia were again increased in size, and nerve fibers invaded blood vessels and other tissues, confirming that the tumor released a diffusible factor that promoted neuronal survival and nerve fiber outgrowth.

Realizing it would be difficult to investigate the nature of the tumor-derived factor using only *in ovo* preparations, Levi-Montalcini traveled to Brazil in the early 1950s to learn neuronal tissue culture, an emerging field that was not yet commonly used by neuroembryologists. In Brazil, Levi-Montalcini learned how to successfully culture dorsal root and sympathetic ganglia as **explants**—individual intact tissues (for example, a ganglion) grown in a cell culture dish. This allowed her to test whether the tumor tissues placed at a distance away from a ganglion directly influenced nerve fiber growth *in vitro*. She found that when a sarcoma tumor segment and a ganglion were cultured in the same dish, nerve fibers grew outward from the ganglion explant, forming what was described as a "halo effect." In control cultures of ganglia cultured with other tumors or various mouse tissues, neurite outgrowth was limited or did not occur at all. Thus, there appeared to be something released by the sarcoma tumors that promoted extensive neurite outgrowth from chick dorsal root and sympathetic ganglion explants.

Levi-Montalcini sent detailed drawings of her findings to Hamburger as she collected data on the outgrowth from ganglia at different developmental stages (**Figure 8.4**). In 1954 they published their first paper using this cell culture method to demonstrate the outgrowth-promoting properties of substances released by the sarcoma tumor cells. Levi-Montalcini's bioassay method soon became a primary means of testing whether different substances promoted outgrowth from ganglia. Such assays remain a popular tissue culture method today.

The factor released by sarcoma 180 is found to be a protein

Upon returning to Hamburger's lab, Levi-Montalcini began collaborating with a postdoctoral fellow, the biochemist Stanley Cohen, who had joined

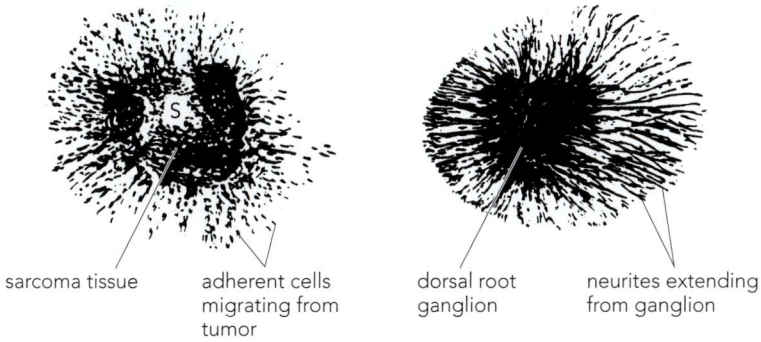

Figure 8.4 First images to reveal the effects of tumor tissue on the outgrowth of dorsal root ganglia *in vitro*. In 1952, Rita Levi-Montalcini developed a tissue culture method to assay outgrowth of dorsal root and sympathetic ganglia. A ganglion was co-cultured with fragments of sarcoma tissue to test for the presence of diffusible, growth-promoting factors. This image is one of Levi-Montalcini's original India ink drawings, illustrating that in the presence of sarcoma 180 tumor (left, S), a dense halo of nerve fiber outgrowth extended from a dorsal root ganglion explant (right) after 24 hours in culture. (From Levi-Montalcini R [1987] *Science* 237:1154–1162.)

Figure 8.5 Snake venom contains a nerve growth factor. In studies designed to test whether extracts of tumor tissue contained nucleic acids, snake venom was added to cultures containing tumor extracts. The snake venom contained phosphodiesterases—enzymes known to break down nucleic acids. (A) In the presence of tumor extract only, chick dorsal root or sympathetic ganglia produced a dense halo of neurite outgrowth. (B) When snake venom was added to the tumor extracts, a dense halo of outgrowth was again noted and appeared greater than that observed in cultures treated with tumor extract only. (C) Surprisingly, the control cultures that contained only snake venom produced greater neurite outgrowth than the cultures treated with tumor extract. Subsequent analyses revealed that the snake venom contained greater nerve-growth-promoting activity than the tumor extract. (D–G) Images from Cohen and Levi-Montalcini's 1956 paper revealing that the nerve-growth-promoting activity from extracts of snake venom was greater than that from control medium or tumor extract. (D) A dorsal root ganglion grown in a standard cell culture medium without additional extracts (control) does not produce neurite outgrowth. (E) Neurite outgrowth occurred in the presence of an unprocessed snake venom sample. (F) Denser and more extensive growth was observed when the snake venom was partially purified. (G) Neurite outgrowth was less dense when grown in the presence of a partially purified sample of sarcoma 180 tumor extract. (D–G, from Cohen S & Levi-Montalcini R [1956] *Proc Natl Acad Sci USA* 42:571–574.)

Hamburger's lab to identify the factor that promotes nerve fiber growth. He subjected extracts of sarcoma 180 tissues to various treatments to characterize the type of factor involved. The first studies, published in 1954, revealed that the substance released by the sarcoma 180 tumor was a nucleoprotein, though at this stage it was unclear if the active substance was a nucleic acid or a protein.

Cohen's studies suggested that the factor might be a protein, because the nerve-growth-promoting activity was destroyed by proteases—enzymes that break down proteins. However, there was also concern that viruses or other contaminating nucleic acids could be in the extracts. A colleague suggested he treat the tumor extracts to remove any DNA or RNA in the samples. To eliminate these potential contaminants, the tumor extracts were treated with snake venom containing phosphodiesterases—enzymes that break down nucleic acids. The snake venom was milked from pit vipers (either rattlesnakes or moccasins) living at the Ross Allen Reptile Institute, a former theme park and educational center located in Silver Springs, Florida. That the scientists had to order snake venom from a reptile institute in order to remove nucleic acids from the tumor extract samples is just one reminder that the reagents and techniques available to biochemists in the 1950s were quite different from those available today.

Explants of chick dorsal root and sympathetic ganglia were cultured with the venom-treated sarcoma extracts and the extent of neurite outgrowth was compared to ganglia cultured with untreated sarcoma extracts. In addition, ganglia were treated with snake venom alone to confirm that the venom was not toxic to the ganglia. Ganglia grown in the presence of venom-treated sarcoma extracts extended nerve fibers, consistent with the hypothesis that the active substance was a protein and not a nucleic acid (**Figure 8.5**A, B). The scientists also observed that ganglia grown in the presence of the venom-treated sarcoma extracts appeared to produce more

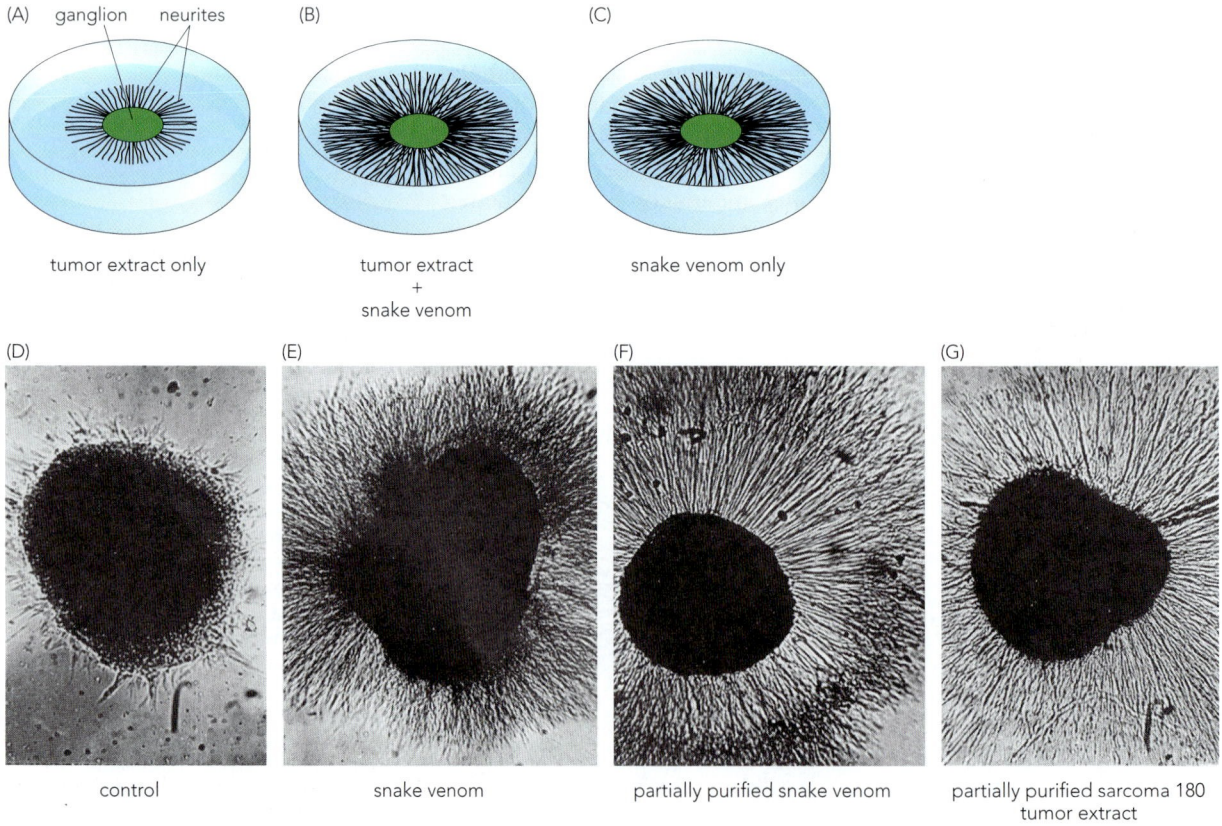

neurite outgrowth than ganglia treated with tumor extract alone (Figure 8.5C). In addition, a surprising and ultimately more influential discovery was made: The ganglia treated solely with snake venom exhibited extensive neurite outgrowth, exceeding the outgrowth produced by tumor extracts alone. The scientists subsequently found that snake venom was the most potent promoter of neurite outgrowth they had found to date (Figure 8.5D–G).

Nerve growth factor is identified in salivary glands

Cohen then sought to purify the active protein. However, protein purification requires large quantities of starting material, and because snake venom glands are very small and not readily available, they were not a good source of material. Cohen recalled that snake venom glands are modified salivary glands. Salivary glands are not only larger, but are more easily obtained from common animal models such as mice. He therefore tested salivary glands from male mice and found that the growth-promoting factor was present there as well. It was fortunate that Cohen started with male mice, because salivary glands from female mice do not produce the same level of the factor. Remarkably, for reasons that still remain unknown, the male mouse salivary gland is the most abundant known source of the nerve-growth-promoting factor.

Using these salivary glands as raw material, by 1956 Cohen was able to purify a protein that they called the nerve-growth-stimulating factor, now called nerve growth factor (NGF). Like the tumor extracts and snake venom, purified NGF promoted the outgrowth of dorsal root and sympathetic ganglia *in vitro* (**Figure 8.6**A). To confirm that NGF is needed for normal development, the scientists developed an antiserum that blocked the activity of NGF. When this antiserum was added to cultures of NGF-treated DRG, neurite outgrowth was inhibited, demonstrating that the antiserum blocked NGF activity (Figure 8.6B). Further, when this antiserum was injected into newborn mice at the time when sympathetic neurons innervate target tissues, the sympathetic ganglia were greatly reduced in size, with very few neurons remaining in any single ganglion (Figure 8.6C). This "immunosympathectomy" was further evidence that NGF normally acts *in vivo* to promote the outgrowth and survival of sympathetic neurons. Additional studies revealed that injections of NGF rescued neurons that

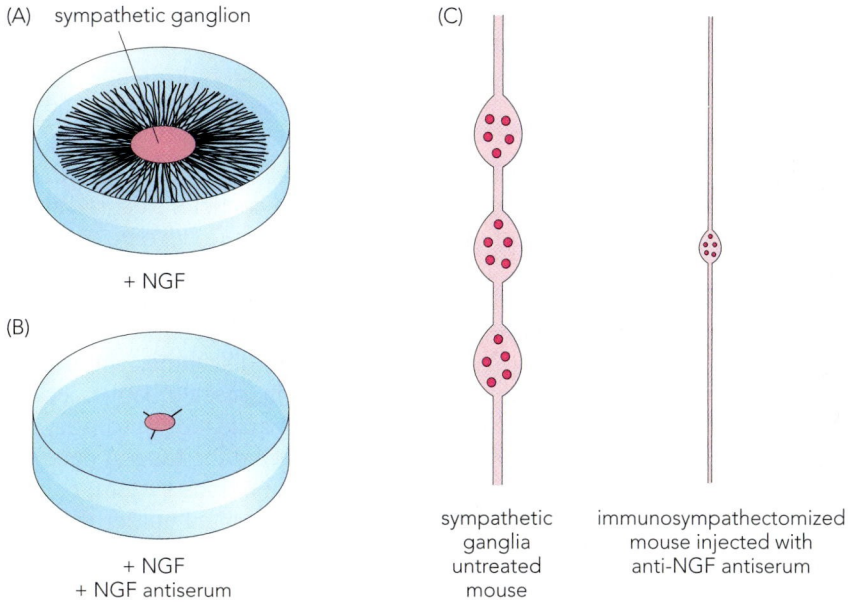

Figure 8.6 Nerve growth factor was purified from mouse salivary glands and found to be required for the development of sympathetic neurons. Using male mouse salivary gland tissue, Stanley Cohen purified nerve growth factor (NGF). (A) Similar to extracts from sarcoma 180 tissue or snake venom glands, NGF promotes a dense halo of neurite outgrowth from dorsal root or sympathetic ganglion explants grown *in vitro*. (B) NGF-induced outgrowth is blocked by the addition of antiserum that inhibits NGF activity. (C) The NGF antiserum also blocks the effects of NGF *in vivo*. When NGF antiserum was injected into newborn mice at the time of target innervation, sympathetic ganglia were greatly reduced in size, with few neurons remaining. These studies demonstrated that NGF is required *in vivo* at the time of target innervation to promote survival of innervating sympathetic neurons. (C, adapted from Levi-Montalcini R & Booker B [1960] *Proc Natl Acad Sci USA* 46:384–391.)

would normally die during development. Thus, *in vitro* as well as *in vivo* manipulations confirmed that NGF was important for normal development of sympathetic and dorsal root ganglion neurons at the time of target innervation. The significance of the discovery of NGF was highlighted in 1986 when Rita Levi-Montalcini and Stanley Cohen received the Nobel Prize in Physiology or Medicine for their work (**Box 8.1**).

Further characterization of NGF became possible as laboratory techniques evolved. In the 1970s, the amino acid sequence of NGF was determined, and in the 1980s, NGF was cloned. In the 1990s, when techniques for producing gene knockout mice were first developed, mice lacking the gene for NGF were generated. These mice had drastically decreased numbers of neurons in dorsal root and sympathetic ganglia, consistent with the various *in vivo* and *in vitro* studies conducted over the years.

Recognition that not all neuronal populations respond to NGF leads to the discovery of brain-derived neurotrophic factor

Once purified NGF was available, researchers from multiple labs were also able to investigate the growth factor's functions in other aspects of neural development. NGF was found to affect not only the neurons in dorsal root and sympathetic ganglia, but also the neurons in most neural-crest-derived ganglia in the PNS and the basal forebrain cholinergic neurons in the CNS. However, NGF did not influence all neuronal populations studied. Motor

Box 8.1 NGF and the intersection of two extraordinary careers

Rita Levi-Montalcini and Stanley Cohen shared the Nobel Prize in Physiology or Medicine in 1986 for their discovery of NGF and its role in neural development. Levi-Montalcini (**Figure 1**) was born in Turin, Italy, in 1909. She received a degree in medicine and surgery from the University of Turin in 1936, where she continued to work as a research scientist until 1938, when laws forbade individuals of Jewish descent from working in universities or the medical profession. In autobiographical writings she described how she was forced into hiding during World War II and conducted her studies in a makeshift lab set up in a bedroom. This research, when later published, motivated Viktor Hamburger to invite her to work in his lab. Originally anticipating a one to two year stay, she ultimately spent 26 years in Hamburger's lab. Levi-Montalcini lived to be 103 years old, and throughout her life she continued to study NGF, first in the United States and then in her native Italy. Over the years, she also wrote extensively about her experiences working with Hamburger, Cohen, and others throughout her career, providing a personal perspective on the process of discovering the first known nerve growth factor.

Figure 1 Rita Levi-Montalcini shown in her office in 1990. (From Aloe L [2004] *Trends Cell Biol* 14:395–399.)

Stanley Cohen (**Figure 2**) was born in 1922 in Brooklyn, New York. He received a bachelor's degree in biology and chemistry from Brooklyn College in 1943, a master's degree in zoology from Oberlin College in 1945, and a Ph.D. in biochemistry from the University of Michigan in 1948. Cohen has made remarkable and diverse contributions throughout his career. In the process of discovering NGF, Cohen injected baby mice daily with a crude extract of adult male mouse salivary glands. He observed that not only did the treated mice develop an enlarged sympathetic nervous system, but also their eyes opened and teeth erupted earlier than untreated controls. In contrast, injections of purified

Figure 2 Stanley Cohen shown talking with students on a visit to Oberlin College. (Courtesy of Oberlin College Archives.)

NGF enlarged the sympathetic nervous system, but did not alter eye or tooth development. Recognizing that the crude extract must contain another active component, he went on to identify epidermal growth factor (EGF) and, later, its receptor (EGFR), which was the first ligand-activated tyrosine kinase to be identified. These discoveries ultimately led to the use of anti-receptor antibodies and tyrosine kinase inhibitors as therapies for cancers that overexpress proteins in the EGF family.

Cohen also investigated **cytokines**, coining the term for this large family of multifunctional molecules. Some cytokines—like ciliary neurotrophic factor (CNTF), discussed later in this chapter—regulate aspects of neural development. Thus, Cohen's valuable contributions to science extended far beyond his discovery of NGF.

neurons in the spinal cord and neurons in most placode-derived PNS ganglia failed to respond to NGF. Such findings suggested that additional growth factors were present in other target tissues to regulate survival and outgrowth of their corresponding neuronal populations.

In the late 1970s, Yves-Alain Barde, Hans Thoenen, and colleagues described a novel factor detected in a glioma-cell-conditioned medium. When glioma cells were grown in cell culture, they secreted proteins into the surrounding cell culture medium. This "conditioned medium" contained a factor that shared some, but not all, properties of NGF. Like NGF, the glioma-cell-conditioned medium promoted growth of dorsal root ganglion explants. In contrast to NGF, however, the biological activity was not blocked by antibodies to NGF (**Figure 8.7**).

The scientists searched for other, more abundant sources of this growth-promoting factor. They found a similar biological activity was present in extracts of whole pig brain and used this tissue to purify a growth-promoting substance that was distinct from NGF. In 1982, more than 25 years after the purification of NGF, Barde, Edgar, and Thoenen reported the discovery of a new growth factor called brain-derived neurotrophic factor (BDNF). As noted previously, it requires large quantities of starting material to purify a protein. This was clearly illustrated in the discovery of BDNF, during which 3 kilograms of pig brain were processed to generate just 2 micrograms of protein.

BDNF was shown to be distinct from NGF, not only because antibodies to NGF failed to inhibit BDNF activity, but also because BDNF promoted

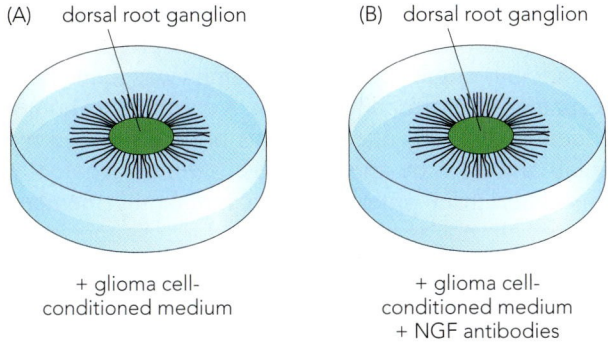

Figure 8.7 Glioma-cell-conditioned medium contains proteins distinct from NGF. (A) A dorsal root ganglion cultured in the presence of growth medium conditioned by glioma cells produces a dense halo of neurite outgrowth. (B) Addition of antibodies that block NGF activity does not limit this neurite outgrowth, suggesting that the glioma cell-conditioned medium contains a nerve-growth-promoting factor distinct from NGF.

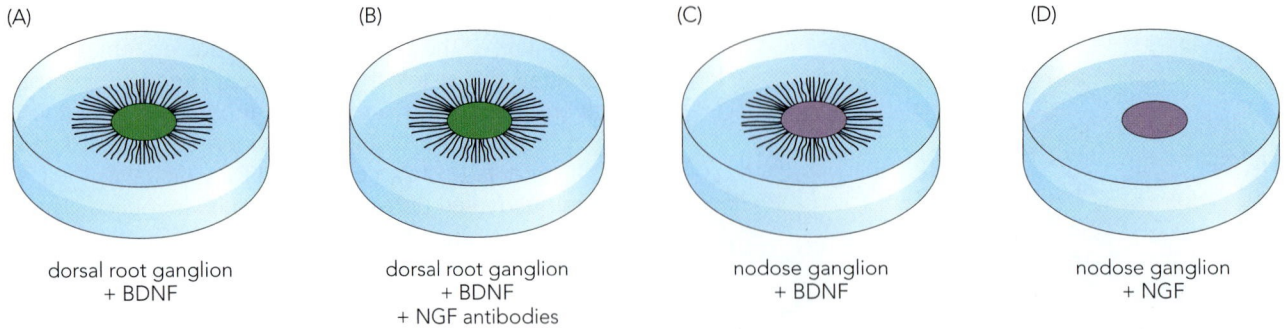

Figure 8.8 The biological activity of brain-derived neurotrophic factor (BDNF) differs from that of NGF. (A) Like glioma-cell-conditioned medium, BDNF promotes neurite outgrowth from a dorsal root ganglion explant. (B) BDNF-induced outgrowth is not inhibited when antibodies specific to NGF are added, indicating that the neurite-outgrowth-promoting properties of BDNF are distinct from those of NGF. (C) BDNF also promotes outgrowth of nodose ganglia, a neuronal population that is unresponsive to NGF (D).

the survival and neurite outgrowth of neuronal populations that were not influenced by NGF. For example, the placode-derived neurons of the nodose ganglia were responsive to BDNF, but not NGF (**Figure 8.8**). Subsequent studies demonstrated that BDNF and NGF have overlapping but distinct spatial and temporal patterns of expression, further emphasizing that they are two different molecules.

Discoveries of other NGF-related growth factors rapidly followed

In 1989, Barde and colleagues determined the amino acid sequence of BDNF and found it was very similar to NGF, sharing about 50% sequence identity and indicating that NGF and BDNF are part of the same gene family. This finding prompted several labs to look for additional members of this family of nerve growth factors.

Whereas it took decades to purify and sequence NGF and several years to purify and sequence BDNF, four labs were able to identify a third member of the NGF family of growth factors within months of the publication of the BDNF amino acid sequence using molecular techniques rather than protein purification methods. The new factor was called various names by the different groups who discovered it, including NGF-2 and hippocampal-derived neurotrophic factor, but is now called neurotrophin-3 (NT-3). The entire family of NGF-related proteins is now known as the **neurotrophins**.

NT-3, like NGF and BDNF, promotes survival and outgrowth of dorsal root ganglia, though different subpopulations of neurons within the ganglia are influenced by the different neurotrophins. The developmental pattern of NT-3 expression also differs from the expression patterns of NGF and BDNF. For example, NT-3 is generally expressed at a higher level during development than in adult tissues, whereas BDNF levels are often lower during development and higher in adult tissues, where it appears to play numerous roles in neuronal maintenance and synaptogenesis.

A fourth member of the neurotrophin family was found in *Xenopus* in 1991 and was named neurotrophin-4 (NT-4). A mammalian homolog of this protein was identified shortly after and was originally called mammalian NT-4 by one lab and NT-5 by another. Now often designated as NT-4/5, these proteins differ from other neurotrophin family members in that NT-4/5 sequences are not as highly conserved across species, explaining why the *Xenopus* and mammalian proteins initially appeared to represent two distinct neurotrophins. Neurotrophin-6 and neurotrophin-7 were discovered in 1994 and 1998, respectively, but have only been detected in fish. To date their functions are not as well characterized as those of the first four neurotrophins.

NGF SIGNALING MECHANISMS AND NEUROTROPHIN RECEPTORS

In addition to characterizing the biological activity and tissue distribution of NGF and the other neurotrophins, scientists also investigated the ways such growth factors exerted their survival-promoting effects. Because neuronal cell bodies are often located at a distance from target tissues, the mechanism by which growth factors reach the cell body to promote cell survival was initially unclear. For many years, scientists suspected that the nerve fibers must express receptors that take up the growth factor and transport it to the cell body. However, this hypothesis could not be tested until appropriate techniques were available.

NGF undergoes retrograde transport from the nerve terminal to the cell body

By the early 1970s scientists had developed a method to evaluate whether NGF was taken up by nerve terminals and transported **retrogradely** to the cell body. In the first study, published in 1974, Hendry and colleagues injected radiolabeled NGF (^{125}I-NGF) into the mouse iris, a tissue that is innervated by sympathetic neurons of the superior cervical ganglia (SCG). If NGF were transported along the nerve fibers to the cell bodies as hypothesized, the ^{125}I-NGF signal would be detected in neurons of the SCG. Using this method, the scientists confirmed that NGF was transported back to the cell bodies of the sympathetic neurons; the radiolabeled NGF was detected in the neurons of the SCG (**Figure 8.9**). Proteins similar in size to NGF were not transported to the cell bodies, nor were NGF proteins whose structure was modified, indicating that the observed retrograde transport was specific for NGF. Later studies found that radiolabeled NGF was also transported retrogradely when injected near sciatic or spinal nerve terminals of adult mice. Thus, both developing and mature nerve fibers expressed unidentified receptors that could specifically bind NGF and initiate transport of the growth factor to the cell body.

The importance of retrograde transport of NGF for neuronal survival was demonstrated *in vitro*. In the mid 1970s, a cell culture technique was developed called the Campenot chamber, named for its inventor, Robert Campenot. With this method, dissociated neurons are placed in a central chamber, and the neurites extend and grow beneath a silicon barrier to reach one of two adjacent chambers (**Figure 8.10**A, B). The central

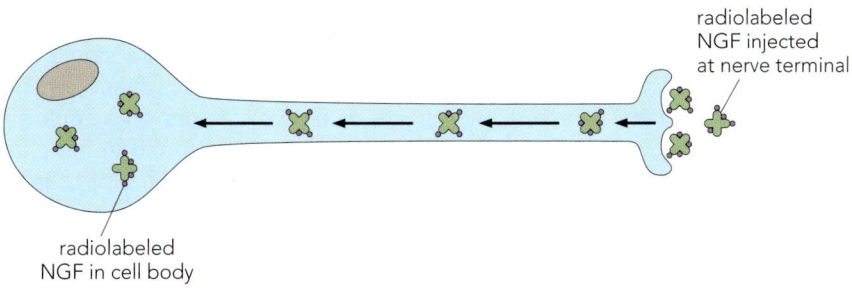

Figure 8.9 NGF is retrogradely transported to the neuronal cell body. When radiolabeled NGF was injected in the region of nerve terminals, scientists observed that the NGF was transported back to the cell body (arrows). These studies demonstrated that NGF did not need to be expressed near the cell body to exert its effects. Instead, NGF produced at sites distant from the cell body could be carried back to the cell body via retrograde transport to influence signal transduction pathways and gene expression.

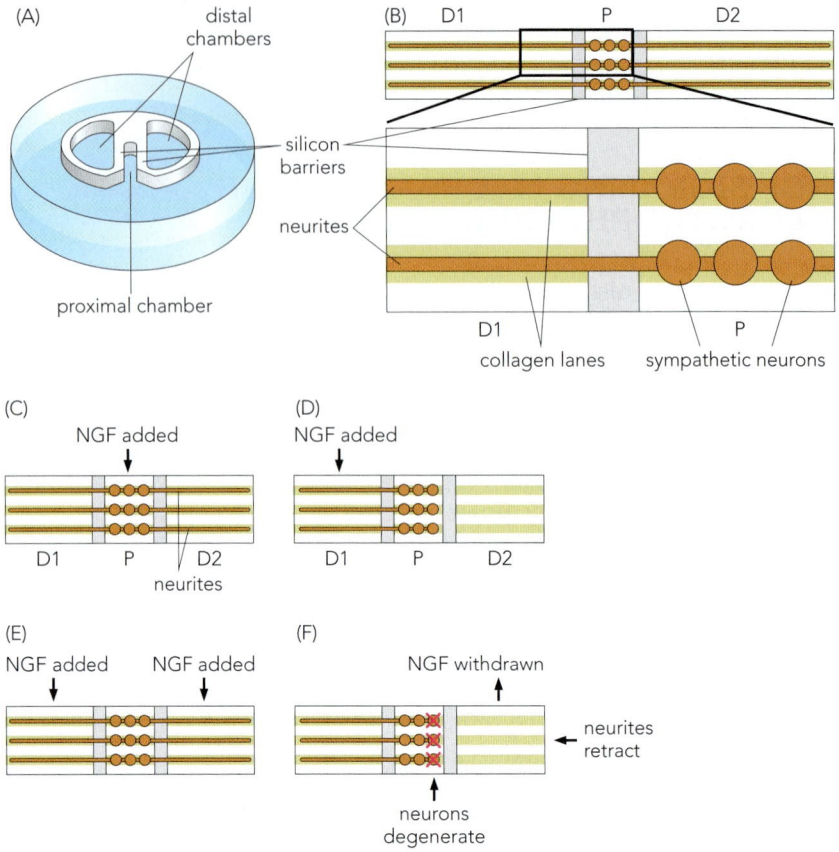

Figure 8.10 ***In vitro*** **studies demonstrate NGF is retrogradely transported to the neuronal cell body and required for neuronal survival.** A cell culture method was developed that allowed scientists to selectively add NGF to cell bodies or nerve terminals. (A) A chamber is added to a petri dish. This chamber consists of a single central chamber (the proximal chamber) and two distal chambers. Silicon barriers separate the proximal and distal chambers. (B) When the cell bodies of sympathetic neurons are placed in the proximal chamber (P) under culture conditions that support neuronal survival and allow neurites to grow, the neurites extend beneath the silicon barriers and extend into each of the distal chambers (D1 and D2). The lower panel is an enlarged view of the boxed area in the upper panel. (C–F) Selective addition or withdrawal of NGF to the different chambers revealed retrograde transport of NGF. (C) When NGF was added to the proximal chamber, neurites extended into both distal chambers (D1 and D2). (D) NGF, added to one distal chamber only (D1), promoted neurite outgrowth to the treated side. However, neurites did not grow to the untreated chamber (D2). (E) When NGF was added to both distal chambers, neurites extended to both D1 and D2. (F) If NGF was subsequently withdrawn from one chamber (D2), the neurites that extended to that chamber retracted and their associated cell bodies degenerated. Together these experiments demonstrated that NGF is retrogradely transported to the neuronal cell body to promote neuronal survival. (Adapted from Campenot RB, Lund K, and Mok S-A [2009] *Nature Protocols* 4:1869-1887.)

(proximal) or adjacent (distal) chambers can be individually treated with NGF to limit growth factor exposure to the cell body or nerve terminals. When NGF was added to the cell bodies of sympathetic neurons located in the proximal chamber, the neurons survived and extended neurites into the distal chambers (Figure 8.10C). Further, the addition of NGF to one or both distal chambers still promoted the survival of sympathetic neurons (Figure 8.10D, E). However, if NGF was withdrawn from one of the distal chambers, the neurites in that chamber retracted and their associated cell bodies degenerated (Figure 8.10F). Together, these latter studies provided additional evidence that NGF must bind to receptors on the nerve terminals and be transported back to the cell body to influence cell survival.

NGF receptors are first identified in the PC12 cell line

Primary cell cultures consist of cells taken directly from an animal model and grown *in vitro*. Examples of primary cell cultures include explants of whole DRG and sympathetic ganglia, such as those developed by Levi-Montalcini, and the dissociated cell preparations used by Campenot and others. **Cell line cultures** are also important experimental tools. A cell line consists of immortalized cells derived from a single cell and therefore contains a homogeneous cell population. This characteristic allows researchers to assess the effects of growth factors on a uniform cell population, rather than the mixed cell populations found in ganglia. Cells lines also have the advantage of providing large numbers of cells to analyze without the need to dissect cells from animals for each experiment. A cell line was identified in the 1970s that proved very useful for studying NGF. PC12 cells, an adrenal chromaffin cell line derived from a rat pheochromocytoma, develop characteristics of sympathetic neurons when

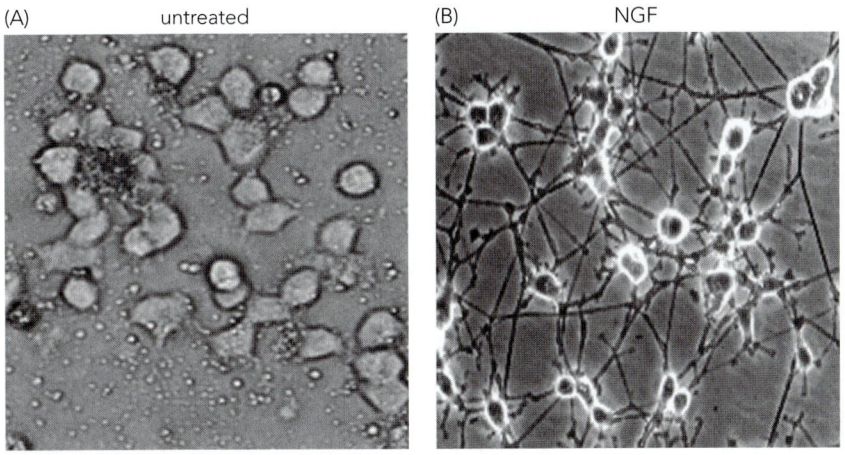

Figure 8.11 The PC12 cell line is used to study NGF signaling mechanisms. The PC12 cell line is an adrenal chromaffin cell line derived from a rat pheochromocytoma that normally lacks characteristics of neurons (A). When treated with NGF, these cells develop characteristics of sympathetic neurons (B). PC12 cells first became an important experimental tool in the 1970s and continue to be used by researchers today. (From Rocco ML, Balzamino BO, Passeri PP et al. [2015] *PLoS One* 10(4): e0124810.)

treated with NGF (**Figure 8.11**). Importantly, like primary sympathetic neuronal cultures, PC12 cells bind radiolabeled NGF. PC12 cells became an important tool for testing the mechanisms by which NGF exerted its effects and for identifying the structure of NGF receptors.

By the mid 1980s, evidence from PC12 cells and primary neuronal cultures suggested there were two types of NGF receptors. One was a low-affinity receptor that released bound NGF quickly, and the other was a high-affinity receptor that released bound NGF slowly. In 1986 and 1987, the laboratories of Moses Chao and Eric Shooter identified a glycoprotein that binds NGF with low affinity. Structural analysis of what was then called the LNGFR (low-affinity NGF receptor) revealed that the receptor lacks an intracellular domain. Thus, the kinetics and structure of the LNGFR suggested that this receptor alone could not be the sole signaling receptor for NGF, even though it clearly bound NGF and was expressed on many NGF-dependent neurons, as well as on Schwann cells and other nonneuronal cells.

Later, following the discoveries of BDNF, NT-3, and NT-4/5, those neurotrophins were also found to bind to the LNGFR with affinity and kinetics similar to NGF. Thus, the LNGFR was not NGF-specific (**Figure 8.12**A). The

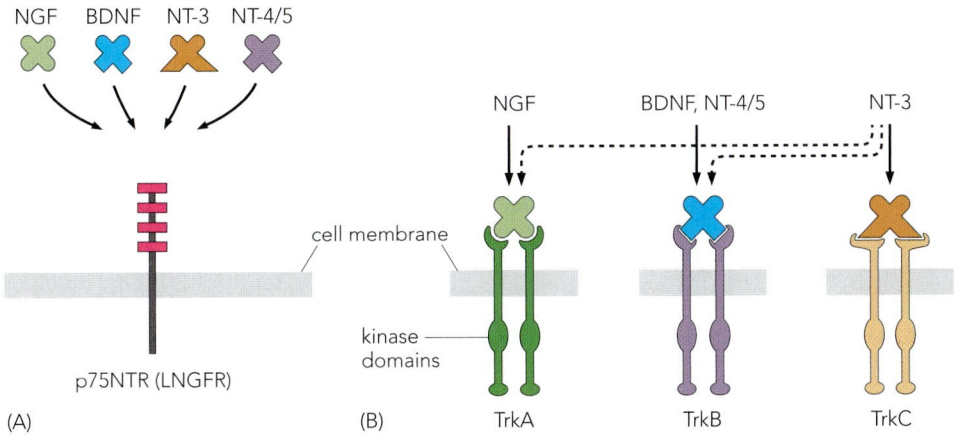

Figure 8.12 All neurotrophins bind to the p75 neurotrophin receptor, but selectively bind to specific Trk receptors. (A) The p75 neurotrophin receptor, initially called the low-affinity NGF receptor (LNGFR), was the first receptor identified that could bind to NGF. It was later found that NGF, BDNF, NT-3, and NT-4/5 all bound to this receptor with equal, low affinity. Thus, the name was changed to the p75 neurotrophin receptor (p75NTR). (B) Each neurotrophin binds to a specific Trk receptor with high affinity (solid arrows). NGF binds with high affinity to TrkA, the first Trk receptor discovered. BDNF and NT-4/5 bind with high affinity to TrkB, while NT-3 binds to TrkC with high affinity. When NT-3 is present at sufficiently high concentrations, it can also bind TrkA and TrkB (dashed arrows).

name of the receptor was subsequently changed to reflect its size; it is now known as the $p75^{LNGFR}$ or p75NTR (p75 neurotrophin receptor). Scientists knew that the p75NTR alone could not mediate the biological effects of NGF, and several studies suggested the presence of an additional high-affinity receptor. As is often the case in science, the identification of the high-affinity NGF receptor came from an unexpected source. In 1986, Mariano Barbacid's laboratory discovered a gene for a tyrosine kinase receptor that was identified in colon carcinoma cells. This **oncogene** encoded a protein with a tyrosine kinase domain coupled to a sequence typically associated with non-muscle tropomyosin, a common cellular component. The tropomyosin sequence was located in place of the extracellular binding domain and this finding led to the name of the receptor—the tropomyosin-receptor kinase (Trk). In noncancerous cells, the gene for this receptor, now called *TrkA*, produces a 140-kD protein that is expressed on restricted neuronal populations as well as on PC12 cells. Subsequent investigations by Luis Parada and others confirmed that TrkA is the high-affinity NGF receptor that scientists sought for so many years.

Following the discovery of TrkA, other members of the Trk family were soon identified, and different binding specificities were observed for each neurotrophin. The second Trk receptor discovered, TrkB, is the high-affinity receptor for BDNF and NT-4/5. TrkC is the high-affinity receptor for NT-3. However, NT-3 can also bind TrkA and TrkB in some contexts (Figure 8.12B). In addition to the full-length tyrosine kinase receptors, TrkB and TrkC also exist in truncated forms that lack the kinase domain.

The distribution of full-length Trk receptors on different neuronal populations matches the known growth factor preferences of the neurons; each Trk receptor is expressed on neurons known to respond to the associated neurotrophin during development or adulthood. In the 1990s, as gene knockout mice became available for the first time, studies confirmed the role of Trk receptors *in vivo*. In the PNS, the deficits found in the *TrkA*, *TrkB*, and *TrkC* knockout mice largely matched those observed in mice lacking the corresponding neurotrophin. For example, mice lacking *TrkA* had significant decreases in the number of sympathetic and DRG neurons that were similar to the losses observed in mice lacking *Ngf*. Mice lacking *TrkB* had a decrease in nodose neurons, as did the mice lacking the TrkB ligands BDNF and NT-4/5 (**Figure 8.13**).

Although the neuronal losses are generally similar in the mice lacking a given Trk receptor and a corresponding neurotrophin, subtle differences are noted in some neuronal populations, thus suggesting the presence of overlapping or redundant mechanisms to prevent the complete loss of a neuronal population in the absence of a single *Trk* gene.

Activation of Trk receptors stimulates multiple intracellular signaling pathways

The Trk receptors function similar to other receptor tyrosine kinases. The neurotrophins are dimers, and each portion of the ligand binds to a corresponding Trk receptor monomer, causing these subunits to approach each other. Once the two receptor subunits are in close contact, they cross-phosphorylate: Each subunit phosphorylates tyrosine residues on the intracellular portion of the other subunit (**Figure 8.14**).

As more was discovered about the structure and function of Trk receptors, a model emerged to explain the mechanisms by which neurotrophins undergo retrograde transport to the cell body. In this model, the ligand binds to the transmembrane receptor protein and is taken up by the cell in a specialized transport vesicle called an **endosome**. The receptor's ligand-binding domain is enclosed within the vesicle, while its kinase

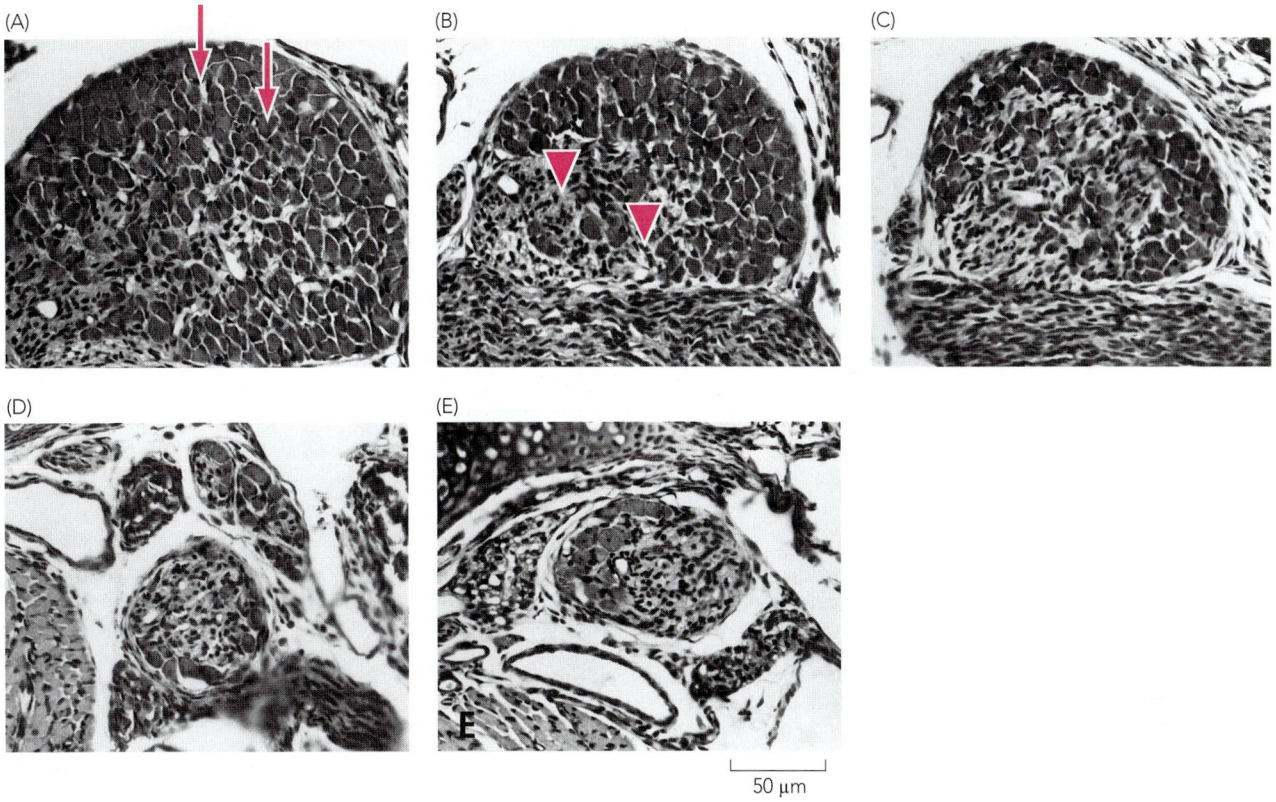

Figure 8.13 **Gene knockout mice confirm a role for Trk receptors *in vivo*.** In the 1990s, mice lacking genes for specific neurotrophins or Trk receptors were generated. The patterns of neuronal deficits in the mice were typically similar in those lacking the receptors and those lacking one or more of the corresponding growth factors, as seen in this example of nodose ganglia from wild-type mice and mice lacking *Bdnf, Nt-4 /5,* or *TrkB*, the receptor for BDNF and NT-4/5. (A) A section through nodose ganglion of wild-type mouse reveals a large ganglion with healthy neurons throughout (arrows). Fewer healthy neurons and more dying (pyknotic neurons, arrowheads B) are observed in the nodose ganglia from mice lacking *Bdnf* (B), *Nt-4/5* (C), both *Bdnf* and *Nt-4/5* (D), or *TrkB* (E). Subpopulations of nodose neurons rely on either BDNF or NT-4/5 and therefore the *Bdnf/Nt-4/5* double knockout mouse has the fewest surviving neurons and most closely resembles the *TrkB* mutation. (From Erickson JT, Conover JC, Borday V et al. [1996] *J Neurosci* 16: 5361–5371.)

domain extends outward from the vesicle surface (**Figure 8.15**). Because the ligand continues to bind the receptor, the receptor remains active during transport, and additional signaling molecules, such as adaptor proteins, are recruited to the endosome. Different proteins associated with endosomes at early and late stages of transport have been noted. Once the endosome complex reaches the cell body, intracellular signaling cascades, such as those described below, are initiated to promote cell survival. While this model is currently proposed as the primary method of transporting neurotrophins to the cell body, other models suggest that the entire ligand–receptor complex does not need to reach the cell body to activate downstream signaling molecules. Transport requirements for neurotrophin signaling remain an ongoing focus of investigation.

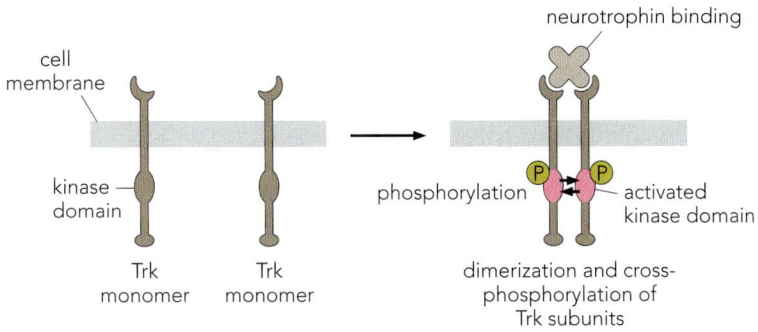

Figure 8.14 **Neurotrophin binding leads to Trk dimerization and autophosphorylation.** Neurotrophins exist as dimers that bind to Trk monomer subunits. The binding of the neurotrophin ligand brings the receptor subunits into close contact, so they cross-phosphorylate each other's intracellular kinase domains, thus initiating signal transduction cascades in the receptor-bearing cell.

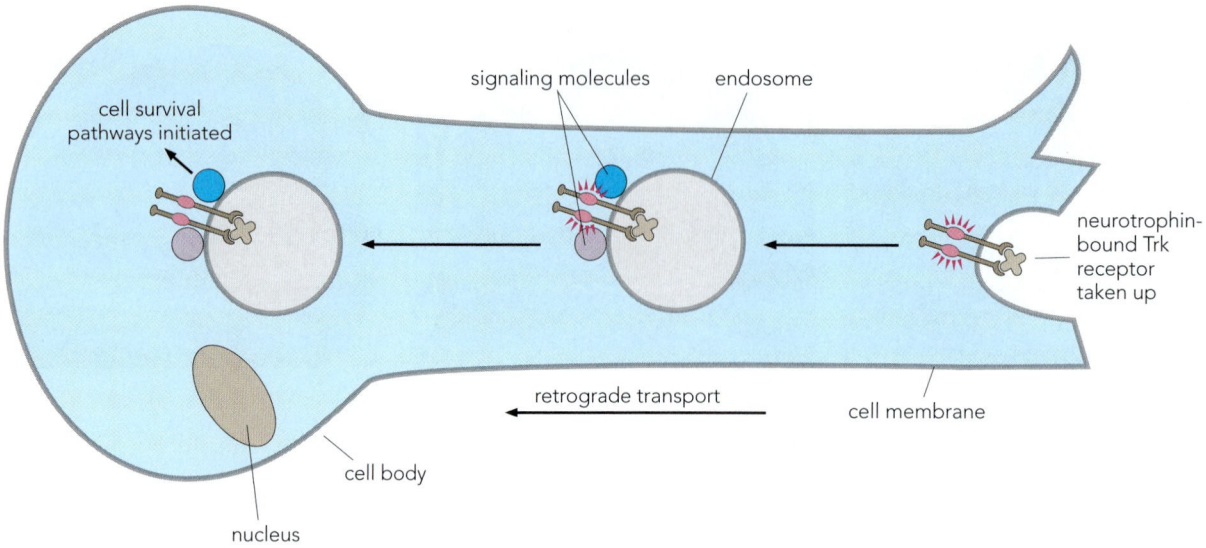

Figure 8.15 A neurotrophin and its bound receptor are retrogradely transported by an endosome. When a neurotrophin binds to its corresponding Trk receptor, the ligand–receptor complex is internalized and transported along the axon to the cell body in a transport vesicle called an endosome. The ligand remains inside the endosome, while the receptor kinase domains extend outward from the vesicle surface. Because the ligand is still bound to the receptor, the receptor remains active during transport and recruits additional signaling molecules to the endosome surface. Once the endosome complex reaches the cell body, it is able to initiate signal transduction cascades associated with cell survival. (Adapted from Miller FD & Kaplan DR [2001] *Neuron* 32:767–770.)

Activation of the Trk receptor's kinase domains leads to phosphorylation of additional tyrosine residues outside the kinase domains. These phosphorylated sites then recruit other proteins that initiate intracellular signaling cascades. Among the proteins that bind to phosphorylated sites on the Trk receptor are adapter proteins such as Shc (C-terminal Src-homology-2 domain) and the enzyme phospholipase C γ (PLCγ). These proteins, in turn, activate molecular pathways leading to various transcriptional events that regulate functions as varied as neurite outgrowth and synaptic plasticity, cell differentiation, and neuronal survival. The precise intracellular effects of the neurotrophins depend in part on the cellular context and stage of development.

The three primary intracellular pathways activated by Trk receptor binding are the PLCγ, Ras (a GTPase initially found in rat sarcoma), and phosphatidylinositol 3-kinase (PI3-K) pathways. In neurons, PLCγ generally influences synaptic plasticity, while Ras regulates the differentiation of cells, and PI3-K promotes cell survival. These pathways often interact through additional proteins, however, and overlapping functions are found in several neuronal contexts. For example, Ras can also influence cell survival pathways.

Figure 8.16 provides an overview of the signaling pathways activated by the binding of neurotrophins to Trk receptors. Activation of the PLCγ pathway occurs when PLCγ associates with phosphorylated sites on the intracellular portion of the Trk receptor. Activation of PLCγ leads to the subsequent activation of the IP3 (inositol 1,4,5-triphosphate) and DAG (diacylglycerol) pathways. IP3 induces release of Ca^{2+} from intracellular stores, while DAG activates PKC (protein kinase C). Both the IP3 and DAG pathways regulate the expression of genes and proteins associated with synaptic plasticity.

Ras pathway activation is initiated with the recruitment of the adaptor protein Shc to the phosphorylated tyrosine residues on the receptor. The Shc protein recruits Grb2 (Growth factor receptor-bound protein 2), which interacts, in turn, with SOS (Son of sevenless) to activate Ras. Ras then activates the MAP (Mitogen-activated protein) kinase signaling cascade,

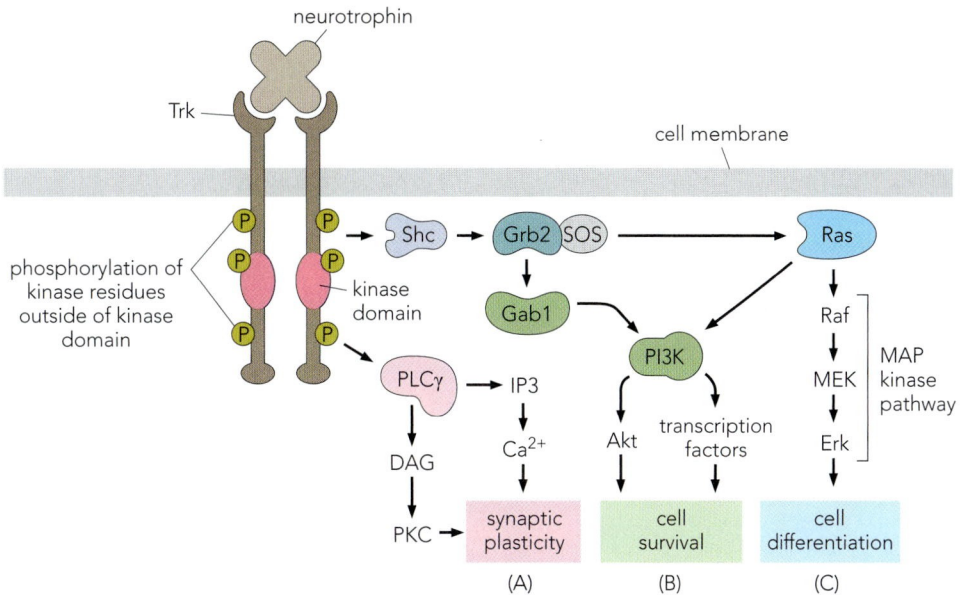

Figure 8.16 Trk phosphorylation initiates signal transduction cascades that regulate synaptic plasticity, cell survival, and cell differentiation. Upon neurotrophin binding, phosphorylation of kinase residues outside of the kinase domains recruits adaptor proteins to the receptor, thereby initiating various signal transduction cascades. Recruitment of PLCγ to the phosphorylated Trk receptor can activate both the IP3 and DAG pathways that influence synaptic plasticity (A). Recruitment of the Shc adaptor protein leads to the subsequent recruitment of Grb2 and ultimately to activation of GAB1 or Ras to influence the PI3 kinase pathway that promotes cell survival (B). Additionally, recruitment of Shc followed by Grb2 interacting with SOS can activate Ras and initiate the MAP kinase pathway to influence cell differentiation (C). Akt, murine thymoma viral oncogene homolog; Erk, extracellular signal-regulated kinase; GAB1, Grb2-associated binding protein 1; Grb2, Growth factor receptor-bound protein 2; MAP, Mitogen-activated protein; MEK, MAP kinase/Erk kinase; PI3-K, phosphatidylinositol 3-kinase; SOS, Son of sevenless. (Adapted from Reichardt LF [2006] *Philos Trans R Soc Lond B Biol Sci* 361(1473):1545–1564.)

which includes the sequential phosphorylation of Raf (rapidly accelerated fibrosarcoma kinase), MEK (MAP kinase/Erk kinase) and Erk (extracellular signal-regulated kinase). Phosphorylated Erk enters the nucleus, where it regulates various genes associated with neuronal differentiation.

The PI3-K pathway can be activated by Ras or by a sequence of adaptor proteins, including Shc, Grb2, and GAB1 (Grb2-associated binding protein 1). PI3-K then activates the protein kinase Akt (named for a murine thymoma viral oncogene homolog) or transcription factors. Activation of either ultimately influences the expression of genes associated with neuronal survival. In addition, Akt can inhibit cell death pathways, as discussed later in the chapter.

Interaction of full-length Trk receptors with truncated Trk receptors or p75NTR further influences cell survival

The response of a neuron to a neurotrophin can further be modified by the presence of truncated, noncatalytic receptors expressed on the same cell as the full-length Trk receptors. As noted above, TrkB and TrkC exist in truncated, noncatalytic forms. Although the function of these receptors remains largely unclear, studies suggest truncated TrkB and TrkC may sequester their corresponding neurotrophins to increase the local concentration of those growth factors, then somehow direct the ligands to full-length receptors on the cells that require them. Alternatively, the truncated receptors may take up and sequester growth factors to prevent activation of a full-length Trk receptor. For example, it has been hypothesized that truncated TrkC could bind local NT-3 to prevent this neurotrophin from activating TrkB at the wrong stage of development (**Figure 8.17**A)

Another noncatalytic receptor, the low-affinity p75NTR, is co-expressed with Trk receptors on many neuronal populations. The function of p75NTR

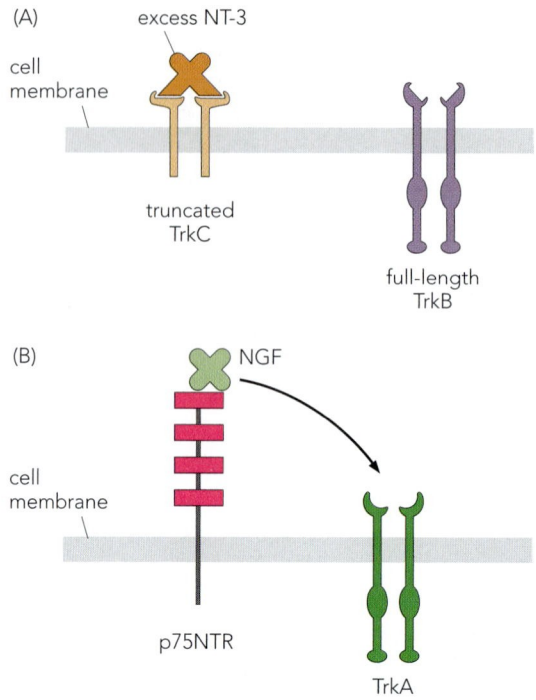

Figure 8.17 Truncated Trk receptors and p75NTR interact with full-length Trk receptors to modify cellular responses. Several models have been proposed to explain how truncated Trk receptors and p75NTR interact with full-length Trk receptors to influence neurotrophin signaling. (A) One suggestion is that truncated TrkC receptors can take up excess NT-3 to ensure the ligand does not activate full-length TrkB receptors at the wrong time in development. (B) p75NTR may bind NGF then release it to nearby, unbound TrkA receptors to enhance the efficiency of NGF signaling.

remained a mystery even as the functions of the various Trk receptors were being documented. The p75NTR binds all neurotrophins with equally low affinity, but because it lacks a catalytic subunit and therefore cannot transduce an intracellular signal, p75NTR's possible function in neurotrophin signaling was not readily apparent.

As scientists continued to investigate the actions of p75NTR and Trk receptors during development, a possible role for p75NTR emerged: The low-affinity receptor might help concentrate the neurotrophins at the target cell and "hand off" mature neurotrophin molecules to the high-affinity Trk receptor. This would presumably optimize signaling efficiency (Figure 8.17B). Evidence to support this hypothesis includes the observation that neurons expressing a high-affinity receptor plus p75NTR can survive and extend neurites at concentrations of neurotrophin lower than the concentrations required by neurons that express the high-affinity receptor alone. For example, *in vitro* experiments showed that DRG neurons isolated from mice lacking *p75NTR* require much more NGF to survive than DRG neurons from wild-type mice expressing both *TrkA* and *p75NTR*. Thus, under some circumstances, p75NTR appears to enhance the response of neurons to neurotrophins.

As studies of p75NTR and Trk receptors continued, an additional, surprising role for p75NTR emerged. Scientists discovered that while p75NTR enhanced Trk receptor signaling to promote neuronal survival in some contexts, it also mediated neuronal cell death in some neuronal populations. The mechanisms by which neurotrophins and p75NTR interact to mediate neuronal death during development are detailed later in the chapter.

Other growth factors also regulate neuronal survival and outgrowth

The preceding descriptions of neuron–target interactions focus on the role of neurotrophins. However, neurotrophins are not the only family of molecules known to influence developing neurons. Many other growth factors have been shown to influence the survival of various neuronal populations, including members of the fibroblast growth factor (FGF) family, the glial-cell-line-derived neurotrophic factor (GDNF) family (GDNF, neurturin,

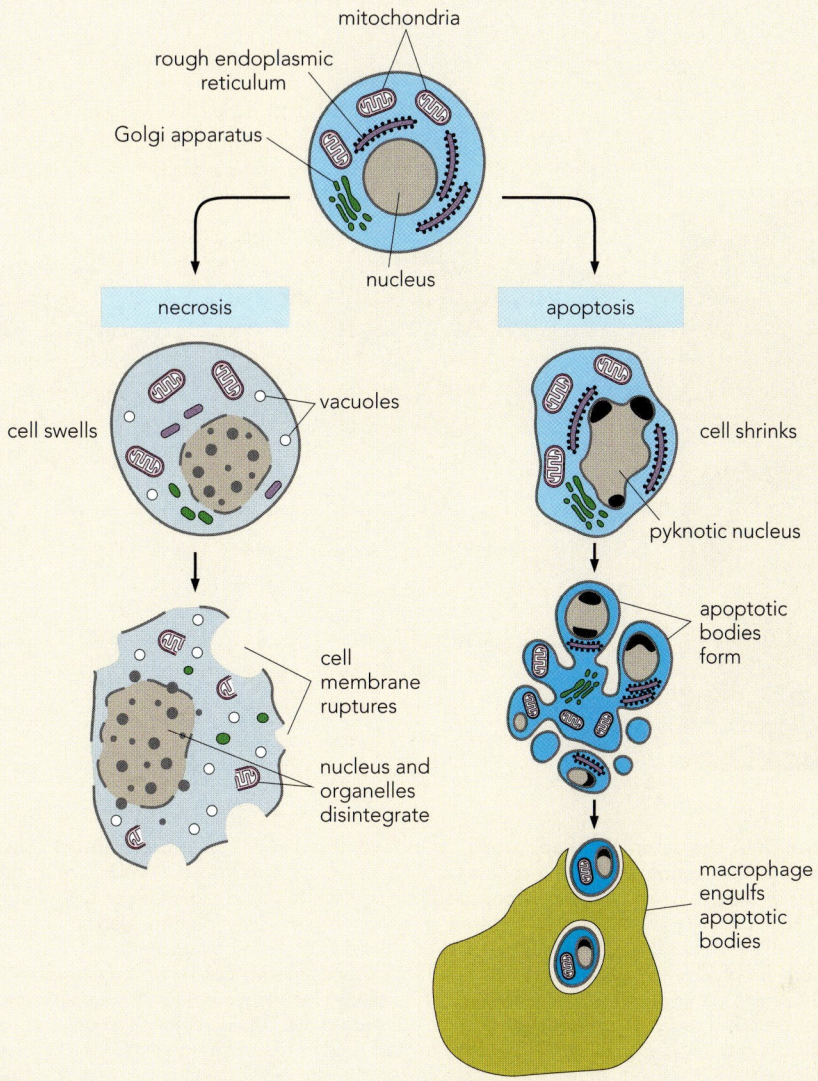

Figure 1 Characteristics of cell death resulting from necrosis and apoptosis. When a normal cell dies, the stages of cell death differ depending on whether the cell undergoes necrosis or apoptosis. At the onset of necrotic death, the cell begins to swell, and vacuoles form in the cytoplasm and organelles. The cell continues to enlarge as the nuclei and organelles disintegrate, until the cell membrane ultimately ruptures. The release of cell contents provokes an inflammatory response. As the cell initiates apoptotic death, its cytoplasm and nuclear chromatin begin to condense. At first, the organelles and cell membrane remain intact, even as the cell body and nucleus begin to shrink. However, the cell surface soon becomes convoluted, and the nucleus and organelles break off into discrete, membrane-bound fragments called apoptotic bodies. These apoptotic bodies are degraded and ultimately phagocytized by nearby macrophages or glial cells. (From Kerr JFR, [1995] *Trends Cell Biol* 5:55–57.)

or cells are treated to rupture the cell membrane and the resulting **lysates** are run on a gel, the fragments are separated on the basis of size and distributed along the length of the gel, resembling the rungs of a ladder (**Figure 2**A). The size of the bands produced is determined by comparing them to marker standards that are run at the same time. Cells dying by necrosis do not undergo orderly DNA fragmentation, so lysates of those cells do not produce the ladder-like pattern on a gel.

To visualize apoptotic death in tissues or individual cells, the TUNEL method (terminal deoxynucleotidyl transferase dUTP nick-end labeling) can be used. The fragmented, nicked ends of DNA can be detected when tissues are incubated with an enzyme (TdT, deoxynucleotidyl transferase) and the nucleotide dUTPs (2′-deoxyuridine 5′-triphosphates). The TdT adds the dUTPs to the nicked ends of the DNA (Figure 2B). The dUTPs are also labeled with a marker that is attached prior to incubation or after attaching to the nicked ends of the DNA. When viewed through a microscope, the markers are visible and reveal which cells have undergone apoptosis (Figure 2C).

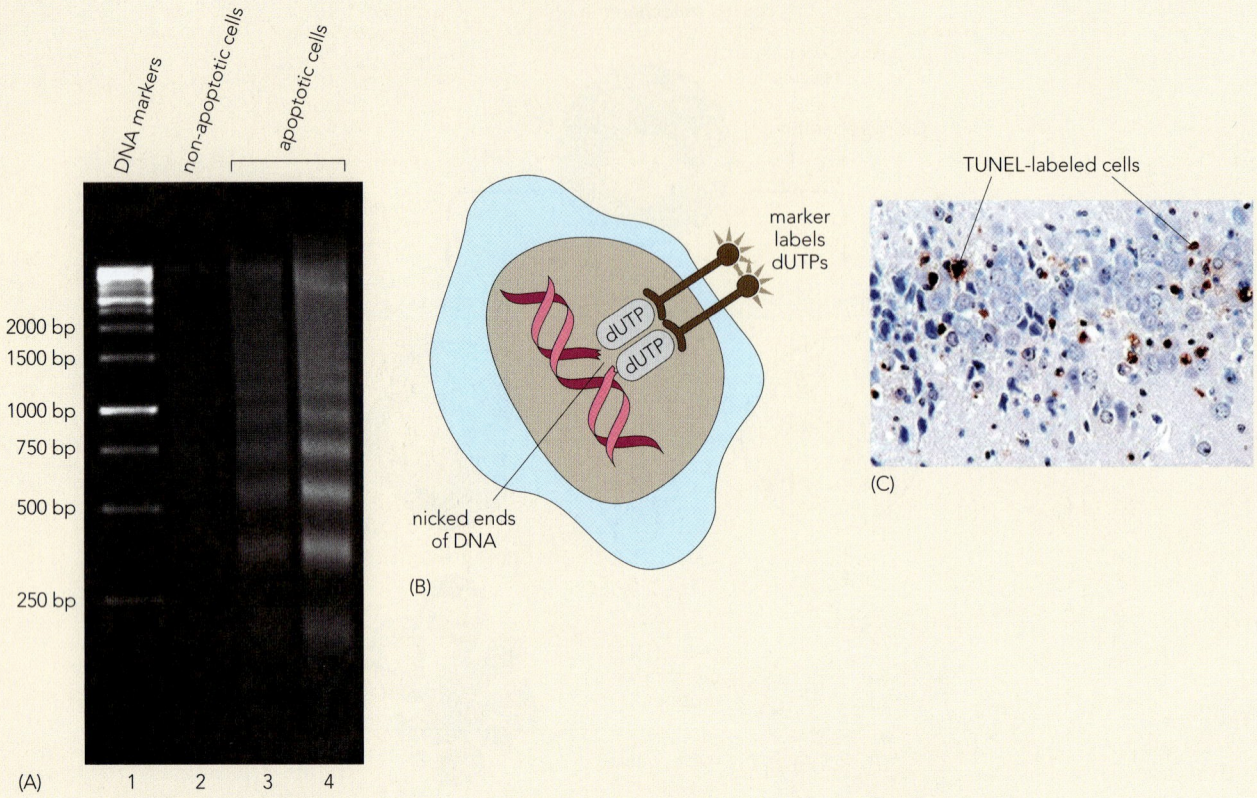

Figure 2 The fragmented DNA characteristic of apoptosis can be identified in cell lysates and intact cells. (A) Separation of DNA by gel electrophoresis distinguishes apoptotic and non-apoptotic cell death. During apoptosis, DNA is broken into fragments of 180 base pairs (bp) or multiples of 180 bp. When cell lysates from apoptotic cells are run on an agarose gel, DNA fragments are separated by size, forming what looks like a ladder. Lysates from non-apoptotic cells do not form such a pattern. Markers standards (1) are run on the gel to estimate the size of the bands detected in the cell lysates. Cells that do not undergo apoptotic cell death do not reveal a ladder-like pattern (2). Lysates from apoptotic cells reveal a ladder-like pattern of bands in sizes consistent with apoptotic DNA fragmentation (3 and 4). (B) The TUNEL (terminal deoxynucleotidyl transferase dUTP nick-end labeling) method allows visualization of apoptotic cells. Fragmented DNA can also be visualized in tissue sections or cell cultures using the TUNEL method. When terminal deoxynucleotidyl transferase is added to tissues in the presence of dUTPs, the enzyme adds dUTPs to the fragmented, or nicked, ends of DNA. In this example, a marker that labels dUTP is subsequently added, and a chemical reaction has taken place to allow visualization of the cells through a microscope. (C) A tissue section from the hamster hippocampus illustrates apoptotic cells (brown) labeled with the TUNEL method. (A from Wang, Q, Wang Y-L, Wang, K et al. [2015] *Oncology Letters* 9:278-282; C from Xiao S-Y, Guzman H, Zhang H, et al. [2001] *Emerging Infectious Diseases* 7:714-721.)

Studies reveal cell death is an active process dependent on protein synthesis

In the 1980s, scientists studying PCD made several observations that suggested PCD was not merely a passive process that resulted from a lack of neurotrophic factor. *In vitro* and *in vivo* studies demonstrated that blocking RNA or protein synthesis prevented cell death that would normally occur. The results indicated that some proteins must be required to initiate cell death.

One such study from the lab of Eugene Johnson, Jr., utilized the established cell culture methods for dissociated sympathetic neurons. The researchers first cultured the neurons for several days in the presence of NGF, then added anti-NGF antiserum to block the NGF effects and thereby induce cell death (**Figure 8.19**A). The time course used in these experiments was designed to correspond to the stages at which neurons would contact target cells and compete for NGF *in vivo*.

Among the results of this study, the researchers noted that cell death typically occurred within 24 to 48 hours following NGF removal

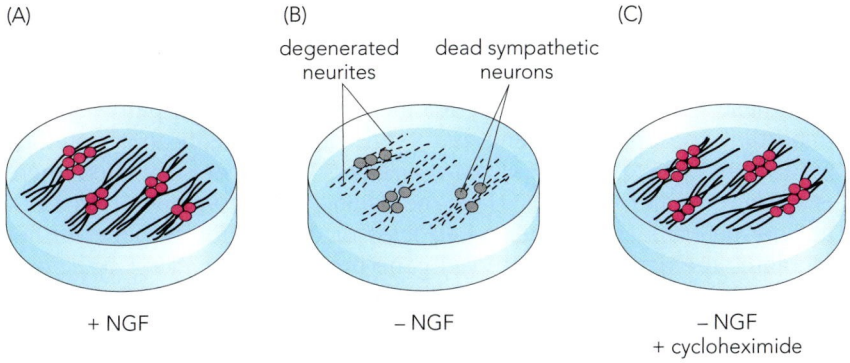

Figure 8.19 Programmed cell death requires the production of new proteins. Scientists cultured sympathetic neurons harvested at the time of normal target innervation to test whether protein synthesis was a necessary step in the induction of cell death. (A) When cultured in the presence of NGF, dissociated sympathetic neurons survived and extended long processes. (B) When NGF was withdrawn and antibodies to NGF were added to remove any remaining growth factor, the neurites began to degenerate and the neurons began to die within 24 to 48 hours. (C) However, if inhibitors of protein synthesis such as cycloheximide were added during the first 24 to 48 hours following NGF removal, the cells survived. These studies indicated that protein synthesis was needed to activate cell death.

(Figure 8.19B). However, cell death was prevented when the protein synthesis inhibitor cycloheximide was added to the cultures lacking NGF (Figure 8.19C). Thus, by preventing the formation of new proteins, the cells were able to survive even in the absence of NGF. These studies confirmed that proteins were necessary to initiate cell death and suggested that the survival effects attributed to NGF may result from the inhibition of such proteins. These discoveries helped shift the focus of research to explore active cell death pathways in cells that undergo PCD.

Cell death genes are identified in *C. elegans*

At the time these *in vitro* studies were being conducted, other scientists were starting to unravel cell death mechanisms in other systems. It soon became apparent that cell death pathways are shared across species. As noted in previous chapters, *C. elegans* provides several advantages as an animal model. The comparatively small number of cells found in specific locations helps scientists track what becomes of individual cells, including neurons. In *C. elegans*, 131 of the 1090 somatic cells undergo PCD. The cells that are removed by PCD are always the same cells, and death always occurs at specific times in development.

Several genes discovered in the 1990s were found to prevent cell death when mutated. Researchers inferred that wild-type alleles of these genes must cause cell death under normal conditions. These findings supported the emerging idea that some forms of cell death were active, genetically controlled processes. The first genes found to regulate cell death in *C. elegans* are known as *ced* (cell death abnormality) genes, because mutations in these genes disrupt the normal patterns of cell death. However, as other members of the *ced* family were identified, scientists recognized that wild-type alleles of some *ced* genes promote cell survival. The *ced* genes that promote cell death are called pro-apoptotic genes, while *ced* genes that promote survival are called anti-apoptotic genes (**Table 8.2**).

Table 8.2 Pro-apoptotic or anti-apoptotic genes in *C. elegans*

C. elegans gene	Role in PCD	Activity in wild-type	Loss-of-function phenotype	Mammalian homologs
ced-3	Pro-apoptotic	Cell death pathway	↓ cell death	*Caspase*
ced-4	Pro-apoptotic	Cell death pathway	↓ cell death	*Apaf-1*
ced-9	Anti-apoptotic	Cell survival pathway	↑ cell death	*Bcl-2*
egl-1	Pro-apoptotic	Cell death pathway	↓ cell death	*Bim, Bad, Bid, Puma, Noxa*

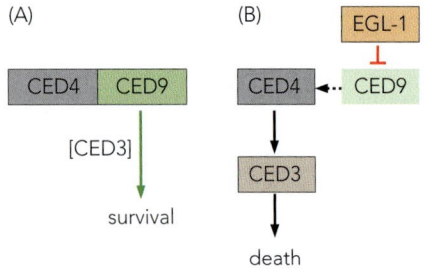

Figure 8.20 Cell death and cell survival pathways in *C. elegans*. Cell survival and cell death are regulated in *C. elegans* through the interaction of several proteins. (A) Cell survival results when the anti-apoptotic protein CED9 forms a complex with the pro-apoptotic protein CED4. This prevents CED4 from interacting with pro-apoptotic CED3 and thus stops cell death. (B) Apoptosis results if the pro-apoptotic EGL-1 inhibits anti-apoptotic CED9. If CED9 cannot maintain a complex with CED4, CED4 is free to dissociate (dashed arrow) from the CED4/CED9 complex and interact with pro-apoptotic CED3, leading to cell death.

Two pro-apoptotic genes are *ced-3* and *ced-4*. If either of these genes is absent, the number of cells undergoing PCD decreases. If both *ced-3* and *ced-4* are absent in *C. elegans*, all PCD is prevented, indicating that these genes are necessary for mediating cell death pathways. An important anti-apoptotic gene, *ced-9*, prevents cell death that would be induced by the expression of *ced-3* and *ced-4*. If *ced-9* is mutated, there is an increase in cell death. Thus, the balance of *ced-9*, *ced-3*, and *ced-4* activity determines whether a cell lives or dies. A fourth gene further regulates the activity of these *ced* gene products. *Egg-laying defective 1 (egl-1)*, first described in 1983, is a member of the large *egl* family of genes that plays multiple roles in *C. elegans* development that extend beyond egg-laying behavior. The *egl-1* gene inhibits the anti-apoptotic actions of *ced-9*, thus allowing pro-apoptotic genes to become active and induce to cell death.

Figure 8.20 illustrates how the various CED proteins interact to promote cell survival or cell death. In the absence of active EGL-1, CED9 forms a complex with CED4, preventing CED4 from interacting with CED3 to initiate the cell death pathway. Thus, when CED9 and CED4 form a complex, the cell survives. However, if EGL-1 interacts with CED9, CED9 is no longer able to complex with CED4. With CED9 inactivated by EGL-1, the pro-apoptotic CED4 can separate from the CED9/CED4 complex to interact with and activate CED3, which then induces cell death.

Homologs of the *C. elegans ced* and *egl* genes contribute to the mammalian apoptotic pathway

Soon after the discovery of the pro- and anti-apoptotic genes in *C. elegans*, scientists identified homologs of the *ced* and *egl-1* genes in vertebrates. These homologs were found to regulate pathways that are similar, though more complex. Scientists also began to explore how trophic factors interact with these pathways to regulate neuronal survival. The homolog of *ced-3* is a caspase (Table 8.2). Caspases are cysteine proteases that cleave target proteins next to an aspartate residue. The specificity of the enzymes, involving a cysteine at the enzyme's active site that cleaves near aspartate, gives rise to the term caspase. The first mammalian caspase related to *ced-3* was originally called ICE (interleukin-1β converting enzyme). This enzyme is now known as caspase-1 and is about 28% similar to the CED3 protein. Since the initial discovery of caspases, it has been determined that *Caspase-9* is the homolog of *ced-3*.

The mammalian homolog of *ced-4* is *Apaf-1* (apoptotic protease activating factor-1). Apaf-1 and other proteins form a complex called an **apoptosome**, a critical component of the cell death pathway. The homolog of *ced-9* is *Bcl-2*, a member of a large family of molecules initially discovered as a part of the B-cell lymphoma-2 family of genes.

In addition to the anti-apoptotic *Bcl-2* gene, the Bcl-2 family includes multiple genes divided into two subfamilies: one anti-apoptotic, the other pro-apoptotic. The best-characterized anti-apoptotic, survival-promoting members of the Bcl-2 family are *Bcl-2* and *Bcl-xL*, while the best-known pro-apoptotic members are *Bax* and *Bak*, which interact with several related genes to promote cell death. The Bcl-2 family is further divided into three classes of proteins based on the number of Bcl-2 homology (BH) domains. The anti-apoptotic proteins, Bcl-2 and Bcl-xL, have four BH domains (BH1, BH2, BH3, and BH4), whereas the pro-apoptotic proteins Bax and Bak have three BH domains (BH1, BH2, and BH3). A second group of pro-apoptotic proteins has only the BH3 domain. These BH3-only proteins include Bad, Bim, Bid, Puma, and Noxa. The *egl-1* gene in *C. elegans* is homologous to the BH3-only proteins in mammals. Similar to the effects of *ced* gene mutations, deleting anti- or pro-apoptotic *Bcl-2*-related genes in mice alters the extent of cell death in the PNS and CNS. When the anti-apoptotic genes *Bcl-2* or *Bcl-xL* were deleted, the number of neurons that died was greater. Conversely, when the pro-apoptotic gene *Bax* was deleted, there was an increase in the number of neurons that survived.

It is now recognized that cleavage of several distinct pro-caspases results in the activation of the corresponding caspases that in turn mediate cell death pathways. One of the first cell death pathways described in mammals involves the release of cytochrome *c* from the mitochondria. When pro-apoptotic signals activate Bax or Bak, these proteins form complexes at the mitochondrial membrane that somehow induce the formation of a pore, allowing release of cytochrome *c* into the cytosol. The released cytochrome *c* then forms a complex with Apaf-1 (**Figure 8.21**, pathway 1). In the presence of ATP, Apaf-1 undergoes a conformational change that allows it to interact with pro-caspase-9, forming an apoptosome. Pro-caspase-9 is cleaved within the apoptosome, and the resulting mature caspase-9 functions as the **initiator caspase** that activates, in turn, subsequent caspases, such as caspase-3, the predominant **effector caspase** in neurons undergoing cell death.

The pathway outlined in Figure 8.21 is often called the mitochondrial pathway or the **intrinsic pathway**. This pathway can be activated by cellular stress, DNA damage, or the loss of trophic factor, such as that observed following the withdrawal of NGF from sympathetic neurons *in vitro* (see Figure 8.19). The removal of growth factor also leads to the activation of BH3-only proteins such as Bim and Puma. These pro-apoptotic proteins either activate Bax directly or inhibit the pro-survival functions of Bcl-2 and Bcl-xL (Figure 8.21).

Trophic factors can increase neuronal survival by inhibiting the apoptotic pathway at various points. For example, trophic factors can increase the synthesis of the Bcl-2 protein that then inhibits the activity of Bax and Bak at the mitochondrial membrane, thereby preventing the release of cytochrome *c* (Figure 8.21, pathway 2). Without cytochrome *c*, the apoptosome does not form and the caspases that induce cell death cannot be activated. In addition, when trophic factors stimulate the PI3-K pathway described earlier (see Figure 8.16), the phosphorylation of Akt not only has the ability to phosphorylate transcription factors associated with cell survival, but can also phosphorylate and inhibit pro-apoptotic proteins such as Bim, Bad, and Bax. Thus, trophic factors not only stimulate survival pathways, but also actively inhibit the cell death cascade.

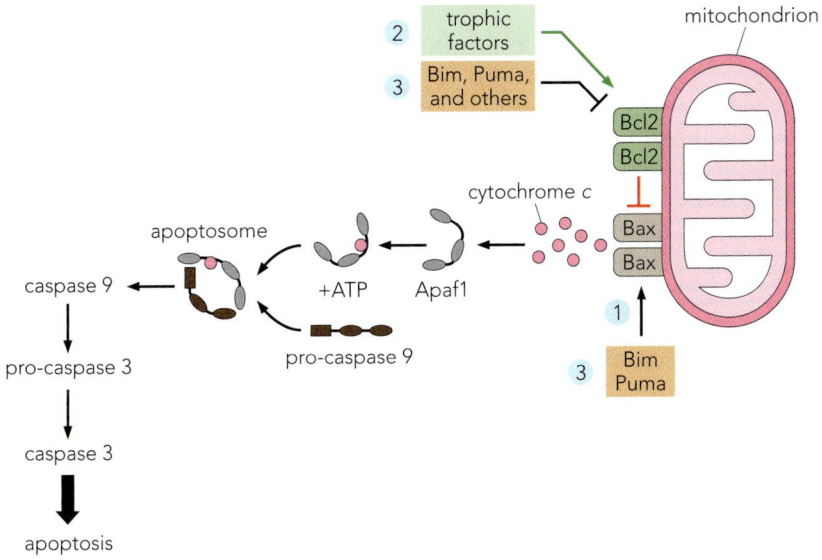

Figure 8.21 Cell death and cell survival pathways in mammals. The figure illustrates some of the ways in which mammalian pro- and anti-apoptotic pathways interact. (1) Activation of the pro-apoptotic protein Bax leads to the release of cytochrome *c* from mitochondria. Cytochrome *c* then interacts with Apaf-1, which in the presence of ATP undergoes a conformational change that allows it to interact with pro-caspase-9, a pro-apoptotic enzyme. Apaf-1, cytochrome *c*, and pro-caspase-9 form a structure called an apoptosome. Pro-caspase-9 is cleaved within the apoptosome, and the resulting caspase-9 cleaves pro-caspase-3. The activated caspase-3 then causes cell death. (2) However, if trophic factors are present, they can activate the anti-apoptotic protein Bcl-2, which in turn inhibits the activity of Bax, thereby preventing the release of cytochrome *c* from mitochondria and blocking the apoptotic pathway. (3) Other pro-apoptotic proteins such as Bim and Puma are activated upon removal of growth factors and can promote apoptosis either by activating Bax or inhibiting Bcl-2. (Adapted from Kristiansen M & Ham J [2014] *Cell Death Differ* 21(7): 1025–1035.)

p75NTR and precursor forms of neurotrophins help mediate neuronal death during development

The final element of cell death pathways to consider is the activation of an **extrinsic pathway** through "death receptors." Scientists made two important observations that helped elucidate how the extrinsic pathway functions in the developing nervous system. One discovery was that p75NTR is related to the tumor necrosis family (TNF) of receptors, a group of receptors that are often called death receptors, because ligands that bind to them initiate intracellular signaling pathways that lead to cell death. In studies utilizing various cell lines, Dale Bredesen and colleagues found that under culture conditions that induced cell death, the extent of cell death was enhanced in cells expressing p75NTR. In addition, Barde and colleagues noted that whereas NGF normally prevented cell death in retinal neurons expressing both TrkA and p75NTR, NGF induced cell death in retinal neurons that expressed only p75NTR.

The second influential discovery regarding the extrinsic pathway of cell death was the finding that different forms of neurotrophins serve distinct roles in the developing nervous system. Neurotrophins are first generated as **pro-neurotrophins**—precursor molecules of approximately 270 amino acids. Pro-neurotrophins are then cleaved to yield mature neurotrophins of roughly 120 amino acids. Initially, it was thought that only the mature forms were important during neural development. In recent years, however, roles for the unprocessed, precursor forms in regulating cell death have been identified.

Whereas the mature, processed neurotrophins bind to p75NTR with low affinity, the unprocessed pro-neurotrophins bind to p75NTR with higher affinity. In the developing nervous system, p75NTR binds to unprocessed pro-neurotrophins, often in the presence of a co-receptor called sortilin. Adaptor proteins are then recruited to the intracellular portion of the receptor complex to initiate the signaling cascades that lead to cell death (**Figure 8.22**). The exact mechanism by which this occurs is still under investigation, but in many instances the *c*-jun N-terminal kinase (JNK) pathway is involved. When NGF is withdrawn from PC12 cells, for example, the pro-survival Erk pathway is no longer activated, but instead the JNK pathway becomes activated. Several targets downstream of JNK then lead to cell death. In some cases, JNK activates c-jun transcription factors that translocate to the nucleus to activate pro-apoptotic genes. In other contexts, JNK activates p53, which in turn activates pro-apoptotic genes such as *Bax* that initiate caspase cascades that lead to cell death (Figure 8.22).

Investigators continue to explore how neurotrophins and pro-neurotrophins interact with p75NTR during normal development to influence appropriate patterns of cell survival and cell death. It appears that co-expression of specific Trk receptors influences how p75NTR responds to different neurotrophins. As noted above, in neurons that only express p75NTR without a Trk receptor, cell death can be initiated in the presence of a pro-neurotrophin. However, when both a Trk receptor and p75NTR are present, cell survival is often enhanced in the presence of the corresponding neurotrophin. In other experiments, scientists have noticed that a noncorresponding neurotrophin can induce cell death rather than survival in neurons expressing both a Trk receptor and p75NTR. For example, during normal development neurons of the sympathetic chain ganglia respond to NGF but not BDNF. If sympathetic neurons expressing TrkA and p75NTR are exposed to NGF, neurons survive. BDNF normally has no effect on these neurons, because it does not bind to TrkA. However, if BDNF is experimentally overexpressed, there is an increase in cell

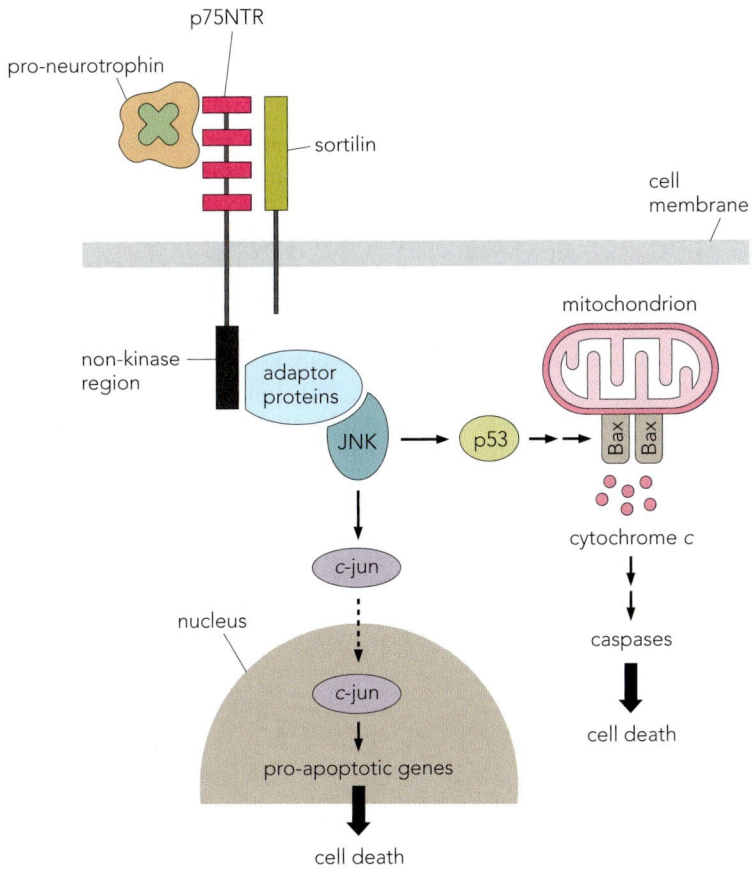

Figure 8.22 Binding of unprocessed pro-neurotrophins to the p75NTR complex activates cell death pathways. When an unprocessed pro-neurotrophin molecule binds to the extracellular region of a p75NTR–sortilin complex, adaptor proteins are recruited to the intracellular, non-kinase region of p75NTR that initiate signaling cascades such as the c-jun N-terminal kinase (JNK) pathway. JNK activation can induce cell death via multiple mechanisms. JNK can activate the transcription factor c-jun, which then translocates to the nucleus (dashed arrow) to activate pro-apoptotic genes. The JNK pathway can also activate p53, which in turn enhances the activity of Bax. The pro-apoptotic Bax causes cytochrome *c* to be released from the mitochondria, thus allowing it to form an apoptosome and activate the caspase cascade (double arrows) that leads to cell death.

death in sympathetic neurons. It is thought that the pro-BDNF that these animals produce acts through the p75NTR receptor to initiate cell death pathways.

The combined activities of neurotrophins, pro-neurotrophins, Trk receptors, and p75NTR are necessary to regulate the balance of cell survival and death *in vivo*. Other signals are also important in ensuring that the proper number of cells survive. For example, inhibitor of apoptosis proteins (IAP) are found in the cytoplasm, where they can inhibit the activity of caspase-3 and caspase-9 and thereby promote cell survival. These IAP proteins can be blocked, in turn, by the mitochondrial protein Smac/Diablo. How IAP proteins function in different cells remains an area of ongoing research. Studies that map the pathways that regulate cell survival and death in various species and cell populations not only provide important information on how cell death occurs under normal conditions, but also generate insights that may ultimately yield interventions to halt cell death in neurodegenerative diseases.

SUMMARY

This chapter has focused on the role of target-derived growth factors and the mechanisms by which they and other growth factors regulate survival and PCD during development. Over the long and interesting history of growth factor research, scientists have come to understand that growth factors not only promote survival by activating specific intracellular pathways, but also inhibit pathways that promote cell death. A delicate balance of growth factor production, receptor expression, and signal transduction cascades determines whether a neuron lives or dies. To ensure proper formation of the nervous system, all of these components must be carefully regulated to make sure the right cells are living or dying at the correct time of development.

Additional roles for the neurotrophins and other growth factors continue to be discovered, and evidence now shows that such factors play multiple roles, such as regulating the expression of neurotransmitters and ion channels at various times during development and adulthood. While their role in the CNS remains less clear than in the PNS, it appears that neurotrophins are often involved in regulating aspects of cellular differentiation, synaptic plasticity, and the maintenance of synaptic connections. These latter topics are discussed in Chapter 10.

As illustrated in many examples above, the importance of trophic factors extends beyond development, and their release is not limited to target tissues. As scientists continue to learn more about trophic factors and their expression patterns, more light is shed on their varied functions. For example, several growth factors are present in adult tissues, and some of these are up-regulated in response to injury. Growth factors are also released by different glial cells and even by neurons themselves. Hope remains that one day growth factors can be delivered effectively to regions of the nervous system that have been damaged by injury or disease to restore nerve cell function or delay neuronal death.

FURTHER READING

Adler R, Landa KB, Manthorpe M & Varon S (1979) Cholinergic neuronotrophic factors: intraocular distribution of trophic activity for ciliary neurons. *Science* 204(4400):1434–1436.

Barde YA, Lindsay RM, Monard D & Thoenen H (1978) New factor released by cultured glioma cells supporting survival and growth of sensory neurons. *Nature* 274:818.

Barde YA, Edgar D & Thoenen H (1982) Purification of a new neurotrophic factor from mammalian brain. *EMBO J* 1(5):549–553.

Bothwell M (1995) Functional interactions of neurotrophins and neurotrophin receptors. *Annu Rev Neurosci* 18:223–253.

Bueker ED (1948) Implantation of tumors in the hind limb field of the embryonic chick and the developmental response of the lumbosacral nervous system. *Anat Rec* 102(3):369–389.

Buss RR, Sun W & Oppenheim RW (2006) Adaptive roles of programmed cell death during nervous system development. *Annu Rev Neurosci* 29:1–35.

Bredesen DE & Rabizadeh S (1997) p75NTR and apoptosis: Trk-dependent and Trk-independent effects. *Trends Neurosci* 20:287–290.

Campenot RB (2009) NGF uptake and retrograde signaling mechanisms in sympathetic neurons in compartmented cultures. *Results Probl Cell Differ* 48:141–158.

Cohen S (2004) Origins of growth factors: NGF and EGF. *Ann NY Acad Sci* 1038:98–102.

Chao MV, Bothwell MA, Ross AH et al. (1986) Gene transfer and molecular cloning of the human NGF receptor. *Science* 232(4749):518–521.

Curtis R & DiStefano PS (1994) Neurotropic factors, retrograde axonal transport and cell signalling. *Trends Cell Biol* 4(11): 383–386.

Davies AM (2003) Regulation of neuronal survival and death by extracellular signals during development. *EMBO J* 22(11): 2537–2545.

Davis S, Aldrich TH, Valenzuela DM et al. (1991) The receptor for ciliary neurotrophic factor. *Science* 253(5015):59–63.

Dekkers MP, Nikoletopoulou V & Barde YA (2013) Cell biology in neuroscience: death of developing neurons: new insights and implications for connectivity. *J Cell Biol* 203(3):385–393.

Frade JM, Rodríguez-Tébar A & Barde YA (1996) Induction of cell death by endogenous nerve growth factor through its p75 receptor. *Nature* 383:166–168.

Hamburger V (1939) Motor and sensory hyperplasia following limb bud transplantations in chick embryos. *Physiol Zool* 12:268–284.

Hashino E, Shero M, Junghans D et al. (2001) GDNF and neurturin are target-derived factors essential for cranial parasympathetic neuron development. *Development* 128:3773–3782.

Hendry IA, Stockel K, Thoenen H & Iversen LL (1974) The retrograde axonal transport of nerve growth factor. *Brain Res* 68(1):103–121.

Johnson D, Lanahan A, Buck CR et al. (1986) Expression and structure of the human NGF receptor. *Cell* 47(4):545–54.

Kaplan DR, Hempstead BL, Martin-Zanca D et al. (1991) The trk proto-oncogene product: a signal transducing receptor for nerve growth factor. *Science* 252(5005):554–558.

Kristiansen M & Ham J (2014) Programmed cell death during neuronal development: the sympathetic neuron model. *Cell Death Differ* 21(7):1025–1035.

Levi-Montalcini R, Meyer H & Hamburger V (1954) In vitro experiments on the effects of mouse sarcomas 180 and 37 on the spinal and sympathetic ganglia of the chick embryo. *Cancer Res* 14(1):49–57.

Levi-Montalcini R (1975) NGF: an uncharted route. In The Neurosciences: Paths to Discovery (Worden FG, Swazney JP & and Adelman G eds), pp 245–265. MIT Press.

Leibrock J, Lottspeich F, Hohn A et al. (1989) Molecular cloning and expression of brain-derived neurotrophic factor. *Nature* 14:149–152.

Lindsay RM, Alderson RF, Friedman B et al. (1991) The neurotrophin family of NGF-related neurotrophic factors. *Restor Neurol Neurosci* 2(4):211–220.

Martin DP, Schmidt RE, DiStefano PS et al. (1988) Inhibitors of protein synthesis and RNA synthesis prevent neuronal death caused by nerve growth factor deprivation. *J Cell Biol* 106(3):829–844.

Martin-Zanca D, Hughes SH & Barbacid M (1986) A human oncogene formed by the fusion of truncated tropomyosin and protein tyrosine kinase sequences. *Nature* 319(6056):743–748.

Oppenheim RW (1996) Neurotrophic survival molecules for motoneurons: an embarrassment of riches. *Neuron* 17(2): 195–197.

Pettmann B & Henderson CE (1998) Neuronal cell death. *Neuron* 20(4):633–647.

Radeke MJ, Misko TP, Hsu C et al. (1987) Gene transfer and molecular cloning of the rat nerve growth factor receptor. *Nature* 325(6105):593–597.

Reichardt LF (2006) Neurotrophin-regulated signalling pathways. *Philos Trans R Soc Lond B Biol Sci* 361(1473): 1545–1564.

Sendtner M, Kreutzberg GW & Thoenen H (1990) Ciliary neurotrophic factor prevents the degeneration of motor neurons after axotomy. *Nature* 345(6274):440–441.

Sendtner M, Schmalbruch H, Stockli KA et al. (1992) Ciliary neurotrophic factor prevents degeneration of motor neurons in mouse mutant progressive motor neuronopathy. *Nature* 358(6386):502–504.

Skaper SD (2008) The biology of neurotrophins, signalling pathways, and functional peptide mimetics of neurotrophins and their receptors. *CNS Neurol Disord Drug Targets* 7(1): 46–62.

Stahl N & Yancopoulos GD (1994) The tripartite CNTF receptor complex: activation and signaling involves components shared with other cytokines. *J Neurobiol* 25(11):1454–1466.

Zigmond RE (2011) gp130 cytokines are positive signals triggering changes in gene expression and axon outgrowth in peripheral neurons following injury. *Front Mol Neurosci* 4:62.

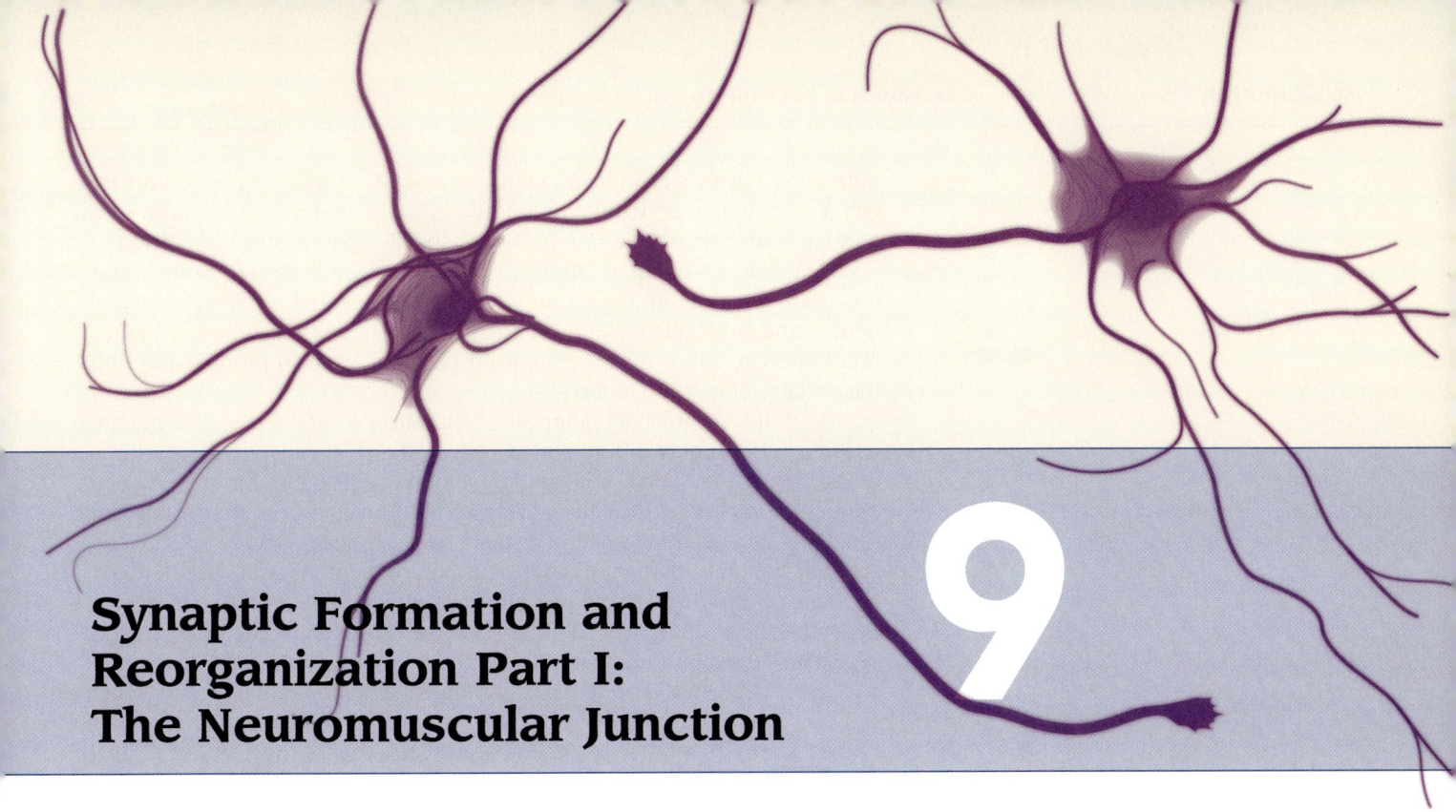

9 Synaptic Formation and Reorganization Part I: The Neuromuscular Junction

This chapter introduces two essential aspects of neurodevelopment: the process of **synaptogenesis**—the formation of new synaptic contacts in the nervous system—and the process of **synaptic reorganization**—the subsequent strengthening or loss of a subset of these connections. The developmental mechanisms that regulate formation and reorganization of the vertebrate **neuromuscular junction** (**NMJ**), the most studied of all the neural connections, are detailed here in Chapter 9. Mechanisms underlying synapse formation and reorganization in the central nervous system (CNS) follow in Chapter 10.

The study of synapses, including how they initially form and later reorganize, has a long history—one often marked by lively debates and strong differences of opinion. As noted in previous chapters, neurobiologists in the late 1800s debated whether neurons communicated through a syncytial network or through connections between individual cells. Evidence ultimately demonstrated that neurons communicated through small spaces—connections now termed **synapses**. In the first half of the twentieth century, scientists engaged in another debate, this one regarding whether the communication at synapses occurred primarily through chemical or electrical signals. These differing opinions were often referred to as the "war of the soups and the sparks," with the "soups" referring to chemical signals and the "sparks" to electrical signals. As scientists tested the two hypotheses, it was eventually determined that the majority of synapses utilize chemical signals in the form of **neurotransmitters** to mediate neural communication. Chapters 9 and 10 therefore focus on the development of structures associated with **chemical synapses**. While most of what is currently known about synaptic development comes from studies of the more prevalent chemical synapses, mechanisms regulating the development of electrical synapses, such as those formed by **gap junctions**, also remain an active area of study, and in some synaptic locations chemical and electrical signals work together to optimize neuronal communication.

The establishment of functional synapses is a dynamic process that takes place over an extended period of time. The initial formation of chemical synaptic connections requires several steps that lead to differentiation of specialized cellular elements in each synaptic partner cell. For example, an extending, motile growth cone must transform into a nerve terminal capable of releasing a specific neurotransmitter, and the target cell must begin to produce the corresponding neurotransmitter receptors. In addition, numerous proteins are required to direct localized neurotransmitter release and organize receptors so they are available for rapid binding. All of these specialized proteins are also produced during the process of synaptogenesis.

Perhaps surprisingly, after all of the effort to convert each synaptic partner into a highly differentiated cell, a subset of synaptic connections is lost during the normal course of development. Thus, rather than induce differentiation in only the subset of synaptic connections required for neural function, the nervous system instead over-produces highly specialized synaptic contacts, then eliminates a portion of them. As will be seen in the examples from the NMJ and CNS neurons, the elimination of connections depends often, though not entirely, on neural activity. The stabilization of synapses occurs in part by maintaining synapses that form functional partners. As the innervating neuron initiates action potentials in the target cell, the cells eventually begin to fire action potentials in synchrony with one another. The phrase "cells that fire together, wire together" describes one common mechanism by which cells maintain selected synaptic connections. Any target cells that do not receive sufficient input, or those that fire asynchronously with the innervating neuron, will eventually lose the synaptic connection; thus, those that are "out of sync, lose their link." This concept was originally introduced by Donald Hebb to explain synaptic changes that occur during learning or memory, and such synaptic changes are therefore referred to as Hebbian modifications or Hebbian plasticity.

There are many shared mechanisms governing synaptogenesis and reorganization in the NMJ and CNS, as well as several aspects that are unique to each synaptic type. This chapter begins with an overview of the synaptic structures found in synapses throughout the nervous system. Next, the specific elements found in the NMJ are described, followed by descriptions of several key experiments that led to the current models of how connections at the NMJ are formed and reorganized. As with other aspects of nervous system development, many of the molecules and signaling pathways identified in synaptic development are conserved across vertebrate and invertebrate animal models.

CHEMICAL SYNAPSE DEVELOPMENT IN THE PERIPHERAL AND CENTRAL NERVOUS SYSTEMS

Numerous studies over the past century have demonstrated how chemical synaptic transmission occurs between two neurons or between a neuron and a specialized cell, such as a muscle fiber or sensory cell. In most chemical synapses, an action potential signals a **presynaptic nerve terminal** to release a neurotransmitter that diffuses across the synaptic space to bind to corresponding receptors on the surface of the **postsynaptic partner cell**, thereby initiating a response in that cell. Charles Sherrington first used the term "synapse" in 1887 to describe the small gap between communicating nerve cells. However, the chemical synapse was not visualized conclusively until the mid-1950s, when techniques for preparing tissues for electron microscopy became sufficiently advanced to provide images of nerve cells. Studies by George Palade and Sanford Palay and Eduardo De Robertis and H. Stanley Bennett in 1955 were the first to prove that neurons were separated by a well-defined **synaptic cleft**. These studies were also the first to reveal the ultrastructural organization of the synaptic elements that are studied in such detail today.

By the late twentieth century, further advances in microscopy—particularly in the areas of time-lapse and fluorescence imaging—and new methods in cellular and molecular biology allowed scientists to identify hundreds of intracellular synaptic specializations. The intricate components of synaptic structures are currently studied at levels of detail unimaginable even 25 years ago.

Reciprocal signaling by presynaptic and postsynaptic cells results in the development of unique synaptic elements

As an extending nerve fiber approaches a target cell, the tip transforms from a motile growth cone to a presynaptic nerve terminal. Rather than producing the cytoskeletal elements necessary for growth cone motility and axon elongation, the presynaptic terminal must now form new structures, such as **synaptic vesicles** that contain the correct neurotransmitter, protein complexes called **active zones** that direct the synaptic vesicles to precise locations along the nerve terminal membrane, and **voltage-dependent calcium (Ca^{2+}) channels** that open in response to an action potential and initiate the release of the neurotransmitter into the synaptic cleft (**Figure 9.1**).

The postsynaptic cell develops its own specialized elements, including **neurotransmitter receptors** that selectively bind neurotransmitters and **scaffolding proteins** to hold the receptors in place directly across from the active zones of the presynaptic terminal (Figure 9.1). Excitatory neurons in the CNS are further characterized by an electron-dense organelle called the **postsynaptic density** (**PSD**) that helps anchor postsynaptic proteins across from the presynaptic terminal. Both pre- and postsynaptic partners produce various **adhesion molecules** to stabilize the newly formed synaptic connections. The precise alignment of pre- and postsynaptic elements ensures neurotransmitters released by the synaptic vesicles reach the postsynaptic receptors as quickly as possible.

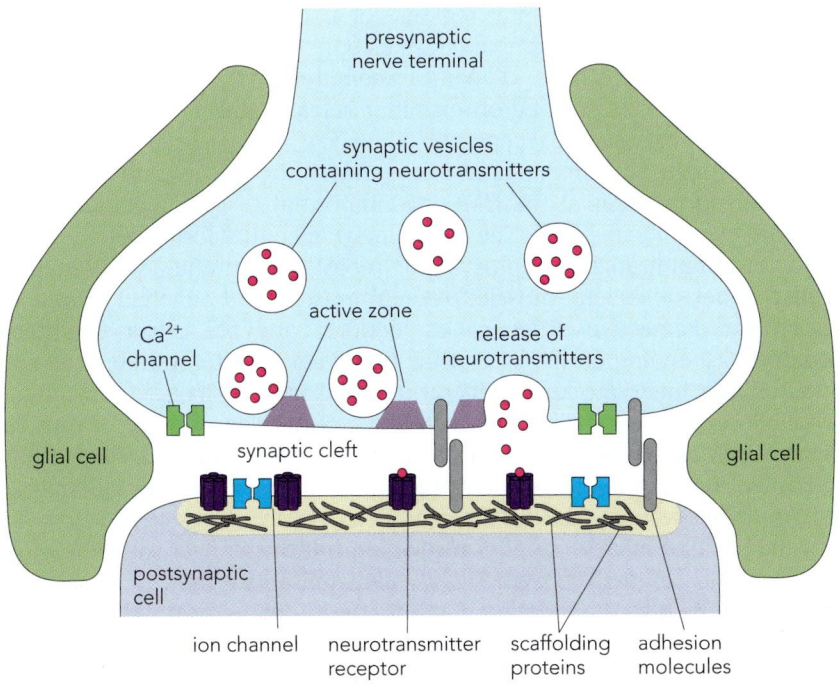

Figure 9.1 Pre- and postsynaptic cells have specialized synaptic elements and are surrounded by glial cell processes. Presynaptic nerve terminals are characterized by an abundance of synaptic vesicles. The synaptic vesicles line up at the active zones—areas of proteins along the presynaptic membrane adjacent to the postsynaptic cell. Voltage-dependent Ca^{2+} channels open in response to an action potential and initiate the release of neurotransmitter into the synaptic cleft. The neurotransmitter molecules bind to neurotransmitter receptors on the postsynaptic cell. Receptors are aligned directly across from the active zones so that neurotransmitter reaches the receptors quickly. Also located in postsynaptic cells are ion channels that open in response to neurotransmitter binding and initiate the action potential in that cell. Scaffolding proteins help anchor the postsynaptic receptors and ion channels across from the presynaptic nerve terminal. Various adhesion molecules are located in pre- and postsynaptic cells that initiate or stabilize the synaptic contacts. Finally, glial cells surround the pre- and postsynaptic contacts to provide signals to maintain the synaptic elements and to remove cellular debris left by retracting nerve terminals. These basic synaptic elements are found at synapses throughout the nervous system.

Vertebrate synapses also include the processes of glial cells. Thus, the vertebrate synapse is referred to as a **tripartite complex**, because it is comprised of a presynaptic cell, a postsynaptic cell, and a glial cell (Figure 9.1). At most synapses in the central and peripheral nervous systems, the glial cells extend cellular processes that surround pre- and postsynaptic cells to help maintain the necessary synaptic elements as well as remove cellular debris generated during synapse elimination.

As scientists documented the formation of the various synaptic elements during the early stages of synaptogenesis, several reports noted that some postsynaptic specializations began to form prior to the arrival of the presynaptic terminal. Such observations suggested that the postsynaptic cell develops first to provide signals that direct the differentiation and connectivity of the presynaptic cell. However, other studies reported that the presynaptic nerve terminal produces some synaptic elements prior to contact with the postsynaptic cell, suggesting that presynaptic differentiation was independent of postsynaptic signals. In many cases, it was simply unclear whether pre- or postsynaptic elements were the first to arise. It now appears that both pre- and postsynaptic cells begin to express immature synaptic elements prior to contact, but only differentiate organized, mature synaptic elements in response to signals obtained from the synaptic partner. Therefore, both pre- and postsynaptic cells provide reciprocal cues that induce maturation of synaptic elements to ensure that a proper synaptic connection is formed, maintained, and reorganized, if necessary.

THE VERTEBRATE NEUROMUSCULAR JUNCTION AS A MODEL FOR SYNAPSE FORMATION

Studied for more than a century, the vertebrate neuromuscular junction (NMJ) is currently the best characterized of all chemical synaptic connections and is described in this section to provide an example of mechanisms that regulate synapse formation and reorganization in the peripheral nervous system (PNS). The ability to study NMJs in such detail is due in part to the relatively large size of these synapses and the ability to isolate individual synaptic regions for experimentation. Among the many influential studies using the NMJ as a model system were those that (1) proved the existence of neurotransmitters, (2) identified synaptic vesicles as the sites of neurotransmitter release, and (3) confirmed the presence of postsynaptic receptors.

In order to understand how synaptogenesis and synaptic reorganization occur at the NMJ, it is important to first understand the cellular elements involved. The NMJ is a synapse that forms between the axon terminal branches of presynaptic motor neuron and a postsynaptic muscle fiber of skeletal muscle. Skeletal muscles are attached to bones, often via connective tissue called tendons, and are under voluntary control. Depending on the size of the particular skeletal muscle, a motor neuron can innervate up to several hundred muscle fibers by extending multiple nerve terminal branches (**Figure 9.2**). The number of muscle fibers that a single motor neuron innervates is called a **motor unit**, because stimulation of that neuron leads to contraction of all the muscle fibers it innervates. Each terminal branch contacts a specialized region on the skeletal muscle called a **motor endplate**, an oval-shaped region of a muscle cell that appears slightly elevated above the rest of the cell surface at the time of innervation. Within the endplate region, the terminal branches end as **synaptic boutons** (terminal end bulbs), where they form a synaptic connection with the muscle fiber. Myelinating Schwann

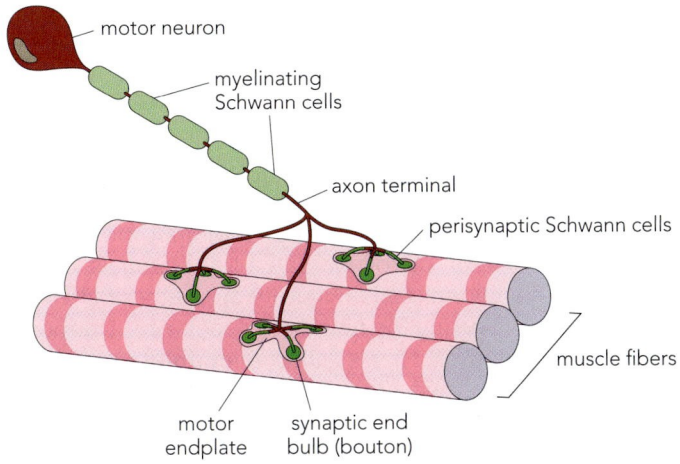

Figure 9.2 Vertebrate motor neurons innervate multiple skeletal muscle fibers. Motor neurons form synaptic connections with skeletal muscle fibers. The sites of synaptic contact are called neuromuscular junctions. In adult vertebrates, each motor neuron extends a myelinated axon that branches to establish contact with several skeletal muscle fibers. Each terminal branch establishes synaptic contacts at a specialized region on the muscle surface called the motor endplate. The synapses form between the synaptic end bulbs (boutons) and the muscle cell membrane in the endplate. Terminal branches are unmyelinated but are covered by specialized Schwann cells called perisynaptic Schwann cells.

cells surround the axons of motor neurons, while specialized perisynaptic Schwann cells surround the terminal branches (Figure 9.2).

At the NMJ, the presynaptic motor axon releases acetylcholine to depolarize the postsynaptic muscle cell

Acetylcholine (ACh) is the neurotransmitter utilized by vertebrate motor neurons. ACh is synthesized in the presynaptic neurons by the enzyme **choline acetyltransferase (ChAT)**, then it is packaged into synaptic vesicles that line up at the active zones. The active zones are protein complexes located in discrete regions of the presynaptic terminal that help anchor the synaptic vesicles directly across from the nicotinic **acetylcholine receptors (AChRs)** located on the postsynaptic skeletal muscle cell. These ionotropic AChRs are ligand-gated cation channels that cluster at the crests of the **postjunctional folds**—invaginations in the skeletal muscle cell membrane (**Figure 9.3**). The localization of the

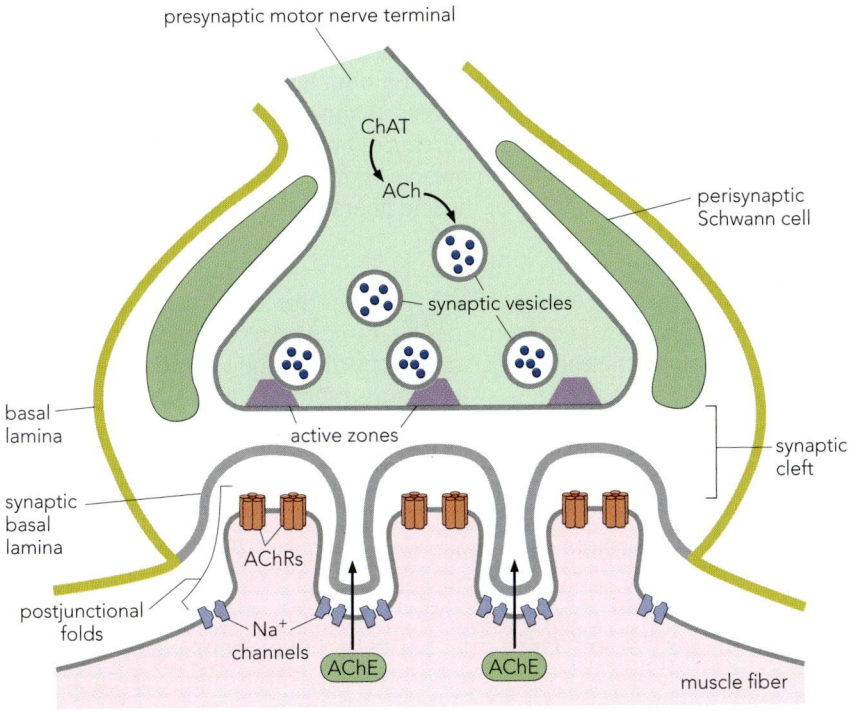

Figure 9.3 Components of the mature neuromuscular junction (NMJ). In the presynaptic motor neuron, acetylcholine (ACh) is synthesized by choline acetyltransferase (ChAT). The ACh is then packaged into synaptic vesicles that line up along the active zones in the presynaptic nerve terminal. The active zone proteins are concentrated opposite the postjunctional folds of the postsynaptic muscle cell. The crests of the folds contain the acetylcholine receptors (AChR), while the troughs contain voltage-gated sodium (Na^+) channels. The postsynaptic cell releases acetylcholine esterase (AChE) into the synaptic cleft, where it breaks down excess ACh. The precise arrangement of these pre- and postsynaptic elements ensures rapid and reliable neurotransmission at the NMJ. Perisynaptic Schwann cells surround the NMJ, and basal lamina covers the presynaptic terminal and muscle surface. In contrast, the synaptic basal lamina is restricted to the synaptic cleft.

AChRs to the crests of the postjunctional folds ensures that the released ACh reaches and binds to these neurotransmitter receptors rapidly.

In response to an action potential initiated in a motor neuron, voltage-gated Ca^{2+} channels accumulated near the active zones in the nerve terminals open, thus leading to an influx of Ca^{2+} ions. Ca^{2+} causes the synaptic vesicles to fuse with the neuronal membrane and release ACh into the synaptic cleft. The binding of ACh to the AChRs leads to an influx of sodium (Na^+) and other cations into the postsynaptic muscle cell and the subsequent rapid activation of voltage-gated Na^+ channels located in the troughs of the postjunctional folds (Figure 9.3). As these channels open, more sodium flows in, causing the muscle to depolarize. This postsynaptic action potential in the muscle fiber is called an **endplate potential (EPP)**. The EPP then propagates away from the endplate in both directions to initiate muscle contraction.

Each muscle fiber and myelinated axon is surrounded by a basal lamina. The synaptic cleft, however, contains the **synaptic basal lamina**, a unique structure with a different protein composition. Among the proteins found in the synaptic basal lamina is **acetylcholine esterase (AChE)**, the enzyme that breaks down ACh. Muscle cells secrete the predominant form of AChE found at the NMJ, and this enzyme becomes concentrated in the synaptic basal lamina. Because of this, any excess neurotransmitter is removed quickly to prevent signaling beyond what is required to initiate muscle contraction. Scientists have developed a number of assays to detect ChAT, ACh, AChRs, and AChE both *in vitro* and *in vivo*. These labeling methods have been extremely useful in identifying many of the developmental events that take place during synaptogenesis of the NMJ.

The distribution of AChRs has been mapped in developing muscle fibers

Skeletal muscle tissue arises from the fusion of mesoderm-derived, spindle-shaped, mononucleated **myoblasts** (**Figure 9.4**A). During embryonic development, several myoblasts fuse to form a primary myotube that eventually extends the length of the future muscle (Figure 9.4B, C). Secondary myotubes arise soon after and fuse to the existing primary myotubes (Figure 9.4D). The myotubes continue to develop and ultimately form a mature **muscle fiber** (or **myofiber**), which is a long, cylindrical, multinucleated cell. Each myofiber is multinucleated because it arises from the fusion of the individual myoblasts (Figure 9.4E). The nuclei of these myoblasts can be visualized at intervals of about 10 μm along the length of the muscle fiber. Mature human muscle fibers can be up to 30 cm long and 10 to 100 μm wide. Bundles of muscle fibers make up the mature skeletal muscle (Figure 9.4F)

As myotubes start to form, they begin to express the genes encoding AChRs. Each myotube expresses AChRs across the cell surface in a generally uniform pattern. However, some newly formed AChRs are located in small aggregates along the cell surface and many are found near the middle third of the cell surface—the area of the future motor endplate. This is a form of prepatterning in which an accumulation of centrally located AChRs is present prior to innervation. The receptors are still able to migrate along the muscle surface, however, and they are not restricted to the site at which they first accumulate (**Figure 9.5**A).

The prepatterning of AChRs is modified once the motor nerve terminals begin to contact their target myotubes. Prior to contact, motor nerve terminals begin to release some ACh from the few synaptic vesicles present (Figure 9.5B). Once the nerve terminal contacts a myotube, there is an increase in ChAT expression in the motor neurons, indicating a corresponding increase in ACh production. The ACh is packaged into

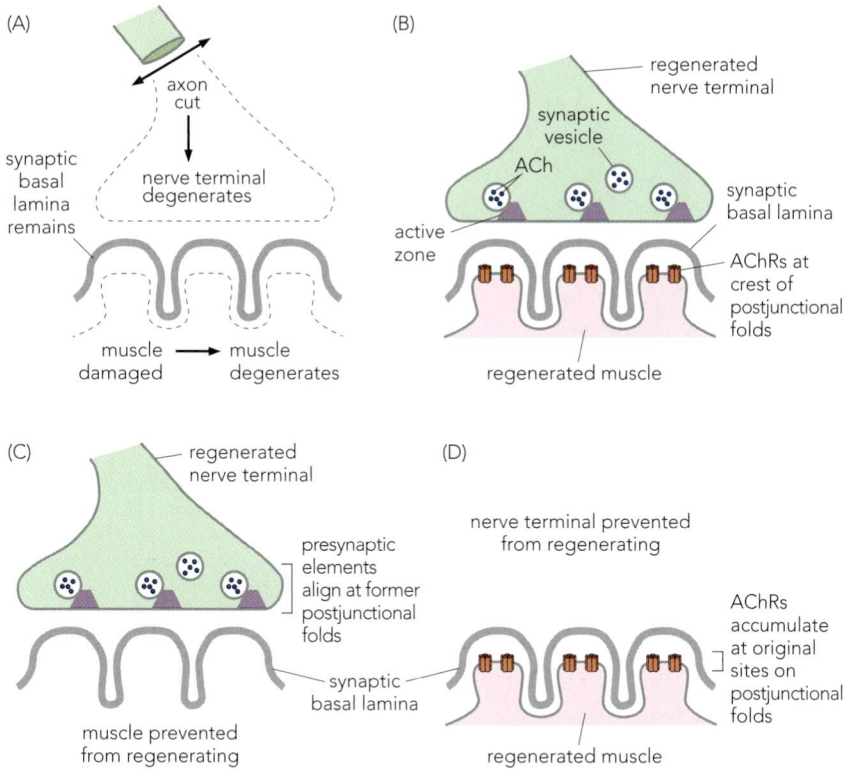

Figure 9.9 Experiments revealed the synaptic basal lamina contains signals to organize pre- and postsynaptic elements. The NMJ of a frog regenerates following damage to motor axons and muscle tissue. (A) When the innervating motor axon is cut and the target muscle is damaged, both the nerve terminal and muscle fiber degenerate. The synaptic basal lamina, however, remains. (B) Following a recovery period, both the axon and muscle fiber regenerate and pre- and postsynaptic elements assemble as usual, creating a new, fully functional synaptic connection. (C) When the muscle fiber was irradiated so that it could not regenerate, the regenerated motor nerve terminal contacted the remaining synaptic basal lamina at the original synaptic site and produced active zone proteins and synaptic vesicles above the regions of the former postjunctional folds. This suggested the synaptic basal lamina contained signals that guided the nerve terminal and induced presynaptic specializations. (D) Conversely, when the motor nerve terminal was prevented from regenerating, but the muscle fiber was allowed to regenerate, the new muscle produced AChRs concentrated in the crests of the postjunctional folds. Like the presynaptic terminal, these new postsynaptic elements were located at the site of the original synapse. This suggested that the synaptic basal lamina also contained signals that organize postsynaptic elements.

contacted the original synaptic sites on the synaptic basal lamina. Further, the presynaptic elements differentiated as expected, with the synaptic vesicles and active zones accumulating directly opposite the folds of the synaptic basal lamina (Figure 9.9C). In other experiments, muscle was allowed to regenerate, but the motor axons were cut and prevented from re-innervating the new muscle tissue. In these preparations, AChRs formed in the new muscle at the site of original innervation, directly below the folds of remaining synaptic basal lamina. These experiments suggested that signals from the synaptic basal lamina remained at the former synaptic site to direct the accumulation of AChRs to the correct location (Figure 9.9D). Together these studies provided the first evidence that the signals regulating synaptogenesis at the NMJ were concentrated in the synaptic basal lamina. Since this initial discovery, a number of molecules needed for both pre- and postsynaptic specializations have been found concentrated in this region.

Because a regenerating nerve fiber growing to a target in an adult animal encounters a very different cellular environment than a newly formed nerve fiber extending in the embryo, signals important during regeneration may be quite different from those required during development. Researchers therefore must test putative signals in both contexts in order to determine the necessity of the signal during development and regeneration. As will be discussed, the signals originally detected in models of NMJ regeneration are also present during NMJ development.

AChRs cluster opposite presynaptic nerve terminals in response to agrin released by motor neurons

Studies to identify the signals derived from synaptic basal lamina first focused on the role of the synaptic basal lamina in inducing postsynaptic AChR aggregation. Two experimental methods greatly aided these investigations. The first was a myotube cell culture assay that was developed to map AChR

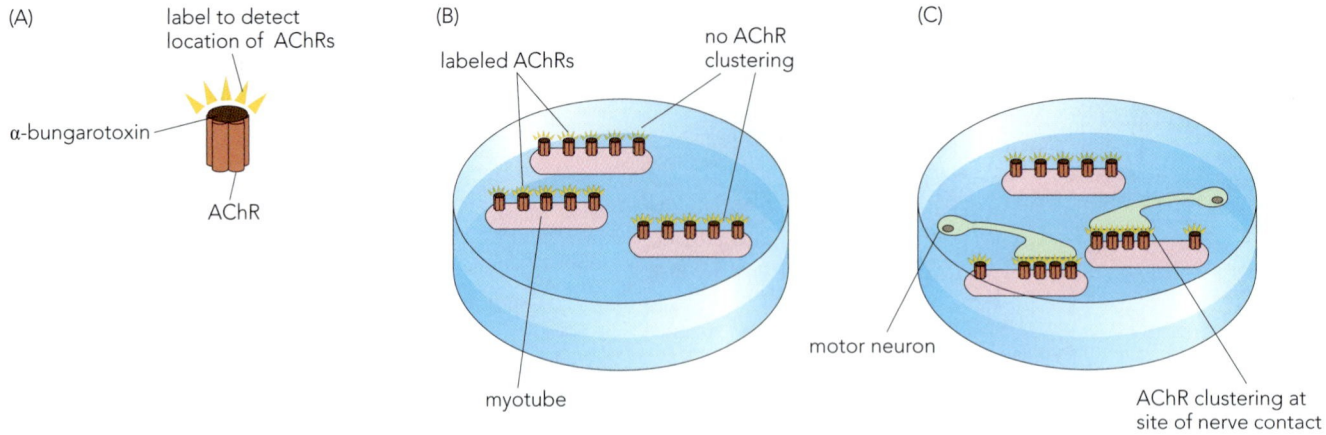

Figure 9.10 Labeling with α-bungarotoxin is used to map the distribution of AChRs on myotubes *in vitro*. (A) α-Bungarotoxin binds to nicotinic AChRs and can be used as an experimental tool to inhibit neural transmission or label receptors. When α-bungarotoxin is conjugated to a radioactive, enzymatic, or fluorescent marker, the location of AChRs along a muscle can be visualized. (B) When mytotubes were grown in cell culture in the absence of motor neurons, unclustered AChRs were observed along the cell surface. (C) When motor neurons were added to the myotube cultures, AChRs clustered at sites of nerve contact.

distribution *in vitro*. In order to visualize the distribution of AChRs along the muscle surface, a snake toxin called α-bungarotoxin, was conjugated to a radioactive, enzymatic, or fluorescent marker (**Figure 9.10**A). α-Bungarotoxin specifically binds to nicotinic AChRs and can be used experimentally to block neural transmission. Additionally, when labeled with a radioactive, enzymatic, or fluorescent marker, α-bungarotoxin can be used to identify the location of AChRs on the cell surface. Using labeled α-bungarotoxin, scientists found, in the absence of motor neurons, that AChRs were distributed across the surface of the myotubes (Figure 9.10B). Although some small clusters of AChRs were noted in the absence of motor neurons, the distribution changed considerably when motor neurons were added to the cultures. In co-cultures of motor neurons and myotubes, the AChRs began to cluster below the areas of nerve fiber contact (Figure 9.10C). Thus, this method provided a means to investigate AChR distribution in response to a variety of *in vitro* experimental manipulations, including the addition of soluble factors and local application of ACh.

A second important experimental approach utilized the electric organ from the ray *Torpedo californica* (**Figure 9.11**A), an animal model first used in 1772. As discovered in the 1930s, the electric organ of this fish contains a remarkably high concentration of cholinergic synapses. AChRs are so abundant in these rays that Christopher Miller described *Torpedo californica* as "essentially a swimming purified AChR." The electric organ shares many features of the NMJ and has been used in multiple studies of synaptogenesis over the years. Importantly, like muscle cells, the cells of the electric organ are covered in basal lamina, thus providing a large source of material to analyze. Without the discovery of the electric organ of *Torpedo* as an abundant source of cholinergic synapses, many of the biochemical and molecular discoveries related to the function of the NMJ would not have occurred as early as they did, if at all.

Using the *Torpedo* electric organ, the McMahon lab extracted insoluble proteins from basal lamina and tested the ability of the protein extracts to cluster AChRs in cultured chick myotubes. With this method, the lab eventually purified a proteoglycan they called **agrin**. When added to cultures of myotubes, purified agrin induced clustering of AChRs (Figure 9.11B). The clustering of AChRs in response to agrin was inhibited when antibodies to agrin were added to cell culture preparations (Figure 9.11C). Further studies indicated that agrin is concentrated in the

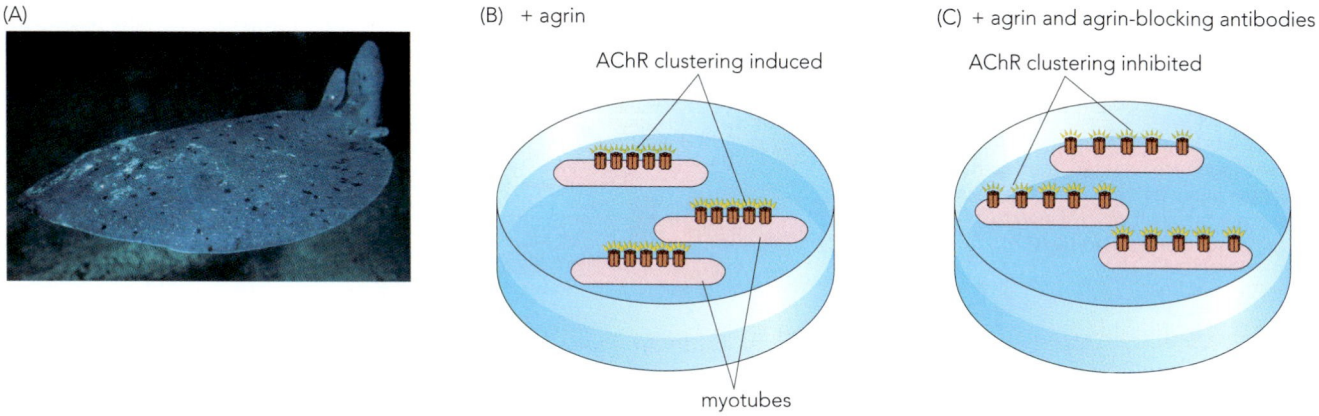

Figure 9.11 Agrin induces clustering of AChRs *in vitro*. (A) Photograph of the ray *Torpedo californica*. The electric organ of *Torpedo californica* shares many characteristics of the NMJ, including having a high concentration of AChRs and cells that are covered in basal lamina. This basal lamina provided an abundant source of material used for isolating proteins, such as agrin, that are necessary for AChR aggregation. (B) Labeled α-bungarotoxin was used to identify the location of AChRs in cultured myotubes. Addition of purified agrin to the myotube cultures induced clustering of the AChRs. (C) When antibodies that block agrin function were added to these cultures, the clustering of AChRs was inhibited. (A, courtesy of Daniel Gotshall/NoAA.)

synaptic basal lamina, produced by motor neurons, and released by the presynaptic nerve terminal. The importance of agrin in NMJ formation was confirmed in knockout mice lacking *agrin*. In these mice, AChRs are present at normal levels but clustering is greatly reduced and stable NMJs fail to form. The mice die at birth due to respiratory failure caused by a lack of functional innervation to the muscles of respiration.

Together these findings led to what was called the **agrin hypothesis**. In this model, first described in the 1990s, agrin released by the innervating nerve terminal becomes concentrated in the synaptic basal lamina, where it signals postsynaptic AChRs to cluster opposite the innervating presynaptic nerve terminal. In the absence of agrin, such receptor clustering does *not* occur and a functional NMJ fails to form. The basic tenants of the agrin hypothesis have been supported, although modifications were added when a role for ACh was later discovered.

The agrin hypothesis is revised based on additional observations

While the initial studies clearly demonstrated that agrin plays a major role in formation of the NMJ and the agrin hypothesis accounts for many aspects of the postsynaptic differentiation observed in muscle fibers, other seemingly conflicting observations led to further evaluation of the proposed role of agrin and ultimately to revisions to the original agrin hypothesis. One observation that first suggested agrin released by the nerve might not be essential for clustering AChRs was the finding that agrin was also produced in muscle tissue and adjacent Schwann cells. This suggested that AChR clustering and NMJ formation could be regulated by any of the cells at the NMJ and may not require nerve-terminal-derived agrin. However, subsequent studies identified different isoforms of agrin. In mammals, the agrins are designated X, Y, or Z to indicate the splice site where amino acids are added in the different isoforms. Nerve terminals release the Z form of agrin (called Z^+ agrin), and this isoform has a much greater clustering effect than muscle-derived agrin that lacks the amino acid insert at the Z site (Z^- agrin). It is the Z^+ isoform that influences AChR clustering at the NMJ. The role of Z^- agrin in muscle is not yet known. It does not induce receptor clustering, even in the absence of Z^+ agrin, and it is localized to many other nonneural tissues, including lung and kidney.

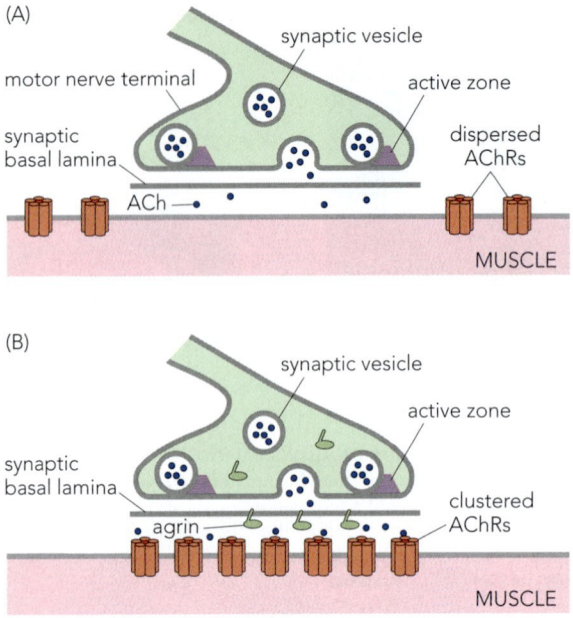

Figure 9.12 The agrin–ACh hypothesis proposed how nerve terminal signals interact to regulate dispersal and clustering of AChRs. (A) ACh released by the presynaptic nerve terminal causes AChRs to disperse away from the site of nerve contact. (B) When both ACh and agrin are released, the AChRs cluster below the nerve terminal. This suggests that the agrin acts to stabilize existing AChRs and prevent dispersal initiated by ACh.

The roles of the Y and X isoforms of agrin are also unclear. Y agrin binds heparin and has been proposed to modify interactions between agrin and other molecules located at the NMJ. However, in the absence of Y agrin, AChR clustering still occurs.

Another important discovery was that the ACh released from the innervating motor nerve terminal led to dispersal of AChR away from the presynaptic terminal (**Figure 9.12**A). Studies in 2005 demonstrated that ACh appears to block the synthesis of new AChR subunits while also initiating the removal of existing AChRs by endocytosis. Thus, the innervating nerve terminal releases signals that regulate both clustering and dispersal of AChRs.

The observation that the presynaptic nerve terminal can both induce AChR clustering and initiate dispersal of these same receptors led to the **agrin–ACh hypothesis**, a proposal that takes into account the release of both agrin and ACh from the innervating nerve terminal. In this model, it is thought that neural agrin functions to stabilize the prepatterned clusters of AChRs at the central region of the muscle fiber by counteracting the effects of ACh. The nerve-derived agrin becomes concentrated in the synaptic basal lamina, where it signals to receptors located on muscle fibers and stabilizes nearby AChRs, thus preventing their dispersal when exposed to ACh (Figure 9.12B). Any AChRs not stabilized by agrin are lost upon exposure to ACh. *In vivo* evidence supporting the agrin–ACh hypothesis was found in mice lacking genes for both agrin and ChAT—the enzyme that synthesizes ACh. When both signals are absent, AChRs still cluster as expected. This indicates that in the absence of ACh, agrin is not required to cluster the AChRs. Further, the lack of agrin does not cause a decrease in AChR clustering, because there is no ACh present to initiate dispersal of the receptors. Thus, the primary role of neural agrin in AChR clustering may be to prevent ACh-induced dispersal.

The receptor components MuSK and Lrp4 mediate agrin signaling

With the initial discovery of agrin, several labs began to search for a receptor on muscle fibers that would mediate agrin-induced clustering of AChRs. These investigations ultimately led George Yancopolous and colleagues to identify a new receptor tyrosine kinase in 1995. The receptor was named MuSK for muscle specific kinase. Subsequent experiments demonstrated that MuSK was an essential component of the agrin-signaling pathway. For example, MuSK is expressed at basal levels in developing muscle cells, particularly in the central region of the muscle fiber. MuSK expression is up-regulated at the time of normal innervation, as well as during re-innervation of denervated muscle. In *Musk* knockout mice, AChRs are present, but no receptor clustering is observed and functional NMJs fail to form. Similar to the *Agrin* knockout mice, these mice die at birth due to respiratory failure. Thus, signaling through the MuSK receptor was necessary for normal clustering of AChRs at the NMJ synapse. Even though experimental evidence indicated MuSK signaling was required for postsynaptic clustering of AChRs, scientists realized that the extracellular portion of the MuSK receptor was unable to directly bind agrin. Therefore, it was proposed that an additional signaling component must be involved. This putative component was referred to as MASC, for myotube-associated specificity component. MASC was presumed to bind agrin before interacting with MuSK to induce clustering of AChRs.

Several labs sought this missing signaling component, and more than 10 years after the discovery of MuSK, a co-receptor was identified. In 2009, low-density lipoprotein receptor-related protein 4 (Lrp4) was found to bind Z^+ agrin directly. Lrp4 and MuSK form a complex in the muscle cell membrane, and the interaction between the two co-receptors is increased upon binding of agrin to Lrp4 (**Figure 9.13**). Agrin binding also leads to an increase in MuSK phosphorylation. Phosphorylated MuSK then recruits and phosphorylates the cytoplasmic adapter protein Downstream of kinase 7 (Dok7). Once activated, Dok7 further increases the phosphorylation of MuSK and recruits additional adapter proteins that ultimately lead to the clustering of AChRs (Figure 9.13). Among the adaptor proteins recruited by Dok7 are CRK (v-crk avian sarcoma virus CT10 oncogene homolog) and CRK-L (CRK-like) that activate Rho GTPase-binding proteins associated with the actin cytoskeleton that helps stabilize AChR clustering. The requirement for Lrp4 and Dok7 in NMJ formation was confirmed in mice lacking either gene. As with mice lacking *Musk*, in mice lacking either *Lrp4* or *Dok7*, aggregation of AChRs was absent, and the NMJ failed to form. The

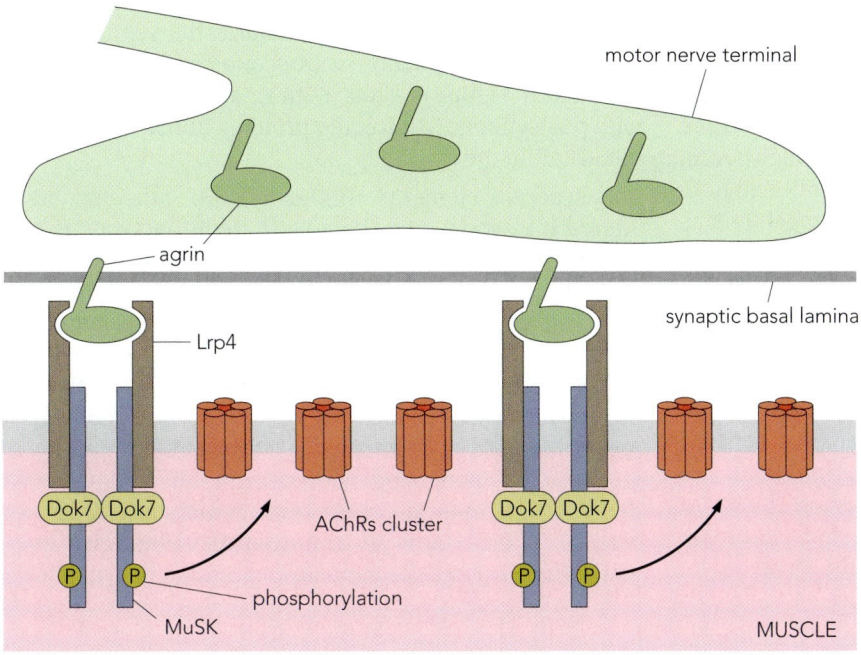

Figure 9.13 Agrin activity is mediated by the Lrp4-MuSK receptor complex. Agrin released by the nerve terminal is concentrated in the synaptic basal lamina. Agrin binds to the Lrp4 receptor that interacts with the receptor tyrosine kinase MuSK. The interaction between Lrp4 and MuSK is further increased upon agrin binding. Agrin binding leads to an increase in MuSK phosphorylation, the recruitment of adaptor proteins such as Downstream of kinase 7 (Dok7), and the clustering of AChRs.

importance of Dok 7 is also revealed in congenital muscular disorders in humans. Human *Dok7* mutations are one cause of congenital myasthenic syndrome (CMS), a condition that causes muscle weakness (myasthenia) and fatigue upon exertion of skeletal muscles.

Additional proteins are also proposed to interact with MuSK to regulate postsynaptic specializations. For example, a number of isoforms of vertebrate Wnt have been found to interact with Frizzled-like regions on the extracellular portion of MuSK. In zebrafish, the Wnt–MuSK interaction appears to influence prepatterning and clustering of AChRs in at least some contexts. In some experiments, the addition of Wnt isoforms to cultures of myotubes increased AChR clustering even in the absence of agrin, suggesting a role for Wnts in NMJ formation. Not all studies found a direct effect of Wnts on clustering, however, and many *in vitro* and *in vivo* studies suggest that the clustering of AChRs increases when both agrin and Wnts are present. The signaling pathways involved in this additive effect are not fully established. MuSK has been found to interact with components of the canonical Wnt signaling pathway (see Figure 4.17), including Dishevelled (Dvl) and adenomatous polyposis coli (APC). This interaction may link MuSK to the cytoskeletal elements associated with stabilizing AChRs at the nerve terminal. The Wnt homolog, *wingless* (*wg*), is also an important component of NMJ formation in *Drosophila* (**Box 9.1**).

Box 9.1 Invertebrate model systems of NMJ formation

As with vertebrate animal models, the mechanisms regulating synaptogenesis in invertebrates remain active areas of research. The NMJs of *Drosophila* and *C. elegans* have provided considerable insight into the mechanisms of synaptic formation across species and both animal models share several common features with the vertebrate NMJ. Similar to vertebrates, studies in these invertebrate models revealed that pre- and postsynaptic elements begin to appear prior to the formation of a synaptic contact, and postsynaptic receptors cluster at that site of the arriving presynaptic nerve terminal. However, there are also many unique aspects of NMJ formation in these species. Whereas the vertebrate NMJ utilizes ACh, the fly NMJ utilizes glutamate, and the worm can use one of three different neurotransmitters at an NMJ—namely, ACh, glutamate, or serotonin. The examples mentioned here illustrate examples of some of the unique mechanisms regulating clustering of neurotransmitter receptors in the postsynaptic membrane of the fly and worm NMJ.

Development of the *Drosophila* NMJ

In the fly NMJ, pre- and postsynaptic elements begin to appear prior to the formation of a synaptic contact. As noted previously, the *Drosophila* NMJ uses the neurotransmitter glutamate. The glutamate binds to the corresponding ionotropic ligand-gated glutamate receptors, the iGluRs. The arrival of the presynaptic nerve terminal induces aggregation of the receptors below the terminal. This clustering depends, at least in part, on Wingless (Wg), a homolog of the vertebrate Wnt proteins. Wg is secreted by the motor neurons and binds to Frizzled (Fz) receptors on the developing muscle cells (**Figure 1**). In the absence of *Wg*, glutamate receptors and scaffolding proteins are disrupted across the muscle surface. Additional signals from the Neto family of proteins (neuropilin and Tolloid-like proteins) have been shown to regulate iGluR clustering. Neto interacts with the subunits of the receptors prior to innervation. Neto appears to stabilize the subunits of the iGluRs by interacting with scaffolding proteins in the muscle (Figure 1). Thus, both Fz and Neto associate with postsynaptic scaffolding proteins to help stabilize the iGLuR in the fly NMJ.

The synaptic elements in the fly NMJ can also be altered in response to changes in neural activity. For example, if synaptic input is reduced, the postsynaptic neuron can increase the number of neurotransmitter receptors and ion channels to return the postsynaptic firing rate to baseline levels. Conversely, if the activity of the postsynaptic glutamate receptors is inhibited or the number of receptor subunits are reduced, the presynaptic cell will compensate for the reduced muscle response by increasing neurotransmitter release. In this way, the muscle contraction remains at a stable level. This is a form of **homeostatic plasticity** in which the partner cells increase expression of synaptic

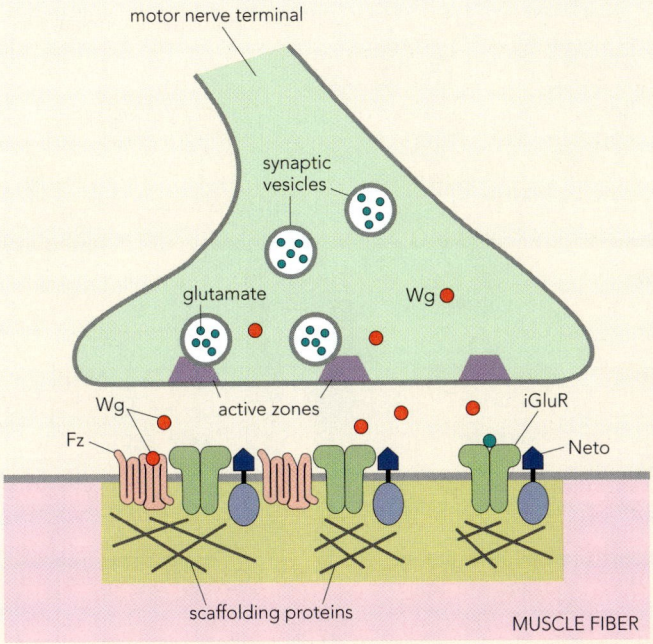

Figure 1 The *Drosophila* NMJ. The *Drosophila* presynaptic motor nerve terminal releases the neurotransmitter glutamate that binds to corresponding ionotropic ligand-gated glutamate receptors (iGluRs) on the muscle fiber. Clustering of the iGluRs below the nerve terminal is influenced by Frizzled (Fz) receptors and the Neto family of proteins that interact with postsynaptic scaffolding proteins. Wingless (Wg) secreted by the motor nerve terminal binds to the Frizzled (Fz) receptors located on the muscle fiber. Signals from the Neto family of proteins interact with the subunits of the iGluRs prior to innervation and appear to stabilize the subunits of the iGluRs. Together, Fz and Neto associate with postsynaptic scaffolding proteins to help stabilize the iGluRs in the fly NMJ.

elements in response to alterations in neural signaling. Homeostatic plasticity is also observed in mammalian NMJ and CNS neurons (see Chapter 10).

Development of the *C. elegans* ACh NMJ

One of the neurotransmitters utilized at the NMJ of *C. elegans* is ACh. Among the receptors present on muscles of the worm body wall are AChRs that bind to the agonist levamisole. These levamisole-sensitive receptors reveal prepatterned clustering in the postsynaptic muscle prior to innervation. The clustering is further regulated in part by signals derived from the muscle cells. Muscle cells secrete a protein called LEV-9 (levamisole resistant-9) that interacts with LEV-10, a transmembrane muscle protein. Another secreted, muscle-derived protein called OIG-4 (one immunoglobulin domain containing protein-4) enhances this clustering. Clustering is further regulated in response to nerve-derived signals. As the presynaptic motor nerve terminal approaches, the differentiating nerve fiber releases the protein MADD-4 (Map kinase-activating death domain). MADD-4 helps cluster the AChRs across from the presynaptic terminal. The intracellular mechanisms by which MADD-4 activates this clustering is not yet known (**Figure 2**).

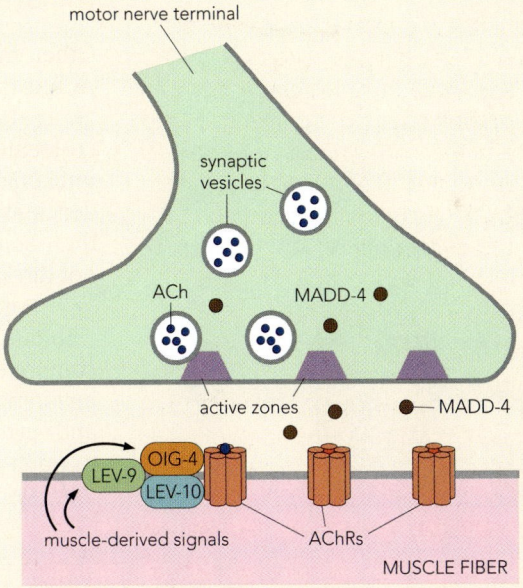

Figure 2 The *C. elegans* NMJ. ACh is one of the neurotransmitters utilized at the *C. elegans* NMJ. Prepatterning of the levamisole-sensitive AChRs is observed in the postsynaptic muscle fiber prior to innervation. The clustering is influenced by a protein called LEV-9 (levamisole resistant-9) secreted by muscle fibers. LEV-9 interacts with a transmembrane muscle protein, LEV-10, and another muscle-derived protein called OIG-4 (one immunoglobulin domain containing protein-4) to enhance this clustering. As the presynaptic motor nerve terminal approaches the differentiating muscle cell, it releases the protein MADD-4 (Map kinase-activating death domain) that further increases the clustering of the AChRs across from the presynaptic terminal.

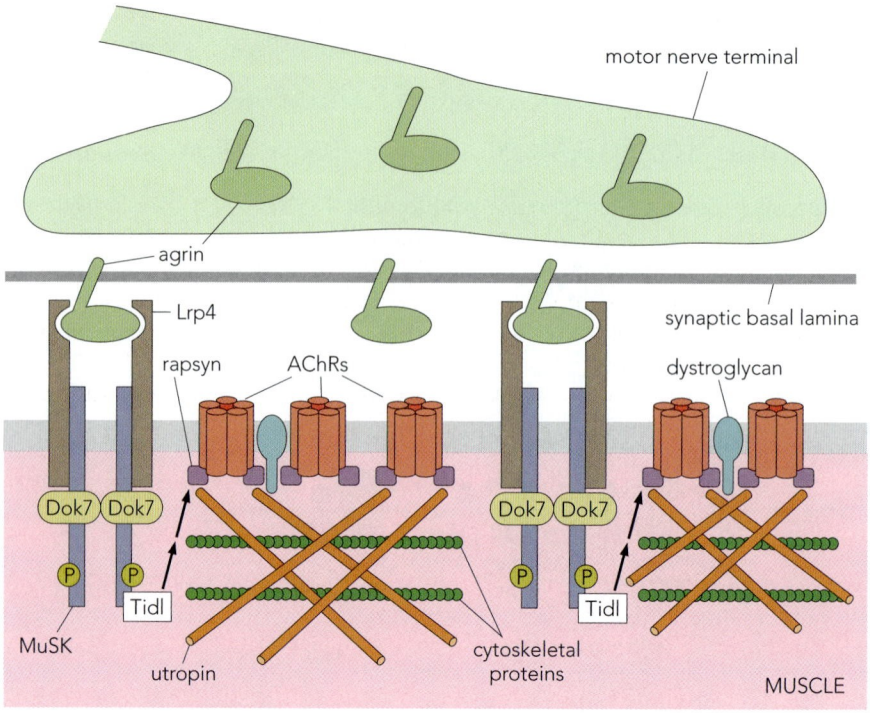

Figure 9.14 Rapsyn interacts with MuSK to anchor AChRs to the cytoskeleton. Rapsyn interacts with AChRs, utropin, dystroglycan, and cytoskeletal proteins to help anchor the AChRs to the postsynaptic site. MuSK activity appears to be linked to rapsyn by a number of downstream signaling pathways. In this example, the binding of agrin to the Lrp4-MuSK receptor complex activates Tid1 (tumorous imaginal disc 1), which interacts, in turn, with rapsyn to stabilize the AChRs in the postsynaptic membrane.

Rapsyn links AChRs to the cytoskeleton

Other molecules also interact with MuSK to cluster and stabilize the AChRs to ensure these receptors align precisely with the presynaptic active zones. Rapsyn, first discovered in the late 1980s, is one such molecule that co-localizes with AChRs. Mice lacking rapsyn fail to cluster AChR. However, unlike *Musk*- or *Agrin*-mutant mice, the unclustered receptors of rapsyn-deficient mice remain near the central region of the muscle fiber, rather than being widely dispersed across the muscle. This suggested rapsyn is not directly influenced by the release of ACh from the nerve terminal.

Rapsyn is now known to function downstream of MuSK to help cluster AChRs—likely by interacting with other proteins to link the receptors to components of the cytoskeletal network. A number of associated proteins—including dystroglycan and utrophin—are thought to interact with rapsyn and cytoskeletal proteins to stabilize AChRs. Additionally, signaling through the vertebrate homolog of the *Drosophila tumorous imaginal disc 1* (*tid1*) gene appears to link MuSK to rapsyn (**Figure 9.14**).

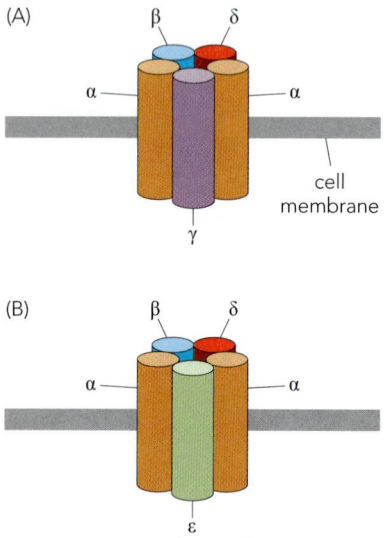

Figure 9.15 Acetylcholine receptor subunit composition changes during development. (A) Embryonic AChRs contain five subunits: two α, one β, one δ, and one γ. (B) Mature AChRs also have five subunits, but the γ subunit is not synthesized in adults. Instead, an ε subunit is synthesized.

AChR subunits are synthesized in nuclei adjacent to the nerve terminal

The clustering of AChR opposite the active zones of the presynaptic nerve terminal is one of many postsynaptic changes that take place in response to the arrival of the motor nerve terminal. Changes are also noted in the location of nuclei that transcribe the subunits of embryonic and adult AChRs. AChRs are pentamers comprised of five subunits. Both embryonic and adult AChRs have four subunits in common: two α, one β, and one δ. However, the fifth subunit in embryonic AChRs is a γ subunit, whereas in adult AChRs it is an ε subunit (**Figure 9.15**). In rodents, the switch in subunit composition takes place during the first postnatal weeks when the expression of embryonic γ subunits is down-regulated and the expression of the ε subunits begins. The reasons for this shift in subunit composition are not entirely clear, but it appears that γ and ε subunits have different functions. For example, the γ subunit in embryonic muscle appears to be more effective in depolarizing the smaller embryonic muscle cells. The embryonic γ subunits may also influence the prepatterning of AChRs prior to the arrival of the nerve terminal. In mice lacking only the γ subunit, the normal AChR prepatterning does not occur. However, once the shift to the production of the ε subunit begins in these mice, the AChRs begin to accumulate near the central region of the muscle.

At both embryonic and adult stages, the genes for the specific AChR subunits are expressed in nuclei concentrated adjacent to the nerve terminal. Thus, the arrival of the nerve terminal not only induces migration and clustering of AChRs, but also stimulates the migration and accumulation of nuclei that express genes for the AChR subunits. The nuclei below the site of nerve terminal contact are called **subsynaptic nuclei**, whereas remaining nuclei outside the region of innervation are called **extrasynaptic nuclei** (**Figure 9.16**). How the nuclei migrate to the site of nerve contact is not yet understood.

As noted previously, the release of ACh by the innervating nerve terminal and the resulting depolarization of the muscle fiber represses AChR subunit transcription in the extrasynaptic nuclei (Figure 9.16). This repression means that AChRs are not replaced during normal receptor turnover and receptors in the extrasynaptic regions of the muscle are

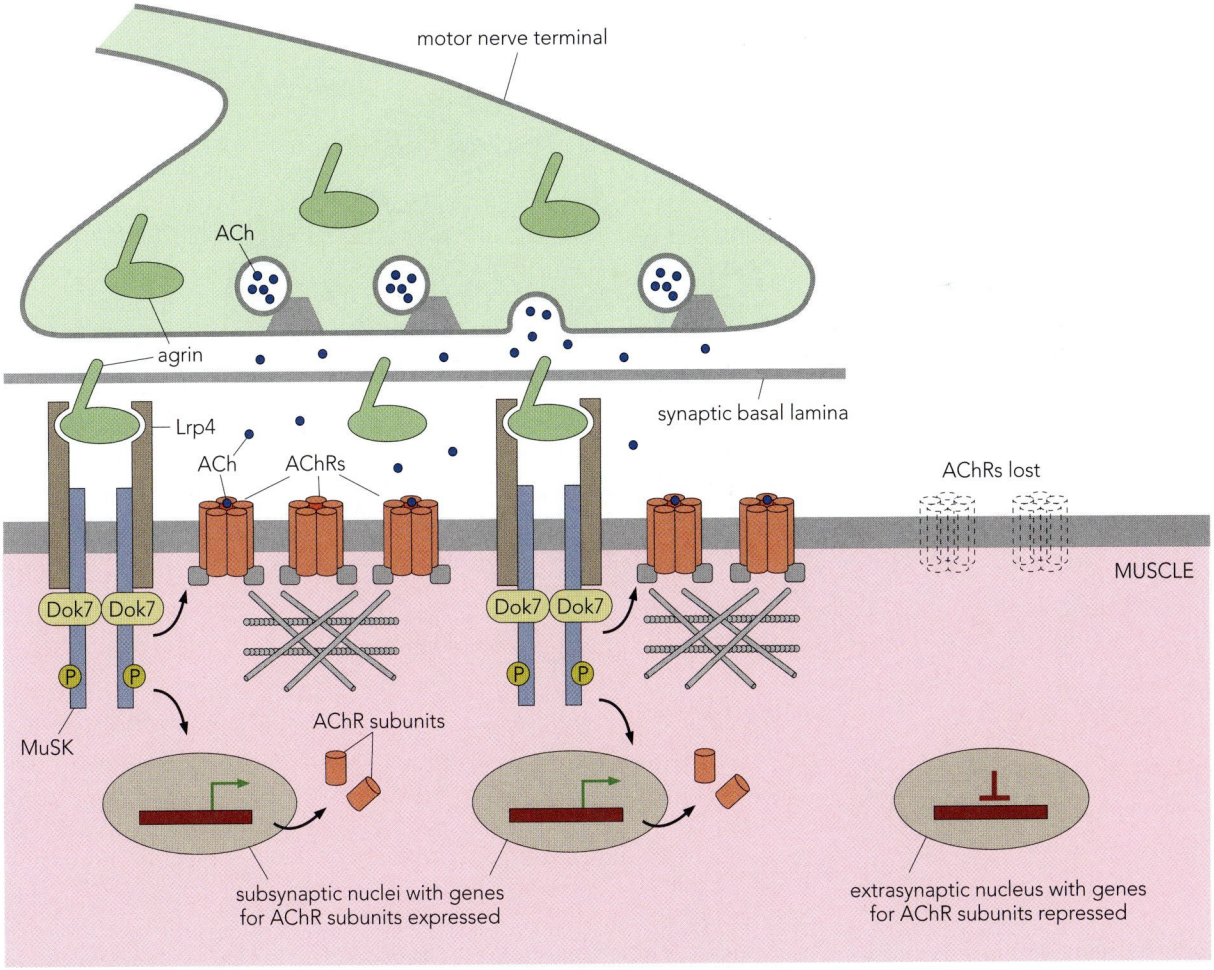

Figure 9.16 Genes for AChR subunits are selectively expressed in subsynaptic nuclei. In addition to agrin's role in clustering AChRs, activation of MuSK also contributes to the synthesis of AChR subunits in subsynaptic nuclei. In contrast, ACh released by the nerve terminal represses transcription of AChR subunit genes in the extrasynaptic nuclei, so receptors located in extrasynaptic regions of the muscle are eventually lost.

eventually lost. However, repression of AChR subunit expression does not occur in the subsynaptic nuclei. Instead, transcription of AChR subunits continues in these regions, leading to the local production of receptors at the nerve terminal. Although the mechanisms governing the accumulation of subsynaptic nuclei at the site of nerve terminal contact are still being investigated, the local production of AChR subunits is currently thought be influenced by the agrin concentrated in the synaptic basal lamina. The agrin signals through Lrp4-MuSK to help mediate gene transcription in the subsynaptic nuclei. Thus, it appears that agrin not only induces receptor clustering and prevents AChR dispersal in response to ACh, but also influences the local production of receptor subunits at the synaptic site.

Other signals may also help regulate synthesis of AChR subunits. One early candidate for regulating synthesis of AChR was isolated from spinal cord and brain extracts in the 1990s. This protein was originally termed ARIA for acetylcholine receptor-inducing activity, and it was subsequently found to be Neuregulin-1 (Nrg1), a protein discussed in Chapter 6 for its roles in glial cell development. ARIA/Nrg1 is produced by the motor neuron and transported down the nerve terminal where it is released. Nrg1 binds to ErbB receptors, which are located on muscle, nerve, and Schwann cells (**Figure 9.17**). Initial *in vitro* studies suggested that Nrg1 increased

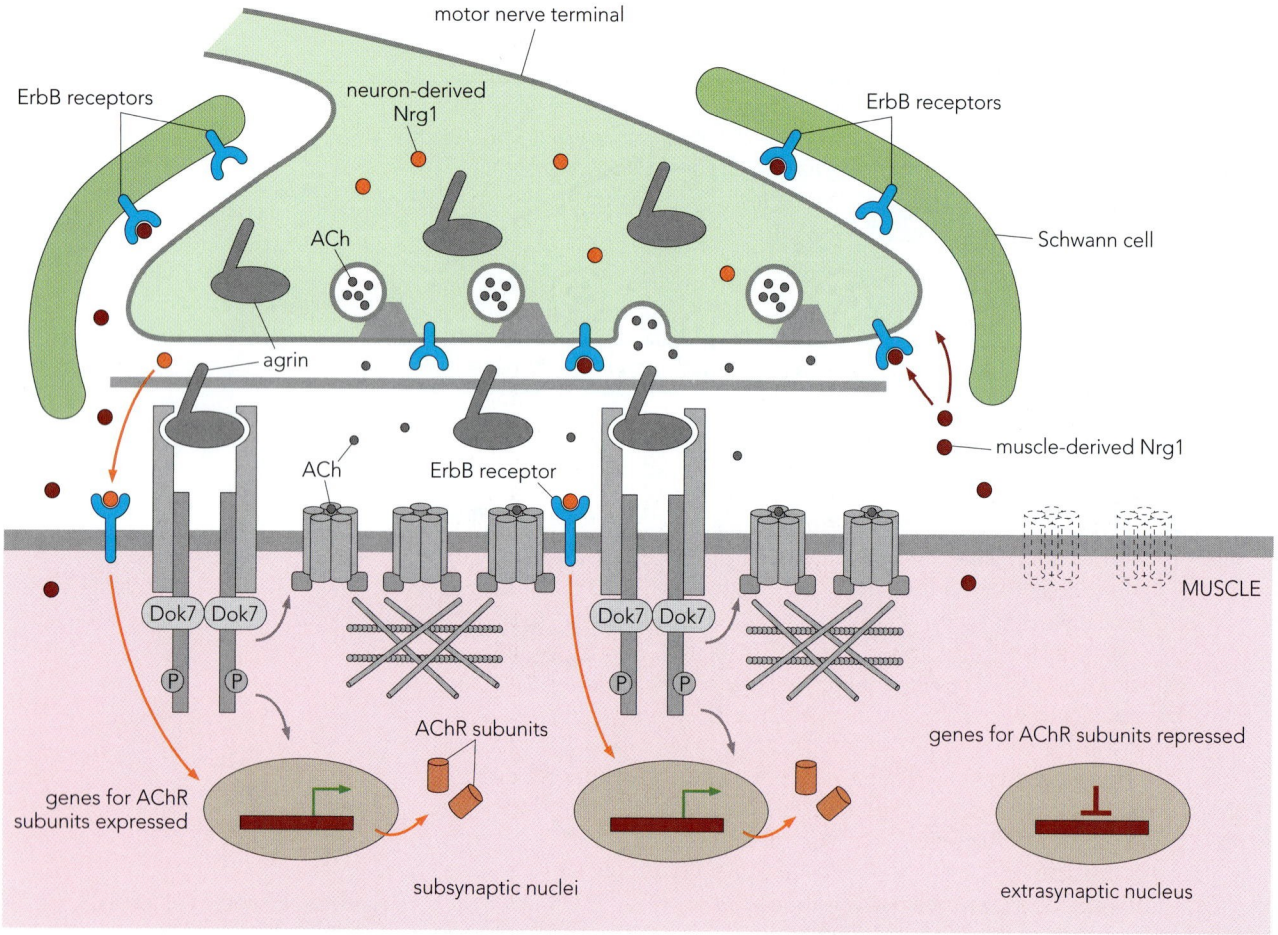

Figure 9.17 Neuregulin-1 may also influence the synthesis of AChR subunits in subsynaptic nuclei. Neuregulin-1 (Nrg1) produced by the motor neuron is believed to bind to ErbB receptors on the muscle fiber and interact with agrin to enhance the synthesis of AChR subunits in subsynaptic nuclei (bright red arrows). However, the mechanism by which this occurs is still not clear. Further, Nrg1 is released by muscle and ErbB receptors are also found on motor neurons and perisynaptic Schwann cells (dark red arrows). Muscle-derived Nrg1 activates ErbB receptors on Schwann cells to promote their survival. Schwann cells, in turn, help maintain the NMJ. Thus, the Nrg1 pathway may mediate multiple aspects of NMJ formation and maintenance.

mRNA transcript levels for the α, δ, and ε subunits of AChR. Thus, for many years Nrg1 was thought to be the nerve-derived signal for the synthesis of AChR subunits. In recent years, however, the proposed role of Nrg1 has been revised.

A role for Nrg1 in AChR synthesis *in vivo* has been more difficult to ascertain, leading to continued debate about its function during normal development. Mice lacking the gene for *Nrg1* are embryonic lethal and die prior to the formation of the NMJ. Thus, the role of Nrg1 during embryonic development cannot be determined in these mice. Therefore, conditional knockout mice were generated in which ErbB receptors were only deleted in muscle, but not other cell types. Neuromuscular synapses still formed in these mice, suggesting that Nrg1 signaling through ErbB receptors on muscle fibers was not required for the synthesis of AChR subunits and NMJ formation. Because the ErbB receptors remained on motor neurons and Schwann cells, Nrg1 released by either motor neurons or muscle cells could have activated these receptors to stabilize the NMJ (Figure 9.17). One hypothesis is that Nrg1 may regulate the survival, proliferation, or migration of Schwann cells, which are also important for NMJ development. Therefore, loss of ErbB receptors only on muscle fibers would not eliminate NMJ formation.

The current working model suggests agrin and Nrg1 somehow interact to modulate transcription of the correct level of AChR subunits. Agrin appears to be the primary mediator of AChR synthesis in subsynaptic nuclei, but alone is insufficient to generate the number of receptors normally found at the NMJ. For example, one set of *in vivo* experiments found that agrin injected into adult muscle, at a site away from the active NMJ, induced ectopic postsynaptic specializations such as AChRs and their associated anchoring proteins. However, the concentration of AChRs produced was lower than at the original synapse. These findings supported the idea that additional signals are needed for full transcriptional activation of AChR subunits. Nrg1 may be one such signal.

Perisynaptic Schwann cells play roles in NMJ synapse formation and maintenance

The first studies of the NMJ focused on how motor neurons and muscle cells interacted to regulate formation and stabilization of pre- and postsynaptic elements. As scientists uncovered various ways that the synaptic partner cells interact with each other, it became apparent that a third element of the synapse was also critical for successful synaptogenesis. The synaptic glial cells—Schwann cells in the PNS and astrocytes in the CNS—are now known to have important roles in synaptogenesis, synapse maintenance, and synaptic remodeling.

In mammals, Nrg1 promotes survival of all Schwann cells, and the survival of **perisynaptic Schwann cells** (**PSCs**) is required for maintenance of a functional NMJ. The PSCs of the NMJ are distinct from the myelinating Schwann cells along the motor axon. The PSCs are non-myelinating glial cells that form a cap around the motor axon terminals (see Figures 9.2 and 9.9). There are typically three or four PSCs at each NMJ, depending on the size of the synapse. Studies in frogs and mammals suggest that the PSCs generally are not needed for the initial guidance of the nerve terminal to the muscle cell. Once the nerve terminal arrives, however, the PSCs appear to provide signals necessary to stabilize and maintain the synaptic connection. For example, studies in frog in which PSCs were selectively ablated revealed that the NMJ could not remain functional for more than several hours. Instead, over the course of a week, the nerve terminal withdrew and the NMJ was lost in the absence of the PSC.

In vitro studies further demonstrated the importance of Schwann cells in maintaining the NMJ. Frog Schwann cells grown in cell culture released factors that induced synaptogenesis when added to co-cultures of frog neurons and muscle cells. The addition of the Schwann-cell-conditioned medium led to an increase in synaptic contacts between the co-cultured motor neurons and muscle cells, as well as increased clustering of AChRs in the muscle cells. Subsequent investigations revealed that the Schwann cells release transforming growth factor β1 (TGFβ1). In both frog and rat, TGFβ1 is detected in Schwann-cell-conditioned medium and expressed in PSCs. Addition of purified TGFβ1 to co-cultures of neurons and muscle cells increased synaptic contacts and AChR clustering similar to Schwann-cell-conditioned medium. Further, these effects on NMJ formation were inhibited when TGFβ1 blocking antibodies or binding proteins were added to Schwann-cell-conditioned medium, or when the receptor for TGFβ1 was blocked with a specific kinase inhibitor. Together, these studies demonstrated the importance of Schwann-cell-derived TGFβ1 in promoting synaptogenesis. TGFβ1 was also noted to increase agrin expression in rat spinal motor neurons, suggesting a possible mechanism underlying the increased AChR clustering seen in the cultured muscle cells (**Figure 9.18**). Thus, PSCs provide essential signals that regulate multiple aspects of synaptogenesis at the NMJ.

The synaptic basal lamina concentrates laminins needed for presynaptic development and alignment with postjunctional folds

As first noted in the studies by McMahon and colleagues in the 1980s, the synaptic basal lamina contains proteins that organize both presynaptic and postsynaptic specializations. While considerable progress has been made in identifying proteins in the synaptic basal lamina that regulate development of postsynaptic elements, comparatively less is known about signals that regulate presynaptic development. This was due in part to the lack of an assay to directly assess development of a particular presynaptic element similar to that developed for monitoring AChR clustering in muscle fibers. In recent years, however, as the number of identified presynaptic proteins has increased, methods for labeling some of these proteins have been developed, thus allowing researchers to establish cell culture assays to monitor how presynaptic elements are formed and maintained in response to signals originating in the synaptic basal lamina.

Like all basal lamina, the synaptic basal lamina contains collagen, entactin, and proteoglycans. Additionally, the synaptic basal lamina

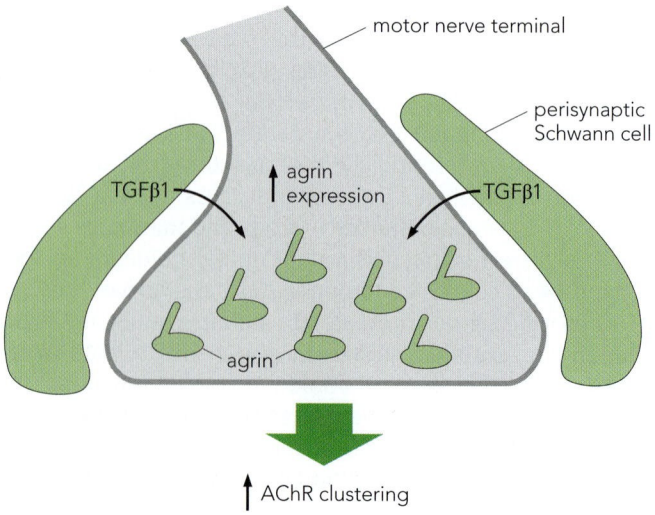

Figure 9.18 TGFβ1 is released by perisynaptic Schwann cells to support the NMJ. The TGFβ1 produced by perisynaptic Schwann cells is thought to increase expression of agrin, thus leading to increased AChRs clustering in the postsynaptic cell and the stabilization of NMJ contacts.

contains AChE, agrin, neuregulin, and specific laminin isoforms that differ from those in the extrasynaptic basal lamina. Laminins are currently among the best-characterized signals that influence presynaptic development. Several distinct isoforms of laminin are prominent in this region, including laminin-4, laminin-9, and laminin-11. As noted in previous chapters, laminins are a large family of proteins that interact through heterophilic binding. Isoforms of the proteins differ in the composition of various α, β, and γ chains. Laminins found in the synaptic basal lamina differ from extrasynaptic laminins based on the composition of their β chains, while the laminin isoforms concentrated in the synaptic basal lamina differ from each other based on their α-chain subunits. For example, the chain composition of extrasynaptic laminin-1, found mainly on the muscle surface, is α1, β1, and γ1. Based on the specific α, β, and γ subunit composition, laminin-1, the first laminin discovered, is now often referred to as laminin-111. Laminins concentrated in the synaptic basal lamina include laminin-9, laminin-11, and laminin-4, which differ in their α subunits. The laminin-9 chain is comprised of α4, β2, and γ1; the laminin-11 chain is comprised of α5, β2, and γ1 subunits; and the laminin-4 chain is comprised of α2, β2, and γ1 subunits. Thus, laminin-9 is also called laminin-421, laminin-11 is called laminin-521, and laminin-4 is called laminin-221.

It now appears that each laminin in the synaptic basal lamina has a specific function to ensure proper alignment of pre- and postsynaptic elements at the NMJ (**Table 9.1**). Laminin-9 is necessary to localize the presynaptic active zones above the postjunctional folds, whereas laminin-11 appears to participate in forming the presynaptic terminals and restricting the growth of PSCs so that their processes do not enter the synaptic cleft. In contrast, laminin-4 is needed for development of the postjunctional folds on the postsynaptic muscle surface.

As noted in previous chapters, laminins can bind to various combinations of integrin receptors located on neuronal cell surfaces. In the early part of the twenty-first century, researchers made the surprising discovery that the laminins containing the β2 subunit can also specifically bind to the pore-forming subunits of the subtypes of voltage-dependent

Table 9.1 Signals for NMJ formation

Signals regulating postsynaptic elements	Localization	Function
Acetylcholine	Released by presynaptic motor nerve terminal	Disperses AChRs
Agrin (Z^+ isoform)	Released by presynaptic motor nerve terminal	Stabilizes clusters of AChR below nerve terminal; regulates synthesis of AChR subunits in subsynaptic nuclei
Lrp4	Located in muscle fiber membrane	Binds agrin
Laminin-4	Concentrated in synaptic basal lamina	Influences development of postjunctional folds
MuSK	Located in muscle fiber membrane	Associates with Lrp4; mediates effects of agrin
Neuregulin	Released by motor nerve terminal and muscle	Influences survival of Schwann cells; plays a role in synthesis of AChR subunits
Rapsyn	Co-localized with AChRs in muscle	Interacts with MuSK; links AChRs to cytoskeleton
Signals regulating presynaptic elements	**Localization**	**Function**
Laminin-9	Concentrated in synaptic basal lamina	Organizes active zones above postjunctional folds
Laminin-11	Concentrated in synaptic basal lamina	Restricts growth of Schwann cell processes; influences growth of presynaptic terminals
Lrp4	Located in muscle fiber membrane	May influence presynaptic differentiation independent of MuSK

Ca^{2+} channels concentrated in the presynaptic motor nerve terminal. Interestingly, these laminins do *not* bind to the subtypes of voltage-dependent Ca^{2+} channels located *outside* of the nerve terminal. Thus, laminins concentrated in the synaptic basal lamina are able to selectively organize and stabilize presynaptic elements. More recently, it has been observed that the intracellular domains of the voltage-dependent Ca^{2+} channels interact with proteins associated with the active zone protein complex. One such protein is Bassoon, a cytosolic scaffolding protein located at the active zone complex. Thus, the synaptic basal lamina-concentrated laminins directly bind to and stabilize the location of the presynaptic voltage-dependent Ca^{2+} channels that are linked, in turn, to the components of the active zone protein complex (**Figure 9.19**). The importance of this laminin-voltage-dependent Ca^{2+} channel signaling was demonstrated in various knockout mice. In mice lacking genes for either β2 laminin or the presynaptic voltage-dependent Ca^{2+} channels, a decrease in the number of active zones in the presynaptic nerve terminal was observed. Because active zones line up adjacent to postjunctional folds, linking voltage-dependent Ca^{2+} channels to active zone proteins ensures that Ca^{2+} influx leads to the localized release of ACh above the AChRs clustered at the top of the postjunctional folds.

Other signals regulate presynaptic differentiation as well. Among the candidate molecules identified to date are members of the fibroblast growth factor family (FGF). FGFs are released by muscle cells and bind to corresponding FGF receptors on the presynaptic terminal, where they appear to regulate aspects of synaptic vesicle clustering. Lrp4 is also expressed on presynaptic terminals, where it may influence synaptic vesicle clustering, although conclusive evidence of this role has not yet been confirmed. As described in Chapter 10, many other molecules have been identified that regulate the differentiation of presynaptic elements in the CNS. Some of these may also play a role in NMJ development.

MODELS OF SYNAPTIC ELIMINATION IN THE NMJ

As mentioned at the beginning of this chapter, the formation of mature synaptic contacts is a prolonged and dynamic process that includes a period in which some synaptic connections are eliminated and others are strengthened. The time course over which this happens for NMJ synapses has been well documented. Initially, during embryogenesis, each motor

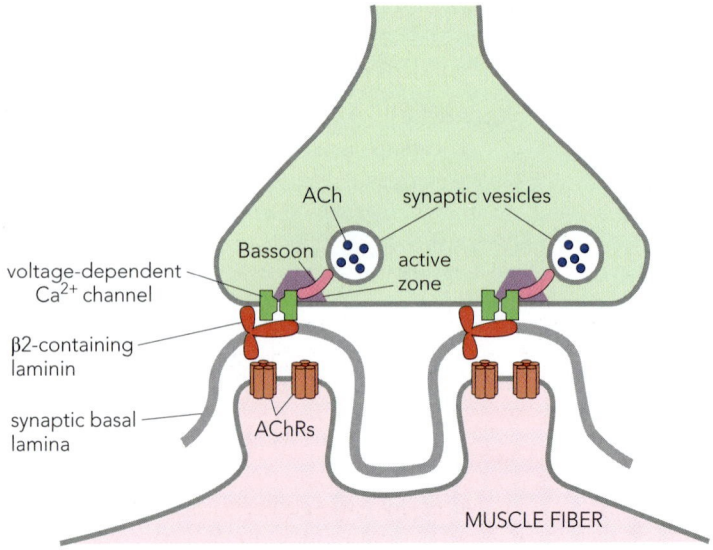

Figure 9.19 Laminins concentrated in the synaptic basal lamina bind to voltage-dependent Ca^{2+} channels in the presynaptic motor nerve terminal. Laminins containing the β2 subunit bind to the pore-forming subunits of voltage-dependent Ca^{2+} channels in the presynaptic nerve terminal. The intracellular regions of the Ca^{2+} channels also interact with the cytosolic scaffolding protein Bassoon that is associated with the active zone protein complex. In this way, the laminins in the synaptic basal lamina are able to organize the presynaptic elements above AChRs clustered at the postjunctional folds on the postsynaptic muscle fiber.

neuron innervates more than one muscle fiber, and each muscle fiber is innervated by multiple presynaptic terminals at a single location (see Figure 9.6). In rodents, the subsequent process of synaptic elimination typically occurs during the first three postnatal weeks. During this stage, all but one of these presynaptic terminals is lost. As in other regions of the nervous system, neural activity in the presynaptic neuron helps regulate which synapses stabilize and which are lost.

While the role of neural activity in regulating synapse stabilization and elimination is now well established, it took many years to determine how this occurred. Since the first studies conducted in the 1970s, several seemingly conflicting results were reported. In some experiments, for example, increased neural activity was observed to increase the extent of synapse elimination. In other experiments, however, increased neural activity led to the stabilization of NMJ connections. In still other experiments, blocking all neural activity stabilized NMJ connections and prevented synapse elimination. Over time, these seemingly conflicting results were reconciled by evidence that synapse elimination depended on *differences* in the relative level of neural activity in presynaptic neurons and muscle fibers.

The relative levels of neuromuscular activity determine which terminal branches remain at the endplate

A number of experiments conducted by Jeff Lichtman and colleagues illustrated how differences in relative levels of neural activity influence synaptic elimination in the rodent NMJ. In one set of studies, the scientists applied the AChR-binding antagonist α-bungarotoxin at a singly innervated NMJ. The toxin blocks neural transmission through AChRs and was applied via microelectrodes to either inhibit neural transmission at a restricted region of the endplate or across the entire endplate (**Figure 9.20**). When this inhibitor was applied to a limited region of the endplate, the motor nerve terminal branches and AChRs at that site were lost, while contacts at the untreated region of the endplate remained intact. In contrast, when the entire NMJ was inhibited, no synapses were eliminated. Thus, synapses were eliminated only when there were differences in levels of neurotransmission between different terminal branches at a single endplate.

Several models have been suggested over the years to account for this effect. Most note there must be different signals released by the pre- or postsynaptic cell in response to neural activity. These are often described as "punishment" or toxic signals and "protective" or trophic signals. For example, one early model suggested that the muscle released a protease, whereas the nerve terminal released a protease inhibitor. If activity in the nerve and muscle occurred at the same time, both signals would be released at the same time. Therefore, the protease inhibitor (protective signal) would block the effects of the protease (punishment signal) and the synapse would remain (**Figure 9.21**A). If the activity of the pre- and postsynaptic cells did *not* match, however, then the signals would be released at different times. If muscle activity was greater, more "punishment" signal would be released and there would not be enough "protective" signal released by the nerve terminal. Thus, the synapse would not be protected and the nerve terminal would withdraw (Figure 9.21B). If neural activity was absent in both cells, then neither signal would be released, and all synaptic contacts would remain (Figure 9.22C).

BDNF and pro-BDNF are candidates for the protective and punishment signals

The protective/punishment model provided a framework for continued studies of how differences in neural activity regulate synapse elimination

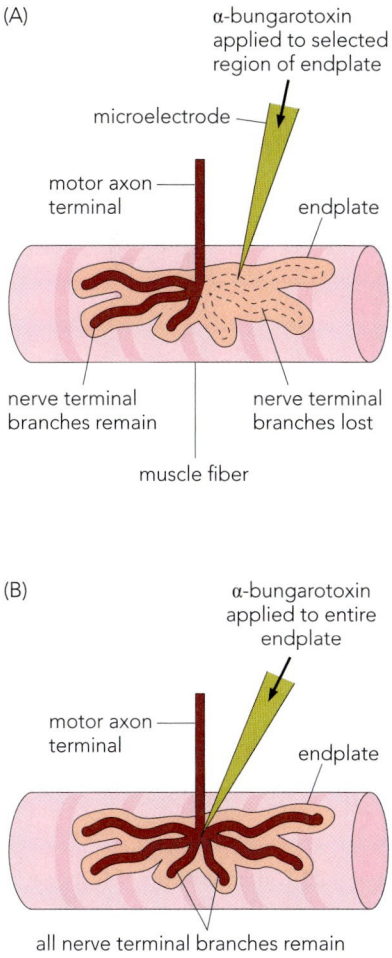

Figure 9.20 Differences in neural activity influence synapse elimination and stabilization. α-Bungarotoxin was used to inhibit neural transmission at the NMJ. (A) When α-bungarotoxin was injected into a small region of the motor endplate, only the nerve terminals at the site of inhibition were lost. (B) When α-bungarotoxin was applied to the entire endplate, all of the nerve terminals remained. This indicated that synapse elimination only occurred when there were differences in the amount of neural transmission between the nerve terminal branches.

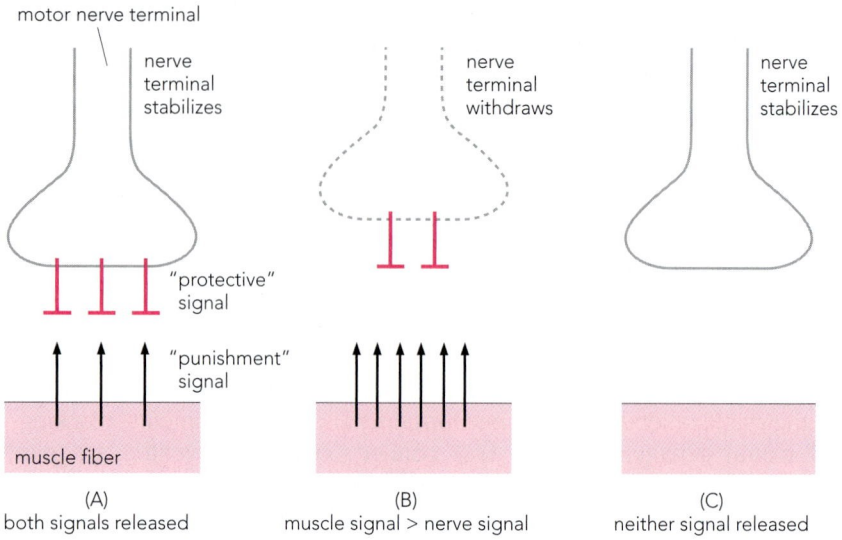

Figure 9.21 Model of activity-dependent synapse elimination in the NMJ. Synapse elimination in the NMJ results when there are differences in the activity levels of the motor neuron and muscle fiber. One model to explain this effect proposed that muscle released a "punishment" signal that could damage the motor nerve terminal and initiate its withdrawal. The motor neuron could prevent nerve terminal withdrawal, however, by releasing a "protective" signal to block the effect of the muscle-derived signal. (A) When both the motor neuron and muscle are active at the same time, both the punishment and protective signals are released, and the motor nerve terminal is stabilized. (B) If the activity levels of the motor neuron and muscle cell are asynchronous, the signals are released at different times. In this example the muscle activity is greater and therefore the amount of punishment signal released is greater. Thus, the protective signal is insufficient to counteract the punishment signal and the nerve terminal withdraws. (C) If the motor neuron and muscle cell are inactive at the same time, neither signal is released. Therefore, the synapse remains.

and stabilization at the NMJ and perhaps other synapses in the PNS and CNS as well. A number of studies in the CNS and PNS have demonstrated that BDNF is released by target cells in response to neural activity. As noted in Chapter 8, neurotrophins are initially synthesized as precursor forms (pro-neurotrophins) that are then cleaved to mature forms. Pro-neurotrophins bind to the p75NTR-sortilin receptor complex, whereas mature neurotrophins bind to Trk receptors. Recent studies in frog and mouse suggest that different forms of BDNF serve as the punishment and protective signals at the NMJ. Based on a series of *in vitro* and *in vivo* experiments, a model emerged in which pro-BDNF is released by muscle cells in an activity-dependent manner and innervating nerve fibers express both p75NTR-sortilin and Trk B receptors. When the released pro-BDNF binds to and signals through the p75NTR-sortilin receptor complex on motor axons, those innervating nerve fibers retract. However, neural activity also initiates secretion of proteases from the nerve terminal that cleave muscle-derived pro-BDNF to its mature form. The mature BDNF then activates the TrkB receptors on the remaining nerve fibers to stabilize those synaptic connections (**Figure 9.22**). Thus, protease release in response to

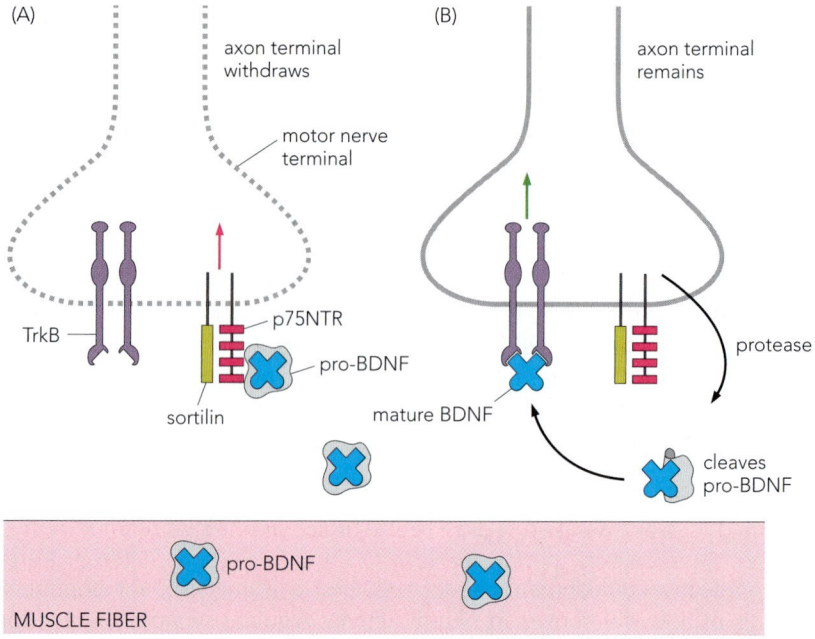

Figure 9.22 Different forms of BDNF may provide both a punishment and protective signal. (A) Pro-BDNF is released by muscle fibers in an activity-dependent manner. When pro-BDNF binds to p75NTR-sortilin receptors on innervating motor nerve terminals, it initiates signal transduction pathways that lead to nerve terminal withdrawal. (B) Neural activity causes proteases to be released by the nerve terminal. These proteases cleave the pro-BDNF to generate mature BDNF that then binds to the TrkB receptors also located on innervating motor nerve terminals. This signaling pathway stabilizes the nerve contact. Thus, the nerve-derived protease converts the muscle-derived punishment signal into a muscle-derived protective signal.

neural activity cleaves the pro-BDNF to mature BDNF. Cleaving pro-BDNF to the mature form changes the muscle-derived pro-BDNF punishment signal mediated by the p75NTR-sortilin receptors to a protective signal that then activates TrkB receptors on the nerve terminals. In this model, the protease serves as the factor that converts the "punishment" signal to a "protective" signal. Those motor axons that do not have the protective protease signal released will be eliminated. In this way, neural activity determines which axons respond to BDNF and therefore, which axons are stabilized.

In vivo studies confirm the importance of these two signaling pathways. In one set of experiments, p75NTR knockout mice were injected with a sortilin propeptide that inhibits the binding of pro-BDNF. This approach rendered both p75NTR and sortilin receptors unable to transduce signals. In the absence of p75NTR-sortilin signaling, more synapses remained. Conversely, in other experiments in which TrkB signaling was inhibited, greater synaptic elimination occurred. Further, inhibiting the cleavage of pro-BDNF by injecting a metalloprotease inhibitor into muscle fibers, also increased synaptic elimination by preventing formation of mature BDNF and the activation of TrkB receptors. Thus, the balance of available pro-BDNF and BDNF signaling pathways regulated by neural activity appears to contribute to the extent of synaptic elimination. Researchers continue to investigate how these events are regulated at the NMJ and whether similar events occur in CNS synapses.

Perisynaptic Schwann cells influence the stability of synaptic connections

In recent years, the role of PSCs was found to be critical to a number of aspects of NMJ synapse formation and refinement. The PSCs not only release signals to support synaptic contacts (see Figure 9.19), but also are active during synapse elimination. PSCs phagocytize and remove the cellular debris associated with eliminated axon terminals. It may be that the PSCs randomly choose terminals to remove or they may be able to detect which synaptic contacts are less active and selectively remove only those synapses. PSCs, like other glial cells, express receptors for neurotransmitters. PSCs express both muscarinic AChRs and purinergic receptors that are activated by the neurotransmitters ACh and adenosine-5′-triphosphate (ATP), respectively. Both ACh and ATP are released during the stages of NMJ formation and elimination. Thus, PSCs have the means to monitor the activity of presynaptic terminals and influence which synapses are stabilized and which are lost.

SUMMARY

To establish a functional vertebrate NMJ, reciprocal signals are exchanged between pre- and postsynaptic cells. Both the presynaptic motor nerve terminal and the developing myotube express synaptic elements prior to making initial contact. Once contact is initiated, however, signals are produced to form a stable, mature synaptic connection. For example, agrin is released from the nerve terminal to bind to the Lrp4 receptor that interacts with the MuSK receptor on the muscle. This leads to clustering or aggregation of AChRs below the presynaptic nerve terminal and prevents ACh-induced dispersal of AChR. Several molecules, including rapsyn, help anchor the receptors in place by linking them to the cytoskeleton. Agrin, signaling through MuSK, also regulates synthesis of AChR subunits in subsynaptic nuclei.

In addition to agrin, the nerve also releases Neuregulin (Nrg-1), which binds ErbB receptors to further contribute to maintenance of the NMJ and possibly to localized synthesis of AChRs. Laminin signals in the nerve terminal help organize the active zones to align directly above the postjunctional folds. Together, all of these events result in a highly organized NMJ, with synaptic vesicles located above the AChRs clustered at the crest of the postjunctional folds. The numerous molecular signals and the precise timing of synaptogenesis events give rise to a synapse structured to provide rapid and reliable neural transmission from motor neurons to skeletal muscle. Not all NMJ connections that form during development remain into adulthood, however differences in neural and muscle activity at individual NMJs appear to regulate signals, such as BDNF, that determine which synapses are lost and which are maintained. PSCs release signals to support the remaining NMJ and also participate in the removal of cellular debris from the lost connections.

Since the initial discoveries of agrin, MuSK, ARIA/Nrg-1, and rapsyn over 20 years ago, there have been considerable advances in our understanding of the signaling pathways used to regulate clustering and stabilization of AChRs in the postjunctional folds of the NMJ. Despite a vast amount of new knowledge, many questions remain about how the different proteins interact at different stages of normal NMJ development and in different disease conditions (**Box 9.2**). Scientists continue to investigate these mechanisms and have already identified additional molecules, beyond those described here, that appear to interact with MuSK and other postsynaptic molecules. Given the numerous signaling pathways identified to date and the availability of well-established assays for studying NMJ differentiation and synapse elimination, scientists will likely continue to expand the understanding of how the vertebrate NMJ is formed and reorganized.

Box 9.2 Post-polio syndrome: When new branches of motor nerve terminals can no longer function

During the 1940s and 1950s, one of the most feared diseases in the United States was poliomyelitis, a viral infection that causes degeneration of motor neurons in the spinal cord and brain stem. With the introduction of the first polio vaccine in 1955, polio was eventually eradicated from the United States. By the 1990s, polio was eradicated from most other countries of the world as well, though several cases continue to be found each year in a limited number of countries.

Although often thought of as a disease that always results in paralysis or death, the outcomes of a polio infection are quite variable. Some individuals experience only a few mild symptoms. Others experience muscle weakness, whereas others develop paralysis. Many of those who initially experience muscle weakness go on to recover muscle function. This recovery appears to be due to the branching of motor nerve terminals from healthy motor neurons to the nearby muscle fibers that lost innervation when infected motor neurons died. Thus, the new branches compensate for the lost motor neurons.

Post-polio syndrome (PPS) is a condition that appears three to four decades after the initial infection with the polio virus. The new symptoms include gradually progressive muscle weakness and muscle atrophy. The cause of PPS is unclear, but it appears the new connections made after the initial infection gradually become overworked and can no longer handle the increased demands. These motor neurons or their connections then begin to degenerate, causing the progressive weakness and atrophy noted by patients. At this time there is no cure for PPS and treatments rely primarily on physical and occupational therapy to maintain functional connections as long as possible. As scientists continue to study NMJ formation and elimination under normal conditions, insights into the causes of PPS may also emerge.

FURTHER READING

Axelrod D (1983) Lateral motion of membrane proteins and biological function. *J Membr Biol* 75:1–10.

Balice-Gordon RJ & Lichtman JW (1993) In vivo observations of pre- and postsynaptic changes during the transition from multiple to single innervation at developing neuromuscular junctions. *J Neurosci* 13:834–855.

Balice-Gordon RJ & Lichtman JW (1994) Long-term synapse loss induced by focal blockade of postsynaptic receptors. *Nature* 372:519–524.

Brown MC, Jansen JK & Van Essen D (1976) Polyneuronal innervation of skeletal muscle in new-born rats and its elimination during maturation. *J Physiol* 261:387–422.

Burden SJ, Sargent PB & McMahan UJ (1979) Acetylcholine receptors in regenerating muscle accumulate at original synaptic sites in the absence of the nerve. *J Cell Biol* 82:412–425.

Chen PJ, Martinez-Pena YVI, Aittaleb M & Akaaboune M (2016) AChRs are essential for the targeting of rapsyn to the postsynaptic membrane of NMJs in living mice. *J Neurosci* 36:5680–5685.

Culican SM, Nelson CC & Lichtman JW (1998) Axon withdrawal during synapse elimination at the neuromuscular junction is accompanied by disassembly of the postsynaptic specialization and withdrawal of Schwann cell processes. *J Neurosci* 18: 4953–4965.

Duxson MJ, Usson Y & Harris AJ (1989) The origin of secondary myotubes in mammalian skeletal muscles: ultrastructural studies. *Development* 107:743–750.

Feng Z & Ko CP (2008) Schwann cells promote synaptogenesis at the neuromuscular junction via transforming growth factor-beta1. *J Neurosci* 28:9599–9609.

Gautam M, Noakes PG, Moscoso L et al. (1996) Defective neuromuscular synaptogenesis in agrin-deficient mutant mice. *Cell* 85:525–535.

Ghazanfari N, Fernandez KJ, Murata Y et al. (2011) Muscle specific kinase: organiser of synaptic membrane domains. *Int J Biochem Cell Biol* 43:295–298.

Glass DJ, Bowen DC, Stitt TN et al. (1996) Agrin acts via a MuSK receptor complex. *Cell* 85:513–523.

Glass DJ, Apel ED, Shah S et al. (1997) Kinase domain of the muscle-specific receptor tyrosine kinase (MuSK) is sufficient for phosphorylation but not clustering of acetylcholine receptors: required role for the MuSK ectodomain? *Proc Natl Acad Sci USA* 94:8848–8853.

Godfrey EW, Nitkin RM, Wallace BG et al. (1984) Components of Torpedo electric organ and muscle that cause aggregation of acetylcholine receptors on cultured muscle cells. *J Cell Biol* 99:615–627.

Gu Y & Hall ZW (1988) Characterization of acetylcholine receptor subunits in developing and in denervated mammalian muscle. *J Biol Chem* 263:12878–12885.

Hallock PT, Xu CF, Park TJ et al. (2010) Dok-7 regulates neuromuscular synapse formation by recruiting Crk and Crk-L. *Genes Dev* 24:2451–2461.

Je HS, Yang F, Ji Y et al. (2012) Role of pro-brain-derived neurotrophic factor (proBDNF) to mature BDNF conversion in activity-dependent competition at developing neuromuscular synapses. *Proc Natl Acad Sci USA* 109:15924–15929.

Je HS, Yang F, Ji Y et al. (2013) ProBDNF and mature BDNF as punishment and reward signals for synapse elimination at mouse neuromuscular junctions. *J Neurosci* 33:9957–9962.

Kummer TT, Misgeld T & Sanes JR (2006) Assembly of the postsynaptic membrane at the neuromuscular junction: paradigm lost. *Curr Opin Neurobiol* 16:74–82.

Lieth E & Fallon JR (1993) Muscle agrin: neural regulation and localization at nerve-induced acetylcholine receptor clusters. *J Neurosci* 13:2509–2514.

Liu Y, Padgett D, Takahashi M et al. (2008) Essential roles of the acetylcholine receptor gamma-subunit in neuromuscular synaptic patterning. *Development* 135:1957–1967.

Lin S, Maj M, Bezakova G et al. (2008) Muscle-wide secretion of a miniaturized form of neural agrin rescues focal neuromuscular innervation in agrin mutant mice. *Proc Natl Acad Sci USA* 105:11406–11411.

McMahan UJ & Slater CR (1984) The influence of basal lamina on the accumulation of acetylcholine receptors at synaptic sites in regenerating muscle. *J Cell Biol* 98:1453–1473.

Meier T, Masciulli F, Moore C et al. (1998) Agrin can mediate acetylcholine receptor gene expression in muscle by aggregation of muscle-derived neuregulins. *J Cell Biol* 141:715–726.

Nishimune H, Sanes JR & Carlson SS (2004) A synaptic laminin-calcium channel interaction organizes active zones in motor nerve terminals. *Nature* 432:580–587.

Nishimune H (2012) Active zones of mammalian neuromuscular junctions: formation, density, and aging. *Ann NY Acad Sci* 1274:24–32.

Nitkin RM, Smith MA, Magill C et al. (1987) Identification of agrin, a synaptic organizing protein from Torpedo electric organ. *J Cell Biol* 105: 2471–2478.

Redfern PA (1970) Neuromuscular transmission in new-born rats. *J Physiol* 209:701–709.

Reist NE, Magill C & McMahan UJ (1987) Agrin-like molecules at synaptic sites in normal, denervated, and damaged skeletal muscles. *J Cell Biol* 105:2457–2469.

Sanes JR, Marshall LM & McMahan UJ (1978) Reinnervation of muscle fiber basal lamina after removal of myofibers. Differentiation of regenerating axons at original synaptic sites. *J Cell Biol* 78:176–198.

Sanes JR & Lichtman JW (1999) Development of the vertebrate neuromuscular junction. *Annu Rev Neurosci* 22:389–442.

Sanes JR & Lichtman JW (2001) Induction, assembly, maturation and maintenance of a postsynaptic apparatus. *Nat Rev Neurosci* 2:791–805.

Shi L, Fu AK & Ip NY (2012) Molecular mechanisms underlying maturation and maintenance of the vertebrate neuromuscular junction. *Trends Neurosci* 35:441–453

Smith MA, Yao YM, Reist NE et al. (1987) Identification of agrin in electric organ extracts and localization of agrin-like molecules in muscle and central nervous system. *J Experimtl Biol* 132:223–230.

Son YJ & Thompson WJ (1995) Schwann cell processes guide regeneration of peripheral axons. *Neuron* 14:125–132.

Tezuka T, Inoue A, Hoshi T et al. (2014) The MuSK activator agrin has a separate role essential for postnatal maintenance of neuromuscular synapses. *Proc Natl Acad Sci USA* 111:16556–16561.

Valenzuela DM, Stitt TN, DiStefano PS et al. (1995) Receptor tyrosine kinase specific for the skeletal muscle lineage: expression in embryonic muscle, at the neuromuscular junction, and after injury. *Neuron* 15:573–584.

Walsh MK & Lichtman JW (2003) In vivo time-lapse imaging of synaptic takeover associated with naturally occurring synapse elimination. *Neuron* 37:67–73.

Witzemann V, Barg B, Criado M et al. (1989) Developmental regulation of five subunit specific mRNAs encoding acetylcholine receptor subtypes in rat muscle. *FEBS Lett* 242:419–424.

Wu H, Xiong WC & Mei L (2010) To build a synapse: signaling pathways in neuromuscular junction assembly. *Development* 137:1017–1033.

Yang X, Arber S, William C et al. (2001) Patterning of muscle acetylcholine receptor gene expression in the absence of motor innervation. *Neuron* 30:399–410.

Zong Y & Jin R (2013) Structural mechanisms of the agrin-LRP4-MuSK signaling pathway in neuromuscular junction differentiation. *Cell Mol Life Sci* 70:3077–3088.

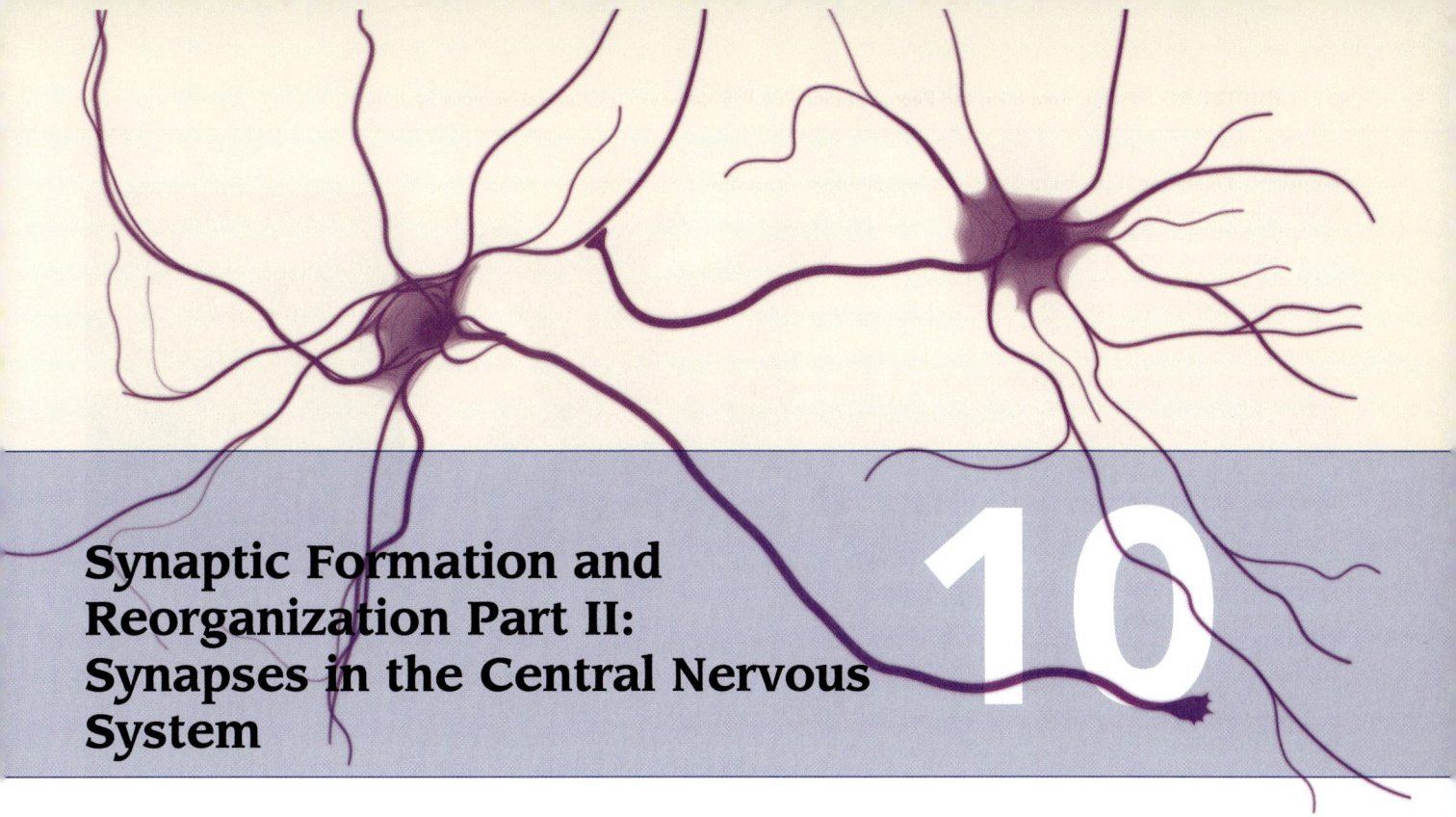

Synaptic Formation and Reorganization Part II: Synapses in the Central Nervous System

10

The formation of synapses in the central nervous system (CNS) begins early in embryogenesis and often continues through adulthood. In many vertebrate species, synapses produced during embryonic stages are later modified or eliminated as a result of intrinsic cues or neural activity. In many regions of the CNS, synapses continue to be reorganized throughout adulthood by a process called **neural plasticity**. Similar to the neuromuscular junction (NMJ) detailed in Chapter 9, synaptic formation and stabilization in the CNS relies on the exchange of reciprocal cues between presynaptic and postsynaptic partner cells.

Investigating synapse formation and reorganization in the brain and spinal cord presents many more challenges than those encountered studying the NMJ. As noted in Chapter 9, there are a number of reasons why the NMJ has proven to be an advantageous experimental system. Its relatively large size, experimental accessibility, structural simplicity, and use of a single neurotransmitter system (acetylcholine) are among the characteristics that have made the NMJ useful for studying synaptogenesis. In contrast, CNS synapses generally lack these same features. Typically, CNS synapses are smaller, less accessible, and more difficult to isolate from surrounding cells. In addition, CNS synapses can utilize one or more of the hundreds of different neurotransmitters and neuropeptides that have been identified. The small size and vast number of neurons in the CNS also make it difficult to isolate individual pre-and postsynaptic cells. In addition, one of the largest obstacles to studying specific synaptic connections in the CNS is that each neuronal cell body and dendrite is almost completely covered with nerve terminals from the numerous synaptic partners. Multiple glial cell processes also contact each neuron. The density of innervation was highlighted in a 1969 study of cat spinal motor neurons. In this study, a series of electron micrographs was collected so that the density of innervation on a single motor neuron could be documented (**Figure 10.1**A). This study was one of the first to clearly illustrate the numerous cellular contacts made at a single CNS neuron. In more recent years, fluorescent microscopy has been used to label synaptic elements and document synaptic contacts on CNS neurons under normal and experimental conditions (Figure 10.1B).

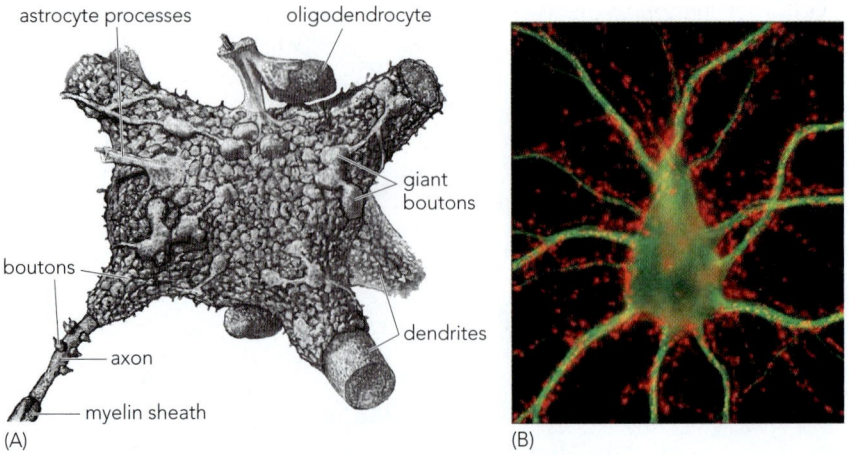

Figure 10.1 Neurons in the central nervous system (CNS) are contacted by multiple synaptic partners. (A) A drawing based on serial electron micrographs of a single adult cat spinal cord motor neuron reveals that the neuronal cell body is covered with multiple synapses (boutons and giant boutons) and processes from astrocytes. Each raised area on the neuron represents a point of synaptic or astrocyte contact. Dendrites and the axon also receive multiple synaptic contacts. This complexity is one reason it is more difficult to study the signals that regulate CNS synaptogenesis. (B) A fluorescent label specific for presynaptic terminals (red) reveals the large number of synaptic connections on the cell body and dendrites (green) of a single motor neuron. The axon is not stained in this image. (A, from Poritsky R [1969] *J Comp Neurol* 135(4):432-451; B, courtesy of Olaf Mundigl and Pietro de Camillo.)

Despite the challenges of isolating individual pre- and postsynaptic regions, a large number of synaptic specializations have been discovered in CNS neurons over the course of the past 25 years. It is quite remarkable that so many individual synaptic elements have been identified in the tiny cellular spaces that comprise the pre- and postsynaptic regions. Identifying these many structural elements and determining their multiple functions have allowed investigators to develop hypotheses to explain how synaptic formation, stabilization, and elimination take place in the CNS. Among the identified synaptic specializations are proteins necessary for clustering synaptic vesicles at the presynaptic active zone and proteins required for clustering neurotransmitter receptors at the postsynaptic membrane. Like the NMJ, having synaptic vesicles and neurotransmitter receptors closely associated ensures rapid and reliable synaptic transmission. In addition to identifying numerous synaptic proteins, scientists have begun to identify how different intracellular signaling pathways intersect to govern synapse formation and reorganization across the lifespan of an animal.

This chapter focuses on some of the common mechanisms that influence the formation and reorganization of synaptic elements in the CNS. Many of the same signals and intracellular signaling pathways used at other stages of neural development are also used during these processes. Among the molecules important for CNS synaptogenesis are cell adhesion molecules, brain-derived neurotrophic factor (BDNF), and members of the Wnt and Ephrin families. These molecules often interact with Rho GTPases or other proteins to influence cytoskeletal elements that in turn impact synapse stabilization or elimination. These signaling mechanisms are further examples of how the nervous system utilizes available signals rather than creates new proteins for each developmental event.

The chapter concludes with descriptions of processes that regulate synaptic reorganization and elimination in the CNS during early postnatal and adult stages. While synapse reorganization and elimination take place throughout the vertebrate CNS, examples from the mammalian visual system are highlighted because development of this system has been described in previous chapters and the system remains one of the most common model systems for studies of synaptic formation and reorganization.

EXCITATORY AND INHIBITORY NEURONS IN THE CENTRAL NERVOUS SYSTEM

Understanding the structure and function of the many different synaptic elements associated with excitatory and inhibitory synapses helps interpret the current models of CNS synapse formation and reorganization. At

excitatory synapses the release of the presynaptic neurotransmitter causes the postsynaptic neuron to depolarize and thereby increases the likelihood that the neuron will fire an action potential. Excitatory synapses are usually formed between a presynaptic terminal and a postsynaptic **dendritic spine**, the small protrusion that extends from the dendritic shaft. A smaller number of excitatory synapses form between the axon terminal and the dendritic shaft. The excitatory synapses that contact dendritic spines primarily form in one of two ways. A **terminal synapse** forms when the presynaptic nerve terminal contacts a dendritic spine. Such synapses are most like those seen in the vertebrate NMJ, in which a nerve terminal contacts a muscle fiber. In contrast, an ***en passant* synapse** forms when a region of the axon shaft enlarges and produces presynaptic specializations that contact a dendritic spine on the postsynaptic cell (**Figure 10.2**). These are similar in morphology to the synapses that form in the invertebrate nervous system.

Glutamate is the primary excitatory neurotransmitter in the CNS. Glutamate binds to several different glutamate receptors, including those that form ion channels (ionotropic) and those that use G proteins (metabotropic). Two ionotropic receptors that have been extensively studied during synaptogenesis are the NMDA (*N*-methyl-D-aspartic acid) and AMPA (alpha-amino-3-hydroxy-5-methyl-4-isoxazolepropionic acid) receptors. These names reflect the name of specific chemical agonists that bind to each of these glutamate receptors.

In addition to glutamate, some excitatory neurons package and release other neurotransmitters, such as acetylcholine (Ach), serotonin (5-hydroxytryptamine, 5-HT), dopamine (DA), or one of the various neuropeptides. In some cases, a single excitatory neuron will release more than one neurotransmitter. This typically involves glutamate and one of the modulatory neurotransmitters such as histamine, 5-HT, or DA. The combination of neurotransmitters helps determine whether an action potential will fire.

Inhibitory synapses are often formed at the dendritic shaft or near the postsynaptic cell body, although some synapses are also located at distal regions or spines of the dendrite. At inhibitory synapses, the release of the presynaptic neurotransmitter hyperpolarizes the postsynaptic neuron, thus decreasing the likelihood that the neuron will fire an action potential.

Most inhibitory neurons package and release GABA (gamma aminobutyric acid) or glycine—neurotransmitters that bind to corresponding receptors on the postsynaptic neuron. As with the excitatory neurons, neuromodulators may also be co-released with inhibitory neurotransmitters to regulate the response of the postsynaptic cell and determine whether an action potential will fire.

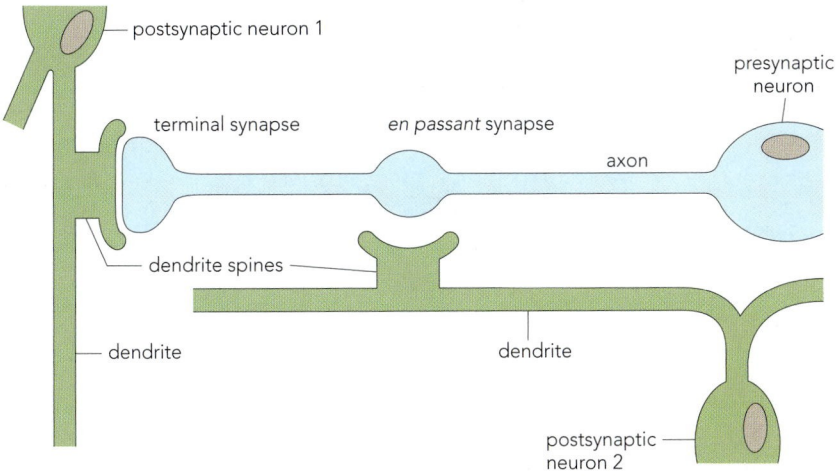

Figure 10.2 Organization of terminal and *en passant* synapses in the CNS. An excitatory terminal synapse forms between the end of the axon of the presynaptic neuron and a postsynaptic dendritic spine as shown at postsynaptic neuron 1. An *en passant* synapse occurs when presynaptic specializations form along the axon shaft and contact a dendritic spine (postsynaptic neuron 2). Inhibitory synapses are structurally similar, but typically form along the dendritic shaft.

Many presynaptic and postsynaptic elements are similar in excitatory and inhibitory synapses

The general organization of pre-and postsynaptic elements is similar for excitatory and inhibitory neurons in the CNS. Like the presynaptic terminal of a motor neuron, the presynaptic terminals of CNS axons contain synaptic vesicles and active zone proteins. These presynaptic elements are concentrated above the corresponding neurotransmitter receptors clustered at the postsynaptic membrane. Other presynaptic elements include voltage-gated ion channels, adhesion molecules, and scaffolding proteins that help anchor the presynaptic elements at the presynaptic site (**Figure 10.3**).

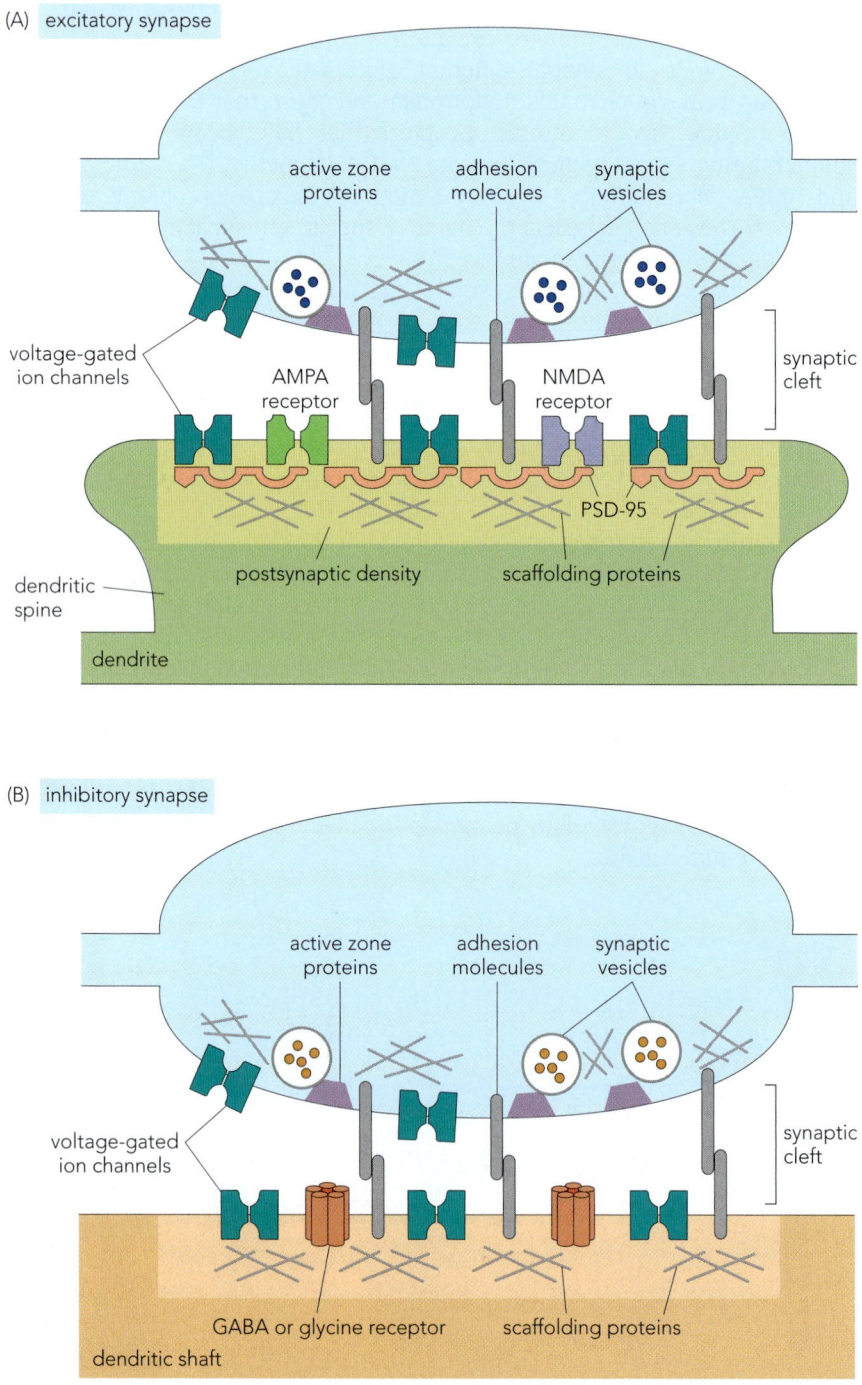

Figure 10.3 Pre- and postsynaptic specializations in excitatory and inhibitory CNS synapses. (A) Presynaptic elements in an excitatory glutamatergic synapse include the glutamate-releasing synaptic vesicles, active zone proteins, voltage-gated ion channels, scaffolding proteins, and adhesion molecules. The postsynaptic dendritic spine is characterized by a prominent postsynaptic density (PSD) that contains the NMDA and AMPA neurotransmitter receptors, voltage-gated ion channels, adhesion molecules, and scaffolding proteins such as postsynaptic density-95 (PSD-95). (B) Similar synaptic elements are found in the presynaptic cells of an inhibitory synapse. GABA (γ-aminobutyric acid) and glycine are common inhibitory neurotransmitters released by different inhibitory neurons. Active zone proteins, voltage-gated ion channels, adhesion molecules, and scaffolding proteins are also found in the presynaptic terminal of inhibitory neurons. Inhibitory synapses typically form along the dendritic shaft that lacks a prominent postsynaptic density. However, scaffolding proteins concentrate at the site of synaptic contact to help anchor GABA or glycine receptors below the presynaptic elements.

The postsynaptic specializations of CNS neurons include neurotransmitter receptors, voltage-gated ion channels, and adhesion molecules. The postsynaptic elements clustered at the site of synaptic contact are also held in place by various scaffolding proteins. The specific scaffolding proteins located in the postsynaptic membrane often differ between excitatory and inhibitory neurons. For example, a protein called PSD-95 (postsynaptic density protein-95) is only found in postsynaptic dendrites of excitatory synapses (Figure 10.3).

Unlike the NMJ, there is no synaptic basal lamina in the synaptic cleft between CNS neurons, although there is a concentration of visible proteinaceous material. Because the synaptic cleft is smaller at CNS synapses, the pre- and postsynaptic cells are close enough that adhesion molecules, such as those described in a subsequent section, can span the cleft and stabilize connections (Figure 10.3).

The postsynaptic density is an organelle found in excitatory, but not inhibitory, neurons

In 1959, soon after the identification of synapses by electron microscopy, E. George Gray reported structural differences in the postsynaptic regions of excitatory and inhibitory synapses. These studies revealed two types of synapses, called type I (asymmetric) and type II (symmetric), based on the appearance of electron-dense material (**Figure 10.4**). In symmetric synapses, the pre- and postsynaptic densities appear similar. With asymmetric neurons, however, a prominent electron-dense region is observed in the postsynaptic cell. This difference is due to the presence of a specialized region in excitatory neurons called the **postsynaptic density (PSD)**. Scientists now know that the asymmetric neurons are excitatory neurons, whereas the symmetric neurons represent inhibitory neurons.

The PSD is a disc-shaped, electron-dense organelle that adheres to the postsynaptic cell membrane of excitatory synapses to anchor proteins directly below the active zones of the presynaptic cell. In the 1970s, scientists began isolating proteins from the PSD and hundreds of proteins have now been identified in this small region of the postsynaptic dendrite. Neurons in different regions of the CNS express different PSD proteins. Among the common PSD proteins are ion channels, cytoskeletal proteins, adhesion molecules, signaling kinases, and scaffolding proteins. The protein composition of the PSD is linked to the function of individual excitatory synapses throughout the lifespan of the animal, and size of the PSD is thought to correlate with the strength of the synaptic connection.

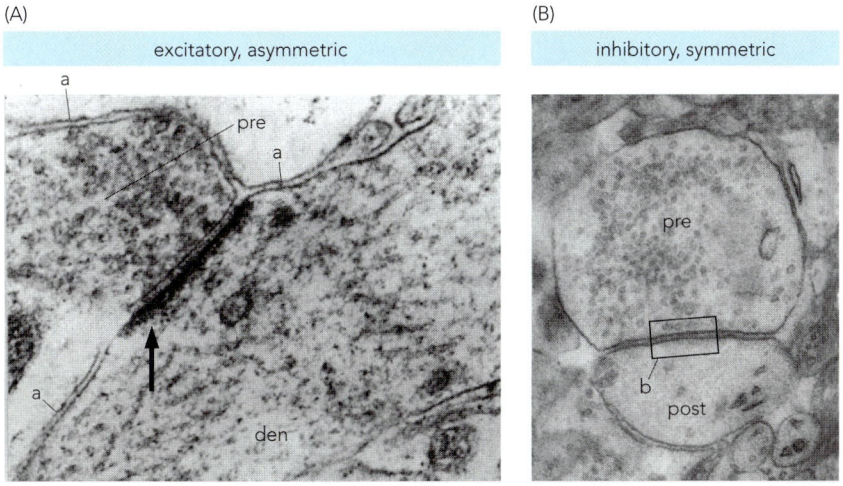

Figure 10.4 Electron micrographs reveal symmetric and asymmetric excitatory synapses. In 1959 George Gray published photomicrographs revealing differences in the postsynaptic densities of excitatory (asymmetric) and inhibitory (symmetric) neurons. (A) The postsynaptic region of the dendrite (den) of excitatory neurons reveals an area of electron-dense material (arrow) that is not observed in other regions of the dendrite (areas labeled a) or in the presynaptic nerve terminal (pre). (B) Because the postsynaptic region at an inhibitory neuron does not contain an electron-dense material, the regions on either side of the synaptic cleft (box, b) appear symmetrical (pre, presynaptic nerve terminal; post, postsynaptic neuron). (From Gray G [1959] *J Anat* 93:420–433.)

Examples of how changes in the protein composition of the PSD occur as synapses are strengthened, weakened, or reorganized during development and adulthood are described in later sections of the chapter.

Of the scaffolding proteins found in postsynaptic excitatory neurons, the PSD-95 (postsynaptic density protein-95) family is one of the best characterized to date (see Figure 10.3A). The PSD-95 family contains four members: PSD-95, PSD-93, and the synapse-associated proteins-97 and -102 (SAP-97 and SAP-102, respectively). Currently, the role of PSD-95 is the best understood.

PSD-95 interacts with other proteins to cluster and anchor those proteins within the PSD. These interactions occur through the PDZ domains found in each of the proteins. PDZ domains were named for the three proteins in which the domains were first identified—the vertebrate PSD-95, the *Drosophila* discs large, and the tight junction protein (ZO-1). PDZ domains are about 90 amino acids in length and are among the most common protein domains identified that interact with other proteins to form large complexes. PDZ-containing proteins are particularly prominent in areas of cell–cell contact, such as synapses. Unlike excitatory synapses, inhibitory synapses do not have a prominent PSD. The postsynaptic membrane of inhibitory synapses is comprised of a simpler protein composition, yet this region accomplishes similar functions, such as clustering neurotransmitters receptors (see Figure 10.3B).

Cell adhesion molecules mediate the initial stabilization of synaptic contacts

In the CNS, it appears that the pre- and postsynaptic cells first establish tentative contacts and then rely on reciprocal signals to determine whether further differentiation into a mature synaptic contact will occur. The initial tentative contacts are established through interactions with various cell adhesion molecules found on each putative synaptic partner. A number of cell adhesion molecules (CAMs) have been identified that influence the stabilization of CNS synapses. Among those identified to date are members of the cadherin and IgG superfamilies. In the presence of calcium, cadherins interact through homophilic binding mechanisms to mediate binding and stabilization of pre- and postsynaptic partners (**Figure 10.5**). Cadherins have also been shown to influence the clustering of glutamate and GABA receptor subunits in postsynaptic dendrites of excitatory and inhibitory synapses, respectively.

SynCAM I (synaptic cell adhesion molecule I) is a member of the IgG superfamily expressed throughout the vertebrate CNS, particularly at excitatory synapses. The SynCAM gene family consists of four genes that are expressed only in vertebrates. SynCAM I is found in several tissues,

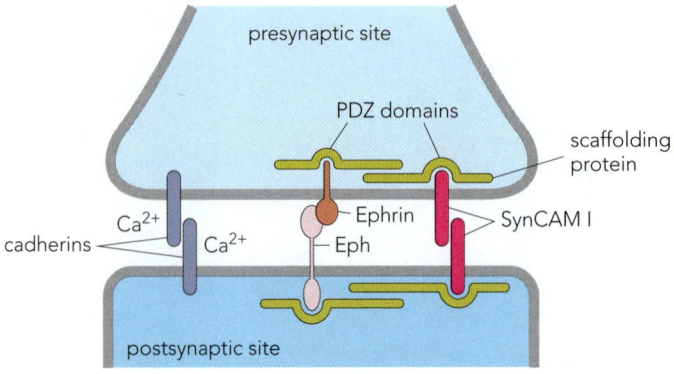

Figure 10.5 Adhesion molecules stabilize contacts between pre- and postsynaptic cells. A number of cell adhesion molecules have been identified that stabilize synaptic contacts. Cadherins bind to one another through homophilic mechanisms that require calcium (Ca^{2+}). Cadherins are thought to stabilize initial synaptic connections. SynCAM I (synaptic cell adhesion molecule I) is a member of the IgG superfamily that also interacts through homophilic binding, but unlike cadherins, SynCAM I does not require calcium. The binding of SynCAM I molecules leads to interaction with scaffolding proteins on pre- and postsynaptic neurons to stabilize cytoskeletal elements. Other nontraditional adhesion molecules also help stabilize synaptic contacts. In some neurons, ephrin B ligands on the presynaptic neuron bind to EphB receptors on the postsynaptic neuron. The cytoplasmic tails of these molecules are also thought to interact with scaffolding proteins to stabilize cytoskeletal elements.

but its highest level of expression is in the CNS. Peak expression levels of SynCAM I correspond to periods of synaptogenesis. SynCAM I homophilic binding does not require the presence of calcium. In fact, SynCAM I was originally discovered in a search for neuronal cell adhesion molecules that did not require calcium. Among the functions associated with SynCAM I are the induction of presynaptic specializations and the stabilization of pre- and postsynaptic elements. Stabilization of synaptic connections occurs because the cytoplasmic regions of SynCAM I interact with the PDZ domains of scaffolding proteins in the pre- and postsynaptic cells (Figure 10.5). The scaffolding proteins help anchor the appropriate synaptic elements and stabilize the synaptic connection.

In addition to traditional cell adhesion molecules, in some CNS neurons interactions between ephrin B ligands and EphB receptors help stabilize newly formed CNS synapses. Thus, in addition to mediating intracellular signaling cascades that influence other aspects of neural development, as noted in previous chapters, EphB/ephrin B interactions may also function as adhesion molecules at synapses. This appears to occur because the cytoplasmic tails of the ephrin B ligands and the EphB2 receptors interact with the PDZ domains of scaffolding proteins that stabilize synaptic connections (Figure 10.5).

Neurexins and neuroligins also induce formation of synaptic elements and stabilize synaptic contacts

Neurexins and neuroligins, first discovered in the 1990s, are a family of cell surface proteins expressed in neurons throughout the CNS. The neurexins are located in the presynaptic cell, while the neuroligins are in the postsynaptic cell. The two molecules attach to each other through heterophilic binding mechanisms and signal bidirectionally so that each induces the synaptic elements in the partner cell. Evidence of this bidirectional signaling was demonstrated in cell culture experiments in which nonneuronal cell lines were generated to overexpress either neurexin or neuroligin. The nonneuronal cells were then co-cultured with CNS neurons. In cultures with nonneuronal cells expressing neurexins, postsynaptic elements were induced in the neurons. In contrast, in cultures in which nonneuronal cells expressed neuroligin, presynaptic elements were induced in the neurons (**Figure 10.6**).

Each binding partner in the neurexin–neuroligin complex has an intracellular domain that interacts with the PDZ domains of local scaffolding proteins. On the presynaptic side, neurexins, of which there are over 3000 potential isoforms, interact with the PDZ domains associated

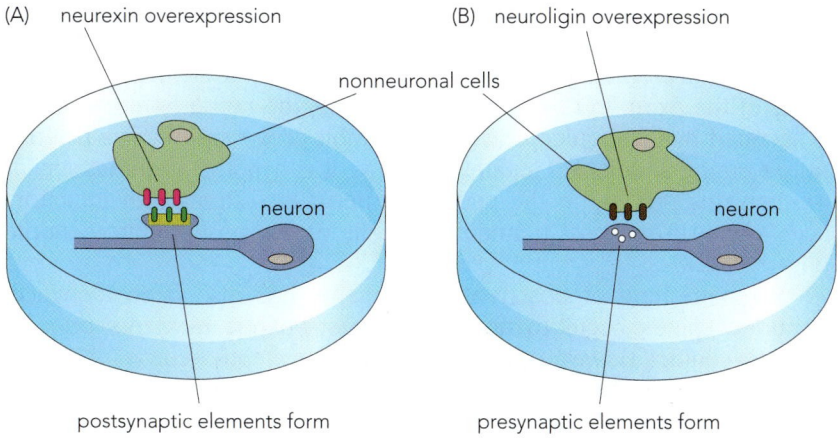

Figure 10.6 ***In vitro*** **studies reveal that neurexin and neuroligin induce different synaptic elements.** (A) Neurexin is a presynaptic protein that binds to postsynaptic neuroligin. Neurexin was overexpressed in nonneuronal cells then cultured with CNS neurons. The neurons that came into contact with the neurexin-expressing nonneuronal cells began to accumulate postsynaptic elements such as neurotransmitter receptors and PSD proteins. (B) In contrast, nonneuronal cells that overexpressed neuroligin induced presynaptic elements such as active zone proteins and synaptic vesicles in the contacting neurons. Together, these experiments demonstrated that the neurexin–neuroligin interactions induce pre- or postsynaptic specializations in partner cells.

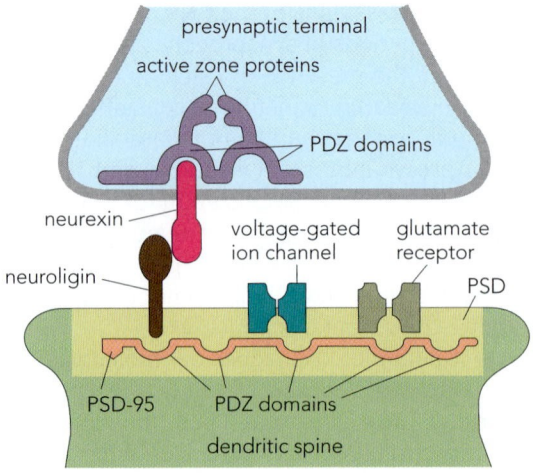

Figure 10.7 Neurexin and neuroligin interact with PDZ domains of scaffolding proteins. In the presynaptic terminal, neurexin interacts with PDZ domains of scaffolding proteins that help anchor associated active zone proteins. On the postsynaptic neuron, neuroligin interacts with a PDZ domain of PSD-95. The multiple PDZ domains of PSD-95 help anchor glutamate receptors and ion channels to the postsynaptic site.

with active zone proteins (**Figure 10.7**). It has been proposed that the numerous potential isoforms of neurexin help establish synaptic specificity in the CNS. However, whether this occurs has not yet been determined.

In excitatory postsynaptic neurons, the cytoplasmic tails of the neuroligins interact with a PDZ binding domain of PSD-95. PSD-95 has multiple PDZ domains that help cluster and anchor ion channels and glutamate receptor subunits (Figure 10.7). Thus, neurexin–neuroligin binding links presynaptic active zone proteins with postsynaptic proteins to ensure that the pre- and postsynaptic elements are closely aligned at the site of contact. One model of how CNS connections form suggests that adhesion molecules such as the cadherins are important in forming initial synaptic connections. Then proteins that interact with scaffolding proteins, such as SynCAM I, EphB/ephrin B, or neurexin/neuroligin stabilize the pre- and postsynaptic elements. Further stabilization and maturation of synaptic elements then occurs as the pre- and postsynaptic cells exchange reciprocal signals.

Reciprocal signals regulate pre- and postsynaptic development

Similar to the NMJ, specific pre- and postsynaptic elements are present before synaptogenesis begins. Many of the necessary proteins are therefore ready to interact with each other as soon as a synaptic contact is made. For example, studies have found evidence of prepatterning in the presynaptic axons of excitatory *en passant* synapses. Initially as an axon extends, synaptic vesicle proteins transiently accumulate at several locations along the axon. Synaptic vesicles then begin to preferentially accumulate at specific sites along the axon shaft. A number of signals likely determine where the presynaptic terminal will form. In some cases, it appears that local inhibitory signals prevent formation of presynaptic elements in certain areas of the axon. In *C. elegans*, for example, the netrin homolog Unc-6 is secreted by local guidepost cells to inhibit presynaptic specializations from forming at adjacent axonal locations. It is hypothesized that similar mechanisms operate in the vertebrate nervous system, with inhibitory cues arising from local nonneuronal cells or the postsynaptic dendrite (**Figure 10.8**).

Prepatterning of postsynaptic elements has also been identified. Dendritic spines begin to accumulate postsynaptic elements prior to synaptic contact. However, contact with the presynaptic neuron is required for further maturation of the postsynaptic site. The immature dendritic spine makes several transient contacts with the presynaptic axon shaft in

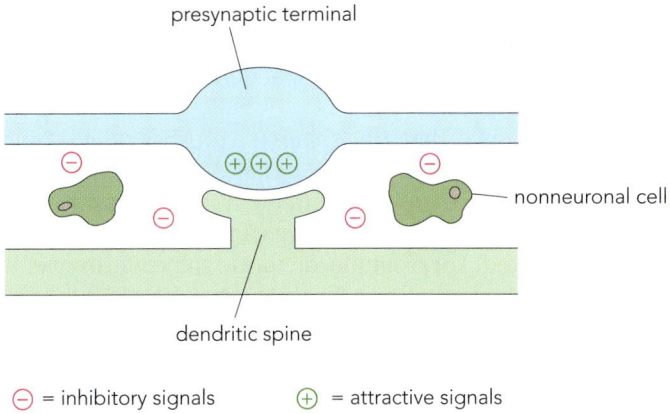

Figure 10.8 Pre- and postsynaptic signals determine where an excitatory synapse will form. Presynaptic specializations only form at restricted regions of the axon shaft. Inhibitory signals from the dendrite shaft and adjacent nonneuronal cells help prevent presynaptic elements from forming at other sites along the axon. The developing presynaptic site releases signals that attract and stabilize dendritic spines to the correct location.

what appears to be an attempt by the dendrite to identify and respond to cues originating in the presynaptic axon. In at least some neuronal populations, the presynaptic neuron may provide attractive signals that direct the initial contact between the immature postsynaptic dendritic spine and the presynaptic axon (Figure 10.8).

Dendritic spines are highly motile and actively seek presynaptic partners

The dendritic spines of excitatory neurons begin as highly motile filopodia that extend from the dendritic shaft (**Figure 10.9**A). These immature spines are slender processes rich in filamentous actin (F-actin) that appear to function much like the filopodia that project from the growth cone at the end of an extending axon (see Chapter 7). The dendritic filopodia eventually alter their morphology and become mature dendritic spines (Figure 10.9B–D).

The morphology of dendritic spines is described based on the overall shape of the neck and head regions. In general, those identified as stubby spines have no obvious neck and a larger head, whereas thin spines have a small head and long, thin neck. Mushroom-shaped spines have a large

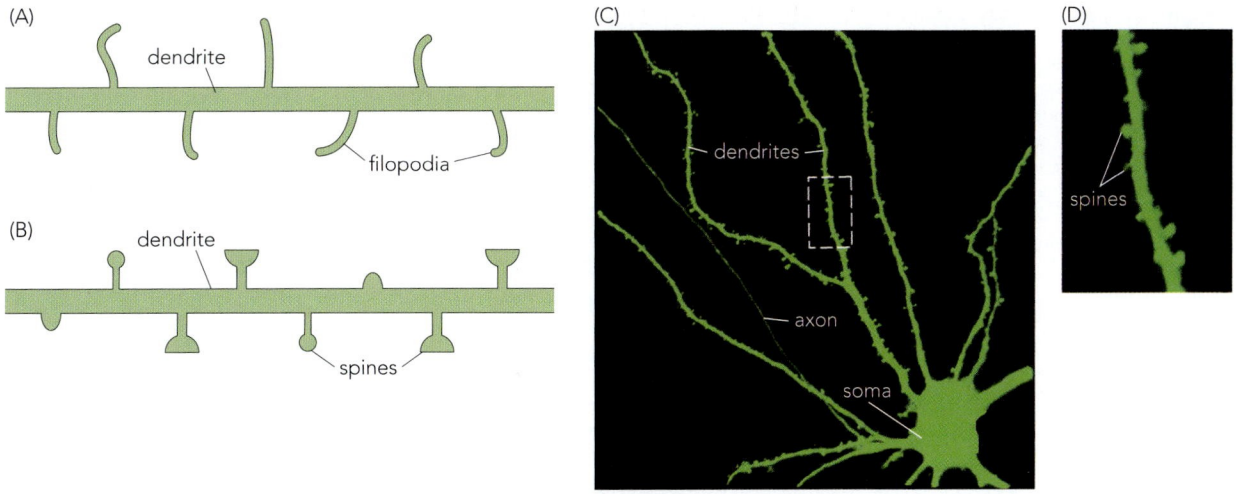

Figure 10.9 Dendritic filopodia are highly motile extensions that later form spines. (A) Dendritic filopodia are actin-rich protrusions that are highly motile prior to synapse formation. As each filopodium establishes a point of synaptic contact, it is converted to a dendritic spine (B). (C) An example of dendritic spines in a hippocampal neuron expressing green fluorescent protein. The boxed region is shown enlarged in panel D. (C and D, from Tolias KF, Duman JG & Um K [2011] *Prog Neurobiol* 94:133–148.)

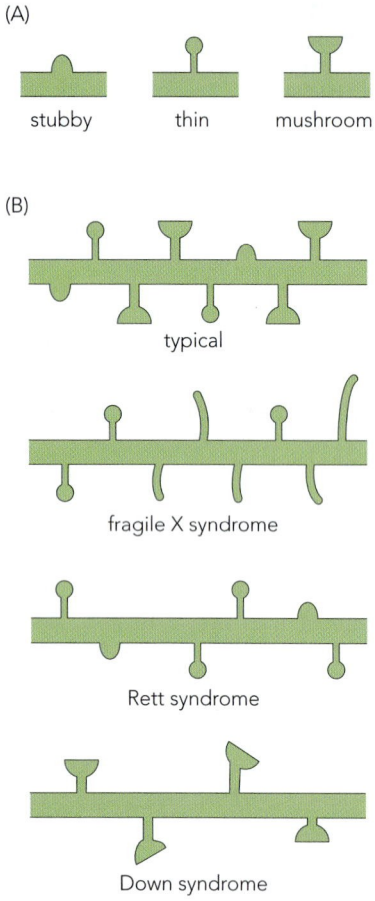

Figure 10.10 Dendritic spine morphology is altered in neurodevelopmental disorders. (A) Mature dendritic spines reveal different morphologies, which are described as stubby, thin, and mushroom-shaped. The size of the spine head is correlated with synaptic strength. (B) Compared to the spine density and morphologies observed in typical brains, those with neurodevelopmental disorders reveal changes characteristic for each disorder. With Fragile X syndrome, spine density is variable and the spines present tend to be longer and immature, often resembling filopodia. In Rett syndrome, there is a decrease in spine density and fewer mushroom-shaped spines, whereas in Down syndrome, there are fewer spines and those present have larger, atypical spine heads. (A, adapted from Yuste R & Bonhoeffer T [2004] *Nat Rev Neurosci* 5:24; B, adapted from Phillips M & Pozzo-Miller L [2015] *Neurosci Lett* 601:30–40.)

head and a thick neck (**Figure 10.10**). The size of the spine head appears to be associated with the synaptic strength, with larger heads being stronger. This is due in part to the increased space available for the postsynaptic neurotransmitter receptors.

The development of dendritic spines is a process called **spinogenesis**. In many ways, spinogenesis appears to be linked to synaptogenesis, because the number of spines is correlated with the number of synapses that form. However, there are differences that suggest the two developmental events are not directly linked. For example, dendritic spines form over a period of minutes, whereas the formation of mature synapses often takes place over a period of days or weeks. Moreover, in at least some cellular contexts, spinogenesis can occur in the absence of axonal contact. In both *Weaver* and *Reelin* mutant mice that are missing presynaptic cerebellar granule cells (see Chapter 5), the dendritic spines on postsynaptic Purkinje cells still develop and appear morphologically similar to wild-type mice. These findings suggest that spines can form independent of synaptic contact. However, it is hypothesized that there are subtle structural changes in postsynaptic specializations due to the loss of reciprocal signals between pre- and postsynaptic partners.

Whether spinogenesis is directly linked to synapse formation or not, the importance of spinogenesis is noted by the large number of neurodevelopmental disorders associated with decreases in dendritic spine number and altered morphology. For example, reduced spine density and alterations in spine morphology have been noted in schizophrenia, autism, Fragile X syndrome, Rett syndrome, and Down syndrome. Studies of these disorders in mouse models or human autopsy specimens have revealed a number of consistent changes associated with each disorder. For example, compared to the spine density and morphology observed in typical adult brains, those with Fragile X syndrome have long spines that resemble filopodia, a morphology that suggests these spines never reach a mature state (Figure 10.10B). The density of spines appears to be variable, at least in those specimens observed to date. In individuals with Rett syndrome, spine density is reduced and there are fewer spines exhibiting a mushroom-shaped morphology. Those with Down syndrome have a decreased number of spines and those present tend to have large heads (Figure 10.10B). Studies of humans with Down syndrome indicate that spine density begins to decrease after 4 months of age, suggesting there is excess pruning of putative synaptic connections during early postnatal development. As discussed later in the chapter, loss of synaptic connections is a normal part of postnatal development, but in those with Down syndrome the loss of dendritic spines is dramatically increased.

BDNF influences dendritic spine motility and synaptogenesis

A number of molecules important for other aspects of neural development, such as brain-derived neurotrophic factor (BDNF), have been found to influence dendritic spine motility and synapse formation. The importance of BDNF signaling in synapse formation is seen in the hippocampus of mice lacking TrkB. These mice have fewer dendritic spines and a reduction in synapse number in this region of the CNS. Studies have also shown that BDNF released by presynaptic neurons activates TrkB receptors on postsynaptic dendrites, leading to increased motility of dendritic filopodia. The increased motility is thought to increase the likelihood that a filopodium will initiate a synaptic connection.

In recent years, mechanisms by which BDNF can influence filopodia motility have been identified. BDNF influences a number of Rho family small

GTPase signaling pathways. Rho family small GTPases, a subfamily of the Ras superfamily of small GTPases, comprise a family of G proteins that are utilized in multiple intracellular signaling cascades and often influence actin and microtubule dynamics (see Chapter 5). Rho GTPases are activated by various guanine nucleotide exchange factors (GEFs) that convert GDP to GTP. The ability of BDNF to influence different GEFs and activate Rho GTPases therefore provides a direct means for this neurotrophin to regulate cytoskeletal dynamics during synaptogenesis.

Several experiments have demonstrated a role for GTPase signaling pathways in synaptogenesis. In cultured rat hippocampal neurons, for example, BDNF was found to activate the GEFs, Vav2 (vav guanine nucleotide exchange factor 2), and Tiam1 (T-cell lymphoma invasion and metastasis inducing protein 1) to influence formation of dendritic spines. In other experiments, synapse formation was induced when BDNF activated the Rho family GTPase Rac1 (ras-related C3 botulinum toxin substrate 1). Thus, BDNF not only influences neuronal survival, but by activating specific GEFs and Rho family GTPases, BDNF-TrkB signaling regulates the motility and stability of the cytoskeletal elements to influence the formation and stability of synaptic connections (**Figure 10.11**).

BDNF is also known to increase the density of postsynaptic receptors in spinal cord inhibitory neurons. The mechanisms by which changes in receptor density occur is not entirely clear, but it seems to involve signaling through the MAP kinase pathway that is activated following the binding of BDNF to TrkB (see chapter 8). Thus, BDNF signaling activates different intracellular signaling cascades to influence multiple aspects of synaptic development in different CNS neuronal populations.

Eph/ephrin bidirectional signaling mediates presynaptic development

Since 2001, a number of studies have revealed that ephrin ligands and their associated Eph receptors influence pre- and postsynaptic development. As noted in earlier chapters, ephrin ligands bind to Eph tyrosine kinase receptors to initiate signal transduction cascades in the receptor-bearing cell through forward signaling. In addition, the binding of the Eph receptors to the membrane-bound ligand can also induce signal transduction cascades in the ligand-bearing cell through the process of reverse signaling.

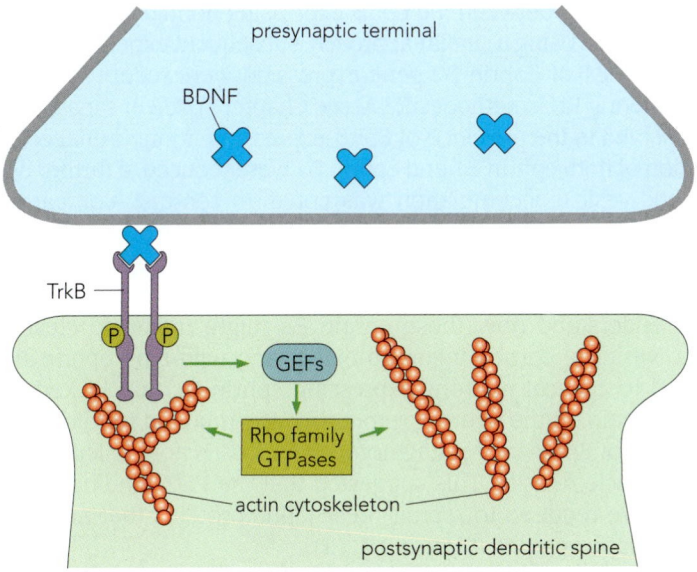

Figure 10.11 BDNF activates Rho family small GTPases to influence cytoskeletal dynamics. BDNF released by the presynaptic terminal binds to TrkB on the postsynaptic neuron. Phosphorylated TrkB can then activate GEFs (guanine exchange factors), which stimulate activity of various members of the Rho family of small GTPases. The Rho GTPases then influence the motility of the actin cytoskeleton to regulate synapse formation and stability.

Both forward and reverse signaling mechanisms are used during synapse development. In some cellular contexts, when Eph receptors in presynaptic neurons are activated by ephrin ligands on the postsynaptic neurons (forward signaling), clustering of presynaptic elements is observed. In other contexts, postsynaptic Eph receptors activate presynaptic ephrin ligands (reverse signaling) to induce clustering of presynaptic elements. Similarly, forward and reverse signaling mediate development of postsynaptic elements. Members of both the EphA/ephrin A and EphB/ephrin B subclasses have been found to influence synaptogenesis. Many of the studies have focused on the role of EphB/ephrin B signaling and this subclass is highlighted in the subsequent discussion to provide just one example of how receptor–ligand interactions commonly used throughout neural development can also impact CNS synaptogenesis.

A number of experiments conducted in the laboratories of Matthew Dalva and Mark Henkemeyer are among those that have contributed to the current models of how ephrin signaling influences synapse formation. For example, activation of EphB receptors by ephrin B ligands (forward signaling) has been shown to influence presynaptic development of hippocampal and cortical neurons in a number of experimental preparations. An *in vitro* assay was developed in which ephrin B3 was expressed in nonneuronal cells. These cells were then co-cultured with hippocampal neurons. A fluorescently labeled marker for excitatory presynaptic neurons (VGLUT1—vesicular glutamate transporter 1—a protein found in the membrane of glutamate-containing synaptic vesicles) was then used to assess whether ephrin B3 induced presynaptic specializations in the hippocampal neurons. The presynaptic marker was only observed in the axons that contacted the ephrin B3-expressing nonneuronal cells. This suggested ephrin B3 mediated presynaptic differentiation of excitatory neurons by binding to endogenous EphB receptors on the hippocampal neurons. This finding was consistent with other studies showing EphB activation leads to presynaptic differentiation. In contrast to VGLUT1 labeling, presynaptic markers of inhibitory neurons were not detected in any of the cultures.

Postsynaptic EphB receptors signaling through presynaptic ephrin B1 and ephrin B2 ligands (reverse signaling) also influence presynaptic development by regulating the clustering of synaptic vesicles and the maturation of active zone proteins. Evidence for ephrin-B-directed synaptic vesicle clustering was found in *in vitro* studies of embryonic rat cortical neurons co-cultured with nonneuronal cells generated to express EphB receptors. These studies noted that clustering of synaptic vesicles occurred at the sites of contact between the EphB-expressing nonneuronal cells and the cortical neurons. Using a similar approach, subsequent experiments revealed that if ephrin B1 or ephrin B2 gene expression were reduced by using the short interfering RNA method (siRNA; see Chapter 4), fewer synaptic vesicles were observed in the presence of EphB-expressing nonneuronal cells. If the expression of both ephrin B1 and ephrin B2 were reduced, a further decrease in synaptic vesicle accumulation was noted. In contrast, cortical neurons with decreased levels of ephrin B3 did not reveal a corresponding decrease in synaptic vesicle clustering. Thus, only ephrin B1 and ephrin B2 appear to influence presynaptic vesicle clustering in cortical neurons.

To understand how these ephrin Bs might influence clustering of synaptic vesicles, cortical neurons expressing wild-type ephrin B1 were compared to cortical neurons expressing ephrin B1 that lacked the PDZ binding domain. In the neurons expressing the mutated form of ephrin B1, the addition of EphB-expressing nonneuronal cells now failed to influence synaptic vesicle density. This suggested that the PDZ binding domains of ephrin B1 are required to recruit presynaptic proteins that are needed to cluster synaptic vesicles (**Figure 10.12**).

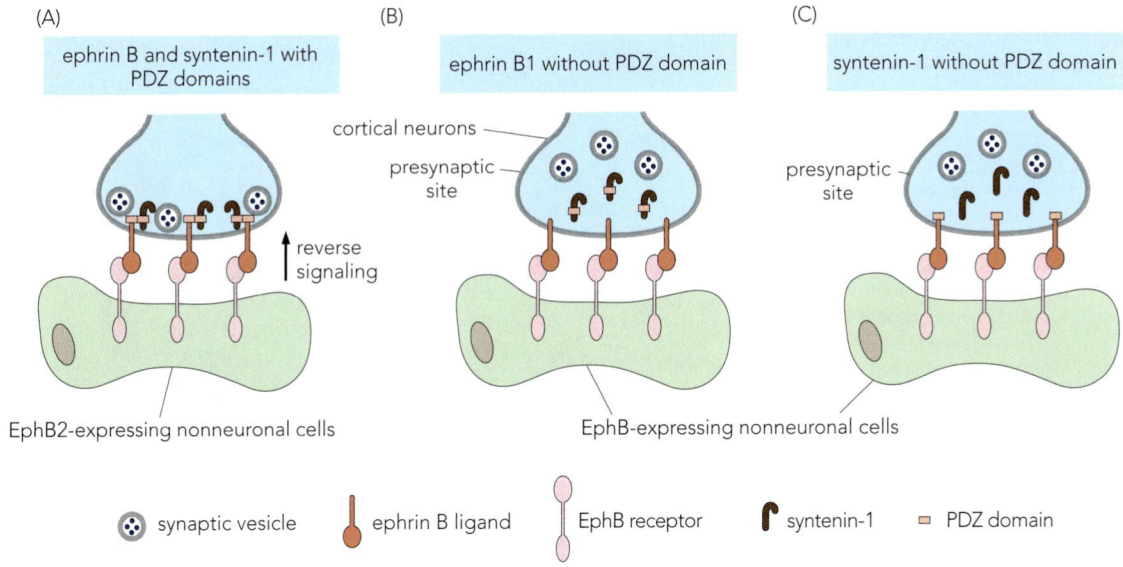

Figure 10.12 EphB receptors signal through ephrin B1 and ephrin B2 ligands to cluster presynaptic vesicles in cortical neurons. *In vitro* experiments were conducted in which cortical neurons were co-cultured with nonneuronal cells expressing EphB receptors. (A) When EphB-expressing nonneuronal cells contact cortical neurons expressing ephrin B1 or ephrin B2, reverse signaling leads to interaction of ephrin B PDZ domains with the PDZ domains of syntenin-1. This interaction induces clustering of synaptic vesicles at the site of contact. (B) Synaptic vesicle clustering did not occur when cortical neurons were generated to overexpress ephrin B1 lacking the PDZ domain. (C) Similarly, when the PDZ domains of syntenin-1 were absent, the synaptic vesicles failed to cluster.

Syntenin-1 is one protein known to interact with ephrin B during presynaptic development. Syntenin-1 is a PDZ-containing adaptor protein involved in a number of cellular processes that require formation of protein complexes. Syntenin-1 was first identified as a melanoma differentiation gene and has since been found to be important not only for tumorigenesis, but also for receptor trafficking and synapse formation. Syntenin-1 is located at both the presynaptic active zone and postsynaptic PSD, indicating roles in assembling presynaptic and postsynaptic signaling molecules.

To investigate the function of syntenin-1 in presynaptic development, syntenin-1 lacking the PDZ domain was overexpressed in cortical neurons. In these neurons, the EphB2-expressing nonneuronal cells were no longer able to induce synaptic vesicle clustering, similar to what was observed when the PDZ binding domain of ephrin B1 was missing (Figure 10.12C). This suggests that syntenin-1 PDZ domains interact with ephrin B1 PDZ domains to cluster the presynaptic vesicles and associated proteins at the proper site. One hypothesis about how this occurs is that the ephrin Bs recruit syntenin-1 to the proper presynaptic location so that syntenin-1 can then anchor synaptic vesicle proteins.

Eph/ephrin signaling initiates multiple intracellular pathways to regulate the formation of postsynaptic spine and shaft synapses

A number of studies in cortical and hippocampal neurons have indicated that ephrin–Eph interactions also influence aspects of postsynaptic development. For example, reverse signaling initiated by presynaptic EphB receptors influences postsynaptic differentiation of ephrin-expressing postsynaptic neurons. These ephrins co-localize with PSD-95 and other PDZ-containing scaffolding proteins located at the postsynaptic cell membrane. Interestingly, ephrin–Eph interactions appear to mediate different aspects of synaptic formation depending on the age and type of CNS neuron, as well as the specific PDZ-containing adaptor proteins utilized. For example, *in vitro* experiments suggest that ephrin B3 interacts

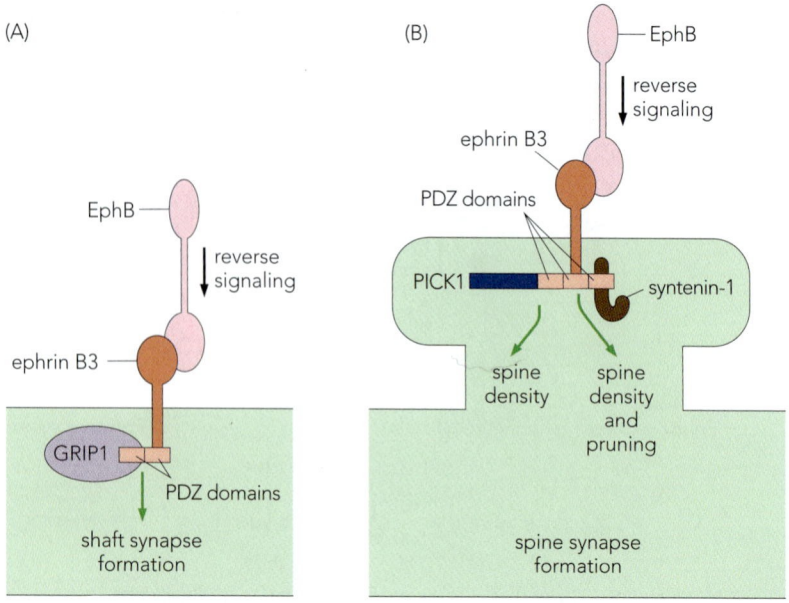

Figure 10.13 Postsynaptic ephrin B3 interacts with different adaptor proteins to regulate excitatory shaft and excitatory spine synapse development. In hippocampal neurons, EphB activates postsynaptic ephrin B3 through reverse signaling. (A) Ephrin B3 interacts with the adaptor proteins GRIP1 (glutamate-receptor-interacting protein 1) to influence formation of shaft synapses, but not spine synapses. (B) The density of spine synapses increases when activated ephrin B3 interacts with the adaptor protein PICK1 (protein interacting with C kinase-1) or with syntenin-1. Syntenin-1 can also regulate the subsequent pruning of excess synapses. Both PICK1 and syntenin-1 interact with cytoskeletal elements in the dendritic spine, thereby linking ephrin B3 signaling to dendritic spine structure and density.

with the adaptor protein GRIP1 (glutamate-receptor-interacting protein 1) to regulate development of excitatory dendritic shaft synapses whereas ephrin B3 interacts with other adaptor proteins to influence development of spine synapses.

In cell culture assays in which an EphB receptor was added to cultures of hippocampal neurons, the resulting ephrin B3 reverse signaling induced clustering of GRIP1 at the postsynaptic membrane of shaft synapses (**Figure 10.13**A). GRIP1 is an adaptor protein that interacts with both ephrin B3 and AMPA receptors. The direct interaction of ephrin B3 and GRIP1 was demonstrated by experiments in which the PDZ domains of ephrin B3 were mutated then overexpressed in hippocampal neurons. The overexpression of the mutated ephrin B3 acted as a dominant negative and prevented normal interactions between the PDZ domains of ephrin B3 and GRIP1. In the absence of functional ephrin B3 PDZ domains, the addition of EphB2 to the cell cultures no longer induced GRIP1 clustering. Thus, under normal developmental conditions, Eph/ephrin B3 reverse signaling appears to recruit GRIP1 to the membrane of postsynaptic shaft synapses where it presumably helps anchor glutamate receptors below the presynaptic terminal.

In other studies of hippocampal neurons, ephrin B3 was found to interact with different PDZ-domain-containing proteins to influence development of dendritic spine synapses. In these experiments, it was found that presynaptic EphB receptors signal through the postsynaptic ephrin B3 to recruit the PDZ-containing adaptor proteins PICK1 (protein interacting with C kinase-1) and syntenin-1. When the PDZ domains of ephrin B3 interacted with the PDZ domains of PICK1, dendritic spine density increased. In contrast, ephrin B3 PDZ domains interacting with syntenin-1 PDZ domains were able to influence spine density as well as the subsequent pruning of excess spines. Thus, postsynaptic modifications that occur through ephrin B signaling appear to take place not only during synaptogenesis, but also during synapse maturation and refinement.

Scientists continue to explore how signals influence synaptogenesis. Evidence suggests that in some cases, Eph/ephrin signaling pathways may intersect with BDNF-TrkB signaling pathways to impact synapse formation, maturation, and refinement. Additionally, similar to BDNF-TrkB signaling, Eph/ephrin signaling also leads to the activation of specific GEFs and

Rho GTPases to mediate cytoskeletal dynamics that in turn regulate spine motility and stabilization. Other signaling pathways have been identified that regulate cytoskeletal dynamics in pre- and postsynaptic neurons. For example, the ubiquitin proteasome system (UPS) has also been shown to influence cytoskeletal dynamics in the presynaptic terminals of invertebrate and vertebrate synapses (**Box 10.1**).

Box 10.1 Ubiquitin-mediated pathways regulate synapse formation

The ubiquitin proteasome system (UPS) was originally discovered in the mid-twentieth century as a system for degrading proteins. Proteins tagged with ubiquitin, a 76-amino acid peptide that attaches to other proteins, are transferred to one of the many **proteasomes** found in the nucleus or cytoplasm of the cell. Only ubiquinated proteins enter the proteasome, where they are degraded into small peptides and released back into the cell for further use. The proteasome is often called the protein-recycling center within a cell.

In addition to the important role in recycling proteins, the UPS regulates several aspects of synaptic development. This first became apparent at the beginning of the current century in studies of invertebrates. It is now recognized that ubiquitination mediates many cellular pathways that ultimately regulate synaptic size, stabilization, and elimination by rapidly degrading various pre- and postsynaptic proteins.

The specific effect of ubiquitin on a target protein depends on the length and configuration of the ubiquitin chains that attach to the targeted proteins. A ubiquitin chain is attached to a target protein following a series of steps involving enzymes called E1, E2, and E3 ubiquitin ligases. E1 ligases are the enzymes that activate ubiquitin then interact with E2 ubiquitin carrier ligases. The E2 ligases are coupled with one of the over 100 specific E3 ligases. Together, this E2–E3 complex transfers ubiquitin to the target protein (**Figure 1**).

One highly conserved E3 ligase was first discovered in invertebrate animal models in 2000. The ligase is called highwire (Hiw) in *Drosophila* and RPM-1 (regulator of presynaptic morphology-1) in *C. elegans*. Within five years, homologs were identified in zebrafish, where it is called Esrom, and in mammals, where it is called Phr1. In all of these animal models, this E3 ligase was found to be important for presynaptic development. Mutations of the E3 ligase gene led to morphological changes in the presynaptic terminals, as well as changes in the number of synaptic contacts that reached the target cell.

Interestingly, the results of a mutation in this E3 ligase gene vary with species. For example, in the *Drosophila* NMJ, loss of *hiw* led to overgrowth of presynaptic axon terminals, whereas a mutation in *C. elegans* or vertebrates resulted in disorganized synapse morphology and fewer contacts at the correct target cell. In all cases, however, development of the correct number of functional synapses depended on the expression of the corresponding *hiw/rpm-1/esrom/Phr-1* gene.

Scientists have now identified several downstream targets of these E3 ligases. In *C. elegans*, RPM-1 interacts with the protein DLK-1 (dual leucine zipper kinase), whereas in *Drosophila* the homologous kinase is called Wallenda (Wd). DLK-1/Wd activates other signaling pathways to ultimately regulate cytoskeletal elements to influence the motility and morphology of the presynaptic terminal.

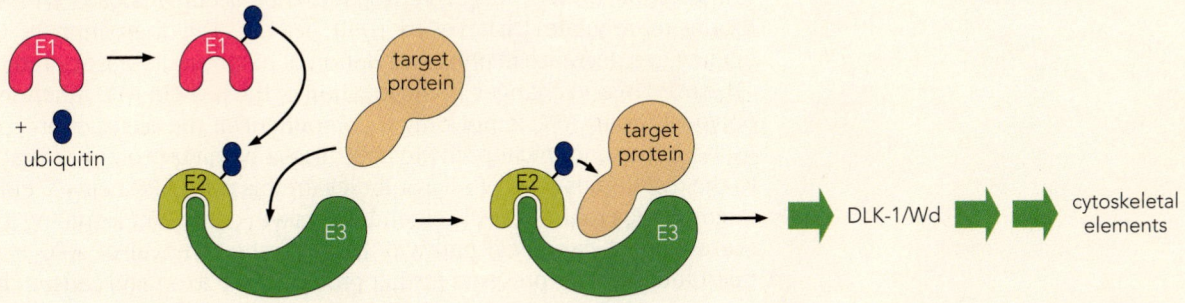

Figure 1 The ubiquitin proteasome system (UPS) regulates proteins required for synapse formation. Ubiquitin is activated by E1 ligase, then transferred to an E2 carrier ligase that is coupled to an E3 ligase. A target protein is recruited to the E2–E3 ligase complex. Once the target protein is ubiquinated, downstream signaling cascades such as the DLK-1/Wd kinases are activated. By activating intracellular pathways that degrade selected cytoskeletal elements, the synapses are rapidly reorganized.

Understanding the role of the UPS in neural development remains an active area of research. Scientists have discovered other E3 ligases that regulate the density of postsynaptic receptors, the stability of postsynaptic scaffolding proteins, and the morphology of dendritic spines. Further, the role of the UPS is not limited to synaptogenesis. Other E3 ligases have been associated with neuronal migration and axonal guidance. Thus, the UPS, the protein-recycling center of a cell, has been adapted to function as a rapid, reliable, and necessary signaling system for regulating many aspects of neural development.

Wnt proteins influence pre- and postsynaptic specializations in the CNS

The Wnt family of proteins also provides bidirectional signals to regulate development of pre- and postsynaptic elements in the CNS. As seen in earlier chapters, Wnt proteins play a variety of roles in nervous system development. During synaptogenesis, Wnt proteins activate specific Frizzled (Fz) receptors to initiate different intracellular signaling cascades that ultimately mediate pre- and postsynaptic development.

A number of *in vitro* and *in vivo* studies in the vertebrate CNS have found that presynaptic development is influenced by Wnt7a secreted by postsynaptic cells. Wnt7a is secreted by postsynaptic neurons in the cerebellum, hippocampus, and spinal cord where it is thought to guide presynaptic terminals expressing the receptor Fz5. Once the presynaptic terminal is at the correct postsynaptic location, Wnt7a then induces clustering of synaptic vesicles and active zone proteins by regulating microtubule stability in the presynaptic terminal.

The importance of this signaling pathway in presynaptic development was seen in experiments in which Wnt7a was added to cell cultures of hippocampal neurons. In these assays, the addition of Wnt7a led to the clustering of presynaptic vesicles and associated proteins. The addition of Wnt7a inhibitors prevented this clustering. Thus, when either soluble Frizzled receptor proteins (sFRPs) or Dickkopf were added to the hippocampal neurons, the Wnt7a-induced clustering of presynaptic proteins was prevented. *In vivo* studies of mice lacking Wnt7a also revealed defects in the clustering of synaptic vesicles and associated presynaptic proteins, further demonstrating the need for Wnt7a in presynaptic development.

The Wnt signaling pathway used during presynaptic development is known as the **divergent canonical pathway** (**Figure 10.14**). This pathway is similar to the canonical pathway (see Chapter 4), but transcription of Wnt target genes does not occur. Instead, Wnt signals locally to regulate Dishevelled (Dvl) activity. Dvl then inhibits GSK3β, leading to a decrease in phosphorylation of microtubule-associated protein 1B (MAP 1B) and changes in the location of the protein APC (adenomatous polyposis coli). APC is not only a component of the destruction complex found in the canonical pathway, but also a binding protein that attaches to the distal (plus) end of microtubules. In this role, APC helps accumulate microtubules to the actively extending growth cone. Wnt signaling through the divergent canonical pathway leads to the removal of APC from the microtubules. This prevents further growth of the axon and causes changes in axon terminal morphology. The changed morphology then allows the stabilized microtubules to attract and attach presynaptic proteins. In this way, Wnt signaling through the divergent pathway regulates cytoskeletal dynamics of presynaptic terminal. To date, the Wnt divergent pathway has been noted in the formation of presynaptic terminal synapses, but it is not known if this pathway is used at all CNS synapses.

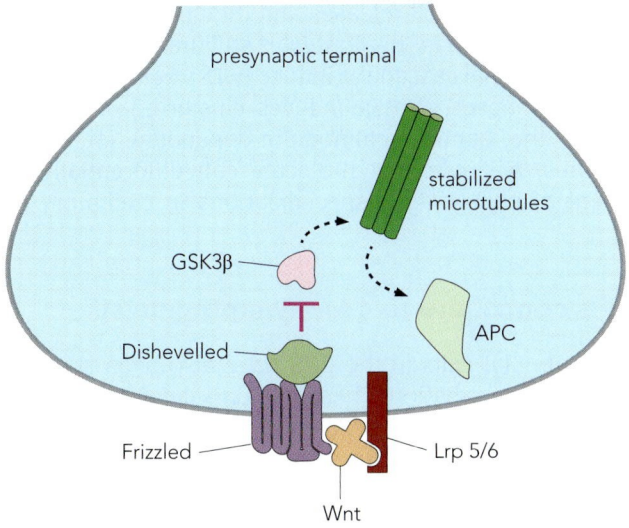

Figure 10.14 Wnt signaling through the divergent canonical pathway regulates presynaptic development. In the divergent canonical pathway, Wnt binding to the frizzled Lrp5/6 receptor complex causes dishevelled to inhibit GSK3β. This leads to a decrease in phosphorylation of microtubule associated protein 1B (MAP 1B, not shown) and relocation of the APC protein that was attached to the distal ends of microtubules. With the removal of APC (dashed arrows), the growth of the axon is stopped and the microtubules stabilize. This enables the microtubules to attract and attach presynaptic proteins.

Different Wnts regulate postsynaptic development at excitatory and inhibitory synapses

Wnt7a is not only released by the postsynaptic cell to influence presynaptic development, but is also secreted by the presynaptic cell to influence postsynaptic specializations at excitatory synapses. This presynaptic Wnt7a influences dendritic spine formation and recruits the PSD-95 protein to the postsynaptic site (**Figure 10.15**). Thus, Wnt7a functions through a bidirectional signaling mechanism to influence the maturation of synaptic elements on both presynaptic and postsynaptic sites of excitatory synapses. However, Wnt7a does *not* appear to influence the development of postsynaptic elements at inhibitory synapses.

Another Wnt family member, Wnt5a, has been observed to influence the development of inhibitory, as well as excitatory, synapses by activating different downstream signaling pathways. At excitatory synapses, Wnt5a clusters the PSD-95 protein in the dendritic spine. In at least some neuronal populations, Wnt5a functions in an **autocrine** manner—that is,

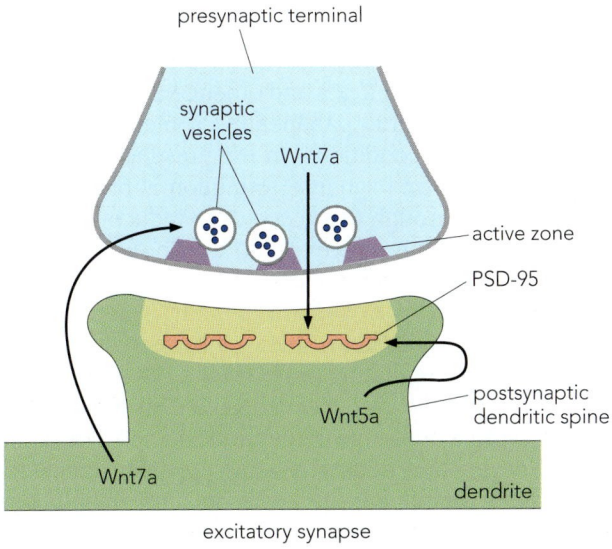

Figure 10.15 Wnt proteins regulate development of presynaptic and postsynaptic elements. Postsynaptic Wnt7a regulates the accumulation of active zone proteins and synaptic vesicles in the developing presynaptic terminal. The presynaptic terminal also releases Wnt7a to recruit PSD-95 into the PSD, thus leading to further maturation of the postsynaptic site. In addition, Wnt5a released by the postsynaptic dendrite functions in an autocrine manner to help cluster the PSD-95 within the PSD.

the same postsynaptic cell releases and then binds Wnt5a. Although the receptor used by Wnt5a in this context is still unknown, Wnt5a appears to activate the JNK signal cascade to induce PSD-95 clustering (Figure 10.15). At inhibitory synapses, Wnt5a induces clustering of $GABA_A$ receptors by activating the CamKII (calmodulin-dependent kinase II) signaling pathway. Thus, a single Wnt molecule can differentially regulate the development of postsynaptic specializations in excitatory or inhibitory synapses.

Glial cells contribute to CNS synaptogenesis

Similar to NMJ, glial cells in the CNS have also been shown to regulate aspects of synapse formation. There is often a delay between the time that CNS axons reach their target cells and the time that synapses form. This period generally coincides with the production of glial cells. Therefore, researchers have long suspected that glial cells release molecules that contribute to synapse formation. Scientists have confirmed that astrocytes release factors that help establish and maintain new synapses.

Among the first studies demonstrating the importance of glial-derived signals in CNS synaptogenesis were those using cultures of retinal ganglion cells (RGCs). *In vitro* studies in the 1990s found that astrocytes induced synaptic elements and helped establish new synapse connections to form between the cultured RGCs. This effect did not require cell contact between astrocytes and neurons, because astrocyte-conditioned medium also induced synaptogenesis when added to cultures of RGCs.

Subsequent studies by Ben Barres and colleagues identified thrombospondin-1 (TSP-1) and thrombospondin-2 (TSP-2) as the proteins produced by astrocytes. There are five members of the TSP glycoprotein family, all of which are produced by multiple cell types in the body to mediate a variety of cell–cell and cell–matrix interactions. In the CNS, TSP-1 and TSP-2 are produced by astrocytes and are located in the developing brain during periods of synaptogenesis.

In vitro studies revealed that synaptogenesis induced by TSP-1 or TSP-2 is similar to that observed with astrocyte-conditioned media. The importance of these proteins was further noted *in vivo*. TSP-1 and TSP-2 are expressed in the developing brain during periods of synaptogenesis, and mice lacking both TSP-1 and TSP-2 form about 30 percent fewer CNS synapses than do wild-type mice. Mice lacking only one TSP gene did not reveal any changes in synaptic number, indicating that these proteins have redundant functions.

How TSPs mediate synaptogenesis is still not entirely clear. A receptor for TSPs, α2δ1, is expressed on RGCs and other CNS neurons. This receptor has a binding site for the drug gabapentin—a medication used to treat epilepsy and other neural conditions. In the presence of gabapentin, the α2δ1 receptor is blocked, and synapse formation is prevented, indicating that TSPs normally act through this receptor to induce synapse formation (**Figure 10.16**). TSPs alone, however, do not induce a fully functional synapse, suggesting that other factors may be necessary for synapse maturation.

TSPs are not the only proteins released by astrocytes. Other growth factors associated with regulating aspects of CNS synaptogenesis such as BDNF and ephrins have also been detected in astrocytes and their corresponding receptors are found on dendritic spines. The other CNS glial cells—oligodendrocytes and microglia—also appear to impact aspects of synaptogenesis. For example, recent studies indicate microglia release the cytokine tumor necrosis factor α (TNFα) that can strengthen newly formed

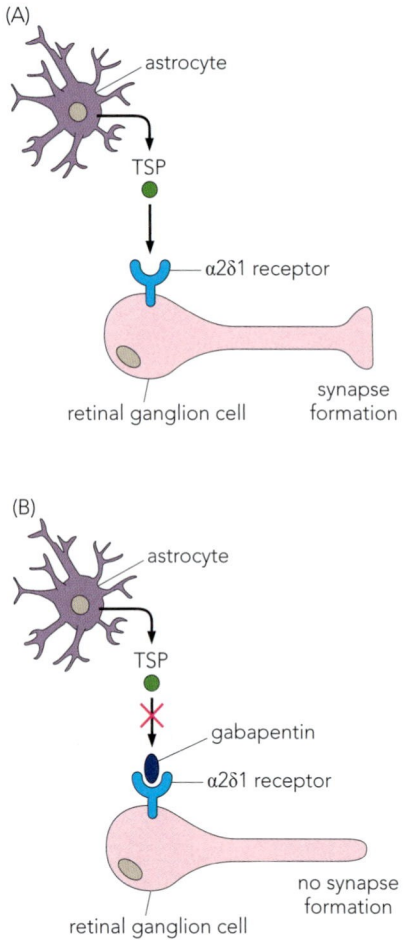

Figure 10.16 Thrombospondins released by astrocytes induce synapse formation. (A) Thrombospondins (TSP) are released by astrocytes and bind to the α2δ1 receptor found on retinal ganglion cell neurons. Both TSP-1 and TSP-2 are released by astrocytes to induce the formation of new synapses between RGCs. (B) The α2δ1 receptor also binds the drug gabapentin. When gabapentin is present, TSPs cannot bind to the receptor, and new RGC synapses are prevented from forming. These studies indicated the necessity of α2δ1 receptor activation in mediating the effects of TSPs on synaptogenesis.

synapses. Thus, a number of factors produced by glial cell populations have the potential to interact and impact synaptogenesis and synaptic maintenance in the CNS.

SYNAPSE ELIMINATION AND REORGANIZATION IN THE CNS

The vertebrate CNS eliminates a large number of recently established synaptic connections during postnatal development. In some regions of the vertebrate CNS, up to 50 percent of synapses are lost. The developmental loss of these synaptic connections is not due to the loss of neurons themselves. Programmed cell death, discussed in Chapter 8, typically occurs during embryonic development, whereas synapse elimination typically begins during early postnatal development. In contrast to normal vertebrate CNS development, a widespread loss of synaptic connections does not take place in the invertebrate CNS.

Similar to the process of synapse elimination in the vertebrate NMJ, the loss of synaptic connections in the vertebrate CNS appears to be due, in part, to differences in the levels of neuronal activity between pre- and postsynaptic partners. The relative level of neural activity between neighboring axons, as well the timing of action potential firing in the postsynaptic cell often determines whether a synaptic connection remains.

While evidence for the importance of neuronal activity in regulating synaptic connections has accumulated for over 60 years, the mechanisms by which neural firing patterns regulate such connectivity are still not fully understood. In some cases, growth factors such as BDNF are released in response to neural activity and may help maintain synapses. Presumably, the synapses that do not have access to adequate levels of these growth factors are the ones that are lost. A number of other growth factors have also been proposed to help stabilize connections or delay elimination. For example, scientists have found stabilization of synaptic contacts, at least transiently, in response to growth factors such as ciliary neurotrophic factor (CNTF), leukemia inhibitory factor (LIF), and glial cell line-derived neurotrophic factor (GDNF). However, whether the levels of these growth factors are directly regulated by neural activity levels remains to be established.

The vertebrate visual system is a popular model to study synapse elimination and reorganization

Much of what is known about the processes underlying synapse elimination and reorganization in the vertebrate CNS was first discovered in studies of the visual system. The visual system is a convenient model system to study synapse formation and elimination for several reasons. The visual system of several vertebrate animal models develops postnatally and the various anatomical regions of the CNS visual pathways are generally accessible. These two features make the vertebrate visual system relatively convenient for experimental manipulations. In addition, the anatomical organization of the central visual pathways has been helpful in designing experiments to determine how synapses are formed and reorganized. In the visual system of mammals that rely on binocular vision, for example, several CNS regions become segregated into layers, with each cellular layer receiving synaptic inputs from either the right or left eye. Such segregation is noted in each **lateral geniculate nucleus (LGN)** of the thalamus, a target of the RGCs, and ultimately in the **primary visual cortex** (**V1**), the target of LGN axons.

The role of neural activity in shaping synaptic connections in the vertebrate visual system pathways has been studied since the mid-twentieth century. As outlined in the next section, scientists have determined over the past several decades that in some regions of the mammalian visual system, synaptic stabilization, elimination, and reorganization are influenced by spontaneous embryonic neural activity, while in other regions these processes are driven by visual stimuli. In both cases, changes in the number of dendritic contacts and postsynaptic receptors are influenced by the amount of neural stimulation received.

Spontaneous waves of retinal activity stabilize selected synapses in LGN layers

The mammalian lateral geniculate nuclei are analogous to the tectal regions previously described in other vertebrates. As in the tectum, the axons of RGCs project to these nuclei in a specific pattern. Each LGN is organized into cellular layers. The number of layers varies with species, ranging from two in rodents to six in primates. In all of these animal models, the RGCs from both eyes initially project axons to postsynaptic neurons in all LGN layers. These overlapping innervation patterns then gradually segregate so that axons that arrive from the ipsilateral eye form stable synaptic connections with neurons in certain layers of the LGN, while axons arriving from the contralateral eye form synapses with neurons in the other layers. For example, in the ferret, axons from the temporal side of the retina project mainly to the ipsilateral LGN and make synaptic connections in layers A1 and C1. In contrast, most axons from the nasal half of the retina cross to the contralateral LGN to contact cells in layers A and C (**Figure 10.17**).

Neural activity generated in the retina is necessary for this synaptic reorganization and segregation to occur in the LGN. If tetrodotoxin (TTX), a drug that blocks action potentials, is infused into the eyes during the period that RGC axons normally segregate into different LGN layers, synaptic reorganization fails to occur. Instead, the axons of RGCs from each eye remain spread across all LGN layers. Interestingly, the process of synaptic reorganization in the LGN takes place before the photoreceptor cells of the retina are fully developed and functional. Thus, visual stimulation is not required for this form of synaptic reorganization. In the late 1980s, scientists proposed that RGCs produce spontaneous action potentials that in turn stimulate neurons in the LGN. In the 1990s, Carla Shatz and colleagues conducted a series of experiments using electrophysiological recordings, calcium imaging, and axonal labeling methods to document the patterns of spontaneous retinal activity in the visual system of the ferret. In the ferret, the formation of the synaptic layers in the LGN occurs from postnatal day

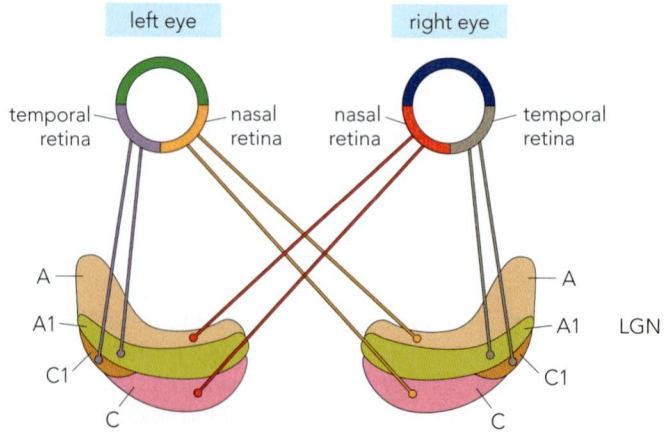

Figure 10.17 Retinal ganglion axons contact distinct layers within the lateral geniculate nucleus (LGN). Retinal axons that project from the temporal (T) or nasal (N) side of the retina become segregated into specific cellular layers in the LGN. In the ferret, temporal axons project mainly to the ipsilateral LGN to contact neurons in layers A1 and C1. The majority of axons from the nasal half of the retina reach the contralateral LGN to contact neurons in layers A and C. (Adapted from Sharma J & Sur M [2014] In: Biology and Diseases of the Ferret, (3rd Ed.); Fox JG & Marini RP (eds) Wiley and Sons, Inc. New Jersey.)

1 (P1) through P21, and eye opening and the onset of vision begin at P30. This provides scientists with a system to study the developmental changes in synaptic connections between RGCs and postsynaptic LGN neurons before and after vision begins.

In these experiments, the scientists demonstrated that RGCs produce spontaneous electrical activity that spreads quickly to adjacent RGCs in waves. The waves do not spread across the entire retina during this short period, but are propagated across a limited segment of the retina. At the completion of the wave, the previously active RGCs remain silent for up to two minutes. After one wave is inactivated, another wave begins at a new, apparently random site in the retina. Therefore, RGCs throughout the retina become active at different times. Additionally, the waves produced in each eye begin at different times and in different retinal locations. As a result, the stimulation that reaches the postsynaptic neurons in the various LGN layers is not synchronous. This asynchronous stimulation creates variable amounts of postsynaptic stimulation to each LGN neuron and ultimately influences axon projection patterns to each layer.

Competition between neurons determines which synaptic connections are stabilized

The repeated stimulation of a postsynaptic LGN neuron by RGCs causes the LGN neuron to begin to fire action potentials. Some of these LGN action potentials will occur at the same time as action potentials from some of the RGC inputs. When this synchronous firing occurs, the synaptic connection between those RGC axons and the LGN neuron becomes strengthened. However, if activation by the RGCs does *not* induce synchronous firing in the postsynaptic LGN neuron, then the presynaptic connections withdraw (**Figure 10.18**). This is an example of the concept of Hebbian plasticity in which "cells that fire together, wire together; those out of sync, lose their link." Thus, the organization of synaptic connections in the layers of the LGN is influenced by patterns of spontaneous electrical activity originating at different locations across each retina.

As in the NMJ, it appears that the differences in level of neural stimulation determine which synaptic connections are stabilized and which are lost. As noted in Chapter 9, muscle fibers initially receive multiple synaptic inputs. When neural activity was blocked at restricted regions along the postsynaptic muscle fiber, the synaptic connections at that region withdrew. However, if the entire NMJ was blocked, synaptic connections remained. Thus, only when differences in the levels of neural activity were present did the synapses reorganize (see Figures 9.20 and 9.21).

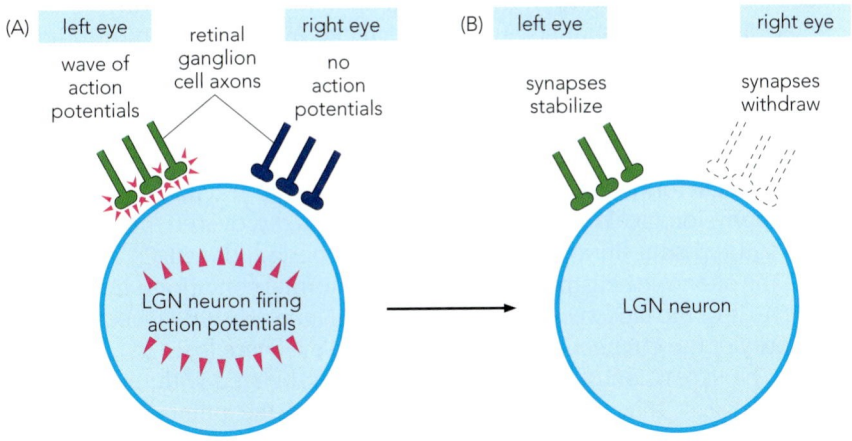

Figure 10.18 Activity levels determine which synapses stabilize in the lateral geniculate nucleus (LGN). Retinal ganglion cells (RGC) produce spontaneous waves of electrical activity that stimulate target neurons in the lateral geniculate nucleus (LGN). Each eye produces waves at different times so that the stimulation to a target neuron varies. (A) At the time a wave of action potentials spreads across RGCs in the left eye in this example, no action potentials are active in the right eye. The action potentials from the left eye are sufficient to activate a neuron in the LGN and that neuron begins to fire action potentials at the same time as the left RGCs. (B) The synchronous firing between RGC inputs from the left eye and the LGN neuron cause those RGC inputs to stabilize. Because synchronous firing did not occur between the RGC inputs from the right eye and the LGN neuron, the contacts from the right eye are withdrawn.

Similar results were found in the studies of RGC projections to the LGN. Experiments were conducted in which drugs that inhibit cAMP were injected into the eyes of ferrets during the period of synaptic reorganization in the LGN. The available levels of cAMP are known to influence the activity of retinal waves. When the drug was injected into both eyes, the RGCs from each eye failed to produce retinal waves. Because the RGC activity in each eye was equally inhibited, there were no differences in neural activity across the LGN layers, and therefore the axons of the RGCs spread throughout all the layers of the LGN. When the drug was injected into only one eye, however, the RGC axons from the treated eye withdrew connections in the LGN, while the axons from the untreated eye maintained connections with neurons in all layers of the LGN. Thus, when differences in neural activity are present, the postsynaptic neurons receiving the greater amount of stimulation—in this case, those innervated by the untreated eye—form stable synaptic connections. The spontaneous temporal and spatial differences in retinal wave activity therefore influence the synaptic organization in LGN cell layers. What is still unclear is whether the retinal waves provide cues to set up initial patterning in the LGN or whether the waves are the instructive cue that establishes the final LGN pattern of layers.

Neural activity resulting from early visual experience establishes ocular dominance columns in the primary visual cortex

Presynaptic neural activity was also found to influence synaptic organization in the primary visual cortex. Whereas neurons in the LGN stabilize synaptic connections independent of visual stimulation, the neurons in the primary visual cortex depend on visual stimulation to maintain synaptic connections.

Studies by David Hubel and Torsten Wiesel begun in the 1960s, for which they won the Nobel Prize in Physiology or Medicine in 1981, were among the first to detail the role of neural activity in establishing mature synaptic connections. As with the LGN, the layers of the primary visual cortex in several mammalian animal models are initially innervated by connections from both eyes. The RGCs project to neurons in the LGN, and those LGN neurons project to the primary visual cortex. In animal models that rely on binocular vision, such as cats and some monkeys, the inputs originating from both eyes initially overlap in layer IV of the primary visual cortex. However, during early postnatal development the projections segregate into distinct regions known as **ocular dominance columns (ODCs)**, which arise from alternating projections originating in the right and left eyes (**Figure 10.19**).

Under normal developmental conditions, visual experience shapes the patterning of the ODCs during early postnatal development. This occurs because each eye views a given visual stimulus from a slightly different angle. Thus, any visual stimulus will activate a different portion of each retina, and it is this difference in stimulation that ultimately leads to the formation of alternating columns in layer IV of the primary visual cortex that persist in adulthood (Figure 10.19A).

Hubel and Wiesel conducted a series of studies to determine how these ODCs formed. Using radiolabeled neuronal tracing methods and electrophysiological recordings, the researchers discovered that if one eye was sutured shut during the early postnatal period, the ODCs associated with the open eye became wider than under normal conditions. In contrast, the ODCs of the closed eye became smaller (Figure 10.19B). Over time, the majority of the connections from the sutured eye were lost.

The expansion of the size of the ODCs associated with the open eye could suggest that visual stimulation increased the number of synaptic connections that formed and stabilized. However, Hubel and Wiesel found

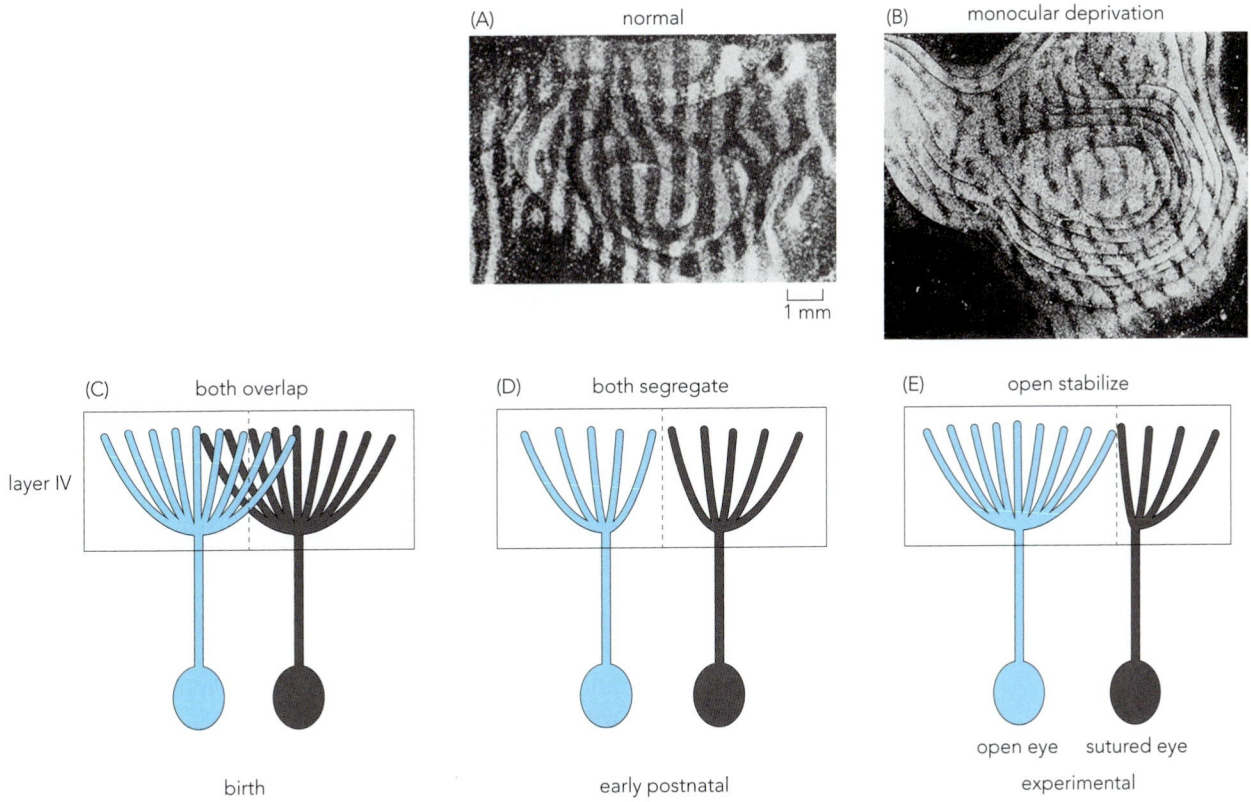

Figure 10.19 Studies reveal activity-dependent changes in synaptic organization of the primary visual cortex. (A) Layer IV of the primary visual cortex of the macaque monkey is normally organized into alternating columns representing input from each eye. In this example, one eye was injected with a radioactive tracer (light bands). The dark bands represent the areas receiving input from the eye that was not injected with the tracer. Each eye projects to alternating columns of equal size. (B) Experiments were conducted in which one eye was sutured shut so that only one eye received visual stimulation during postnatal development. When the radioactive tracer was injected into the open eye, it was found that the layer IV ocular dominance columns were larger (light bands) than those of the sutured eye (dark bands). This was found to be due to stabilization of existing contacts from the open eye and the loss of some connections from the sutured eye. (C–E) Schematic representation of the changes that occur. (C) At the time of birth, axonal projections from both eyes intermingle in layer IV of the visual cortex. (D) During postnatal development, the eyes normally segregate into ocular dominance columns, with each eye represented by alternating columns. (E) When one eye is sutured shut, the activity from the open eye becomes sufficient to maintain initial contacts. Thus, the columns for the open eye appear expanded compared to the sutured eye. (A and B, from Hubel DH, Wiesel TN & LeVay S [1977] *Philos Trans R Soc Lond Biol* 278:377–409.)

that the increased size of a cortical column was due to the maintenance of connections that would normally have been eliminated during development, rather than the formation of new connections. These results indicated that because the sutured eye no longer received visual stimulation, it was no longer able to provide competing neural activity to the ODC neurons it normally innervated. Therefore, the preexisting connections from the other, open eye were able to stabilize (Figure 10.19C–E). Similar to studies in the LGN, when TTX was injected into both eyes to prevent activation of the RGCs, the reorganization of ODCs failed to occur. Together, these experiments were among the first to demonstrate the importance of differing levels of neural activity in regulating the stabilization of synaptic connections.

Another major finding from these studies was that the effect was reversible during a limited period in development. If the sutured eye was re-opened and the previously opened eye sutured shut during the period when synaptic connections were still forming, the columns associated with the previously shut eye widened. Similarly, the columns from the previously open eye became smaller. Thus, synaptic reorganization in response to visual stimulation was plastic during a limited period of postnatal development. This window of plasticity is now called a **critical period**—a time in neural development during which synaptic reorganization can take place. After this period, changes to synaptic connections are less likely to occur and any changes that do take place are much more limited compared to changes that occur prior to the close of the critical period. In the studies by Hubel and Wiesel, for example, when the sutured and opened eyes were reversed later in development or in adulthood, the width of the cortical columns no longer reversed as they did during the critical period.

The molecular mechanisms underlying activity-dependent changes in CNS synaptic connections are beginning to be elucidated. It appears that the expression of many proteins used during the initial formation of synaptic contacts, such as BDNF and ephrins, are also influenced by neuronal activity. The expression of these proteins can then modify synaptic structure and may further enhance neural transmission by influencing the number of receptors anchored at the postsynaptic membrane.

Homeostatic plasticity contributes to synaptic activity

An additional form of plasticity also contributes to the structural and functional changes in synapses that accompany normal postnatal development. **Homeostatic plasticity** occurs when pre- and postsynaptic cells modify synaptic elements and synaptic output in response to changes in the overall level of neural activity. This form of plasticity occurs in the CNS and NMJ of vertebrates and invertebrates (see also Box 9.1) and has become a particularly active area of research in the past decade. Whereas Hebbian modifications take place within a period of minutes, homeostatic changes occur over a period of hours to days. Hebbian and homeostatic plasticity appear to balance one another to maintain stable levels of neural firing. Maintaining action potential firing patterns is necessary to sustain a stable functional network of neurons as other synaptic changes, such as those associated with learning and memory, take place.

Examples of the principles of homeostatic changes are illustrated in **Figure 10.20**. The desired baseline level of neural firing and the expression of postsynaptic neurotransmitter receptors and ion channels are shown in the first panel. If the overall level of neural stimulation is then lowered over an extended period, the postsynaptic cell will compensate for the reduced activity by increasing the number of neurotransmitter receptors and ion channels expressed. Conversely, if neural stimulation is maintained at a high level, homeostatic changes take place to reduce the levels of postsynaptic receptors and ion channels. In both examples, these changes allow the neurons to adjust to an altered environment and return firing patterns to the baseline level. This form of homeostatic plasticity, also called **synaptic scaling**, therefore maintains the desired Hebbian modifications, but prevents further widespread changes in synaptic connectivity that would be detrimental to nervous system function.

Other structural changes are also observed during homeostatic plasticity. As noted above, the synapse size and spine density are correlated with

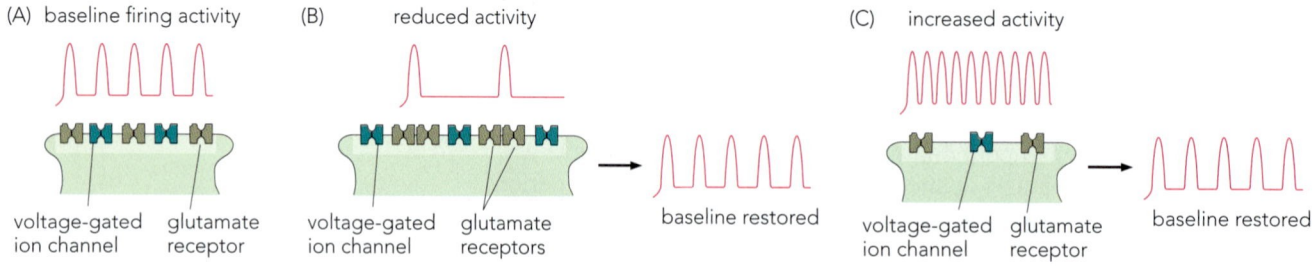

Figure 10.20 Homeostatic plasticity ensures stable neural firing patterns. (A) Neurons have a desired baseline level of neural firing activity. Each postsynaptic neuron expresses a given number of neurotransmitter receptors and ion channels. (B) If neural activity is reduced for a prolonged period of time, the postsynaptic neurons will increase the number of postsynaptic neurotransmitter receptors and ion channels to restore neural firing patterns to baseline levels. (C) If neural activity is increased over a prolonged period of time, the postsynaptic cells will decrease expression of neurotransmitter receptors and ion channels to restore neural activity to baseline levels.

synaptic strength. If the synaptic signaling is altered experimentally through sensory deprivation or decreasing the number or function of postsynaptic neurotransmitter receptors, subsequent changes in dendritic spines will occur. In response to decreased activity, the number of spines increases and the size of the associated PSDs increase, thus providing additional space for postsynaptic elements to accumulate. In addition, the presynaptic neurons will increase synaptic vesicle release and the size of the active zone in response to sensory deprivation to further offset the effects of reduced input.

Homeostatic changes also occur at inhibitory synapses (**Box 10.2**). In the visual system, experimentally induced lesions made at restricted sites of the retina led to a decrease in synaptic activity initially, but over a period of 48 hours, neural activity returned to baseline as the size of dendritic spines on the excitatory pyramidal neurons of the visual cortex increased and the number of spines on inhibitory neurons decreased. These results suggest that inhibitory neurons help restore baseline firing by reducing their inhibitory signaling as neural activity decreases.

A number of studies have also indicated that the amount of synaptic elimination during postnatal development appears to decrease in conditions of sensory deprivation, presumably in an effort to reduce the extent of synaptic connections lost.

Intrinsic and environmental cues continue to influence synapse organization at all ages

It is rather difficult to establish when development of the nervous system ends. This book focuses on the events that occur from neural induction through the initial stages of synaptogenesis and synaptic reorganization. However, changes in synaptic connections continue throughout the lifespan of an animal. While some neuronal populations exhibit the greatest extent of synaptic refinement primarily during early postnatal stages, others continue to modify synaptic connections until puberty. In many vertebrate species, the changes in hormone levels that accompany puberty are thought to limit the ability of some synapses to further modify connections. Thus, hormonal changes stabilize or "crystalize" synapses, often by regulating the stability and organization of postsynaptic elements such as neurotransmitter receptors. In humans, for example, the ability to acquire new language skills or a second language begins to decrease following puberty. Examples of synapse crystallization are also seen in male songbirds. Once testosterone levels begin to increase during puberty, the birds lose the ability to learn new songs. This has been shown to be a result of the inability of the birds to reorganize synaptic connections in the CNS regions associated with song learning

In other areas of the nervous system, ongoing changes in synaptic connections occur throughout adulthood. For example, the neural processes underlying learning and memory require changes in synaptic connections. Numerous studies in the hippocampus, the area of the brain most associated with learning and memory, have provided several examples of how cellular mechanisms regulate the selective strengthening and weakening of synapses in response to various types of stimulation. At excitatory synapses, strengthened synapses have an increased dendritic spine size and more AMPA receptors in the postsynaptic membrane. These changes allow for enhanced neural transmission and are the basis of the process called long-term potentiation (LTP). The opposite changes are seen as synapses weaken. Weakened synapses lose dendritic spines and have fewer AMPA receptors in the postsynaptic membrane, leading to reduced efficiency of neural transmission. These changes are part of the process called long-term depression (LTD). Recent experiments suggest that pro-BDNF signaling through the p75NTR receptor may lead to a reduction in spine number and

Box 10.2 Developing Neuroscientists: Changes in ion channel expression following the onset of hearing

Briana Carroll graduated from Oberlin College in 2011, where she majored in neuroscience. She completed a Ph.D. at Florida State University with Richard Hyson, where she studied developmental changes in the auditory system following hearing onset. She is currently a postdoctoral fellow at the University of Chicago.

Synaptic plasticity refers to changes in the connections made between two neurons. Another form of plasticity occurs during development known as homeostatic plasticity, which involves changes in pre- and postsynaptic elements in a single neuron. Homeostatic intrinsic plasticity occurs within a single neuron and refers to changes in the expression or biophysical properties of ion channel proteins that characterize a given neuron. By altering the expression or properties of ion channels, the behavioral characteristics of a neuron are also altered, ultimately leading to changes in behavioral outcomes. In many cases, changes in the number of synaptic inputs, or their firing patterns, contribute to intrinsic plasticity. The chick auditory system has been used as a model system for understanding the normal development of CNS connections and the effects of external stimuli on both synaptic and intrinsic neuronal properties.

The cochlear nuclei are the site of termination for auditory nerve fibers of the eighth cranial nerve (CN VIII), which connects to hair cells of the inner ear (see Chapter 6). The chick cochlear nuclei are divided into three regions designated nucleus magnocellularis (NM) and nucleus angularis (NA), which receive input from the ipsilateral CN VIII and nucleus laminaris (NL), which receives bilateral input from the NM (shown ipsilaterally in **Figure 1**). The neurons from these nuclei send ascending branches to the superior olivary nuclei (SON), which are found on each side of the brainstem. In addition, descending fibers from SON provide inhibitory connections to the three regions of the cochlear nuclei. This descending inhibitory firing is activated in response to loud or high-frequency sounds and is thought to help shape the response of the ascending signals by suppressing auditory neuron output. Thus, the SON neurons provide a type of gain control to suppress auditory input when necessary.

If the intrinsic plasticity of SON neurons is important for maturation of the gain control, then the properties of ion channels in SON neurons would be expected to change following the onset of hearing. To investigate these possibilities the properties of SON neurons were evaluated at different developmental time points prior to and after hearing onset.

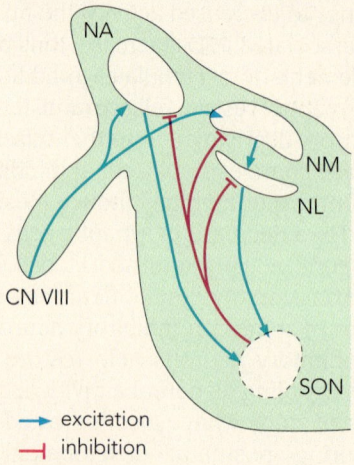

Figure 1 The cochlear nuclei in the chick auditory brainstem receive excitatory and inhibitory input. The cochlear nuclei of the chick auditory brainstem are divided into the nucleus angularis (NA), nucleus magnocellularis (NM), and nucleus laminaris (NL). Excitatory input originates from the eighth cranial nerve (CN VIII). These nuclei send ascending projections to the superior olivary nuclei (SON). The neurons in the SON also send descending, inhibitory fibers to all three regions of the cochlear nuclei. These inhibitory fibers are activated in response to loud or high-frequency sounds to suppress auditory output when needed. Ion channel expression increases after the onset of hearing to allow this inhibitory control to occur.

The ion channel properties of single SON neurons were evaluated using the patch-clamp recording technique. Over the course of normal development, the neurons became more excitable following the onset of hearing, indicating greater ion channel expression. In a subset of recordings, drugs that target sodium and potassium channels were applied. Altering the conductance of these channels had little effect prior to hearing onset, further suggesting that ion channel expression was more limited prior to the onset of hearing. Together, the studies suggest that an increased expression of ion channels facilitates the increased excitability observed in SON neurons following hearing onset.

It is still unclear whether hearing onset itself triggers increased ion channel expression. Whatever the mechanism, increased ion channel expression is an important developmental step that provides mature neurons in the SON the ability to exert gain control in the lower NM, NA, and NL auditory nuclei.

a weakening of synaptic transmission in at least some CNS neurons. This would be similar to the differential effects of BDNF-TrkB and pro-BDNF-p75NTR signaling noted in the NMJ (see Figure 9.22).

In addition to the normal changes in synaptic connections that take place during neural development and adulthood, there are a number of conditions that lead to a loss or disorganization of synapses. For example, altered synaptogenesis may underlie neurodevelopmental disorders such as autism and schizophrenia. Additionally, environmental factors, such as prenatal exposure to drugs (**Box 10.3**) or early postnatal exposure to life stress (**Box 10.4**), can also alter normal synaptic connections.

Box 10.3 Developing Neuroscientists: The effects of prenatal cocaine exposure on developing and adult neural connections

Zeeba Daruwalla Kabir graduated from Oberlin College in 2005 with a major in neuroscience. She then completed her Ph.D. in Neuroscience at the Weill Cornell Medical College. She is currently a postdoctoral research fellow at Weill Cornell in the Department of Pediatric Neurology.

As the central nervous system develops, synapse formation and plasticity can be altered in response to substances in the environment. Environmental changes that occur during prenatal or early postnatal development often have a lasting impact on CNS function. One example of this is the use of certain drugs during pregnancy. During the 1980s, crack cocaine became a common drug of abuse in the United States. During this period, approximately 7.5 million children in the United States were exposed to cocaine prenatally. Children exposed to cocaine prenatally often have deficits in learning and short-term memory in adulthood. Over the years, studies have also shown that the effects of prenatal cocaine on behavior are often subtle, but become increasingly apparent in stressful or demanding situations.

Several animal models have been developed to evaluate the effects of prenatal cocaine exposure on neural development. There is a consensus that prenatal cocaine exposure leads to significant alterations in the structure and function of several brain regions, including the prefrontal cortex (PFC). The PFC is the region of the brain associated with executive functions as well as short-term memory. When cocaine is administered to pregnant mice, it readily crosses the placental barrier and binds the dopamine, serotonin, and noradrenergic transporters in the fetal brain. Under normal conditions these receptor transporters are used to clear excess neurotransmitter from the synaptic cleft. However, the binding of cocaine to the transporters interferes with this mechanism and leads to an increase in the synaptic concentrations of neurotransmitters. Binding of cocaine to the dopamine transporter has been shown to be the main contributor to the drug's effects on the PFC, a region that receives strong dopaminergic input.

In recent years, studies have suggested that many of the cellular changes seen in the PFC following *in utero* cocaine exposure are due to changes in the expression of brain-derived neurotrophic factor (BDNF). This altered BDNF expression leads to changes in neuronal migration and synapse formation. For example, in the developing forebrain, prenatal cocaine exposure impairs both the radial migration of the excitatory glutamatergic neurons and the tangential migration of GABAergic interneurons in the PFC. This ultimately causes an imbalance of excitatory–inhibitory innervation to this region. In the fetal mouse brain, intrauterine cocaine exposure also results in a significant, albeit transient, decrease in BDNF protein levels that appears responsible for the deficit in tangential migration of GABAergic interneurons observed in these animals.

The altered dopaminergic transmission in the cortex of prenatal-cocaine-exposed animals has also been thought to contribute to alterations in the expression of markers of synaptic plasticity, specifically via impaired BDNF signaling. In the adult medial PFC of prenatal-cocaine-exposed animals, there is no change in the constitutive expression of BDNF protein, but there is a reduction in its release in response to neural activity. This change is thought to be a consequence of decreased constitutive binding of certain transcription factors at the promoters of activity-driven regions of the BDNF gene. These findings provide an example of how BDNF expression is regulated by environmental–genetic interactions.

Recently, it has also been shown that adult mice that were exposed to cocaine *in utero* have a deficit in the ability to recall a cue-induced fear memory as a result of decreased BDNF protein levels in the medial PFC. However, if BDNF was infused into the medial PFC region at the time when the memory would normally be extinguished, the mice were then able to respond to the fear-inducing cue. These findings highlight the persistent impact of prenatal cocaine exposure on the development and function of the PFC and shed light onto possible therapeutic strategies for affected individuals.

Box 10.4 Developing Neuroscientists: Understanding genetic susceptibility to mood disorders following early life stress

After graduating from Oberlin College in 2014 with a major in biology, Joshua Kogan became a research assistant at Rockefeller University in the laboratory of Bruce McEwen, assisting with research investigating the impact of stress hormones on brain plasticity. He now attends medical school.

Early life stress during childhood, such as that caused by abuse or neglect, leads to an increased risk of developing mood disorders in adulthood. As demonstrated in animal models, early life stress (ELS) alters synaptic plasticity, dendritic spine number, and neuronal excitability. However, not all individuals who experience ELS develop mood disorders in adulthood. Researchers are exploring how gene and environment interactions determine why some individuals are more susceptible to mood disorders following ELS.

Studies in rodents have demonstrated that chronically elevated cortisol, such as that which occurs following ESL, can contribute to depression-like and anxiety-like behaviors and impaired hippocampal-dependent memory tasks. Some populations of hippocampal neurons show morphological changes, such as decreased spines, dendritic retraction, and changes in neuronal activity. Humans with histories of stress and depression exhibit smaller hippocampal volume and alterations in the ratio of gray matter matter to white matter, as revealed through brain imaging studies. Since many parts of the hippocampus show morphological and functional changes in response to different types of stress, this region is a promising target for exploring the changes in gene expression that can link changes in neuron structure to observed behavioral changes in mood disorders.

Depression and mood disorders are heterogeneous in their causes and symptomology, making it difficult to predict which treatments will be successful for different individuals. Selective serotonin reuptake inhibitors (SSRIs) are the most commonly prescribed drugs for depression and other mood disorders. Although these drugs have been on the market for decades, their mechanism of action remains unclear. One theory is that antidepressants work by opening "windows of plasticity," making a person more likely to benefit from behavioral therapy and a positive environment. It is also thought that some individuals carry genes that make it more likely they will respond to treatment.

One example is the gene encoding a serotonin transport protein. The transporter helps clear excess neurotransmitter from the synaptic cleft and is the target of SSRIs. There are two different variants in the promoter region of this serotonin transport gene termed long (L) and short (S). The long allele leads to increased expression of the protein. Researchers have found that individuals with the L allele who have experienced childhood abuse or neglect have a decreased probability of a major depressive episode in adulthood, compared to those carrying the S allele. Carrying the S allele or exposure to stress may not be enough on their own to cause depression, but the risk is much higher when these factors are combined. Other work has found that in a nurturing environment, individuals with the S allele can actually excel and do better than their L-allele-carrying peers. These types of alleles are said to be "context sensitive." Just like antidepressants, these genes may increase the plasticity of the brain, allowing for a positive environment to limit the chances of developing depression.

This research emphasizes the importance of early intervention in people at high risk for the development of mood disorders. Early treatment may be able to reprogram neural circuitry and allow individuals to successfully adapt to subsequent stressful situations. Scientists are using various animal models to continue to explore how gene–environment interactions impact the likelihood that any given individual will develop a mood disorder. This research may ultimately define the key mechanisms underlying the development of mood disorders, and perhaps lead to the identification of novel treatments.

SUMMARY

Synaptogenesis between neurons in the CNS is more difficult to study than in the NMJ for a number of reasons, including the size of the neurons and the complexity of the many excitatory and inhibitory synapses that form on each individual neuron. While mechanisms of synaptogenesis in the CNS are less well understood than those in the NMJ, remarkable progress has been made in recent years, particularly in identifying signaling molecules that induce the formation of pre- and postsynaptic specializations and mediate synaptic stabilization.

Initial synaptic contacts in the CNS are often induced by interactions between cell adhesion molecules (CAMS). PDZ-containing proteins that interact with local scaffolding proteins then help stabilize select connections. In both the NMJ and CNS there are reciprocal signals exchanged between synaptic partners. These signals lead to the induction of active zone proteins and the clustering of synaptic vesicles and ion channels in the presynaptic terminal. Reciprocal signals also mediate the clustering and stabilization of neurotransmitter receptors and ion channels at the postsynaptic membrane. Additionally, both NMJ and CNS synapses rely on signals from glial cells to form and maintain synapses. Perhaps surprisingly, some of the proteins known to have essential roles in the formation of the NMJ do not appear to function similarly during CNS synaptogenesis. For example, although agrin and neuregulin are detected in many CNS neurons, they are not required for synapse formation. Moreover, rapsyn, which is the protein used for clustering acetylcholine receptors (AChR) at the NMJ, is not used in the CNS, even at central cholinergic synapses. Thus, the identification of a synaptic protein in one region of the nervous system does not always indicate the same function at another synaptic site. In the CNS, BDNF/TrkB, Eph/ephrin, and Wnt/Fz signaling pathways are among the many identified to date that induce the clustering of synaptic specialization in pre- and postsynaptic neurons.

Scientists continue to study how individual connections are selected for elimination during neural development. In most cases, synapse elimination does not appear to result from any inherent defect in the connections that are lost. It has also been noted that synapse elimination does not lead to a net decrease in synaptic effectiveness. Instead, it appears that the remaining connections become more efficient at stimulating the postsynaptic cells.

CNS synapses are often modified throughout adulthood. This plasticity can be influenced by a combination of environmental factors and neural experience. In some cases, intrinsic or extrinsic factors lead to disordered synaptogenesis. One long-term goal of studying how synapses form is to develop therapeutic interventions to promote or restore malformed synaptic connections.

FURTHER READING

Ahmad-Annuar A, Ciani L, Simeonidis I et al. (2006) Signaling across the synapse: a role for Wnt and Dishevelled in presynaptic assembly and neurotransmitter release. *J Cell Biol* 174(1):127–139.

Aoto J, Ting P, Maghsoodi B et al. (2007) Postsynaptic ephrinB3 promotes shaft glutamatergic synapse formation. *J Neurosci* 27(28):7508–7519.

Beekman JM & Coffer PJ (2008) The ins and outs of syntenin, a multifunctional intracellular adaptor protein. *J Cell Sci* 121(9):1349–1355.

Biederer T, Sara Y, Mozhayeva M et al. (2002) SynCAM, a synaptic adhesion molecule that drives synapse assembly. *Science* 297(5586):1525–1531.

Calabrese B, Wilson MS & Halpain S (2006) Development and regulation of dendritic spine synapses. *Physiology* 21:38–47.

Carrasco MA, Castro P, Sepulveda FJ et al. (2007) Regulation of glycinergic and GABAergic synaptogenesis by brain-derived neurotrophic factor in developing spinal neurons. *Neurosci* 145(2):484–494.

Cherra SJ 3rd & Jin Y (2015) Advances in synapse formation: forging connections in the worm. *WIREs Dev Biol* 4(2):85–97.

Christopherson KS, Ullian EM, Stokes CC et al. (2005) Thrombospondins are astrocyte-secreted proteins that promote CNS synaptogenesis. *Cell* 120(3):421–433.

Davis GW (2013) Homeostatic signaling and the stabilization of neural function. *Neuron 80*(3):718–728.

DeFelipe J (2015) The dendritic spine story: an intriguing process of discovery. *Front Neuroanat* 9:14.

Farias GG, Alfaro IE, Cerpa W et al. (2009) Wnt-5a/JNK signaling promotes the clustering of PSD-95 in hippocampal neurons. *J Biol Chem* 284(23):15857–15866.

Flavell SW & Greenberg ME (2008) Signaling mechanisms linking neuronal activity to gene expression and plasticity of the nervous system. *Annu Rev Neurosci* 31:563–590.

Giachello CN, Montarolo PG & Ghirardi M (2012) Synaptic functions of invertebrate varicosities: what molecular mechanisms lie beneath. *Neural Plast* 2012:670821.

Goodman CS & Shatz CJ (1993) Developmental mechanisms that generate precise patterns of neuronal connectivity. *Cell* 72 Suppl:77–98.

Graf ER, Zhang X, Jin SX et al. (2004) Neurexins induce differentiation of GABA and glutamate postsynaptic specializations via neuroligins. *Cell* 119(7):1013–1026.

Hamilton AM & Zito K (2013) Breaking it down: the ubiquitin proteasome system in neuronal morphogenesis. *Neural Plast* 2013:196848

Hruska M & Dalva MB (2012) Ephrin regulation of synapse formation, function and plasticity. *Mol Cell Neurosci* 50(1): 35–44.

Keck T, Keller GB, Jacobsen RI et al. (2013) Synaptic scaling and homeostatic plasticity in the mouse visual cortex in vivo. *Neuron* 80(2):327–334.

Keith D & El-Husseini A (2008) Excitation Control: Balancing PSD-95 Function at the Synapse. *Front Mol Neurosci* 1:4.

McClelland AC, Sheffler-Collins SI, Kayser MS & Dalva MB (2009) Ephrin-B1 and ephrin-B2 mediate EphB-dependent presynaptic development via syntenin-1. *Proc Natl Acad Sci USA* 106(48):20487–20492.

McClelland AC, Hruska M, Coenen AJ et al. (2010) Trans-synaptic EphB2-ephrin-B3 interaction regulates excitatory synapse density by inhibition of postsynaptic MAPK signaling. *Proc Natl Acad Sci USA* 107(19):8830–8835.

Mulligan KA & Cheyette BN (2012) Wnt signaling in vertebrate neural development and function. *J Neuroimmune Pharmacol* 7(4):774–787.

Munno DW & Syed NI (2003) Synaptogenesis in the CNS: an odyssey from wiring together to firing together. *J Physiol* 552(1):1–11.

Opperman KJ & Grill B (2014) RPM-1 is localized to distinct subcellular compartments and regulates axon length in GABAergic motor neurons. *Neural Dev* 9:10.

Park TJ & Curran T (2014) Neurobiology: Reelin mediates form and function. *Curr Biol* 24(22):R1089–1092.

Penn AA, Riquelme PA, Feller MB & Shatz CJ (1998) Competition in retinogeniculate patterning driven by spontaneous activity. *Science* 279(5359):2108–2112.

Petzoldt AG & Sigrist SJ (2014) Synaptogenesis. *Curr Biol* 24(22):R1076–1080.

Scheiffele P (2003) Cell-cell signaling during synapse formation in the CNS. *Annu Rev Neurosci* 26:485–508.

Sheng M & Kim E (2011) The postsynaptic organization of synapses. *Cold Spring Harb Perspect Biol* 3(12):a005678.

Sharma J & Sur M (2014) The Ferret as a Model for Visual System Development and Plasticity. In Biology and Diseases of the Ferret, 3rd ed. (Fox JG & and Marini RP eds), pp 711–734. John Wiley and Sons, Inc.

Tian X & Wu C (2013) The role of ubiquitin-mediated pathways in regulating synaptic development, axonal degeneration and regeneration: insights from fly and worm. *J Physiol* 591: 3133–3143.

Xu NJ, Sun S, Gibson JR & Henkemeyer M (2011) A dual shaping mechanism for postsynaptic ephrin-B3 as a receptor that sculpts dendrites and synapses. *Nat Neurosci* 14: 1421–1429.

Yogev S & Shen K (2014) Cellular and molecular mechanisms of synaptic specificity. *Annu Rev Cell Dev Biol* 30:417–437.

Ziv NE & Smith SJ (1996) Evidence for a role of dendritic filopodia in synaptogenesis and spine formation. *Neuron* 17:91–102.

Glossary

acetylcholine (ACh) Neurotransmitter released by a motor neuron at the neuromuscular junction in vertebrates and some invertebrates. It is also used in the vertebrate CNS, as an excitatory or modulatory neurotransmitter, and in the autonomic nervous system. In some invertebrates, it is the major excitatory neurotransmitter of the CNS.

acetylcholine esterase (AChE) Enzyme that catalyzes the breakdown of acetylcholine. At the neuromuscular junction, an isoform of the enzyme is secreted by muscle cells and becomes concentrated in the synaptic basal lamina.

acetylcholine receptor (AChR) Receptor for the neurotransmitter acetylcholine. The nicotinic AChRs (nAChRs) are nonselective cation channels; they are the postsynaptic receptor at the vertebrate neuromuscular junction and function as excitatory receptors at some CNS synapses. The metabotropic AChRs (muscarinic AChRs or mAChRs) are G-protein-coupled receptors that play a modulatory role.

action potential An elementary unit of nerve impulses that axons use to convey information across long distances. It is all-or-none, regenerative, and propagates unidirectionally in the axon.

activin Protein of the transforming growth factor β (TGFβ) superfamily. One of its many roles during embryogenesis is the induction of mesoderm.

adherens junction Protein complex that forms connections between cells by linking the actin filaments of one cell to those of an adjacent cell. Examples include the sites of attachment between apical progenitor cells and the ventricular surface of the developing neural tube.

adhesion molecule Protein located on a cell surface that mediates binding between cells or between a cell and the extracellular matrix.

agrin Protein secreted by motor neurons that induces aggregation of acetylcholine receptors in the muscle.

agrin hypothesis Idea that agrin released by an innervating motor nerve terminal becomes concentrated in the synaptic basal lamina to induce the clustering of acetylcholine receptors in the muscle directly opposite the nerve terminal.

agrin–ACh hypothesis Idea that both agrin and acetylcholine (ACh), released by an innervating motor nerve terminal, regulate the clustering and dispersal of acetylcholine receptors, respectively. Agrin is thought to stabilize the receptors in the central region of a muscle fiber to prevent them from being displaced upon exposure to ACh.

anencephaly A severe type of neural tube defect that results when the anterior portion of the neural tube fails to close; the cranial portions of the brain and the back of the skull fail to fuse. The condition is fatal.

animal pole The portion of the egg that has the least amount of yolk, located opposite the vegetal pole. It is the site of origin for cells that form ectoderm and mesoderm.

anterior neural border (ANB) In fish, the anterior-most region of the neural ectoderm that becomes a signaling center that secretes molecules to influence the patterning of forebrain regions. Called the anterior neural ridge in other vertebrates.

anterior neural ridge (ANR) The anterior-most region of the neural ectoderm that becomes a signaling center that secretes molecules to influence the patterning of forebrain regions in mice, chick, and other vertebrates. Called the anterior neural border in fish.

anterior visceral endoderm (AVE) In mice, a population of extraembryonic cells that lies below the anterior neural plate. The AVE serves as a signaling center to pattern anterior portions of the neural tube as well as to specify cells that form the anterior neural ridge. Similar to the hypoblast in other species.

anterior–posterior (anteroposterior) axis Refers to the main body axis that extends from head to tail and the neuraxis that extends from the brain to the spinal cord. Also called the rostral–caudal axis.

apoptosis A form of programmed cell death characterized by distinct cellular changes, including the fragmentation of DNA. A naturally occurring developmental process initiated by extracellular signals or intrinsic genetic mechanisms; mediated by intracellular signaling pathways involving enzymes called caspases.

apoptosome Heptamer of Apaf1 proteins that forms on activation of the intrinsic apoptotic pathway; it recruits and activates initiator caspases that subsequently activate downstream effector caspases to induce apoptosis.

apoptotic body Discrete, membrane-bound structure containing condensed fragments of organelles and the nucleus; buds off from a cell undergoing apoptosis.

astrocyte Star-shaped glial cell of the vertebrate central nervous system that plays a variety of roles in neuronal development, synaptic maintenance, and neuronal communication.

astrotactin A glycoprotein found on the surface of migrating neurons in the cerebellum and cerebral cortex; facilitates attachment of the neurons to radial glial cells.

autocrine A type of cellular communication in which a cell releases a signaling molecule that then binds to a corresponding receptor on the surface of that same cell.

autonomic nervous system (ANS) The collected parts of the nervous system that regulate the function of internal organs, including the contraction of smooth and cardiac muscles and the activities of glands.

axon terminal The distal, slightly enlarged end of an axon that forms the presynaptic connection at a synapse.

axon A long, thin process of a neuron, it often extends far beyond the soma and propagates and transmits signals to other neurons or muscle at its presynaptic terminals.

Bergmann glia Specialized radial glia found in the cerebellum that extend multiple, parallel cell processes which help guide cerebellar granule cells from the external to the internal granule cell layer.

blastocoel The central, fluid-filled cavity of an early-stage embryo.

blastocyst The early-stage, pre-implantation embryo of mammals. The structure consists of blastomeres surrounding a large blastocoel cavity.

blastoderm A group of cells (blastomeres) that forms above the central cavity (blastocoel) or yolk of the early-stage embryo.

blastodisc The flat, disc-shaped, early-stage embryo that forms two cell layers (epiblast and hypoblast) above the blastocoel cavity; characteristic of birds, reptiles, and some mammals.

blastomere A cell produced in the zygote during cell division.

blastopore The opening on the outer surface of an early-stage embryo through which cells migrate during gastrulation.

blastula A spherical-shaped, early-stage embryo consisting of blastomeres surrounding the blastocoel cavity; characteristic of amphibians and similar to the blastocyst in mammals. Often used to refer to embryos at similar stages of development in multiple species.

bone morphogenetic protein (BMP) Protein of the transforming growth factor β (TGFβ) superfamily; isoforms play multiple roles in embryogenesis, including induction of epidermal ectoderm and patterning along the dorsal–ventral axis of the neural tube.

Cajal–Retzius (CR) cell Cell located in the marginal zone of the developing neocortex and hippocampus; secretes the protein Reelin.

canonical pathway The established signaling cascade for a given molecule. When referring to Wnt signaling, the Wnt-β-catenin pathway that leads to the transcription of target genes.

caspase Intracellular protease that is involved in mediating the intracellular events of apoptosis.

cell autonomous Of a gene, acting in the cell that produces the gene product.

CDK inhibitory protein A group of small proteins that halt progression of the cell cycle at different stages by blocking the activity of specific cyclin-dependent kinases (CDKs). Primarily involved at the gap 1 (G1) and synthesis (S) phases.

cell cycle The sequence of steps through which a cell progresses prior to cell division. The stages include gap 1 (G1), synthesis (S), gap 2 (G2), and mitosis (M).

cell fate The final cell type established from a precursor cell during the course of development.

cell fate determination The process by which a cell becomes committed to forming a particular cell type. Once determined, no further developmental or experimental manipulations can alter the type of cell that forms.

cell line culture A group of homogenous cells maintained *in vitro* that descend from a single cell and therefore have the same genotype and phenotype.

cell marker A protein or gene expressed in a limited number of cell types and used to distinguish the identity a particular cell.

cellular blastoderm In *Drosophila*, the stage of development when the nuclei located around the cortex of the blastoderm become separated by cell membranes to create individual cells. Forms from the syncytial blastoderm.

cellular determination See *cell fate determination*.

cellular differentiation The process of acquiring the characteristics of a particular type of cell. Cellular differentiation follows cellular determination.

central domain (C-domain) The region of the growth cone that lies closest to the neurite shaft and contains microtubules and organelles. Also called the central mound.

central mound The region of the growth cone that lies closest to the neurite shaft and contains microtubules and organelles. Also called the central domain (C-domain).

central nervous system (CNS) The brain and spinal cord in vertebrates; the brain and nerve cord in some invertebrates.

cephalic Relating to the head.

Cerberus A secreted protein present in the organizer and anterior visceral endoderm that blocks the activity of bone morphogenetic proteins and Wnt. In the anterior neural tube, Cerberus acts to prevent the posteriorization of forebrain regions.

chemical synapse A site at which information is transferred from a neuron to another cell across a synaptic space (cleft), where presynaptic nerve terminal specializations release chemical neurotransmitters and postsynaptic specializations specifically bind the neurotransmitter.

chemoaffinity hypothesis The concept that there are matching proteins or "chemical tags" located on the surface of axonal growth cones and their target cells. Originally proposed by Roger Sperry to describe how the axons of retinal ganglion cells reached their target cells in the optic tectum.

chemotropism The directed extension of a growth cone up or down a chemical gradient.

choice point An intersection of two cellular surfaces on which an elongating axon has the potential to extend.

choline acetyltransferase (ChAT) Enzyme needed to synthesize the neurotransmitter acetylcholine.

chordin One of the first neural inducers identified; promotes formation of neural tissue by blocking the epidermal-inducing actions of bone morphogenetic proteins (BMPs).

chordamesoderm A band of axial mesoderm that forms the notochord and related structures in vertebrate embryos.

ciliary neurotrophic factor (CNTF) A cytokine of the gp130 family of proteins that promotes survival and neurite outgrowth of several neuronal populations in the developing and adult nervous system. Initially identified in the chick eye.

commissure A bundle of nerve fibers that connects the two sides of the brain or spinal cord.

conditional knockout mouse The process of disrupting a gene in a specific spatiotemporal pattern or an animal in which a gene has been disrupted in a specific spatiotemporal pattern. The most common strategy for generating conditional knockouts in mice uses Cre/*loxP*-based recombination. It usually involves inserting a pair of *loxP* elements in introns that flank (an) essential exon(s) of a gene of interest. The gene of interest is only disrupted in cells in which Cre is active or in cells derived from a progenitor in which Cre was active.

contralateral Of the other side of the midline. For example, a contralateral axonal projection is an axon that crosses the midline and terminates on the side of the nervous system opposite the soma.

cortical plate (CP) A transient layer of cells in the developing neocortex that forms between the marginal zone and preplate. The cells ultimately form layers II through VI of the adult neocortex.

cranial neural crest Neural crest cells that originate from the dorsal surface of the neural tube between the midbrain and rhombomere 6. Gives rise to structures associated with the head and neck, including cranial ganglia and skeletal structures.

critical period A sensitive period during development when experience plays an important role in shaping the wiring properties of the brain.

cyclin-dependent kinase (CDK) Protein kinase that has to be complexed with a cyclin protein in order to act. Different cyclin–CDK complexes trigger different steps in the cell division cycle by phosphorylating specific target proteins.

cyclin Protein that periodically rises and falls in concentration in step with the eukaryotic cell cycle. Cyclins activate crucial protein kinases (called cyclin-dependent protein kinases or CDKs) and thereby help control progression from one stage of the cell cycle to the next.

cytokine Extracellular signal protein or peptide that acts as a local mediator in cell–cell communication.

delaminate To separate into different layers. When referring to neural crest cells, the process by which the cells separate from the neuroepithelium of the neural tube.

dendrite Neuronal process that is typically branched with tapered ends; receives synaptic input.

dendritic spine A small protrusion on a dendrite of certain neurons that receives synaptic input from a partner neuron. The thin spine neck creates chemical and electrical compartments for each spine such that it can be modulated independently from neighboring spines.

dermamyotome A segment of a somite that later gives rise to dermis, skeletal muscle, and vascular tissue.

Dickkopf (Dkk) A secreted protein that binds to the LRP5 and LRP6 subunits of the Wnt receptor complex to block Wnt signaling. In the anterior neural tube, Dkk acts to prevent the posteriorization of forebrain regions.

diencephalon One of the five secondary brain vesicles; arises from the posterior region of the prosencephalon and forms structures that include the thalamus, hypothalamus, and optic cup.

divergent canonical pathway A signaling cascade other than the one previously established for a given molecule. When referring to Wnt signaling, the pathway that leads microtubule stabilization.

DNA (deoxyribonucleic acid) Long double-stranded chains of nucleotides. The nucleotides consist of the sugar deoxyribose, a phosphate group, and one of four nitrogenous bases: adenine (A), cytosine (C), guanine (G), or thymidine (T).

DNA ladder The pattern of DNA fragments characteristic of apoptotic cells, as visualized by gel electrophoresis.

dominant negative A type of mutation in which the resulting gene product interferes with the function of the normal gene product.

dorsal blastopore lip (DBL) The dorsal margin of the opening (blastopore) that forms on the outer surface of the amphibian embryo. This region of the blastopore becomes a signaling center critical for directing ectoderm to a neural fate.

dorsal–ventral axis (dorsoventral axis) The body axis that runs from the back (Latin, *dorsum*) to the belly (Latin, *venter*).
downstream In a signal transduction pathway, molecules that are activated after receptor binding or after activation of an earlier step in the pathway.
dynamic microtubule In a growth cone, a motile microtubule (a hollow tube comprised of the protein tubulin) that extends past the stable microtubules of the central domain to influence the direction of growth cone extension.
ectoderm Embryonic epithelial tissue that is the precursor of the epidermis and nervous system.
ectopic Arising or placed at a location where a tissue is not normally found.
effector caspase Apoptotic caspases that catalyze the widespread cleavage events during apoptosis that kill the cell.
electroporation A procedure in which DNA containing a transgene is introduced into cells by applying electrical current to facilitate the transfer of negatively charged DNA molecules into the cells. In animals, this can be achieved by placing a micropipette containing the DNA near the cells of interest and applying electrical current.
embryonic shield In zebrafish, the organizer region that is analogous to the dorsal blastopore lip of the amphibian.
embryonic stem (ES) cell Pluripotent cell derived from early embryos that can be propagated indefinitely *in vitro* and that can give rise to all cell types of an embryo *in vivo*.
***en passant* synapse** Synaptic connection that forms when presynaptic elements concentrate along the axon shaft rather than at the axon terminal.
endoderm The innermost of the three germ cell layers; gives rise to the gut and related structures.
endosome A membrane-enclosed organelle produced by endocytosis. It carries newly internalized extracellular materials and transmembrane proteins.
enteric nervous system A division of the autonomic nervous system that is associated with the gastrointestinal tract and that regulates digestion rather independently of the rest of the autonomic nervous system.
endplate potential (EPP) Depolarization produced in a postsynaptic muscle cell by acetylcholine released from a presynaptic motor neuron in response to an action potential.
ependymal cell A type of glial cell that lines the ventricles and spinal canal.
Eph receptor Receptor tyrosine kinase that binds ephrin with its extracellular domains. Two Eph receptor subtypes, the EphA and EphB receptors, typically bind ephrin As and ephrin Bs, respectively, but this specificity is not absolute. They can also serve as ligands during reverse signaling.
epiboly The movement and thinning of cell layers. In zebrafish, several stages of development are distinguished based on the percentage of the yolk that is surrounded by the blastoderm, reported as percent of epiboly.
ephrin Cell surface protein that usually acts as a ligand for Eph receptors to mediate repulsion during axon guidance. The ephrin family consists of two subfamilies: ephrin As are attached to the extracellular face of the plasma membrane by GPI, and ephrin Bs are transmembrane proteins. They can also serve as receptors during reverse signaling.
enhancer A short segment of DNA that, when bound by transcription factors, increases the likelihood that a gene will be transcribed.
epiblast The outer of the two layers of the blastodisc that gives rise to all future cell layers (ectoderm, mesoderm, and endoderm). It is roughly equivalent to the animal pole of the amphibian blastula.
epigenetic Molecular modifications to DNA and chromatin, such as DNA methylation and various forms of post-translational modification of histones. They do not modify the DNA sequence but can alter gene expression.
excitatory synapse Site where the release of a presynaptic neurotransmitter depolarizes the postsynaptic cell, thereby increasing the likelihood that the cell will fire an action potential.
exosome Small, membrane-bound vesicles that incorporate proteins, mRNAs, and microRNAs; can be packaged for release into the extracellular space.
explant A piece of tissue placed in cell culture.
extraembryonic tissue Tissues that contribute to the maintenance of an embryo, but that do not become part of the embryo proper.
extrasynaptic nucleus In the multinucleated muscle fiber (myofiber), a nucleus that is not located below the innervating motor nerve terminal.
extrinsic pathway Pathway of apoptosis triggered by extracellular signal proteins binding to cell surface death receptors.
fasciclin A cell surface glycoprotein of the IgG superfamily expressed on axons. An ortholog of neural cell adhesion molecule (NCAM) that relies on homophilic binding and often functions to direct the growth of other axons.
fasciculation The bundling of axons; the tendency for growing axons to bind to one another to form a fascicle.
fibroblast growth factor (FGF) A member of a family of secreted growth factors that act as morphogens to pattern early embryos during development.
filopodium Slender process arising from the growth cone of a neurite that actively samples the local environment to select the most favorable substrate. Helps to direct, or steer, the growth cone to a particular pathway.

floor plate A wedge-shaped region of glial cells that forms at the ventral surface of the neural tube; serves as a signaling center for neuronal patterning and axonal guidance.
folic acid Vitamin B_9, or folate.
follistatin An activin-binding protein that was one of the first neural inducers identified; promotes formation of neural tissue by blocking the epidermal-inducing actions of bone morphogenetic proteins (BMPs).
G1 (gap 1) phase Phase of the eukaryotic cell division cycle, between the end of mitosis and the start of DNA synthesis.
G2 (gap 2) phase Phase of the eukaryotic cell division cycle, between the end of DNA synthesis and the beginning of mitosis.
ganglion Clusters of neurons located in the peripheral nervous system. Plural ganglia.
gap gene In *Drosophila* development, a gene that is expressed in specific broad regions along the anteroposterior axis of the early embryo, and which helps designate the main divisions of the insect body.
gap junction Communicating channel-forming cell–cell junction present in most animal tissues that allows ions and small molecules to pass from the cytoplasm of one cell to the cytoplasm of the next.
gastrulation The process by which an embryo is transformed from a ball of cells into a structure with three distinct layers: ectoderm, mesoderm, and endoderm.
gene A segment of DNA that carries the instructions for how and when to make specific RNAs and proteins.
gene knockout mouse A mouse generated using genetic engineering to inactivate a specific gene. It is usually achieved by homologous recombination in embryonic stem cells to create a mutation in the target gene. The resulting mutant mouse is called a knockout mouse for that particular gene.
gene regulation The process by which a cell determines which genes will be expressed (turned on) and which will be repressed (turned off).
germ cell layer One of the three primary tissue layers (endoderm, mesoderm, and ectoderm) of an animal embryo; also called germ layer.
glia Nonneuronal cells of the nervous system, they play essential roles for the development and function of neurons.
gliogenesis The process of generating new glial cells, the nonneuronal cells of the nervous system.
gp130 cytokine A member of a family of peptides that signal through a common receptor system that includes the gp130 subunit. Ciliary neurotrophic factor and leukemia inhibitory factor are among the members of this family of cytokines.
granule cell layer The innermost layer of the adult cerebellum that contains granule neurons as well as interneurons.
growth cone A dynamic structure at the tip of a developing neuronal process, it enables the extension of the process and guides its direction.
Hensen's node In birds, a group of cells at the anterior (cephalic) end of the primitive streak. Analogous to the dorsal blastopore lip of amphibians and node of mammals.
homeobox A 180-base-pair segment of DNA that is found in genes such as those of the *Hox* family. Disruption of this sequence leads to transformations of one body part into another.
homeodomain The 60 amino acid protein product encoded by the homeobox; common to Hox and other transcription factors.
homeostatic plasticity The modification of pre- and postsynaptic elements in response to changes in the overall level of neural activity; used to maintain a stable level of neural firing.
homologous recombination Exchange of nucleotide sequences between two identical or highly similar DNA molecules. It occurs naturally in certain cells due to its role in specific biological processes, such as in germ-line cells during meiotic crossing over. It is also used experimentally for genome engineering, such as the generation of knockout and knock-in alleles.
Hox code The combination of *Hox* genes used to define the development of regions along the anterior–posterior axis.
***Hox* gene** A member of a family of evolutionarily conserved genes that are arranged in genomes in clusters and encode homeodomain-containing transcription factors. *Hox* genes define the anterior–posterior body axis of vertebrates and most invertebrates and contribute to neuronal fate at later stages of development.
hyperdorsalized embryo A *Xenopus* embryo experimentally treated with lithium chloride that displays an expansion of dorsal tissues, including an increase in the size of the nervous system.
hyperplasia An increase in the size of a tissue.
hypoblast The inner of the two layers of the blastodisc. Lying just above the yolk, it is eventually displaced by the definitive endoderm and ultimately forms extraembryonic tissues. It is roughly equivalent to the vegetal pole of the amphibian blastula.
hypodermis The outermost cell layer of the *C. elegans* embryo; equivalent to the ectoderm of other species.
hypoplasia A decrease in the size of a tissue.
imaginal disc Pockets of tissue found in the larval stages of some insects. The tissues are segmentally organized and attach to the inside of the larval epidermis. Although the discs play no role in larval development, they later evert during metamorphosis to form adult structures of the head, thorax, legs, and wings.

immunocytochemistry A staining method that uses antibodies to visualize the distributions of proteins in cells. See *immunohistochemistry*.

immunohistochemistry A staining method that uses antibodies to visualize the distributions of proteins in fixed tissues. The most common form uses sequential application of two antibodies: a primary antibody that binds the protein of interest and a fluorescence- or enzyme-conjugated secondary antibody that binds to the primary antibody. Protein distribution can be visualized by fluorescence or a color substrate produced by the enzyme conjugated to the secondary antibody.

***in situ* hybridization** A method for determining mRNA distribution in tissues by hybridizing labeled gene-specific nucleic acid probes to fixed histological sections or whole-mount tissues.

inhibitory synapse Site where the release of a presynaptic neurotransmitter hyperpolarizes the postsynaptic cell, thereby decreasing the likelihood that the cell will fire an action potential.

initiator caspase Apoptotic caspases that begin the apoptotic process, thus activating the effector caspases.

interkinetic nuclear migration In the developing nervous system, the movement of the nucleus of a neural progenitor cell through its cytoplasmic processes. The nucleus travels from the ventricular (apical) surface of the neural tube to the pial (basal) surface then back to the ventricular surface, where the cell divides.

intermediate target A tissue that helps direct axons toward their final target cells by transiently expressing guidance cues to which those axons respond.

intrinsic pathway Pathway of apoptosis activated from inside the cell in response to stress or developmental signals; depends on the release into the cytosol of mitochondrial proteins normally resident in the mitochondrial intermembrane space.

ipsilateral Of the same side of the midline. For example, an ipsilateral axonal projection is an axon that does not cross the midline and therefore terminates on the same side of the nervous system as the soma.

isthmic organizer (IsO) A signaling center located between the mesencephalon and metencephalon (the isthmus) that is important for the development of the future midbrain, pons, and cerebellum.

isthmus The region of the central nervous system located between the mesencephalon and metencephalon; also called the midbrain–hindbrain border (MHB).

labeled pathway hypothesis Concept that axons express particular surface molecules ("labels") that serve as directional cues for later extending axons.

lamellipodium A veil-like, motile area of a growth cone that contains a meshwork of filamentous (fibrillar) actin (F-actin) and subunits of globular actin (G-actin); located between the filopodia.

lateral geniculate nucleus (LGN) A thalamic nucleus that receives visual input from retinal ganglion cell axons and sends output to the primary visual cortex.

lateral inhibition The process by which neighboring cells are prevented from adopting identical fates through cell–cell interactions, such as those mediated by Notch/Delta.

ligand A molecule that binds to its receptor.

lissencephaly A brain disorder in which the outer surface of the brain is characterized by a smooth surface lacking typical gyri and sulci. The disorder results from defects in neuronal migration that cause cortical neurons to stack up below one another.

lysate The material produced when cell membranes are broken apart.

M phase The mitosis phase of the cell cycle; results in two daughter cells.

marginal zone (MZ) In the developing neocortex, the layer of cells closest to the pial surface that contains the Cajal–Retzius cells; forms layer I of the adult cortex.

medial–lateral axis The body axis that extends from the midline to the sides.

mesencephalon One of the three primary brain vesicles that forms the midbrain region of the neural tube; gives rise to the superior and inferior colliculi (tectum) and midbrain tegmentum.

mesoderm The middle of the three germ cell layers that gives rise to muscle, bone, connective tissues, and the cardiovascular and urogenital systems.

metencephalon One of the five secondary brain vesicles; forms the cerebellum and pons.

microglia A glial cell that functions as the resident immune cell of the nervous system. It engulfs damaged cells and debris. It influences the stability of synaptic connections under some circumstances.

midbrain–hindbrain border (MHB) The region of the central nervous system located between the mesencephalon and metencephalon; also called the isthmus.

molecular layer The outermost of the three layers in the adult cerebellum. It contains various interneurons as well as the elaborate dendritic tress of the Purkinje cells.

morphogen Diffusible signal molecule that can impose a pattern on a field of cells by causing cells in different places to adopt different fates.

motor endplate A specialized, oval-shaped region of skeletal muscle that receives innervation from the terminal branches of a motor neuron.

motor unit A motor neuron and the set of muscle fibers it innervates.

multipotent cell Cells that can develop into more than one specific cell type, but cannot develop into all cell types of the body.

muscle fiber A muscle cell; a long cylindrical multinucleated cell arising from myotubes; also called a myofiber.

myelin The sheath surrounding some axons that is formed by the cellular processes of oligodendrocytes or Schwann cells. It improves the speed of action potential conduction.

myelencephalon One of the five secondary brain vesicles; gives rise to the medulla.
myoblast Mononucleated, undifferentiated muscle precursor cell. A skeletal muscle cell is formed by the fusion of multiple myoblasts.
myofiber A muscle cell; a long cylindrical multinucleated cell arising from myotubes; also called a muscle fiber.
neocortex Outer region of the cerebral cortex, consisting of six cell layers.
neural crest cell Cell originating along the dorsal neural tube in the vertebrate embryo. These cells migrate to give rise to a variety of tissues, including neurons and glia of the peripheral nervous system, pigment cells of the skin, and the bones of the face and jaws.
neural fold The raised area at the lateral regions of the neural plate that curl over to form the neural tube.
neural groove The central indentation along the length of the neural plate.
neural induction The process by which external signals promote the formation of neural tissue from ectoderm.
neural keel In zebrafish, the solid group of cells that forms at the midline of the neural plate. It later gives rise to the neural rod.
neural plasticity Changes of the nervous system in response to experience and learning.
neural plate The layer of ectodermal cells overlaying the notochord that invaginates and gives rise to the neural tube during neurulation.
neural rod In zebrafish, the solid central region that forms from the neural keel and separates from the surface epidermal ectoderm. The central cells later migrate away in order to form the hollow center of the neural tube.
neural tube The hollow tube that is the embryonic precursor to the brain and spinal cord. Comprised of neuroepithelial cells, it is derived from neural plate ectoderm.
neural tube defect A birth defect resulting from failure of the neural tube to close in the cephalic (brain) or spinal cord regions.
neural cell adhesion molecule (NCAM) A glycoprotein expressed on the surfaces of neurons and neurites; cells adhere to one another via homophilic binding mechanisms.
neurite An axon or dendrite; often used to describe neural processes extending in cell culture when it is not clear if the process is an axon or dendrite.
neuraxis The main axis established by the central nervous system (the brain and spinal cord); also called neuroaxis.
neuroblast A neuronal progenitor cell.
neurogenesis The process of generating new neurons.
neuromere Transient segment of the developing neural tube that provides a restricted area for further development; called prosomere in the developing forebrain and rhombomere in the developing hindbrain.
neuromuscular junction (NMJ) The synapse between a motor neuron's presynaptic terminals and a skeletal muscle cell.
neuronal progenitor Precursor cell that has the capacity to develop into a restricted number of neuronal types, but has not yet fully differentiated.
neuronal specificity The process by which progenitor (precursor) cells are directed to express specific genes and adopt a particular neuronal fate. Can also refer to the mapping of axonal projections to the correct target cells.
neuron An electrically excitable cell that receives, integrates, propagates, and transmits information as the working unit of the nervous system.
neurotransmitter receptor A protein that binds and responds to a specific chemical (neurotransmitter) released by a presynaptic cell.
neurotransmitter Molecule that is stored in synaptic vesicles (or dense-core vesicles in the case of neuropeptides) in the presynaptic terminals, is released into the synaptic cleft triggered by presynaptic depolarization, and activates ionotropic or metabotropic receptors on a postsynaptic target cell.
neurotrophic hypothesis Proposal that target tissues release limited quantities of survival-promoting proteins at the time of innervation to ensure that only the correct number of innervating neurons survive.
neurotrophin A family of secreted signaling proteins that regulate the survival, morphology, and physiology of target neurons through binding to specific receptors on those neurons. Mammalian neurotrophins include nerve growth factor (NGF), brain-derived neurotrophic factor (BDNF), neurotrophin-3 (NT-3), and neurotrophin-4/5 (NT-4/5).
node A group of cells at the anterior (cephalic) end of the primitive streak in mammals; analogous to the dorsal blastopore lip of amphibians and Hensen's node of birds.
noggin One of the first neural inducers identified; promotes formation of neural tissue by blocking the epidermal-inducing actions of bone morphogenetic proteins (BMPs).
notochord In the embryo, a midline, mesodermal structure located along the ventral surface of the neural plate and neural tube. It extends from the mesencephalon to the spinal cord, where it serves as a source of signals for patterning ventral regions of the nervous system. It is later incorporated into the vertebral column of most vertebrates.
nucleation The formation of a small complex of approximately three G-actin monomers; the first step in action polymerization.
ocular dominance column (ODC) Preference for receiving and/or representing visual input from one eye over the other eye. In the primary visual cortex of some mammals, such as cats and monkeys, cells in the same vertical columns share the same ocular dominance, thus producing ocular dominance columns.

oligodendrocyte precursor cell (OPC) Cells that originate in the ventricular zone of the third ventricle and migrate to the optic nerve, where they differentiate into mature oligodendrocytes.

oligodendrocyte A glial cell in the CNS that wraps axons with its cytoplasmic extension to form myelin sheath.

oncogene An altered gene whose product can act in a dominant fashion to help make a cell cancerous. Typically, an oncogene is a mutant form of a normal gene (proto-oncogene) involved in the control of cell growth or division.

organizer A group of cells that induces formation of neural tissue; refers to the dorsal blastopore lip (Spemann's organizer) in amphibians, Hensen's node in birds, and the node in mammals.

orthologs Genes or proteins from different species that are similar in sequence because they are descendants of the same gene in the last common ancestor of those species.

pair-rule gene In *Drosophila* development, a gene expressed in a series of regular transverse stripes along the body of the embryo and which helps to determine its segments.

paralogs Genes or proteins that are similar in sequence because they are the result of a gene duplication event occurring in an ancestral organism. Those in two different organisms are less likely to have the same function than are orthologs.

peripheral domain (P-domain) The distal, leading edge of a growth cone consisting of lamellipodia and filopodia.

peripheral nervous system (PNS) Collections of neurons (ganglia) that lie outside of the brain and spinal cord.

perisynaptic Schwann cell (PSC) Specialized, non-myelinating Schwann cell that surrounds the terminal branches of a motor neuron; a component of the neuromuscular junction tripartite complex.

periventricular heterotopia (PH) A brain disorder in which neurons fail to migrate out of the ventricular zone. Instead, the neurons, which appear to be highly differentiated, cluster in nodules near the ventricles.

pioneer axon The first axon to extend in the direction of a target tissue. It expresses guidance cues along its axon shaft to help guide subsequent axons toward a target tissue.

placode In vertebrates, an ectodermal thickening that gives rise to some cranial ganglia and sensory structures.

postsynaptic After, or downstream of, a synapse.

postsynaptic cell A cell that receives and responds to input from another neuron (the presynaptic neuron); a cell specialized for neurotransmitter reception, located after the synaptic cleft.

postsynaptic density (PSD) An electron-dense organelle found in excitatory synapses of the central nervous system; helps anchor certain proteins of the postsynaptic cell across from the presynaptic terminal.

prechordal plate Located at the anterior end of the epiblast, the region where surface ectoderm tightly adheres to the underlying endoderm.

precursor cell Cell with the ability to give rise to a limited number and type of specialized cells, in a state between a stem cell and differentiated cell; also called a progenitor cell.

preplate (PP) A transient layer of cells in the developing neocortex located below the pial surface; contains the first cells to migrate out of the ventricular zone. As additional cells leave the ventricular zone, the preplate separates into two layers—the marginal zone and subplate.

presynaptic cell A cell specialized to release a signal to initiate a response in a partner cell (the postsynaptic cell); the cell located before the synaptic cleft.

presynaptic nerve terminal A structure at the end (or along the trunk) of an axon that is specialized for releasing neurotransmitters onto target cells.

primary cell culture Cells harvested from an animal and maintained *in vitro*.

primary cell layers The three tissue layers (endoderm, mesoderm, and ectoderm) that form during the process of gastrulation; also called germ layers or germ cell layers.

primary visual cortex (V1) The visual cortical area that receives direct input from the lateral geniculate nucleus.

primitive streak The axial thickening in ectoderm that is the site of indentation and migration for cells during gastrulation in birds, reptiles, and some mammals; similar in function to the blastopore of amphibians and embryonic shield of fish.

progenitor cell Cell with the ability to give rise to a limited number and type of specialized cells, in a state between a stem cell and differentiated cell; also called a precursor cell.

programmed cell death (PCD) A form of cell death in which a cell kills itself by activating an intracellular death program.

proneural cluster (PNC) A group of cells expressing proneural genes. Each cell in the cluster has the potential to form a neural precursor.

proneural gene Gene that provides a cell with the potential to form a neural precursor.

pro-neurotrophin Precursor protein that is cleaved to yield the mature form of a neurotrophin; thought to play a role in cell death pathways.

prosencephalon One of the three primary brain vesicles; forms the forebrain region of the neural tube and brain.

prosomeres Transient segments (neuromeres) in the embryonic forebrain that provide restricted regions for further cellular development. Numbered from 1–6, beginning at the mesencephalon and progressing to the telencephalon.
postjunctional fold Invagination in the cell membrane of skeletal muscle at the site of synaptic contact.
proteasome Large protein complex in the cytosol with proteolytic activity that is responsible for degrading proteins that have been marked for destruction by ubiquitination or by some other means.
Purkinje cell layer The middle cell layer in the adult cerebellum, it is characterized by a single layer of the Purkinje neuron cell bodies; also contains cerebellar interneurons.
pyknotic A shrunken nucleus with condensed chromatin which is characteristic of cells dying by apoptosis.
radial glial (RG) cell Progenitor cell in the ventricular zone that extends processes to the ventricular and pial surfaces. The processes provide surfaces for neuronal migration.
receptor A protein that binds and responds to a specific signaling molecule.
Reelin A secreted glycoprotein that influences neuronal migration and synaptic connections in the nervous system; first identified in Cajal–Retzius cells, where it helps direct the migration of cells entering the cortical plate.
resonance hypothesis Proposal that target cells establish connections with innervating axons based on matching electrical activity.
retinal progenitor cell (RPC) Multipotent cell that can give rise to any of the six neuronal populations, as well as the Müller glia, of the retina.
retinoblastoma protein (Rb) Tumor suppressor protein involved in the regulation of cell division; mutated in the cancer retinoblastoma, as well as in many other tumors. Its normal activity is to regulate the eukaryotic cell cycle by binding to and inhibiting the E2F proteins, thus blocking progression to DNA replication and cell division.
retinoic acid (RA) A vitamin A derivative.
retinula cell (R cell) A photoreceptor cell of the *Drosophila* ommatidium.
retrograde From the axon terminal to the cell body.
rhombencephalon One of the three primary brain vesicles that forms the hindbrain region of the neural tube.
rhombomere Transient segment (neuromere) in the embryonic hindbrain that provides a restricted region for further cellular development.
roof plate A wedge-shaped region of glial and neuroepithelial cells that forms on the dorsal surface of the neural tube; serves as a signaling center for neuronal patterning.
rostral–caudal (or rostrocaudal) axis Of a body axis, from head to tail.
S phase The phase of the cell cycle during which DNA is synthesized.
sacral neural crest Neural crest cells that originate from the dorsal surface of the neural tube posterior to somite 28 and give rise to parasympathetic and enteric ganglia.
satellite cell Glial cell of the vertebrate peripheral nervous system that surrounds neuronal cell bodies. Thought to function similarly to astrocytes of the central nervous system.
scaffolding protein Protein that anchors other proteins to a specific location in the cell.
Schwann cell A glial cell in the PNS that wraps axons with its cytoplasmic extension to form myelin sheath.
sclerotome A segment of a somite that later gives rise to cartilage and bone of the axial skeleton and rib cage.
secreted Frizzled-related protein (sFRP) Small, soluble protein that sequesters Wnt to inhibit it from signaling through the Frizzled receptors. In the anterior neural tube, these proteins act to prevent the posteriorization of forebrain regions.
segment polarity gene In *Drosophila* development, a gene involved in specifying the anteroposterior organization of each body segment.
segmentation gene Gene expressed by subsets of cells in the embryo that refines the pattern of gene expression so as to define the boundaries and ground plan of the individual body segments.
sensory organ progenitor (SOP) In *Drosophila*, a precursor cell whose asymmetric divisions give rise to different cells (a socket cell, hair cell, sheath cell, and sensory neuron) in the external sensory organ.
signal transduction Process by which signals originating outside of a cell are conveyed to cytoplasmic components or the nucleus to influence cell behavior.
somal translocation A form of radial migration in which the neuronal cell body travels through its cytoplasmic process toward the pial surface; often used by the earliest generated neurons in the cerebral cortex.
somite One of the paired blocks of mesoderm tissue found alongside the neural tube; comprised of dermamyotome and sclerotome segments.
Sonic hedgehog (Shh) A morphogen that determines cell fate by regulating the expression of specific transcription factors in many developmental contexts. For instance, floor-plate-derived Shh is responsible for determining the different fates of neuronal progenitors located at different positions along the dorsal–ventral axis of the ventral spinal cord. It is also used as a midline attractant for commissural axons.
short interfering RNA (siRNA) Synthetic double-stranded RNAs that bind to complementary mRNA and transiently block gene transcription.

Spemann's organizer A group of cells in the dorsal blastopore lip of amphibians that induces formation of neural tissue and organizes the body axis; discovered by Hans Spemann and Hilde Mangold.

spina bifida A neural tube defect of the spinal cord that results from incomplete closure of the neural tube and overlying vertebrae. The outcomes range from mild to severe, depending on the degree of spinal cord exposure.

spinogenesis The formation of spines on the dendrites of neurons.

stem cell Undifferentiated cell that can continue dividing indefinitely, producing daughter cells that can either commit to differentiation or remain a stem cell (in the process of self-renewal).

subplate (SP) A transient layer of cells in the developing neocortex that is created when the preplate is divided into two layers by the cells of the cortical plate; located between the cortical plate and intermediate zone.

subsynaptic nuclei In the multinucleated muscle fiber (myofiber), nuclei that are clustered below the innervating motor nerve terminal.

subventricular zone (SVZ) A layer of proliferative cells in the neocortex that forms between the ventricular and intermediate zones. It continues to produce neurons and glial cells after the ventricular zone is gone; it persists in regions near the lateral ventricles into adulthood.

sulcus limitans A longitudinal groove along the inner surface of the neural tube that provides an anatomical landmark separating the alar and basal plates; it extends from the midbrain to the spinal cord.

superplate An aggregate of cells below the pial surface in the cerebral cortex of mice with the *Reeler* mutation; contains Cajal–Retzius cells and the neurons that would normally form the subplate.

synapse Communicating cell–cell junction that allows signals to pass from a nerve cell to another cell. In a chemical synapse, the signal is carried by a diffusible neurotransmitter. In an electrical synapse, a direct connection is made between the cytoplasms of the two cells via gap junctions.

synaptic basal lamina The specialized extracellular matrix found in the synaptic cleft between a motor nerve terminal and a muscle fiber. Has a unique protein composition that distinguishes it from other basal lamina.

synaptic bouton The slightly enlarged region of an axon terminal at the site of synaptic contact.

synaptic cleft A 20–100 nm gap that separates the presynaptic terminal of a neuron from its target cell.

synaptic reorganization The strengthening or loss of subsets of existing neuronal connections.

synaptic scaling The modification of postsynaptic elements in response to changes in the overall level of neural activity; used to maintain a stable level of neural firing; a form of homeostatic plasticity.

synaptic vesicle A small, membrane-enclosed organelle (typically about 40 nm in diameter) enriched at the presynaptic terminal. Each vesicle is filled with neurotransmitters and, upon stimulation, fuses with the plasma membrane to release neurotransmitters into the synaptic cleft.

synaptogenesis The formation of new synaptic contacts in the nervous system.

syncytial blastoderm In *Drosophila*, the stage of development when the nuclei of a fertilized egg have migrated to the cortex of the blastoderm and share a continuous cytoplasm.

syncytium A group of cells in which the cytoplasm is continuous.

tangential migration Mode of neuronal migration in which cells travel on existing cells or axons along routes parallel to the ventricular surface.

telencephalon The anterior-most of the five secondary brain vesicles; gives rise to the cerebral cortex, hippocampus, basal ganglia, basal forebrain nuclei, olfactory bulb, and lateral ventricles.

temporal identity factors (TIFs) Transcription factors expressed in a sequential manner that together influence the fate of a cell.

terminal synapse The type of synaptic connection that forms between a presynaptic nerve terminal and a postsynaptic cell.

topographic map The consistent, primarily invariant projection of axons from one region of the nervous system to their target cells in another region.

transcription factor code The combination of transcription factors in a precursor cell that leads to the expression or repression of genes that determine the subtype of cell that forms.

transcription factor A DNA-binding protein that regulates transcription of target genes.

transforming growth factor β (TGFβ) Large family of structurally related secreted proteins that act as hormones and local mediators to control a wide range of functions in animals, including during development.

tripartite complex The three cells that comprise the vertebrate synapse (a presynaptic neuron, a postsynaptic cell, and a glial cell).

trunk neural crest Neural crest cells that originate from the dorsal surface of the neural tube between somites 8 and 28 to give rise to dorsal root ganglia, portions of the sympathetic and parasympathetic ganglia, and adrenal chromaffin cells.

vagal neural crest Neural crest cells that originate from the dorsal surface of the neural tube from the posterior hindbrain to somite 7 to give rise to some cranial ganglia as well as sympathetic, parasympathetic, and enteric ganglia of the trunk.

vegetal pole The portion of the egg that has the most yolk; located opposite the animal pole; site of origin for endoderm.

ventral nerve cord (VNC) An invertebrate CNS structure posterior to the brain. It is analogous to the vertebrate spinal cord.

ventralized embryo A *Xenopus* embryo experimentally treated with UV light that develops without a nervous system or other dorsal structures.

ventricular zone (VZ) A transient single cell layer of neuroepithelial cells adjacent to the lumen of the neural tube; the site of proliferation for the first cells generated in the central nervous system.

visceral endoderm In the mouse embryo, the cell layer that becomes displaced by definitive endoderm; gives rise to extraembryonic tissues.

voltage-dependent calcium (Ca^{2+}) channel An ion channel that allows selective passage of Ca^{2+} and whose conductance is regulated by the membrane potential.

zona limitans intrathalamica (ZLI) A signaling center in the neural tube located at the junction of the prechordal and notochordal regions, thought to influence development of the diencephalon.

zygote A fertilized egg cell.

Index

A

acetylcholine (Ach) 263–264, 272
acetylcholine esterase (AChE) 264
acetylcholine receptors (AChRs) 263–266
  aggregation 269–274, 275
  Dok7 273–274
  invertebrates 275
  Lrp4 273–274
  muscle fiber development 264–266
  MuSK 272–274, 276
  prepatterning 264–265
  Rapsyn 276
  subunits 276–279
Ach *see* acetylcholine
achaete-scute complex (*AS-C*) 148–150
AChE *see* acetylcholine esterase
AChRs *see* acetylcholine receptors
actin
  growth cone motility 188–190
  subunits 188–189
  zipcode binding proteins 209–211
actin-binding proteins 189–190
action potentials 5
active zones 261, 282, 292
activin 38–40, 46, 97, 99
adenomatous polyposis coli (APC) protein 101, 304
ADF (actin depolymerisation factor) 189–190
adherens junctions 116–117
adhesion
  CAMs (cell adhesion molecules) 66, 294–295
  fasciculation 194–198
  motor axon growth 201–202
  Rho GTPases 122
  synapses 261
adrenergic neuron neurotransmitter switch 163–165
aggregation, acetylcholine receptors 269–274
agrin 269–274, 276, 279
agrin hypothesis 271–272
agrin-ACh hypothesis 272
AGS3 (activator of G-protein signaling 3) 118
alar plate 83, 132
ALS *see* amyotrophic lateral sclerosis
amacrine cells 179–181
amino-Sonic hedgehog (Shh-N) fragments 90
amphibians
  blastula formation 30
  developmental stages 13–15
  gastrulation 31–32, 34–35
  lateral inhibition 150–151
  as model organisms 34
  neural induction 36–41
  notochord signaling 85–86
amyotrophic lateral sclerosis (ALS) 246
anatomical landmarks, D/V axis 82–85
ANB *see* anterior neural border
anencephaly 53
animal cap assays 36–42
animal pole 8, 30–31
annexin-6 141
ANR *see* anterior neural ridge
antagonistic signaling, D/V patterning 102–105
anterior neural border (ANB) 58
anterior neural ridge (ANR) 56, 58
anterior neural tube, D/V patterning 104–109
anterior visceral endoderm (AVE) 33, 57–58
anterior–posterior (A/P) axis 7
  neural tube formation 52–57
  segmentation 51–78
    brain vesicles 54–55
    Cdx 76–78
    early boundary formation 54–56
    extraembryonic tissue-derived signals 57–58
    FGF (fibroblast growth factor) 62–63, 76–78
    forebrain 57–60
    hindbrain 65–76
    homeobox genes 69–78
    isthmus 55–56, 61–63
    mesencephalon/metencephalon 60–65
    midbrain 60–65
    patterning signals, interactions 56–57
    retinoic acid 72–76
    segmentation genes 68–78
    spinal cord 76–78
    Wnt activation, midbrain 63–65
    Wnt inhibition, forebrain 58–59
anti-apoptotic genes 251–253
APC *see* adenomatous polyposis coli
apical proteins
  invertebrate neurons 153–154
  vertebrate neurons 116–118
aPKC *see* atypical protein kinase C
ApoER2 *see* apolipoprotein E receptor 2
apolipoprotein E receptor 2 (ApoER2) 130
apoptosis 248–253
apoptosomes 252–253
Arp2/3 complex 189–190
astrocytes 6, 176–168, 306–307
astrotactin 125,135
asymmetric cleavage 117–118
Atonal homolog 1 (Atoh1) 176–177
atonal1 172–173
atypical protein kinase C (aPKC) 153–154
auditory system 143, 175–178, 314
autocrine signaling 305–306
autophosphorylation, Trk dimerization 241–242
autoproteolysis, Sonic hedgehog 90
AVE *see* anterior visceral endoderm
avians
  blastula formation 30
  developmental stages 11–12
  gastrulation 32–33
axons 4–5
  fasciculation 194–198

growth cones 185–202
labeled pathway hypothesis 196–197
local guidance cues 197–202
pathfinding 187–211
pioneer 193–194
retinotectal system 211–223
self-avoidance 222–223
zipcode binding proteins 209–211

B

BAF complex 160–161
basal plate 83
basal proteins
invertebrate neurons 153–154
vertebrate neurons 116–118
Bassoon 282
Bazooka 153–154
Bcl-2 family of proteins 252–253
BCl-2 homology (BH) domains 252–253
Bergmann glia 135–136
bioassays, nerve growth factors 231–233
bipolar cells 179–181
birds
blastula formation 30
forebrain patterning 58
gastrulation 32–33
midbrain–anterior hindbrain patterning 61–62
retinoic acid 74, 77
blastocoel 30
blastocyst 8, 30
blastoderm 8, 30
blastodisc 8, 30
blastomere 8, 30
blastopore 31
blastula 8, 30–31
BMPs *see* bone morphogenetic proteins
bone morphogenetic proteins (BMPs)
anterior D/V patterning 106–107
class A interneuron patterning 97–100
dorsal neural tube patterning 96–100
inhibitors 42–46
neural crest cell induction 138
neural induction 42–48
receptor binding 38–46
signal transduction 46–48, 99–100
SMADs 46–48, 99–100
sympathetic neuron fates 164
boss see bride of sevenless gene
boundary formation
forebrain 57–60
hindbrain rhombomeres 65–68, 71–76
midbrain–anterior hindbrain 64–65
brahma 160
brain vesicles 10, 54–55
brain-derived neurotrophic factor (BDNF)
beta actin translation 211
central nervous system synapses 298–299
discovery 234–236
granule cells, cerebellum 156–157
neuromuscular junction 283–285
neuronal survival 235–236
nodose neurons 236, 240–241
prenatal cocaine exposure 315
receptor binding 239–241, 283–285
Brat proteins 153–154
bride of sevenless gene (*boss*) 171–172, 174
Brn2 158
α-bungarotoxin 270

C

C. elegans see Caenorhabditis elegans
C-domain *see* central domain
Ca^{2+} channels *see* voltage-dependent calcium channels
cadherins 194, 294–295
Caenorhabditis elegans (*C. elegans*) 18–20, 204–206, 251–252, 275, 298
Cajal–Retzius (CR) cells 123, 128–130
CAMs *see* cell adhesion molecules
canonical pathway, Wnt 101–102
capping proteins 189–190
carboxyl-Sonic hedgehog (Shh-C) fragments 90
Carnegie stages (CS) 11, 13
caspases 253
Castor 155
Casz1 181
β-catenin pathway 100–102
caudal 69, 77
Cdc42 190
Cdk inhibitory proteins 121, 168–169
Cdks *see* cyclin-dependent kinases
Cdx, *Hox* gene regulation 76–78
cell adhesion molecules (CAMs) 66, 294–295
cell cycle
neurogenesis 114–115
regulation 118–122
cell death
apoptosis 248–253
embryogenesis 228
extrinsic pathway 254–255
genes 251–253
intrinsic pathway 253
necrosis 249
programmed 248–255
target tissue size 228–229
cell division, cleavage directions 115–116
cell fate determination *see* cellular determination
cell line cultures 238–240
cell lineage mapping in *C. elegans* 18–20
cell-autonomous receptor functions 94
cell-specific markers 36
cell–cell contacts, *Drosophila* eye development 170–175
cellular blastoderm 43–45
cellular determination 112, 147–181
C. elegans 19–20
central nervous system, vertebrate 155–159
epigenetic factors 159–161
inner ear 176–178
invertebrates 151–155
lateral inhibition 148–151, 176–178
neural crest-derived neurons 161–164
oligodendrocytes 167–169
parasympathetic neurons 161–163
retina 178–181
Schwann cells 166
specialized sensory cells 170–181
sympathetic neurons 161–165
temporal expression of transcription factors 154–159
cellular retinoic acid-binding proteins (CRABPs) 74

central domain (C-domain) 187
central nervous system (CNS) 4
  astrocytes 6, 306–307
  cadherins 116–117, 294–295
  cellular migration 122–136
  Drosophila, fate determination 153–155
  early vertebrate development 10
  en passant synapses 291, 296–297
  Eph receptors/ephrins 67, 218–222, 299–303
  excitatory synapses 5, 291–294, 296–297, 305, 312–313
  inhibitory synapses 5, 291–293, 313
  neurexins/neuroligins 295–296
  oligodendrocytes 6, 167–169
  origins 8–10
  postsynaptic density 293–294
  postsynaptic differentiation 295–299, 301–306
  presynaptic differentiation 295–297, 299–301, 303–306
  reverse signalling, ephrins 299–301
  synaptic elimination 307–310
  synaptic reorganization 307–317
  synaptogenesis 290–307
  thrombospondins 306
  ubiquitin-mediated pathways 303–304
  vertebrate cellular determination 155–161
  Wnt 58–59, 63–65, 304–306
Cerberus 59
cerebellum
  adult organization 132
  D/V patterning 105
  fibroblast growth factors 63–64
  neuronal migration 131–136
  Reeler mutation 136
  Zic 106–107
cerebral cortex
  adult organization 123
  neuronal migration 124–131
  stage-dependent fate options 157–159
  temporal identity factors 159
  transient layers 123–124
cerebral spinal fluid (CSF) production 106
CG *see* ciliary ganglion
checkpoint proteins 121–122
chemoaffinity hypothesis 214–223
  axonal self-avoidance 222–223
  ephrins 218–222
chemokines 131
chemotropism 203
chick embryos
  developmental stages 11–12
  forebrain patterning 58
  gastrulation 32–33
  hypoblast 32–33, 58
  midbrain–anterior hindbrain patterning 61–62
  motor neuron innervation 199–200
  retinoic acid 74, 77
chick-quail chimera 61–62, 162–163
choice points 197
cholinergic neurons
  neuromuscular junctions 259–288
  postnatal development 163–165
  see also acetylcholine, motor neurons
chondroitin-sulfate proteoglycans 143
chordamesoderm 32
chordin 41–46, 106
choroid plexus 106
chromatin 159–160
ci *see* cubitus interruptus proteins
ciliary ganglion (CG) neurons 245, 247
ciliary neurotrophic factor (CNTF) 164, 245–247
class A interneurons 97–100
class B interneurons 97–98
class I transcription factors 91
class II transcription factors 91
cleavage, neuronal proliferation/migration 115–116
clinical outcomes
  neuronal migration disorders 127–128, 143
  spina bifida 53–54
  myelin loss 169
closure, neural tube 53–54
clutch proteins 193
CMS *see* congenital myasthenic syndrome
CNS *see* central nervous system
CNTF *see* ciliary neurotrophic factor
CNTF receptor 246–247
co-linearity, *Hox* genes 71, 74, 77–78
cocaine, prenatal exposure 315
cochlear anatomy 175–176
cofilin 189–190
Cohen, Stanley 234–235
collagen 126, 192–193
comm gene 208–209
commissural interneurons 196–197, 202–211
compound eyes, *Drosophila* 170–175
conditional knockout mice 24–25
cone cells 179–180
congenital malformations
  D/V patterning 108–109
  migration-related 127–128, 143
  neural tube defects 53–54
congenital myasthenic syndrome (CMS) 274
conservation, *HOM-C/Hox* genes 69–71
cortical interneuron tangential migration 130–131
cortical plate (CP) 123
  inside-out patterning 126–127
  Reeler mutation 128–129
COS cell lines 88–89, 195
CP *see* cortical plate
CR cells *see* Cajal–Retzius cells
CRABPs *see* cellular retinoic acid-binding proteins
cranial neural crest cells 138–140
Cre recombinase 24–25
critical period 311–312
cross-repression in Shh signaling 91
CS *see* Carnegie stages
cubitus interruptus (ci) proteins 91–94
Cux1 (cut like homeobox 1) 158
cyclin-dependent kinases (Cdks) 119–120, 168–169
cyclins 119–120, 168–169
cyclohexamide 251
cyclopamine 108
cyclopia 108
Cyp26 75–76, 78
cytokines 164–165, 235, 246–247, 306
cytoskeleton
  central nervous system, vertebrate 122
  growth cone 187–191
    actin 187–189

dynamic microtubules 187–188
Rho GTPase effects 190–191
stable microtubules 187–188
substrate binding interactions 187–189, 193
zipcode binding proteins 209–211
neural crest cells 138

D

DBL *see* dorsal blastopore lip
DE-RAREs *see* distal enhancer retinoic acid response elements
decapentaplegic gene (*dpp*) 43–45
defects
D/V axis patterning 108–109
neuronal migration 127–130
neural tube 53–54
delamination of neural crest cells 138
Delta 148–151
demyelinating diseases 169
dendrites 4–5
dendritic spines 291, 295–299, 301–306
deoxyribonucleic acid (DNA) 3
gene regulation 20–22
genes 21
genetic manipulation 23–25
homologous recombination 24
labeling 22
ladders 248–249
methylation 159
depolymerization of actin 189–190
dermamyotome 142
developmental staging
amphibians 14–15
C. elegans 18–20
chick 11–12
Drosophila melanogaster 15–16
human 11–13
mouse 11–13
zebrafish 13–14
Dickkopf (Dkk) 59–60, 304
diencephalon 10, 54–56
differentiation 112
cerebral cortical neurons 157–159
granule cells 155–157
motor neuron precursors 89–95
oligodendrocytes 167–169
parasympathetic neurons 161–163
specialized sensory cells 170–181
sympathetic neurons 161–164
dimerization of Trk receptor 241–242
Dishevelled (Dvl) 304–305
disorganized neurons, Reeler mice 128–129, 136
distal enhancer retinoic acid response elements (DE-RAREs) 72–73
divergent canonical pathway, Wnt 304–305
Dkk *see* Dickkopf
DLK-1 (dual leucine zipper kinase-1) 303–304
DNA *see* deoxyribonucleic acid)
Dok7 *see* Downstream of kinase 7
dominant negative mutations 38–39
dorsal blastopore lip (DBL) 32, 34–36, 56
dorsal interneurons
BMP-related induction 97–100
roof plate signals 97
dorsal patterning
posterior neural tube 95–102
transforming growth factor β-related signals 96–97
roof plate
anterior neural tube 105–106
posterior neural tube 96–97
dorsal root ganglia
migration 141–142
outgrowth 192, 195
survival 229–236
dorsalin 97
dorsal–ventral (D/V) axis 7
patterning 81–110
A/P positioning effects 107–108
antagonistic signaling 102–105
anterior neural tube 104–109
bone morphogenetic proteins 96–100, 106–107
canonical pathway, Wnt 101–102
defects 108
fibroblast growth factor 95
floor plate 82–95
motor neurons 85–95
notochord 82, 84, 85–91, 108
posterior neural tube
anatomical landmarks 82–85
dorsal 95–104
ventral 85–95
retinoic acid 95
roof plate 82–84, 95–102, 105–106, 108–109
Sonic hedgehog (*Shh*) 89–95, 105, 107–108
transforming growth factor β-related signals 96–100
Wnt 100–102, 108
Zic 106–107
Down syndrome 298
downstream activation 21
Downstream of kinase 7 (Dok7) 273–274
dpp see decapentaplegic gene
Dreher mutants 98, 105
Drosophila melanogaster
D/V axis patterning signals 103–104
developmental stages 15–16
ectoderm derivatives 15–18
eye development 170–175
ganglion mother cells 153–155
head patterning genes 58, 68–69
hedgehog gene 87
HOM-C genes 69–71
neural induction 43–45
neuromuscular junction 274–275
ommatidia 170175
proneural cluster 148–150
segmentation genes 68–71, 77
sensory organ progenitors (SOPs) 17, 151–153
ventral nerve cord (VNC) 16–17, 153
Dscams 222–223
Dvl *see* Dishevelled
dynamic microtubules 187–188

E

E3 ligases 303
early life stress (ELS) 316
early neural enhancer retinoic acid response elements (ENE-RAREs) 72–73
ECM *see* extracellular matrix

ectoderm 9
formation, invertebrate 43–45
formation, vertebrate 31–33
neural induction 31–48
ectopic motor neurons from grafting studies 86
EDNRB (endothelin receptor type B) 143
effector caspases 253
EGL *see* external granule layer
electrical activity, motor neuron guidance 202
electroporation 23, 94–95
ELS *see* early life stress
embryonic shield 31
embryonic stem (ES) cells 156–157
en passant synapses 291, 296–297
En2 see Engrailed 2
endoderm 9, 31–33
endosomes 240–242
endothelin3 143
endplate potentials (EPPs) 264, 266–267
ENE-RAREs *see* early neural enhancer retinoic acid response elements
Engrailed 2 (*En2*) 63, 65
Enhancer of split (*E(spl)*) 149
enhancers, *Hox* gene expression 71–76
environmental differentiation cues 161–163
ependymal cells 6, 83
Eph receptors 67–68
adhesion molecules 295
motor neurons 201–202
neural crest cells 143
presynaptic development 299–301
retinotectal system 218–222
rhombomeres 67–68
ephrins 67–68, 143
adhesion molecules 295
central nervous system synapses 299–303
forward signaling 67, 299–300
motor neuron guidance 201–202
neural crest cells 143
reverse signaling 67, 222, 299–301
retinotectal system 218–219
topographic map formation 220–222
epiblast 30–33
epiboly 13–14
epidermal induction 42–46
epigenetic factors in neuronal determination 159–161
ε-subunits of acetylcholine receptors 276–277
ErbB receptors 166, 278–279
Erk (extracellular signal-regulated kinase)
cerebellum formation 63–64
fibroblast growth factor receptor (FGFR) signaling 64
neural induction 47–48
Trk receptor signaling 242–243
ES cells *see* embryonic stem cells
E(spl) see Enhancer of split
excitatory synapses 5, 291–294, 296–297, 305, 312–313
exosomes 221–222
explants 231
expression
ciliary neurotrophic factor 246
forebrain genes 58
Hox genes 71–73
segmentation genes 68–70
Shh-responsive genes 90–91
external granule layer (EGL) 133–135, 156–157
extracellular matrix (ECM) 125–126, 141, 192–193
extraembryonic tissue 32–33, 57–58
extrasynaptic nuclei 277–278
extrinsic pathway, cell death 254–255
eyes
Drosophila 170–175
vertebrate 178–181

F

F-actin 297
fasciclins 197–198
fasciculation 194–198
FGFs *see* fibroblast growth factors
fibroblast growth factor receptors 63–64
fibroblast growth factors (FGFs)
cerebellum development 63–65
forebrain patterning 58
hindbrain patterning 76–78
midbrain–anterior hindbrain patterning 62–65
neural crest cell induction 138
neural induction 37, 47–48
presynaptic differentiation 282
spinal cord patterning 76–78
ventral patterning 95
fibronectin 126
filopodia 187–191, 297–299
fish
anterior neural border 58
developmental stages 13–14
forebrain patterning 58
spinal cord patterning 76–78
floor plate
commissural interneurons 203–204
induction 85–89
neuronal patterning 82–95
Sonic hedgehog 86–89
ventral patterning 89–95
floxed sites 24–25
folic acid 53–54
follistatin 40–46, 106
forebrain
A/P patterning 57–60
D/V patterning 107–108
extraembryonic tissue signaling 57–58
gene expression 58
Wnt inhibition 58–60
formins 189–190
forward signaling, ephrins 67, 200–300
FoxP2 (Forkhead box protein 2) 158
Fragile X syndrome 298
Frazzled (Frl) receptors 206
Frizzled (Frz) receptors 59, 274, 304
Frl *see* Frazzled receptors
frogs
developmental stages 14–15
lateral inhibition 150–151
neural induction 36–41
FRS2 (FGF receptor substrate 2) 64
Frz *see* Frizzled

G

G-actin subunits 189–190
G1 phase 114–115, 121–122
G2 phase 114–115
GABAergic neurons 130–131, 291
γ-subunits of acetylcholine receptors 276–277
ganglia 4
ganglion mother cells (GMCs) 153–155
gap genes 69–70
gap junctions 5
gastrulation 9, 30–35
GATA3 (GATA binding protein 3) 164
Gbx (gastrulation brain homeobox) genes 62
GDF7 (growth differentiation factor 7) 97, 99, 156
GEFs *see* guanine nucleotide exchange factors
gene duplications, *Hox* genes 71
gene knockout mice 23–25
genes
  definition 21
  forebrain patterning 58
  labeling 22
  naming 87
  programmed cell death 251–253
  proneural 116, 148–150
  regulation 20–26
genetic manipulations, mice 23–25
genetic markers, neural tube progenitor cells 84–85
genetic susceptibility, mood disorders 316
germ cell layers 30–33
  see also primary cell layers
GGF (glial growth factor) *see* Neuregulin 1
glia 4–7
  anterior roof plate 105–106, 108–109
  astrocytes 6, 135, 306–307
  Bergmann-type 135–136
  central nervous system synapses 306
  ependymal cells 6, 83
  floor plate, D/V axis patterning 82–95
  gliogenesis 111–113, 166
  internal clocks 168–169
  Neuregulin signaling 135–136, 166
  neuromuscular junctions 262–263, 279–281, 285
  oligodendrocyte development 167–169
  origins 112–113
  peripheral nervous system synapses 262–263, 279–280, 285
  posterior roof plate 82–84, 95–102
  radial migration 124–126, 135–136
  Schwann cells 6, 166, 262–263, 279–281, 285
  synapses 261–262
glial growth factor (GGF) *see* Neuregulin 1
gliogenesis 111–113, 166
  see also neurogenesis
glutamate 274–276
glycine 291
glycogen synthase kinase-3β (GSK-3β) 48, 101, 304–305
glycoproteins 125, 128–130, 136
GMCs *see* ganglion mother cells
gp130 cytokines 165
gradients
  bone morphogenetic proteins (BMPs) 92–101
  cyp26 75–78
  decapentaplegic 103–104
  dorasal protein 103–104
  ephrins, retinotectal system 218–219
  fibroblast growth factor (FGF) 76–78
  retinoic acid 73–78
  Shh 89–95, 104–105
grafting 34, 66, 85–86
Grainyhead 155
granule cells
  adult organization 132
  cerebellar migration 132–136
  differentiation 156–157
  external layer 133–134
  proliferation 133–134
Grb2 (growth factor receptor bound protein 2) 64
GRIP1 (glutamate receptor interacting protein 1) 302
growth cones
  actin-binding proteins 189–190
  cytoskeletal elements 187–191
  extracellular matrix binding 192–193
  identification 186
  morphology 191
  motility 185–191
  pathfinding 187–211
  Ramón y Cajal, Santiago 187
  Rho GTPases 190–191
  substrate preferences 191–202
  zipcode binding proteins 209–211
growth factors, nerve
  bioassays 231–233
  neuronal survival 227–247
  see also individual growth factors
GSK-3β *see* glycogen synthase kinase-3β
guanine nucleotide exchange factors (GEFs) 122, 299

H

hair cells 175–178
Hairy/Enhancer of split-1 (*Hes1*) 177
Hamburger and Hamilton (HH) stages 11
hearing 143, 175–178, 314
hedgehog (hh) 86–87, 91
Hensen's node 33
hepatocyte growth factor (HGF) 201
Hes1 see Hairy/Enhancer of split-1
HGF *see* hepatocyte growth factor
hh *see* hedgehog
HH stages *see* Hamburger and Hamilton stages
highwire (Hiw) 303
hindbrain
  Cyp26 75–76
  ephrin signaling 67–68
  Hox code 71–73
  retinoic acid 72–76
  rhombomeres 65–68
  segmentation genes 68–78
histones 159–161
Hiw *see* highwire
holoprosencephaly (HPE) 108–109
HOM-C genes *see homeotic complex* genes
homeobox (*Hox*) genes
  Cdx 76–78
  co-linearity 71, 74, 77–78
  conservation 69–71
  Cyp26 75–76
  enhancers 71–76

fibroblast growth factor 76–78
hindbrain patterning 69–78
regulation 72–78
retinoic acid responsivity 72–76
rhombomere expression patterns 71–76
spinal cord expression patterns 76–78
homeodomains 70
homeostatic plasticity 274–275, 312–313
homeotic complex (*HOM-C*) genes 69–71
see also homeobox (*Hox*) genes
homologous recombination 24
homophilic binding 194, 197
horizontal cleavage 116
Hox genes *see* homeobox genes
HPE *see* holoprosencephaly
humans, developmental stages 11, 13
Hunchback 155
hyperdorsalized embryos 39–41
hyperplasia 228–229
hypoblast 30, 57–58
hypodermis 19
hypoplasia 228–229

I

IAP *see* inhibitor of apoptosis proteins
Id proteins *see* Inhibitors of differentiation proteins
Id4 168–169
IGL *see* internal granule layer
iGluRs *see* ionotropic ligand-gated glutamate receptors
Ikaros 159, 181
imaginal disk differentiation 16, 170
immunocytochemistry 22, 36
immunohistochemistry 22
immunolabeling 22
in situ hybridization 22, 36
in vitro experiments
acetylcholine receptor clustering 270–271
granule cell differentiation 156
growth cone substrate preferences 191–192
neural induction 36–42
neuronal survival 231–236, 238–239, 245, 250–251
Shh signaling 88–89
synaptic contacts 300–302
in vivo experiments
granule cell differentiation 156–157
neuronal survival 233–234, 240–241
Shh signaling 89
synaptic contacts 309–312
ind (intermediate neuroblast defective) 103–104
induction
anterior neural tube 56–57
floor plate 85–89
neural 29, 33–48
neural crest cells 138
roof plate 96–97
inhibition
lateral 148–151, 172
migratory neural crest cells 142–143
neural tube segmentation 56–57
Wnt 58–60
inhibitor of apoptosis proteins (IAP) 255
Inhibitors of differentiation (Id) protein 168–169, 177
inhibitory synapses 5, 291–293, 313
initiator caspases 253
injury and sympathetic neuron responses 165
inner ear development 175–178, 314
innervation of skeletal muscle 263, 266–267
"inside-out" patterning 126–127
integrin receptors 125–126, 193
integrin subunits 125–126
interkinetic nuclear migration 114–115
intermediate targeting of commissural interneurons 202–211
intermediate zone, spinal cord 83
internal clocks, oligodendrocyte precursors 168–169
internal granule cell layer (IGL) 134
interneurons
BMP-related induction 97–100
roof plate signaling 97
tangential migration 130–131
intracellular signaling
cerebellar induction 63–64
cytoskeletal dynamics 191
neurotrophin activated 242–243
Trk receptor pathways 240–243
intrinsic pathway, cell death 253
intrinsic signaling, midbrain–anterior hindbrain 61–62
invertebrates
cellular determination 151–155
developmental stages 15–18
neural induction 42–46
neuromuscular junctions 274–275
ionotropic ligand-gated glutamate receptors (iGluRs) 274–275
IsO *see* isthmic organizer
isthmic organizer (IsO) 55–56
isthmus 55–56, 61–63

J

jagged 1 (Jag1) 134, 156
jagged 2 (Jag2) 177

K

knockout mice 23–25
Kriesler I mutants 68
Krox20 67–68
Krüppel 155

L

labeled pathway hypothesis 196–197
labeling of genes 22
lamellipodia 187–191
laminin-like cues 204–205
laminins 126, 280–282
lateral ganglionic eminence (LGE) 131
lateral geniculate nucleus (LGN) 307–310
lateral inhibition 150–151, 172
leukemia inhibitory factor (LIF) 164–165
LEV-9/LEV-10 275
levamisole 275
Levi-Montalcini, Rita 234
LGE *see* lateral ganglionic eminence
LGN *see* lateral geniculate nucleus
LIF *see* leukemia inhibitory factor
ligands 21

lipid tethering, Shh 90
lissencephaly 127–128
LNGFR *see* low-affinity NGF receptor
local guidance cues for motor neurons 197–202
long-term depression (LTD) 313
long-term potentiation (LTP) 313
Lou Gehrig's disease 246
low-affinity NGF receptor (LNGFR) 239–240
  see also p75 neurotrophin receptor
low-density lipoprotein receptors 130
loxP (locus of x-over P) 24–25
lozenge (*lz*) 174
Lrp4 (low density lipoprotein receptor-related protein 4) 272–274, 278
Lrp5/6 (low density lipoprotein receptor-related proteins 5 and 6) 59–60, 101–102, 304–305
LTD *see* long-term depression
LTP *see* long-term potentiation
lysates 249
lz see lozenge

M

M phase 114–115, 121
MafB (musculoaponeurotic fibrosarcoma B) 68
MAG *see* myelin-associated glycoprotein
mammalian Inscuteable (mInsc) protein 117
mammals
  gastrulation 32–33
  homeobox genes 71
MAP 1B *see* microtubule-associated protein 1B
Map kinase-activating death domain-4 (MADD-4) 275
maps, retinotectal 214
marginal zone (MZ) 123, 126–130
Mash1 180
Math1 156
  see also Atonal homolog 1
Math3 180–181
Math5 180–181
maturation of neuromuscular junction 267
medial ganglionic eminence (MGE) 131
medial–lateral axis 7
MEK (MAP kinase/Erk kinase) 64, 243
melanocyte migration 143
mesencephalon 10, 54–55, 60–65
mesoderm 9, 31–33
  inducers 37–39
metencephalon 10, 60–65, 132
methylation 53–54, 159
MGE *see* medial ganglionic eminence
MHB *see* midbrain–hindbrain border
mice
  anterior visceral endoderm 57–58
  developmental stages 11, 13
  Dreher mutants 98, 105
  forebrain patterning 57–58
  gastrulation 33
  gene knockout 23–25
  Reeler mutants 128–130, 136
  retinoic acid 73–75
  Weaver mutants 136
microglia 6, 306–307
microtubule-associated protein 1B (MAP 1B) 304
microtubules, dynamic 187–188
microtubules, stable 187–188
midbrain–hindbrain border (MHB) 55–56, 61–62
  see also isthmus
middle interhemispheric (MIH) variant of holoprosencephaly 108–109
midline targets, spinal commissural neurons 202–211
migration
  cellular cleavage direction 115–116
  central nervous system 122–136
  cerebellum 131–136
  clinical human syndromes 127–128, 143
  cortical plate layering 126–127
  defects, human 127–128, 143
  extracellular matrix 125–126
  gastrulation 32–33
  granule cells 134–136
  hindbrain development 65–67
  interkinetic nuclear 113–114
  melanocytes 143
  neural crest cells 140–143
  peripheral nervous system 136–143
  radial glial cells 124–126, 135–136
  Reelin 128–130, 136
  Rho GTPases 122, 138
  somal translocation 124–125
  tangential 130–131
  transient neocortical layers 123–124
  trunk neural crest cells 141–143
  see also pathfinding
MIH *see* middle interhemispheric variant of holoprosencephaly
mInsc *see* mammalian Inscuteable protein
Miranda 154
mitotic spindle orientation 117–118
MKp3 (mitogen-activated protein kinase phosphatase 3) 64
model organisms 11–19
  see also amphibians; avians; *Drosophila*, *C. elegans*, mammals; mice
molecular layer of cerebellum 132
mood disorders 316
morphogens 57
morphological changes at neuromuscular junctions 267
mosaic analysis 171–172
motility
  dendritic spines 297–299
  growth cones 187–191
motor endplates 262, 263–266
  see also neuromuscular junctions
motor neurons
  agrin 269–274, 276, 279
  axonal guidance cues, vertebrate 197–202
  induction 86–95
  innervation 263, 266–267
  invertebrate 274–275
  laminins 281–282
  Neuregulin-1 (Nrg1) 278–279
  neuromuscular junctions 259–288
  pro-BDNF 283–285
  Schwann cells 262–263, 279–282, 285
  subtype induction 89–96
  survival 247
  synaptic elimination 282–285
motor units 262
msh (muscle segment homeobox) 103–104

Müller glial cells 179–180
multipotent cells 147, 156–157
muscle fibers 264–267
MuSK (muscle specific kinase) 272–274, 276
myasthenia 274
myelencephalon 10, 54–55
myelin 6, 166, 169
myelin-associated glycoprotein (MAG) 169
myelination 6, 166, 169
myoblasts 264–265
myosin II 189
myotube cell culture assays 269–270
MZ *see* marginal zone

N

N-cadherin 141, 194
N-WASP (neural-Wiskott-Aldrich syndrome) 190
naming of genes 87
nasal innervation, olfactory system 219–220
NCAM *see* neural cell adhesion molecule
necrosis 249
neocortex, transient layers 123–124
nerve growth factor (NGF)
  experimental discovery 229–234
  identification 233–234
  retrograde transport 237–238, 240–242
  signaling mechanisms 237–240
Neto proteins 264–265
netrins 204–206
neural canal 83
neural cell adhesion molecule (NCAM) 36
  fasciculation 194
  motor neuron guidance 201–202
  neural crest cells 141
  sialylation 194, 201–202
neural crest cells 9, 82
  cranial 138–140
  delamination 137–138
  determination and differentiation 161–164
  migration 140–143
  origin 137–138
  peripheral nervous system migration 136–143
  sacral 139
  trunk 138–139
  vagal 138–139
neural ectoderm 32–33
neural folds 9, 52–53
neural grooves 9, 52
neural induction 29–49
  activin 38–40, 46
  animal cap assays 36–42
  bone morphogenetic proteins 42–48
  early discoveries 33–36
  fibroblast growth factors 37, 47–48
  gastrulation 30–33
  invertebrates 42–46
  mesoderm inducers 37–39
  signaling 35–48
  SMADs 46–48
  transforming growth factor β-related signals 37–39
  Xenopus 36–41
neural keel 9
neural plate 9
  formation 51, 52–53
  lateral inhibition 150–151
neural rod 9
neural tube 9
  A/P segmentation 51–80
    defects 53–54
    early segmentation 54–56
    forebrain specification 57–60
    formation 52–57
    hindbrain 65–76
    mesencephalon/metencephalon 54 –55, 60–65
    midbrain–hindbrain border 55–56, 61–63
    neural plate formation 52–53
    prosomeres 54–55, 58
    rhombomeres 65–67
    spinal cord 76–78
  D/V axis patterning 81–110
    anatomical landmarks 82–85
    anterior 104–109
    posterior dorsal 95–104
    posterior ventral 85–95
  floor plate, posterior neural tube patterning 82–95
  motor neuron patterning 85–95
  roof plate
    anterior patterning 105–106, 108–109
    posterior patterning 82–84, 95–102
  sulcus limitans 83
neural tube defects (NTDs) 53–54
Neuregulin 1 (Nrg1) 135–136, 166, 278–279
neurexins 295–296
neurite outgrowth inhibitor (NOGO) 169
neurites
  axonal self-avoidance 222–223
  definition 185
  extracellular matrix interactions 192–193
  fasciculation 194–198
  growth cone motility 185–191
  labeled pathway hypothesis 196–197
  local guidance cues 197–202
  outgrowth 185–225
  pathfinding 187–223
  pioneer axons 193–194
  retinotectal system 211–223
  Rho GTPases 190–191
  spinal commissural neuron midline targets 202–211
  substrate binding 187–189
  substrate preferences 191–202
neuroblasts 16–17
  D/V patterning signals 103–104
  differentiation 153–155
neurogenesis 111–122
  adherens junctions 116–117
  cell cycle regulation 118–122
  cleavage directions 115–116
  interkinetic nuclear migration 114–115
  polar protein differences 116–118
neurogenins (ngns) 150–151, 163
neuroligins 295–296
neuromeres 54–55
neuromuscular junction (NMJ) 259–288
  agrin 269–274, 276, 279
  brain-derived neurotrophic factor 283–285
  developmental changes 266–267

Dok7 273–274
extrasynaptic nuclei 277–278
invertebrate models 274–275
laminins 280–282
Lrp4 273–274
maturation 267
MuSK 272–274, 276
Nrg1 278–279
pro-BDNF 283–285
Rapsyn 276
Schwann cells 262–263, 279–282, 285
subsynaptic nuclei 277–279
synaptic elimination 282–285
neuronal progenitors
adherens junctions 116–117
apical/basal migration 113–114
definition 81
genetic markers 84–85
interkinetic nuclear migration 114–115
origins 112–113
polar proteins 116–118
neuronal specificity
definition 81
dorsal–ventral axis patterning 81–110
genetic markers 84–85
neuronal survival 227–257
apoptosis 248–253
brain-derived neurotrophic factor 234–236
ciliary ganglion neurons 247
ciliary neurotrophic factor 245–247
extrinsic pathway 254–255
genes 251–253
growth factors 227–247
intrinsic pathway 253
motor neurons 247
nerve growth factor 229–234
neurotrophins 236–244, 254–255
p75 neurotrophin receptor 239–240, 243–244
pro-neurotrophins 254–255
programmed cell death 248–255
target tissue size 228–229
Trk receptors (tropomyosin-receptor kinases) 240–244
neurons 4
fate determination 147–183
migration
central nervous system 122–136
peripheral nervous system 136–143
neurogenesis 111–122
origins 112–113
proliferation 113–122
radial glial cell translocation 124–126, 135–136
somal translocation 124–125
neuropilin 195, 264–265
neurotransmitter receptors 5, 261
see also individual neurotransmitter receptors
neurotransmitters 5, 163–165
see also individual neurotransmitters
neurotrophic hypothesis 228, 230, 246
neurotrophins 236–244
discovery 236–237
oligodendrocyte precursor cell proliferation 167
p75 receptor 239–240, 243–244
retrograde transport 237–238, 240–242
synaptogenesis 283–285, 298–299
Trk receptors 240–244
NgR *see* NOGO-66 receptor
nicotinic acetylcholine receptors 270
NMJ *see* neuromuscular junction
nodal 95, 107
noggin 40–46, 97, 106
NOGO *see* neurite outgrowth inhibitor
NOGO-66 receptor (NgR) 169
noncatalytic receptors 243–244
nonneuronal cells, lateral inhibition 148–151
Notch receptors
granule cell proliferation 156
inner ear development 176–178
invertebrate differentiation 151–153
lateral inhibition 148–151
parasympathetic/sympathetic neuron determination 163
proliferation 117–118
retinal development 180
notochord 56
D/V patterning 82, 84, 85–91, 107–108
motor neuron induction 85–86
ultimate fate 84
Nrg1 *see* Neuregulin 1
NTDs *see* neural tube defects
nucleation of actin polymerization 189
Numb 118, 152–154

O

ocular dominance columns (ODCs) 310–312
ODCs *see* ocular dominance columns
OIG-4 275
olfactory sensory neurons (OSNs) 219–220
olfactory bulb, innervation 219–220
oligodendrocyte precursor cells (OPCs) 167–169
oligodendrocytes 6, 167–169
ommatidia 170–175
oncogenes 240
OPCs *see* oligodendrocyte precursor cells
optic cup 178–179
optic nerve 167–168
optic vesicles 178
organ of Corti 176–178
organizer 34–35, 40–41
orientation of mitotic spindles 117–118
OSNs *see* olfactory sensory neurons
Otd genes 58
otic vesicle 175–178
otocysts 175–178
Otx genes 58, 62, 65

P

P-domain *see* peripheral domain
p27 168–169
p57 168–169
p75 neurotrophin receptor 239–240, 243–244, 254–255, 284–285
pair-rule genes 69–70
par complex proteins 117–118, 153–155
paralogous groups, *Hox* genes 71
parasympathetic neurons 161–163
patched (Ptc) receptor 91–94

pathfinding 187–211
axonal self-avoidance 222–223
chemoaffinity hypothesis 214–223
extracellular matrix 192–193
fasciculation 194–198
filopodia 187–188
growth cones 187–191
labeled pathway hypothesis 196–197
local guidance cues 197–202
motor neurons 197–202
netrins 204–206
pioneer axons 193–194
retinotectal system 211–223
Robo signaling 207–209
semaphorins 195
Slit proteins 206–208
spinal commissural neurons 202–211
substrate binding 187–189
substrate preferences 191–202
topographic maps 214, 220–222
zipcode binding protein 1 209–211
Pax genes 58, 65
PC12 cell line 238–240
PCD *see* programmed cell death
PCL *see* Purkinje cell layer
Pdm 155
PDZ domains 294–296, 301–302
peanut agglutinin-binding glycoprotein 143, 201
peripheral domain (P-domain) 187
peripheral nervous system (PNS) 4
Drosophila 151–153
neural crest cell migration 136–143
origins 8–20
perisynaptic Schwann cells (PSCs) 279–281, 285
periventricular heterotopia (PH) 128
PFC *see* prefrontal cortex
PH *see* periventricular heterotopia
phosphatidylinositol 3-kinase (PI3-K) pathway 242–243
phosphorylation
cyclin-dependent kinases 119–120
retinoblastoma protein 120–121
SMADs 46–48
Smo proteins 92
photoreceptors of *Drosophila* 170–175
Phox 164
PICK1 302
pillar cells 175–176
pioneer axons 193–194, 196–197
placodes 9, 139–140
plasticity
central nervous system 289
homeostatic 274–275, 312–313
superior olivary nuclei 314
vertebrate visual system 307–313
platelet-derived growth factor (PDGF) 167
PLC_{γ} pathway 242–243
PNS *see* peripheral nervous system
Pokémon 87
polarity proteins 116–118, 153–154
polymerization of actin 189–190
polysialic acid (PSA) 194, 201–202
PON (partner of numb) protein 154
posterior neural tube
anatomical landmarks 82–85
dorsal neuronal patterning 95–104
ventral neuronal patterning 85–95
postjunctional folds 263–264
postnatal development
auditory system 314
cholinergic neurons 163–165
vertebrate visual system 307–313
post-polio syndrome 286
postsynaptic cells 5
central nervous system 292–293
neuromuscular junctions 263–264, 272–279
postsynaptic density (PSD) 261, 293–294
postsynaptic density-95 (PSD-95) 294, 305–306
postsynaptic differentiation 295–299, 301–306
postsynaptic partner cells 260–261, 292–294
POU3f2 (POU class 3 homeobox 2) 158
PP *see* preplate
prechordal plate 33
precursors *see* neuronal progenitors
prefrontal cortex (PFC) 315
prenatal cocain exposure 315
prepatterning
acetylcholine receptors 264–265
dendritic spines 296–297
preplate (PP) 123
presynaptic cells 5
central nervous system 292–293
neuromuscular junctions 260–261, 280–282
presynaptic differentiation, central nervous system 295–297, 299–301, 303–306
primary cell cultures 238
primary cell layers 30–33
see also germ cell layers
primary visual cortex (V1) 307–308, 310–313
primitive streak 31–33
pro-apoptotic genes 251–253
pro-BDNF 283–285
pro-neurotrophins 254–255
programmed cell death (PCD) 248–255
apoptosis 248–253
extrinsic pathway 254–255
genes 251–253
intrinsic pathway 253
necrosis 249
proliferation
cell cycle 114–115, 118–122
cellular cleavage direction 115–116
cerebellar neurons 132–134
granule cells 156–157
neurogenesis 113–122
oligodendrocyte precursor cells 167
promoters, retinoic acid response elements 72–75
proneural cluster (PNC) 148–150
proneural genes 116, 148–150
prosencephalon 10, 54–55
prosomeres 54–55, 58
Prospero 153–154
proteasomes 303
Prox1 118
PSA *see* polysialic acid
PSCs *see* perisynaptic Schwann cells
PSD-95 *see* postsynaptic density-95

Purkinje cell layer (PCL)
adult organization 132
origins 132–133
proliferation 132–133
Reeler mutation 128, 136
pyknotic nuclei 229

R

R cells *see* retinula cells
RA *see* retinoic acid
radial glial (RG) cells 124–126, 135–136
Raf 64
Raldh2 *see* retinaldehyde dehydrogenase 2
Rapsyn 276
RAREs *see* retinoic acid response elements
Ras/Erk cascade
cerebellar induction 64
Drosophila eye development 173–174
neural induction 47–48
neurotrophin signaling 242–243
Rb *see* retinoblastoma protein
receptor tyrosine kinases
Eph family 67–68
FGF receptors 63–64
Trk family 240–244
receptors
autophosphorylation 241–242
cell-autonomy 94
cytokine 165, 246–247
extrinsic pathway, cell death 254–255
neurotrophin signaling 239–244, 254–255
noncatalytic 243–244
retrograde transport 237–238, 240–242
signal transduction 21
truncated 38–39, 243–244
recycling, actin 189–190
Reeler mutants 128–130, 136
Reelin protein 128–130, 136
regeneration
ciliary neurotrophic factor 246
inhibition in central nervous system 169
motor neurons 268–269
retinal ganglion cells 212–214
skeletal muscle 268–269
regionalization
co-linearity 71, 74, 77–78
hindbrain 65–78
mesencephalon/metencephalon 60–65
neural crest cells 138–139
regulation
acetylcholine receptor synthesis 278–279
cell cycle 118–122
cerebellar gene expression 64
gene expression 20–26, 64, 72–78
Hox genes 72–78
neuronal survival 227–237
repression, Shh signaling 90–91
resonance hypothesis 212
response elements, retinoic acid 72–75
Reticulon 4 receptor (RTN4R) 169
retinal ganglion cells (RGCs) 178–180, 212–214, 308–310
retinal progenitor cells (RPCs) 180–181
retinaldehyde dehydrogenase 2 (Raldh2) 74
retinas, vertebrate development 178–181
retinoblastoma protein (Rb) 120–121
retinoic acid (RA)
Cdx expression 76–78
Cyp26-mediated degradation 75–76
FGF interactions 76–78
Hox gene regulation 72–78
posterior patterning 73–78
ventral patterning 95
retinoic acid response elements (RAREs) 72–75
retinotectal maps 214
retinotectal system 211–223
chemoaffinity hypothesis 214–223
ephrins 218–222
maps 214
stripe assays 215–218
retinula cells (R cells) 170–175
retrograde transport 237–238, 240–242
Rett syndrome 298
reverse signaling 67, 220–222, 299–301
RG *see* radial glial cells
Rho GTPases
adhesion 122
central nervous system synapses 298–299
cytoskeletal dynamics 190–191
migration 122
neural crest cell induction 138
rhombencephalon 10, 54–55
rhombomeres 65–78
early boundary formation 55
Hox code 71–73
regionalization 65–67
segmentation genes 68–78
ribonucleic acid (RNA) 3, 23, 99–100
Robo signaling 207–209
ROCK (Rho-associated coiled-coil-containing protein kinase) 190–191
rod cells 179–180
roof plate
anterior patterning 105–106, 108–109
induction 96–97
interneuron patterning 97–99
posterior patterning 82–84, 95–102
rostral–caudal axis 7
see also anterior–posterior axis
Rough 172–173
RPCs *see* retinal progenitor cells
RPM-1 (regulator of presynaptic morphology-1) 303
RTN4R *see* Reticulon 4 receptor

S

S phase 114–115, 121
sacral neural crest 139
salivary glands 233–234
sarcoma 180 tumors 230–232
Satb2 (SATB homeobox 2) 158
satellite cells 6–7
scaffolding proteins 261, 282, 293–296, 299
Schwann cells 6
myelinating 166
Neuregulin signaling 166
neuromuscular junctions 262–263, 279–281, 285
perisynaptic 279–281, 285
sclerotome 142–143

secreted Frizzled-related proteins (sFRP) 59, 306
sef-b (similar expression to FGF genes-b) 64
segment polarity genes 69–70
segmentation
  anterior–posterior axis 51–80
  genes 69–70
  hindbrain 65–78
  neural tube 54–56
selector genes 24
self-avoidance, axonal 222–223
semaphorins 194–195, 201–202
Senseless 172–173
sensory cells
  differentiation 170–181
  Drosophila 17, 151–153, 170–175
  hair cells 175–178
  olfactory neurons 219–220
  retinal cells 180–181
sensory epithelia of inner ear 175–178
sensory organ progenitors (SOPs) 17, 151–153
serine threonine kinase receptor transduction 38–46, 99–100
serine/threonine protein kinases, cell cycle 119–120
serotonin transporter variants 316
seven up (*svp*) 174
sevenless gene 87, 171–172, 173–174
sFRP *see* secreted Frizzled-related proteins
short of gastrulation gene (*sog*) 43–45
short interfering RNAs (siRNAs) 23, 99–100, 300
signal transduction
  basic concepts 21
  fibroblast growth factor receptors 47, 64
  serine/threonine kinase receptors 43–45, 99–100
  Trk receptors 243
siRNAs *see* short interfering RNAs
Six genes, forebrain patterning 58, 60
skeletal muscle 262–285
  agrin 269–274, 276, 279
  development 264–266
  Dok7 273–274
  innervation 263, 266–267
  Lrp4 273–274
  MuSK 272–274, 276
  pro-BDNF 283–285
  Rapsyn 276
  regeneration 269
  synapse elimination 282–285
  synaptic basal lamina 267–274
  Slit proteins 206–208
SMADs 46–48, 87, 99–100
smell 219–220
Smoothened (Smo) 91–94
snake venom 232–233
sog see short of gastrulation gene
somal translocation 124–125
somites 11, 142, 200–202
SON *see* superior olivary nuclei
Sonic hedgehog (*Shh*)
  anterior neuronal patterning 105, 107–108
  cerebellum 134, 156
  cross-repression 91
  cyclopamine 108
  floor plate-derived, 88–90
  motor neuron precursor induction 89–95
  notochord-derived 86, 88–90
  Ptc receptor 91–94
  ventral patterning 86–95, 107–108
  zipcode binding protein phosphorylation 209–211
sortilin 284–285
SOS (son of sevenless) 64
Sox2 (sry-box 2) 176–177
SP *see* subplate
Spemann's organizer 34–35
Sperry, Roger 2
spina bifida 53–54
spinal cord 7, 10
  anatomical landmarks 83–85
  axonal bundles 198–199
  Cdx 76–78
  commissural interneurons 202–211
  dorsal–ventral axis patterning 81–104
  lateral inhibition 150–151
  transcription factors 76–78
spinogenesis 298
spongioblast networks 112–113
sprouty 64
stress and development 316
stripe assays 215–218
stripe formation by lateral inhibition, neural plate 150–151
subplate (SP) 123
substrate preferences, growth cones 191–202
subsynaptic nuclei 277–278
subventricular zone (SVZ) 123–124
SuFu *see* Supressor of Fused
sulcus limitans 83
superior olivary nuclei (SON) 314
superplate formation 128
Supressor of Fused (SuFu) 91–92
svp see seven up
SVZ *see* subventricular zone
symmetric cleavage 118
sympathetic chain ganglia 141–142
sympathetic neurons
  environmental differentiation cues 161–163
  migration 141–142
  neurotransmitter phenotypes 163–165
  programmed cell death 250–251
  survival 230–234, 238
synapses 5
  central nervous system 290–317
  neuromuscular junction 259–288
synaptic basal lamina 263–264, 267–274
  agrin 269–274
  laminins 280–282
synaptic boutons 262
synaptic cleft 5, 260–261
synaptic elimination 282–285, 307–310
synaptic reorganization
  central nervous system 307–317
  hearing onset 314
  neuromuscular junctions 259
  visual system 307–313
synaptic scaling 312
synaptic vesicles 5, 261
synaptogenesis 259
  AChR prepatterning 264–265
  agrin 269–274, 276, 279
  astrocytes 306–307

basal lamina 267–274
cadherins 294–295
cell adhesion molecules 294–295
central nervous system 290–307
dendritic spines 291
Dok7 273–274
en passant synapses 291, 296–297
Eph receptors/ephrins 299–303
excitatory synapses 291–294, 296–297, 305, 312–313
inhibitory synapses 291–293, 313
laminins 280–282
Lrp4 273–274
MuSK 272–274, 276
neurexins/neuroligins 295–296
neuromuscular junctions 260–282
Rapsyn 276
reciprocal signaling 296–297
reverse signaling 299–301
Schwann cells 262–263, 279–280
terminal synapses 291
thrombospondins 306
transforming growth factor β 280
ubiquitin-mediated pathways 303–304
Wnt 304–306
synCAM 294–295
syncytial blastoderm 43–45
syncytium 112–113
syntenin-1 302

T

tail structures, induction 56
tangential migration 130–131
target tissue size and neuronal survival 228–229
targeting
axonal self-avoidance 222–223
chemoaffinity hypothesis 214–223
commissural interneurons 196, 202–211
fasciculation 194–198
labeled pathway hypothesis 196–197
motor neurons 197–202
netrins 204–206
retinotectal system 211–223
Robo signaling 207–209
semaphorins 195
Slit proteins 206–208
topographic maps 214, 220–222
zipcode binding protein 1 209–211
Tbr1 (T-box brain 1) 158
telencephalon 10, 55–56, 105–106
temporal expression patterns
invertebrate determination 154–155
retinal development 180–181
vertebrate cortical differentiation 157–159
temporal identity factors (TIFs) 154–155, 181
terminal deoxynucleotidyl transferase dUTP nick-end labeling (TUNEL) method 249–250
terminal synapses 291
tethering, Shh 90
tetrodotoxin (TTX) 308
TGFβ *see* transforming growth factor β
thrombospondins 306
Tid1 (tumorous imaginal disc 1) 276
TIFs *see* temporal identity factors
tissue manipulations 23
TNFα *see* tumor necrosis factor α
topographic maps 214, 220–222
Torpedo californica 270–271
transcription factor code 81, 84–85
transcription factors 21
Cdx (caudal type homeobox) 76–78
Class I 90–91
Class II 90–91
cerebral cortical neuron differentiation 157–159
Drosophila eye development 172–175
granule cells 156–157
homeobox genes 69–78
Kreisler/MafB 68
Krox20 67–68
neural crest cell induction 138
Shh activation 90–91
SMADs 46–48
temporal expression patterns 154–155
transforming growth factor β (TGFβ) 37–39, 96–100, 280
transforming signals, neural tube segmentation 56–57
transient layers of neocortex 123–124
tripartite complexes 246–247, 261–262
tropomyosin-receptor kinases (Trk) 240–244, 254–255, 284–285
trunk neural crest cells 138–139, 141–143
TTX *see* tetrodotoxin
tumor formation 121, 229–230
tumor necrosis factor α (TNFα) 306
TUNEL method *see* terminal deoxynucleotidyl transferase dUTP nick-end labeling method

U

ubiquitin proteasome system (UPS) 303–304
Unc-5 (uncoordinated-5) 205–206
Unc-6 (uncoordinated-6) 204, 296
Unc-40 (uncoordinated-40) 205–206
UPS *see* ubiquitin proteasome system
utrophin 276

V

V1 *see* primary visual cortex
vagal neural crest cells 138–139
VE *see* visceral endoderm
vegetal pole 8, 30–31
ventral nerve cord (VNC) 17, 153
ventral patterning, anterior neural tube 104–109
ventral patterning, posterior neural tube 85–91
ventralized embryos 39–41
ventricular zone (VZ) 83, 123–124
Purkinje cell proliferation 132–133
vertebrates
cellular determination 155–161
developmental stages 11–15
inner ear development 175–178
motor neuron guidance cues 197–202
retina development 178–181
visual system 307–313
vertical cleavage 116
very-low-density lipoprotein receptor (Vldlr) 130
vesicles
neural tube segmentation 54–55
optic 178
otic 175
synaptic 5, 261

visceral endoderm (VE) 33
visual system 178–181, 307–312
  plasticity 307–313
  homeostatic plasticity 312–313
  ocular dominance columns 310–312
Vldlr *see* very-low-density lipoprotein receptor
VNC *see* ventral nerve cord
vnd (ventral nervous system defective) 103–104
voltage-dependent calcium (Ca^{2+}) channels 261, 282
VZ *see* ventricular zone

W

Waardenburg syndrome 143
WAVE (WASP-family verprolin-homolog protein) 190
Weaver mutants 136
Wingless (Wg) 264–265
Wnt
  anterior D/V patterning 106–107
  canonical pathway 101–102
  central nervous system synaptogenesis 304–306
  divergent canonical pathway 304–305
  dorsal neural tube patterning 100–102
  forebrain inhibition 58–60
  Frizzled receptors 59
  midbrain–anterior hindbrain patterning 62–63, 65
  neural crest cell induction 138
  neural induction 47–48
  neuromuscular junctions 274

X, Z

Xenopus laevis 14–15, 36–41, 150–151
ZBP1 *see* zipcode binding protein 1
ZBTB7 (zinc finger and BTB protein 7) gene 87
zebrafish 13–14, 74, 77
Zic (zinc finger of the cerebellum) 106–107
zipcode binding protein 1 (ZBP1) 209–211
zona limitans intrathalamica (ZLI) 56, 58
zygote 8, 30